MUSCLE FOODS

Muscle Foods

Meat

Poultry and

Seafood

Technology

Edited by
Donald M. Kinsman
Anthony W. Kotula
Burdette C. Breidenstein

CHAPMAN & HALL
New York • London

First published in 1994 by
Chapman & Hall
One Penn Plaza
New York, NY 10119

Published in Great Britain by
Chapman & Hall
2–6 Boundary Row
London SE1 8HN

Printed in the United States of America

Library of Congress Cataloging in Publication Data

Muscle foods: Meat, Poultry and Seafood technology / edited by Donald M. Kinsman, Anthony W. Kotula, Burdette C. Breidenstein.
p. cm.
Includes bibliographical references and index.
ISBN 0-412-98641-8
1. Meat. 2. Poultry. 3. Fish as food. 4. Wildlife as food.
I. Kinsman, Donald Markham, 1923– . II. Breidenstein, Burdette C.
III. Kotula, Anthony W.
TX556.M4M89 1994
641.3′06—dc20 94-7189
CIP

British Library Cataloguing in Publication Data available.

Please send your order for this or any other Chapman & Hall book to **Chapman & Hall, 29 West 35th Street, New York, NY 10001, Attn: Customer Service Department.** You may also call our Order Department at 1-212-244-3336 or fax your purchase order to 1-800-248-4724.

For a complete listing of Chapman & Hall's titles, send your request to **Chapman & Hall, Dept. BC, One Penn Plaza, New York, NY 10119.**

Contents

Dedication

This book is dedicated to all peoples of all nations. Recognizing that hunger and malnutrition are the root of the world's problems, it is the fervent hope of the editors that this publication will help to resolve this universal dilemma and contribute in some way to aiding greater global understanding and world peace and prosperity.

> It is not what we give, but what we share,
> The gift without the giver is bare;
> Who gives himself with his alms feeds three,
> Himself, his hungering neighbor, and me.
>
> James Russell Lowell
> in "The Vision of Sir Launfal"

Preface

Traditionally, in the food industry, there has been a distinction made among meat, poultry, seafood, and game. Meat has historically been defined as the edible flesh of animals. This basically referred only to the red meats, namely, beef, lamb, pork, and veal, including both fresh and processed products as well as variety or glandular meats.

It has been recognized more recently that all foods derived from muscle, or muscle foods, have basically the same or similar characteristics in physical and chemical properties. Therefore, it is logical to examine and consider all muscle foods under one cover. This book, therefore, is an attempt to address the various attributes of red meat, poultry, fish, and game under the single heading of muscle foods and to note any differences where they might occur.

It is of interest that of the 10 top U.S. meat companies in 1990, 8 of them were dealing with poultry as well as red meats and that 4 of the 10 were also involved with seafoods. This lends impetus to the inclusion of all three in a book such as this. Furthermore, the rapid increase in consumption of poultry meat to approximately 30 kg (65 pounds) per capita and seafoods to 7 kg (16 pounds) per capita compared to beef at 34 kg (75 pounds) and pork at 30 kg (65 pounds), whereas veal and lamb/mutton represent only 0.5 kg (1 pound) each, all on a carcass basis, makes it all the more logical to study all of these entities under a single heading. Hence, the development of this book—*Muscle Foods*.

The editors are especially pleased to have the cooperation and co-authorship of these knowledgeable meat scientists whose experience with industry, teaching, research, and public service have kept them on the cutting edge of this field and particularly of their specific interest areas. These enthusiastic colleagues have all attained advanced degrees. A number have gained industry experience. All have established outstanding reputations in their specialties. They bring a fresh, highly scientific yet practical perspective to their respective chapters. It is hoped that the users of this book, be they students, industry personnel, academicians, researchers, state or federal regulatory agencies, consumers, or nutritionists, will find it useful and informative. It has been devised with a wide spectrum of people in mind, yet of great enough depth and perspective to contribute to their needs and applications.

The editors, as well as the authors, recognize that each chapter could be expanded to a greater extent but that space does not permit an exhaustive treatment of each topic. Therefore, an attempt has been made to provide the essentials and concepts and direct the reader to other sources for greater depth and insight on those topics of special interest to the reader.

Acknowledgments

A publication of this nature represents the dedicated input of the respective chapter authors. They in turn are grateful for the critique of their writings for scientific corrections as well as organization and readability. We have been extremely fortunate to have had the interest, support, and professional review of each chapter by at least two scientists who are very highly regarded, particularly in the specific areas that they reviewed. We want to publicly thank and acknowledge their expertise and contributions: Chapter 1, Dr. Robert Bray and Dr. James Kemp; Chapter 2, Dr. Donald Beermann and Dr. Fred Parrish; Chapter 3, Dr. Bruce Marsh and Dr. Joseph Sebranek; Chapter 4, Dr. Brad Berry and Mr. Dale Graham; Chapter 5, Dr. Norman Marriott, Dr. Michael Moody, Dr. Glen Schmidt and Dr. Donn Ward; Chapter 6, Mr. Bill Dennis and Dr. Herbert Ockerman; Chapter 7, Dr. Tom Carr and Dr. Jim Wise; Chapter 8, Dr. Donald Kinsman and Dr. Anthony Kotula; Chapter 9, Dr. John Marcy and Dr. Bruce Tompkin; Chapter 10, Dr. Joseph Hotchkiss and Dr. Frederick Oehme; Chapter 11, Dr. Donald Kropf and Dr. Roger West; Chapter 12, Dr. Karen Bett and Dr. Edgar Chambers; Chapter 13, Dr. Robert Cassens and Dr. Ray Field; Chapter 14, Dr. Gary Shults and Dr. Carl VanderZant; Chapter 15, Dr. Brad Berry and Dr. Jane Bowers; Chapter 16, Dr. Dennis Miller and Dr. Joyce Nettleton; Chapter 17, Dr. Steve Goll, Dr. William Schwartz and Dr. Robert Terrell; Chapter 18, Dr. Aaron Brady, Mr. Richard Perdue and Mr. Owen Schweers; Chapter 19, Dr. Richard Epley and Mr. David Small; Chapter 20, Dr. Christopher Canale and Dr. Kimiaki Maruyama.

Additionally, the editors are most appreciative of the great assistance provided by Mrs. Juanita McIntosh of the University of Connecticut Animal Science Department secretarial staff in handling the correspondence and numerous communications involved over a period of 3 years in bringing this publication to reality. Also, Mr. David Schreiber, Jr. of this department contributed immensely to the word processing and standardizing of format as well as the merging of glossaries and indexing of the entire book.

Furthermore, we recognize and offer thanks to our colleagues and to our wives, Helen Kinsman, Joan Kotula and Shirley Breidenstein for their patience, understanding, and encouragement throughout this entire process.

Contributors

Dr. Sam ANGEL
The Volcani Center
Agricultural Research Center
P.O. Box 6, Bet Dagan 50250, Israel

Dr. Blaine B. BREIDENSTEIN
136 ABLMS, Animal Science Dept.
1354 Eckles Ave., Univ. of MN,
St. Paul, MN 55108

Dr. Burdette C. BREIDENSTEIN
8230 Springbrook Drive
Oklahoma City, OK 73132

Dr. Robert E. CAMPBELL
Dr. P. Brett KENNEY
Animal Science Dept., Meat Science Lab
Weber Hall, Kansas State University
Manhattan, Kansas 66506-7196

Dr. James R. CLAUS
Dr. Jhung-Won COLBY
Dr. George J. FLICK
Food Science Dept., Room 101
V.P.I. & S.U
Blacksburg, VA 24061

Dr. Cameron FAUSTMAN
Animal Science Department
University of Connecticut
Storrs, CT 06269-4040

Dr. J. Samuel GODBER
Food Science Dept.
Louisiana State University
Baton Rouge, LA 70803-4200

Dr. Daniel Scott HALE
Animal Science Dept.
348 Kleberg Center
Texas A&M University
College Station, TX 77843-2471

Dr. Larry W. HAND
Animal Science Dept.
104B Animal Science
Oklahoma State University
Stillwater, OK 74078-0425

Dr. Joseph H. HOTCHKISS
Food Science Dept.
Room 119 Stocking Hall
Cornell University
Ithaca, NY 14853

Dr. Jennifer L. JOHNSON
USDA, BARC-East 322
FSIS-Microbiology
10300 Baltimore Ave
Beltsville, MD 20705-2538

Dr. Robert G. KAUFFMAN
Meat and Animal Science Dept.
1805 Linden Drive
University of Wisconsin
Madison, Wisconsin 53706

Dr. Donald M. KINSMAN
Animal Science Department
University of Connecticut
Storrs CT 06269-4040

Dr. Anthony W. KOTULA
Animal Science Department
University of Connecticut
Storrs, CT 06269-4040

Dr. Riëtte L.J.M. van LAACK
Meat Science Research Lab
USDA, ARS, Bldg. 201, BARC-East
Beltsville, MD 20705-2350

Dr. Richard J. McCORMICK
Animal Science Dept.,
Box 3684 University Station
University of Wyoming
Laramie, WY 82071

Dr. Rhonda K. MILLER
Animal Science Dept.
348 Kleberg Center
Texas A&M University
College Station, TX 77843-2471

Dr. William A. MOATS
Meat Science Research Lab
USDA, ARS, Bldg 201, BARC-East
Beltsville, MD 20705-2350

Dr. Anna V. A. RESURRECCION
Food Safety & Quality Enhancement Lab
Univ. of Georgia Experiment Station
Griffin, GA 30223-1797

Dr. Morse B. SOLOMON
Meat Science Research Lab
USDA, ARS, Bldg. 201, BARC-East,
Beltsville, MD 20705-2350

1

Historical Perspective and Current Status

Donald M. Kinsman

From Adam to America

From time immemorial mankind's basic needs and concerns have been for food and shelter. Without food and adequate nutrition all else pales in the struggle for life, for without good health, which stems principally from a balanced diet, one would not continue to need the other elements. Thus, food per se is of prime importance in the well-being and progress of the human race.

Accepting this fact, one may then pose the question, "Which foods?" With a myriad of foods of both plant and animal origin there has been and continues to be a wide selection from both sources. As prehistoric people migrated to new locations and lived as nomads, it became necessary to develop a mobile food supply based largely on animals and their products that could travel with them for sources of draft power, food, and clothing.

A more stable agriculture developed as people established communities and settled more permanently in locations that offered fertile soil, water, and some protection from the elements. Man evolved from food gathering to food cultivation approximately 10,000 to 16,000 years ago. Table 1.1 shows the sequence in which the domestication of animals took place.

With more permanent agriculture established, the production of livestock for draft purposes, followed by their use for food and clothing, became a logical sequence. Early man, especially with the advent of fire for heat and cooking, had come to realize that the flesh of animals was not only good to the taste but somehow fulfilling to the body as a source of sustenance. Fish and fowl were caught or hunted and consumed, as well. Thus, muscle foods became a routine portion of the diet, whether from domestic animals or game or wildlife.

In the course of evolution, during some 200 to 400 centuries, humans learned, by both instinct and need, that food from animals was both flavorful and essential

Table 1.1. Domestication of Animals

Species	Years Ago	Where	Why	From
Dog	8500–9000	Old and New Worlds	Pet, companion	Wolf/jackel
Goat	8500–9000	Old and New Worlds	Food, milk, clothing	Wild goat
Pig	8000–9000	Old and New Worlds	Food, sport	Euopean wild boar
Sheep	6000–7000	Old and New Worlds	Food, milk, clothing	European mouflon and Asiatic urial
Cattle	6000–5000	Old and New Worlds	Religious reasons	Auroch
Chickens	5000–5500	India, Sumatra, Java	Cockfights, shows, food, religion	Jungle fowl
Horse	4000–5000	Old World	Transportation	Wild horse
Fowl: Ducks		China	Food and feathers	Wild duck
Fowl: Geese		Greece/Italy	Food and feathers	Wild goose
Fowl: Turkeys		Mexico and North America	Food and feathers	Wild turkey

Source: Adapted from Campbell and Lasley, 1985.

to well-being. In addition to vegetables, grains, fruits, and nuts in various forms and combinations, it naturally evolved that foods, of animal origin, including what we now refer to as meat, poultry, and seafoods, were all available, desirable, palatable, and nutritious. Thus, muscle foods, became a stable and essential facet of a balanced and healthful diet. The agrarian era developed independently and at different times and places globally. In general, it progressed in the great river valleys of the world, most notably, the Amazon, Danube, Indus, Mississippi, Nile, Rhine, Tigris-Euphrates, and Yellow. Also, the great plains of the world with adequate rainfall, temperature, and climatic ranges yielded great food supplies of both plant and animal sources. Not only did the peoples of these regions subsist on those foodstuffs in season but they also developed methods of preserving surplus foods for consumption in times of less plentiful supply. Today a year-round supply of food is assured in fresh, processed, or preserved forms.

From Corned Beef to Boxed Beef

The Native Americans preceding the European settlers in North America relied on wildlife for much of their sustenance. Indigenous species such as deer, elk, antelope, moose, bison, turkey, small game, finfish, and shellfish were plentiful and widely hunted. For long journeys, pemmican, a mixed, pulverized combination of dried meat, fruits, and berries with added fat provided a concentrated, lightweight, stable food while on the move. Dried strips of bison (buffalo) meat, perhaps salted as well, known as "jerky," were also commonly used, especially by southwestern Indians.

The early colonists learned a great deal from the Native Americans with whom they came in contact, particularly about utilizing native animal and plant life for food and preserving surplus for the winter months. They soon learned to preserve meat by "packing" it in crocks or barrels in layers sprinkled with salt. The moisture from the meat in combination with the salt formed a brine which proved to be a good means of preserving the meat. Extreme saltiness could be controlled by adding water. As overseas trade developed, this "brined meat" found wider demand both as a source of meat for seagoers and as a commodity. By 1635, Captain John Pynchon of Springfield, Massachusetts was shipping barrels of "corned beef" and is today recognized as the first U.S. "meat packer." Thus, an industry evolved that dealt with fresh meat in season and preserved meat the rest of the year. Natural refrigeration, although uncontrollable, did prolong the edible stage of meat, poultry, and seafoods to some extent, namely, short term at low temperatures (0–4°C /32–40°F) to long term at frozen temperatures (−10°C/0°F or lower). These seasonal vagaries continued with the use of natural ice until the development of artificial refrigeration by use of ammonia systems about 1860. These, in turn, were adapted to railroad cars about 1875, which were previously "iced" at specified stops from the Midwest to the east coast. Soon "packing towns" developed along the routes of the railroads, pushing west from the eastern U.S. seaboard. Buffalo, Cincinnati, and, eventually, Chicago became the sites of large meat-packing plants as livestock were driven in "droves" to these centers for processing into fresh meat and processed meat for further distribution to the developing urban population to the east. Cincinnati was popularly known as "Porkoplis" and Carl Sandburg's poem "Chicago" referred to that city as "Hog Butcher for the World." Rail heads emerged as the transcontinental railroad pushed farther west and the long overland cattle droves terminated at such points, so that "terminal markets" were established either to ship livestock farther east by rail or to supply animals to packing plants which were built at those locations. A whole new industry grew, based on meat animals—cattle, sheep, and swine—being raised largely in the midwest and Southwest and then processed and shipped by rail and more recently by truck to the major consuming areas.

The concentration of livestock production by species within the United States is shown in Fig. 1.1. Additionally, the leading poultry processors are listed. It should be noted that some of these are also listed among the leading beef and/or pork processors. More recently, many packers have built their cattle processing plants in the regions of concentrated production near the feedlot areas. There the source of livestock is plentiful to supply these huge new plants that are dressing between 1000 and 4000 head of cattle per day at a single plant. For example, in a 100-mile radius of Amarillo, Texas there are, at any one time, approximately 2,500,000 cattle in feedlots, with packing plants in that region having the capacity to process that much beef. The beef plants are principally in the west central to southwest region, with Nebraska, Kansas, and Texas

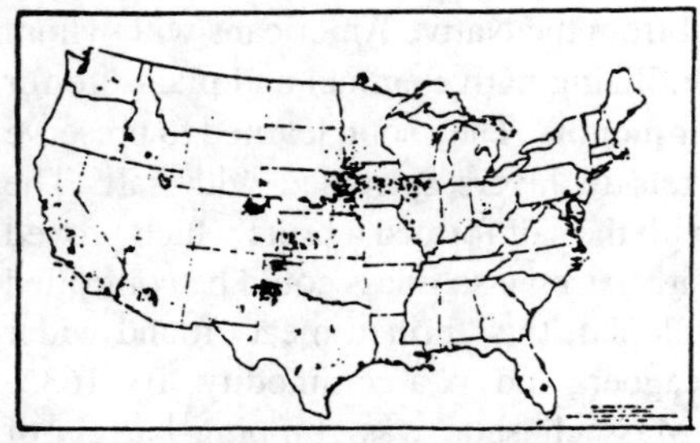

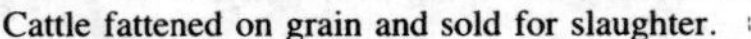

Cattle fattened on grain and sold for slaughter.

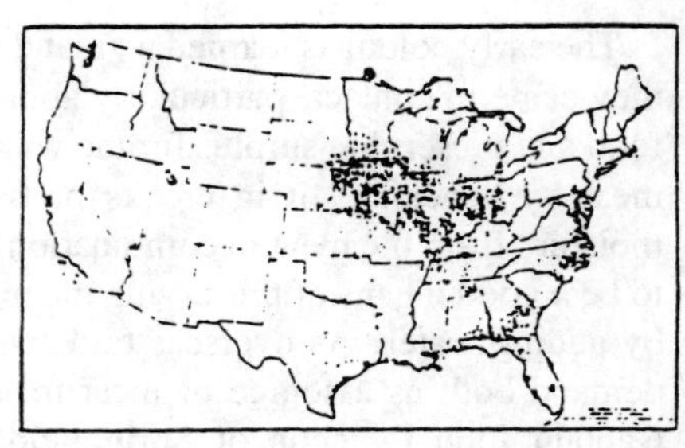

Hogs and pigs.

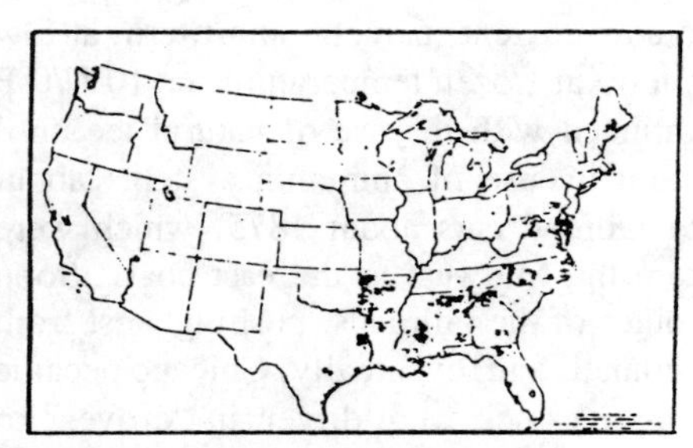

Broilers and other meat-type chickens.

Top Ten Poultry Processors/ Further Processors

Top 100 Rank	Company	Sales (Mil$)
1	ConAgra, Inc.	12,500
4	Tyson foods, Inc.	3,825
11	Perdue Farms, Inc.	1,250
13	Gold Kist, Inc.	1,040
19	Pilgrim's Pride Corp.	721
24	Seaboard Corp.	557
25	WLR Foods, Inc.	494
26	Wayne Poultry	450
28	Townsends, Inc.	400
30	Rocco, Inc.	300

From 1991 Top 100 MEAT & POULTRY, *July 1991*

Source: AMI Meat & Poultry Facts, 1992.

Figure 1.1. Regional concentration of livestock production.

accounting for 55% of the nation's beef slaughter (Fig. 1.2). At major pork plants located mostly in the hog growing areas of the Midwest and north central states, 10,000 to 20,000 hogs per day are dressed. Iowa, Illinois, and Minnesota account for nearly 50% of the U.S. hog slaughter (Fig. 1.2). The larger sheep plants are in the mountain states, notably Colorado, and also in west Texas.

The poultry industry has grown tremendously especially with the development of the broiler industry, first along the eastern seaboard from Maine to the Delaware–Maryland–Virginia area (Delmarva peninsula), and then to Georgia and Alabama, and on to Arkansas. The seafood industry grew, naturally along the seacoast, first on the eastern seaboard and eventually the Gulf Coast and the West Coast. New England was an early seafood location then trailing south along the coast to the Carolinas which have become large seafood processors. The Gulf states concentrated on the shrimp industry, whereas the cold water fish and lobsters dominated the Northeast and salmon and tuna have become a northwest U.S. specialty item. Most recently, freshwater fish production has made a strong

Top 10 Cattle Slaughtering States, by Percentage of Commercial Cattle Slaughter, 1991

TOP 10 CATTLE SLAUGHTERERS IN 1990 (RANKED ALPHABETICALLY)

American Foods Group
BeefAmerica Operating Co., Inc.
ConAgra, Inc.
Excel Corp.
Hyplains Beef Co., Inc.
IBP, Inc.
John Morrell & Co.
National Beef Packing Co.
Packerland Packing Co., Inc.
Sara Lee Corp.

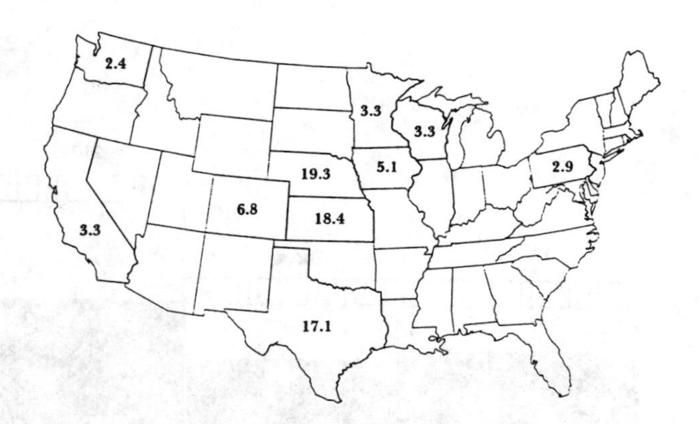

Note: Illinois not included in Top 10 Cattle Slaughtering States because slaughter data not available.

Top 10 Hog Slaughtering States, by Percentage of Commercial Hog Slaughter, 1991

TOP 10 HOG SLAUGHTERERS IN 1990 (RANKED ALPHABETICALLY)

ConAgra, Inc.
Excel Corp.
Farmland Foods, Inc.
FDL Foods, Inc.
Geo. A. Hormel & Co.
IBP, Inc.
John Morrell & Co.
Sara Lee Corp.
Smithfield Foods, Inc.
Thorn Apple Valley, Inc.

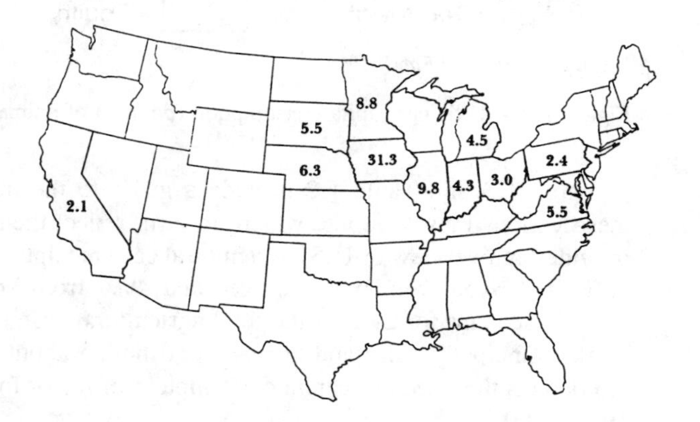

Source: U.S. Department of Agriculture

Figure 1.2. Major cattle and hog slaughtering operations in the United States.

thrust with "fish farms" developing in the southern states, producing huge amounts of catfish in particular, especially in the Mississippi River Delta region.

Thus, we have witnessed in the United States a gradual development of a muscle food industry paralleling the "founding of a nation": from east to west; from rural to urban; from horse and buggy to rail, truck, and air transportation; from hand-to-mouth to ready-to-eat foods; from agrarian to industrial culture; from workaday to leisure pursuits and from high energy to "lite and lively" food requirements. A nation's development and evolution of life-style is reflected in such eating habits and trends.

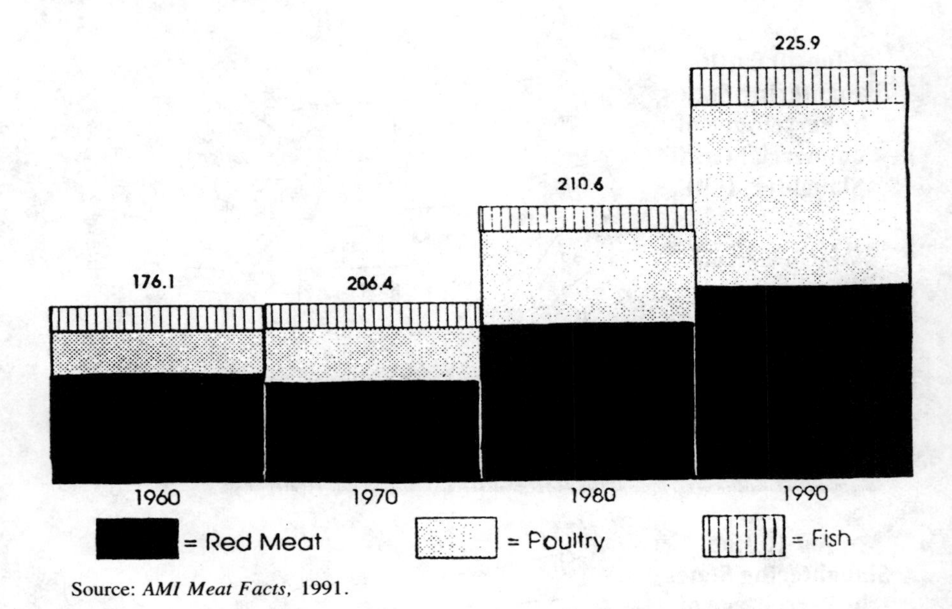

Source: *AMI Meat Facts*, 1991.

Figure 1.3. U.S. per capita consumption (pounds) of animal protein products.

Figure 1.3 and Table 1.2 provide a guide to the trends in consumption of these various muscle foods, which, in turn, reflect their production. Figure 1.4 provides an overview of U.S. agricultural cash receipts for 1991. Approximately 46% is derived from plant sources and 49% from animal production. Meat animals account for 29% of the total agricultural cash receipts, with other livestock, principally dairy and horses, contributing about 20% of the total. Table 1.3 depicts the U.S. per capita consumption of major foods, in 5-year intervals, 1970–1990.

The leading beef and hog slaughtering states are depicted in Fig. 1.2 which also cites the top 10 companies involved in processing these two species. They also represent the leading boxed meat sales companies.

Table 1.2. American Consumption of Animal Protein Products

Type	Eaten Daily	Eaten Weekly	Never Eaten
Red meat (beef, pork, lamb)	15%	85%	3%
Chicken or turkey	5%	89%	2%
Fish or seafood	2%	68%	7%

Respondents claiming to be vegetarians: 5%
(of that 5%, 70% say they sometimes eat meat)

Source: National Restaurant Association Gallup Survey, June 1991.

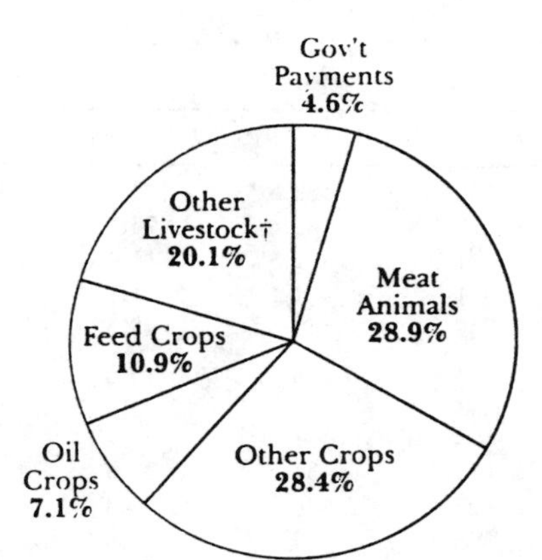

†Includes dairy products, poultry and eggs
Source: U.S. Department of Agriculture.

Figure 1.4. Cash receipts from farming, 1991.

The relatively recent and innovative concept of "boxed beef" has made its presence known in the muscle food industry with the advent of "further processing" of beef initially at the packing plant. This method involves breaking, boning, and trimming the beef carcass at the slaughter plant, vacuum-packaging those trimmed boneless cuts, and boxing and shipping them all over the nation. This has revolutionized the industry so that much less waste (bone and fat) is shipped. These by-products can be better utilized in quantity, in a fresher more sanitary manner, and with greater flexibility at the plant. Furthermore, the cuts are "knife-ready" and, therefore, need less labor at the retail end, require less storage space, lend themselves to better quality assurance and inventory control with less waste disposal, and a much longer shelf-life. Additionally, it permits shipment of preferred cuts to those regions demanding specific cuts without the necessity of shipping full carcasses. Boxed beef has grown steadily, replacing "hanging beef" in quarters, to the extent that it now commands approximately 85% of the quality or retail market for beef. Efficiency, convenience, sanitation, and economy are all served well by the boxed beef concept, which is being utilized by the rest of the muscle food industry to an increasing degree.

Truly we have evolved "from corned beef to boxed beef" in our 500-year development from a colonial upstart to a global leader.

From Exporter to Importer

As the United States developed and gradually moved from an agrarian to an industrial nation and from a predominantly rural to urban population, it naturally

Table 1.3. What Consumers Eat. Per Capita Consumption of Major Foods, 1970 to Present (in Pounds)

Commodity	1970–1974	1975–1979	1980–1984	1985–1989	1989	1990
Meats	130.2	128.6	123.8	120.0	115.9	112.3
Beef	79.1	82.8	73.1	70.5	65.4	63.9
Veal	1.7	2.3	1.4	1.3	1.0	.9
Lamb and mutton	1.9	1.1	1.1	1.0	1.1	1.1
Pork	47.7	42.4	48.3	47.1	48.4	46.3
Fish and shellfish	12.1	12.8	12.9	15.4	15.6	15.4
Poultry Products						
Eggs	37.9	34.5	33.5	31.6	29.9	29.6
Chicken	27.9	30.7	36.3	43.2	47.3	49.4
Turkey	6.8	7.2	8.6	11.7	13.6	14.5
Fluid whole milk	205.2	167.9	135.4	110.7	97.6	90.3
Fluid lowfat milk	46.3	79.5	83.9	99.5	106.5	100.3
Cheese	12.9	16.0	19.5	23.5	23.9	24.7
Butter	5.0	4.4	4.6	4.6	4.4	4.4
Margarine	11.0	11.4	10.9	10.7	10.2	10.9
Fresh fruit						
Citrus	27.1	26.3	25.5	24.3	24.1	21.0
Noncitrus	48.6	54.3	61.2	69.1	71.5	67.2
Fresh vegetables	82.6	83.9	87.0	97.0	103.8	102.3
Potatoes						
Fresh	53.3	47.5	46.5	46.5	47.7	43.6
Frozen	14.9	20.2	19.8	22.9	23.3	24.9
Dry edible beans	6.5	6.3	5.9	6.3	5.5	6.0
Sugar (refined)	100.5	91.5	74.7	61.9	62.5	64.2
Corn sweeteners	21.9	31.3	48.5	68.5	70.3	71.9
Low-calorie sweeteners	5.4	6.6	10.8	18.9[a]	—	—
Flour and cereal products	134.6	141.5	148.2	167.5	175.0	185.4
Rice, milled	7.2	7.4	10.1	12.9	15.6	16.6
Pasta	8.5	10.1	10.2	11.7	12.6	—
Coffee (gallons)	33.1	29.0	26.3	26.8	26.9	26.7
Soft drinks (gallons)	26.2	32.2	35.4	38.6	41.8	42.5

[a] 1985–1988.

Note: Some remarkable changes in eating habits in recent years. Less red meat especially beef. Fewer eggs but a lot more chicken and turkey. Less milk overall and a lot of substitution . . . low-fat milk for whole milk. Less butter but more cheese . . . and more yogurt. Much higher consumption of fresh fruits and vegetables. Significant jumps in cereal, flour, pasta, and rice. Sugar slipping, corn sweeteners, and low-calorie sweeteners soaring.

Source: USDA (1992) Agr. Statistics Board Reports

followed that the urban, industrial portion of the population became more dependent on the agricultural sector to provide its food requirements. This trend continued, with an increasing national population of approximately 260 million, to the point that we as a nation now import agricultural products from many continents. However, the United States exports more agricultural products than it imports, giving it a positive export:import balance of $19.7 billion. This includes large quantities of meat in several forms. In 1991, the United States exported 1.7 billion pounds of meat and poultry to more than 40 countries, with a value of $2.5 billion. It imported 2.6 billion pounds of meat from 29 countries. Five countries accounted for approximately 90% of the imports.

For many years, the United States has participated in activities of the Codex Alimentarius Commission, an international organization which develops food standards to protect consumers and promote fair trade. Codex has 138 member nations and operates under the auspices of the Food and Agriculture Organization and the World Health Organization of the United Nations. Table 1.4 gives a global overview of the consumption of red meats and poultry, by country, and supports the theory that people consume mostly those foods that they produce in greatest abundance. For example, on a per capita basis, beef is consumed in greatest amounts in Argentina and Uruguay; the leading consumers of pork are Hungary, Denmark, Austria, Germany, and Spain; whereas lamb, mutton, and goat meat is most popular in New Zealand, Uruguay, Australia, and Greece. Broiler consumption is greatest in the United States, Hong Kong, Saudi Arabia, Israel, and Canada. Turkey meat is most frequently consumed in Israel, France, Canada, and Italy.

Imports

As of 1991, there were 1,370 foreign plants authorized to export meat and poultry to the United States, of which 637 plants were Canadian. Australia followed with 134 and Denmark with 129. More than 2.6 billion pounds of meat and poultry entered the U.S. in 1990 under federal inspection. The United States is also the world's second largest importer of fishery products.

Australia shipped the most meat, 846 million pounds, principally boneless cow beef for the processing trade. Canada shipped 676 million pounds, primarily pork; and New Zealand shipped 435 million pounds, largely as beef and lamb. The vast majority (81%) of all meat and poultry imports arrived as fresh or fresh frozen commodities, whereas only 17% was processed meats.

Canned hams and other canned pork products imported into the United States originate primarily in Denmark, Hungary, Poland, and The Netherlands. Only canned, thermally treated products are permitted into the United States from countries experiencing hoof and mouth disease (aftosa) in their cattle population or vesicular diseases in their hogs. Thus, the United States imports only canned corned beef or cooked meat from Argentina and other Latin American countries.

Table 1.4. Per Capita Meat Disapperance in Selected Countries,[a] 1992

	Beef and Veal	Pork	Lamb, Mutton and Goat	Broilers	Turkey
	Pounds per Person[b]				
North America					
Canada	78.5	73.4	N.A.	51.7	11.0
Mexico	42.5	20.9	2.4	24.1	1.3
United States	96.2	68.4	1.6	66.8	18.0
South America					
Argentina	152.1	N.A.	4.9	35.5	N.A.
Brazil	49.6	15.2	N.A.	34.8	.8
Colombia	46.5	7.7	N.A.	N.A.	N.A.
Uruguay	133.4	N.A.	N.A.	N.A.	N.A.
Venezuela	39.5	10.4	N.A.	32.1	N.A.
Western Europe					
Belgium-Luxembourg	49.6	107.4	3.7	27.5	4.7
Denmark	49.2	145.1	2.6	21.3	3.4
France	63.5	81.1	12.8	24.3	12.9
Germany	43.2	107.1	2.2	16.2	7.3
Greece	43.0	47.6	12.1	30.7	1.1
Ireland	39.5	84.4	18.1	28.2	8.8
Italy	57.1	69.7	4.0	24.5	10.2
Netherlands	41.7	93.5	2.4	30.2	4.5
Portugal	36.2	54.9	8.4	42.2	6.5
Spain	30.4	110.0	15.4	48.0	2.4
United Kingdom	45.2	51.6	15.0	39.8	8.4
Austria	47.4	112.7	N.A.	18.2	N.A.
Finland	45.9	45.9	N.A.	14.5	N.A.
Sweden	38.1	120.8	N.A.	N.A.	N.A.
Switzerland	57.8	69.0	N.A.	N.A.	N.A.
Eastern Europe					
Bulgaria	27.8	91.9	16.1	N.A.	N.A.
Former Czechoslovakia	60.2	112.9	.9	22.2	N.A.
Hungary	15.0	106.3	.4	34.1	N.A.
Poland	37.8	116.4	.9	10.7	.7
Romania	32.8	51.4	9.3	N.A.	N.A.
Former Yugoslavia	30.2	69.4	5.5	15.9	1.1
Former U.S.S.R.	56.9	39.9	6.4	13.0	N.A.
Middle East					
Israel	36.6	N.A.	N.A.	58.0	22.5
Saudi Arabia	9.7	N.A.	41.4	69.0	N.A.
Africa					
Egypt	21.2	N.A.	3.3	5.6	N.A.
South Africa	39.0	N.A.	12.8	26.1	N.A.

continued

Table 1.4. (Continued)

	Beef and Veal	Pork	Lamb, Mutton and Goat	Broilers	Turkey
	Pounds per Person[b]				
Asia					
China	3.1	49.4	2.4	3.8	N.A.
Hong Kong	25.4	80.9	N.A.	78.2	N.A.
Japan	20.5	36.8	2.0	29.3	N.A.
Korea	15.9	34.4	.2	N.A.	N.A.
Taiwan	6.6	86.6	N.A.	41.9	N.A.
Philippines	4.5	23.4	N.A.	N.A.	N.A.
Oceania					
Australia	77.2	40.8	43.0	49.2	N.A.
New Zealand	65.3	30.9	59.3	N.A.	N.A.

[a] Preliminary data for countries other than United States.

[b] Carcass weight basis.

N.A.—Not available.

Source: AMIX 1993 Meat and Poultry Facts

Central and South American countries do ship a great deal of fresh and frozen beef to the United Kingdom and Europe. Australia and New Zealand export a large amount of beef, lamb, and mutton to the United States. More recently, the demand from Japan for beef and the demand from the Middle East for lamb and mutton have attracted those products from Australia and New Zealand.

Table 1.5 reflects the principal commodities imported by the United States from specific countries. Figure 1.5 depicts the types of meat products imported by the United States. Fresh meat predominates (83%), with processed meat next (16%). Very little poultry (0.6%) is imported. The principal countries from which the United States imports these meats are shown in Fig. 1.6. In 1991, five countries were responsible for 90% of those products, with Australia, Canada, and New Zealand accounting for 75% of the total U.S. meat imports.

With a human population of 15 million people and a livestock population of

Table 1.5. U.S. Imports of Selected Red Meats, by Product and Country

Boneless Beef	Canned Corned Beef	Fresh or Frozen Pork	Canned Hams and Shoulders
Australia	Brazil	Canada	Denmark
New Zealand	Argentina	Denmark	Poland
Canada	Uruguay		Yugoslavia
Costa Rica			Hungary
			Netherlands
			Romania

Source: U.S. Department of Commerce.

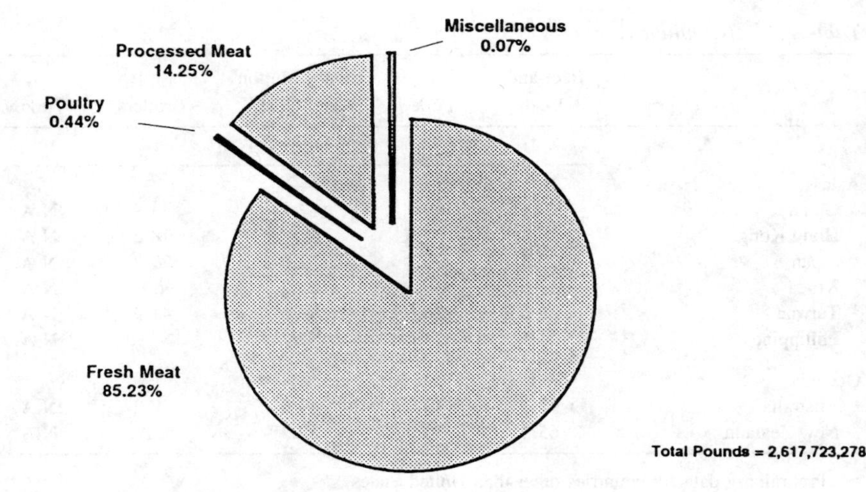

Source: USDA, FSIS, Meat and Poultry Inspection, 1992.

Figure 1.5. Types of products imported into the United States.

25 million cattle and 165 million sheep, Australia is the leading international exporter of meat. The United States, Japan, Korea, and Canada are its major markets. In the United States in 1990 there were 51 importers of Australian beef and in Canada there were 25. The bulk of this is lean, grass-fed manufacturing grade product, usually boneless and frozen. Traditionally there has been a 20:80 ratio between cuts and manufacturing beef from Australia to the United States, although chilled, fresh beef is a growing segment. For 1991, Australian lamb and mutton imported into the United States remained stable (18 million pounds). Air-freighted chilled Australian lamb is evident at the consumer level in supermarkets and the HRI (hotel, restaurant, institution) trade.

Exports

The largest buyer of U.S. meat products is Japan which purchased 813 million pounds valued at $1.4 billion in 1990. This represented 44% of U.S. meat exports. Mexico followed with 25%. The former U.S.S.R. and Eastern Europe were significant markets for U.S. poultry (19.9 million pounds) worth $89.6 million. Japan and Hong Kong were the largest importers of U.S. poultry, with Mexico following closely. The United States is the world's largest exporter of fishery products.

The United States is attempting to increase its export of meat and poultry to the European Common Market but has had difficulty in doing so, as the European Community seeks to resolve its own 15-nation market relationships and inherent problems of national protectionism relative to agricultural products and econo-

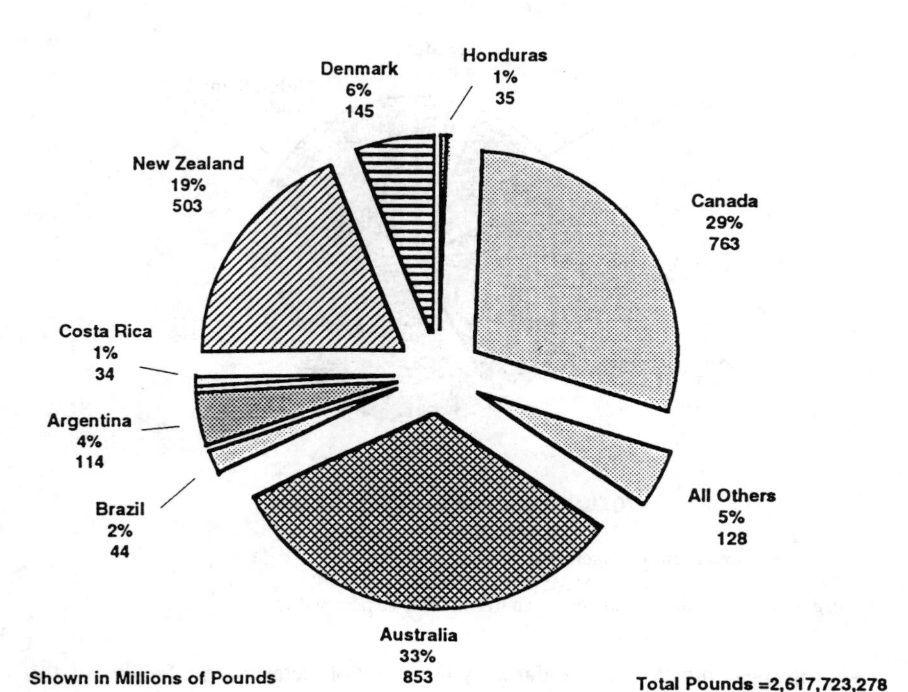

Source: USDA, FSIS, Meat and Poultry Inspection, 1992.

Figure 1.6. Source of product imported into the United States.

mies. The European market continues to be a good one for U.S. variety meats or edible offal, such as liver, heart, tongue, kidneys, sweetbreads, and other glandular meats, and edible fats and oils derived from animal products.

The total U.S. animal related export value for 1991 was $7.025 billion. Red meat accounted for 30%, poultry 15%, and variety meats 7%, or, combined, about 52% of that total. Hides and leather, fats, live animals, and other animal products constituted the remainder (Fig. 1.7). Figure 1.8 identifies the leading markets, by country and trade, for U.S. meat exports. Japan accounts for 44%, Mexico 25%, Canada 13%, and Korea 7% for a total of 89% of U.S. meat exports. International trade in meat, poultry, and seafoods is an important area and will continue to expand with the increasing world population, the global political repositioning, and the emergence of more developing countries. The U.S. export growth in fresh, chilled, and frozen red meats is noticeable in Fig. 1.9 comparing 1991 to 1987. This is further demonstrated by the expansion of U.S. poultry exports to world markets, especially Japan, Hong Kong, Russia, and Mexico (Fig. 1.10). Turkey meat, in particular, has become a very popular export item to Mexico and more recently to Korea (Fig. 1.11). The economy

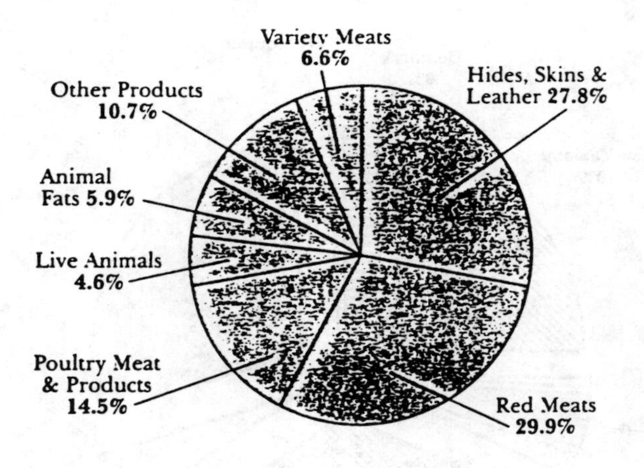

Source: U.S. Department of Commerce

Figure 1.7. Livestock and meat, share of total export value.

and improved nutritional standards will be major determining factors in this trend. Distribution is no longer a major problem, because of modern transportation and preservation systems coupled with advances in food microbiology and packaging methods. With food and proper nutrition as important ingredients in international well-being and ultimate world peace, muscle foods will continue to play a vital role globally.

Muscle Food Facts

One-half of the total food supply worldwide is from mammalian, avian, and aquatic sources. U.S. farmers, representing considerably less than 1% of the world's population, annually produce 29% of the meat, 46% of the milk, and 51% of the eggs consumed in the world. The average U.S. consumer spends 5–6% of total income, or 25% of food budget, for meat. Two-thirds of U.S. agricultural lands are devoted to livestock and animal feeds. An Arabian proverb states: "A land poor in livestock is never rich, a land rich in livestock is never poor." Some countries are more efficient in their food production systems than others. Soil, climate, population, mechanization, agricultural systems, transportation, marketing methods, food preservation capabilities, and population density are some of the major reasons for the disparities; for example, the percentage of population in agriculture (United States = 2% versus China = 71%), percentage of personal spending for food (United States 13.3% versus China = 60.0%),

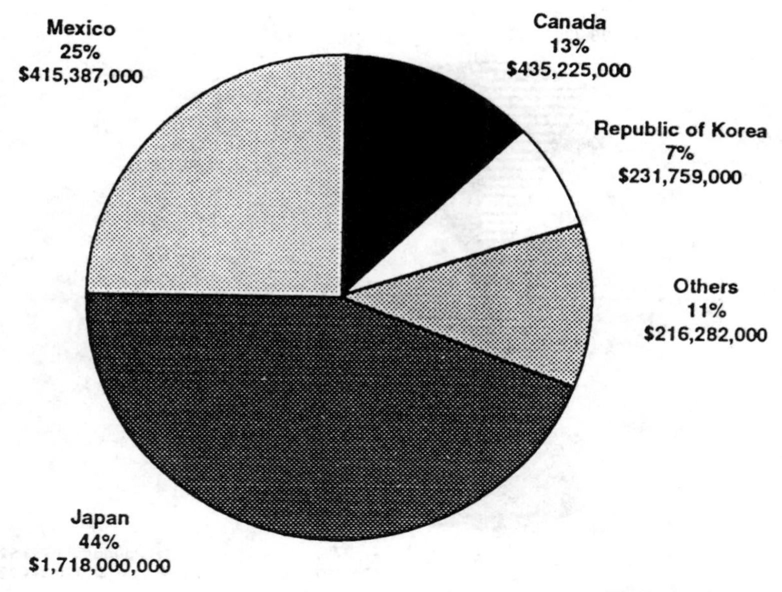

Source: USDA, FSIS, Meat and Poultry Inspection, 1992.

Figure 1.8. Major receivers of U.S. meat exports.

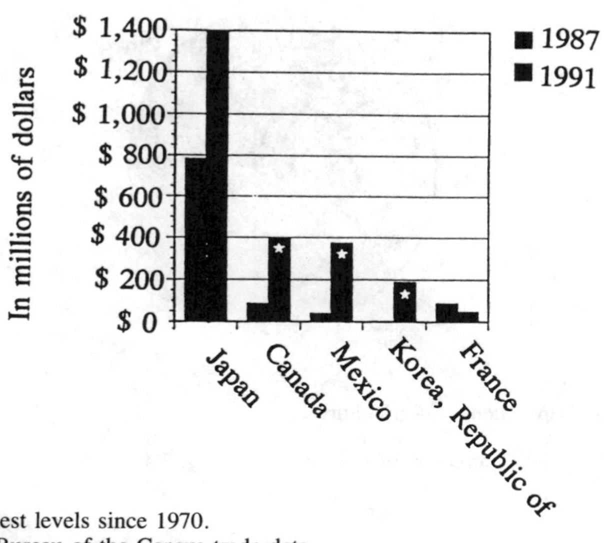

*Denotes highest levels since 1970.
Source: U.S. Bureau of the Census trade data.

Figure 1.9. Export growth: fresh, chilled, and frozen red meats.

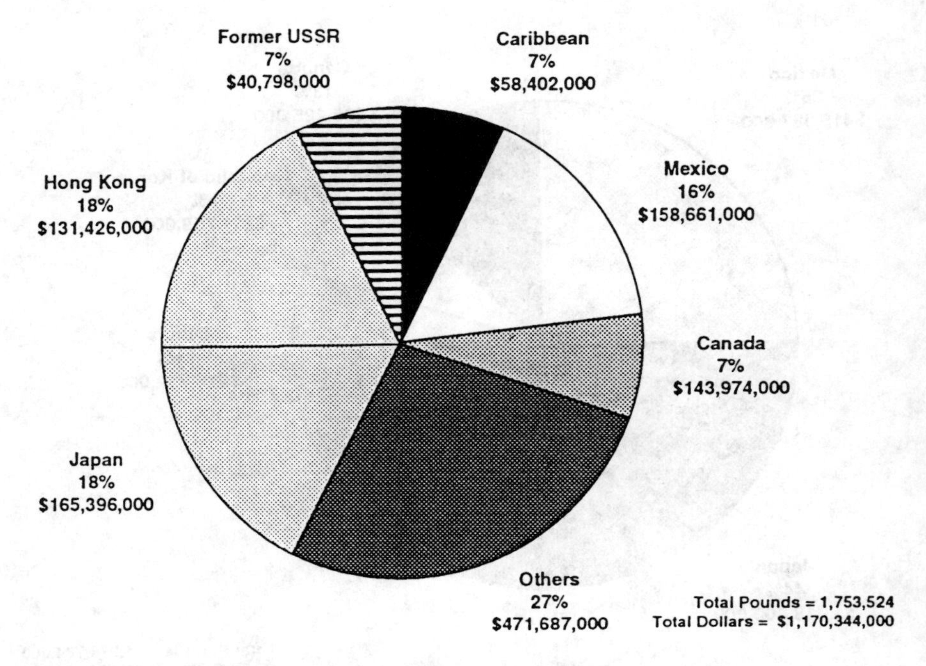

Source: USDA, FSIS, Meat and Poultry Inspection, 1992.

Figure 1.10. Major receivers of U.S. poultry exports.

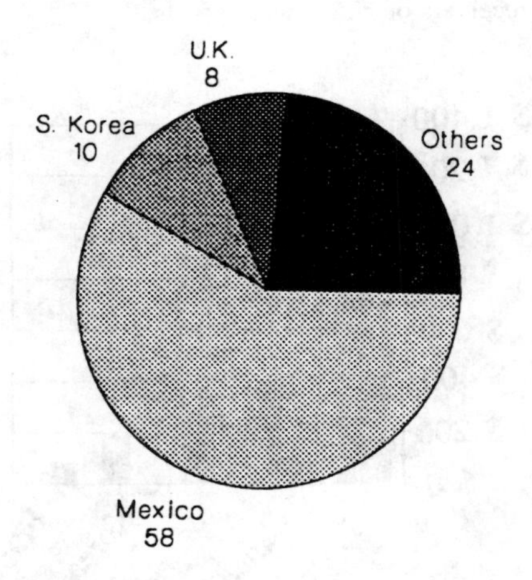

Source: U.S. Department of Agriculture

Figure 1.11. Major importers of U.S.-grown turkey.

food/fiber trade balance (United States = +19.7% versus Japan = −27.4%), minutes worked to earn 1 pound of meat (Canada = 8, China = 180). The aforementioned factors all impact on these statistics. With food surpluses in some areas and great deficiencies in others, distribution becomes a major obstacle requiring appropriate transportation, coordination of program, storage facilities, pest control (insects and rodents), and sociological and ethnic differences relative to preparation and acceptance of "different" foods.

Table 1.2 compares the American consumption of animal protein products (muscle foods) by type and frequency of consumption, thereby providing an indication of choice or preference by the consumer. Twenty-two percent appear to consume at least one of the muscle foods daily and practically all do so in the course of a week. More respondents (7%) never eat fish or seafood, whereas 3% and 2% respectively never eat red meat or poultry.

Figure 1.3 graphically portrays the U.S. per capita consumption of muscle foods at the start of each of the most recent decades, comparing red meats with poultry and fish.

Some interesting trends appear in the data presented in Table 1.3 and later in Table 1.7:

1. Per capita beef consumption, which peaked in the mid-1970s (about 86 lbs), is approximately the same now as in the early 1960s (about 63 lbs).
2. Lamb and veal, although not major meats, have declined from two to three pounds per capita each to about 1 lb each during this period.
3. Pork has remained about the same, with some fluctuations, based largely on its competition with beef.
4. Total per capita edible red meat consumption today, on a national basis, is not greatly different from 1960 (112 lbs versus 116 lbs, a 3.2% decrease).
5. Per capita poultry consumption has escalated dramatically with a 40-lb or 163.9% increase in this 30-year period. Broiler and turkey consumption have both tripled during that period of time.
6. Per capita fish consumption has climbed from 10 to 16 lbs per capita since 1960 (a 55.3% increase) and is predicted to be 20 lbs by the year 2000.
7. Total "white meat" (poultry and fish) per capita consumption has increased tremendously (by 45.3 lbs per capita) for a total of 130% since 1960.
8. Total muscle food per capita intake in the United States increased in the past 30 years, from 176.1 lbs to 218.8 lbs or 23.7% on a retail weight basis.

9. The "white meat" (poultry and fish) percentage of total muscle foods increased from 25% to 45%, whereas the red meat consumption expressed as a percentage of total muscle foods declined accordingly, to 55%.

Consumption Patterns of Muscle Foods

Eating habits change with national development and with differing life-styles. Where once people were basically "meat and potato" oriented and high caloric intake consumers in keeping with heavier physically demanding occupations, they have evolved through greater mechanization and automation to more sedentary, or at least, less strenuous work. Our protein needs may have remained much the same, but our energy requirements have lessened on a national basis. Health factors in face of changing work and life-styles, environmental and welfare concerns, and availability of wider food choices have also influenced our eating habits and needs. These trends reflect the many factors influencing our modern food, nutrition, and health concerns especially impacted by speculation about saturated fats and cholesterol and their relation to health. There is much yet to be learned about the intricate interrelationships of these many factors and their effects on human health, especially in view of our changing life-styles. The former Surgeon General of the United States, C. Everett Koop, M.D. summarized the situation by the following statement (October 1991):

> We have become more concerned about what we should *not* eat than what we should. Food fads come and go, but the basics of a good diet are actually very simple.

His daily dietary recommendations include

- Four to five servings of fruits, vegetables, and grains
- Four servings of bread, pasta, rice, or cereal
- Two servings of milk, cheese, ice cream, or yogurt
- Two servings of meat, fowl, eggs, fish, nuts, or legumes.

Concerning the issue of cholesterol, he notes that the entire population should not overreact to a problem that affects but 25% of all people. In addition, because cholesterol is manufactured in the body naturally, the cholesterol in one's diet does not have the direct relationship to one's blood cholesterol level as many assume.

Table 1.6 shows the U.S. consumer per capita spending for beef, pork, and chicken over the past 15 years. Beef's portion has decreased approximately 7% (from 60% in 1975 to 53% in 1990), pork has increased 3% (from 26% to 29%), and chicken has increased about 6% (from 12% to 18%) when compared on a

Table 1.6. Consumer Per Capita Spending. Beef, Pork, and Chicken

Year	Beef $Spent	Pork $Spent	Chicken $Spent	Total $Spent
1975	134.24	57.88	25.58	217.70
1976	137.69	60.97	25.71	224.36
1977	133.64	58.94	26.70	219.28
1978	155.56	67.49	31.28	254.33
1979	173.69	77.38	34.51	285.58
1980	178.70	79.93	36.29	294.93
1981	181.19	83.36	38.32	302.88
1982	183.33	86.12	38.18	307.64
1983	183.53	87.62	39.35	310.50
1984	184.16	83.11	45.44	312.70
1985	180.37	83.75	44.41	308.53
1986	178.04	87.06	49.60	314.70
1987	175.22	92.32	49.69	317.23
1988	180.97	95.73	55.63	332.33
1989	184.40	94.56	63.83	342.79
1990	189.39	105.66	64.37	359.42
1991[a]	193.50	108.14	67.61	369.26

[a] Projected. Bulletin

Note: Per capita consumer expenditures for meat are projected to increase 2.7% in 1991, but expenditures for chicken are projected to increase 5%, nearly twice as fast as the spending for beef and other meats.

Source: *Cattle-Fax.*

dollar spent basis. These differences are still relevant even though inflation and price increases are reflected in these figures.

The market share of these three major and competing muscle foods is evident in the pie graphs of Fig. 1.12, wherein beef has fluctuated from 42% to 46% to 31% during the past 25 years and is projected at 27% by the year 2000. Pork has been quite constant at 31, 26, and 28%, with 28% projected. Poultry has increased from 23% to 24% to 40%, with 44% projected.

These several methods of analyzing muscle food consumption patterns for the United States all clearly show a definite trend of more seafood and poultry and less red meat. In total, however, recognizing that these three commodities are highly competitive, note that total muscle food consumption (Table 1.7) on a retail weight basis has increased (from approximately 176 to 219 lbs) 43 lbs, or a 24% increase from 1960 to 1991. These "shares" for 1991 are seen in Fig. 1.13 with red meats at 55%, poultry at 38%, and seafood at 7%.

Other factors influencing national eating patterns include ethnic or religious traditions. In a transient society, ethnic preferences tend to get lost with succeeding generations or at least blend into the new society where they relocate. Fortunately, however, most ethnic cuisines are perpetuated especially where

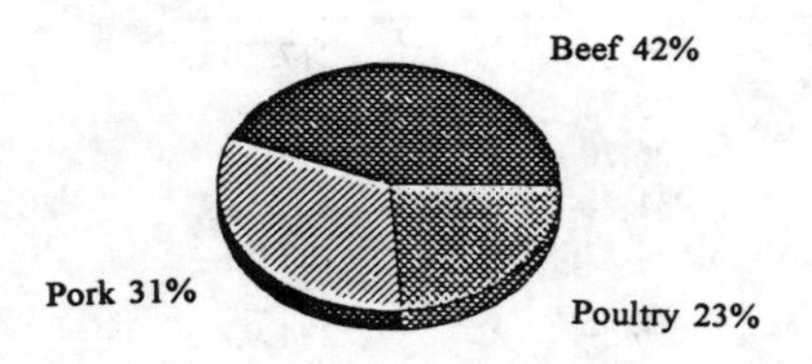

Market Share - 1976

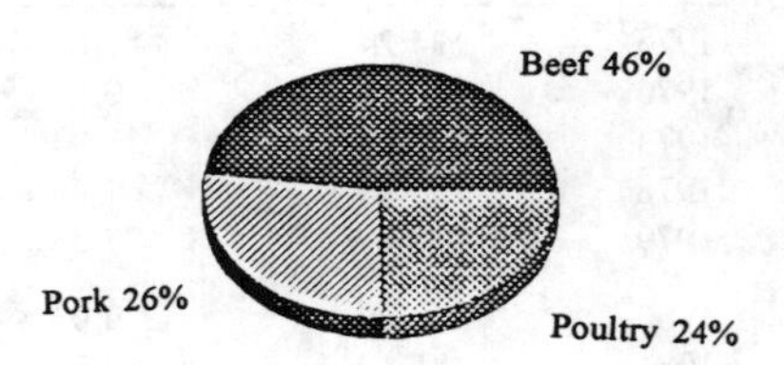

Market Share - 1990

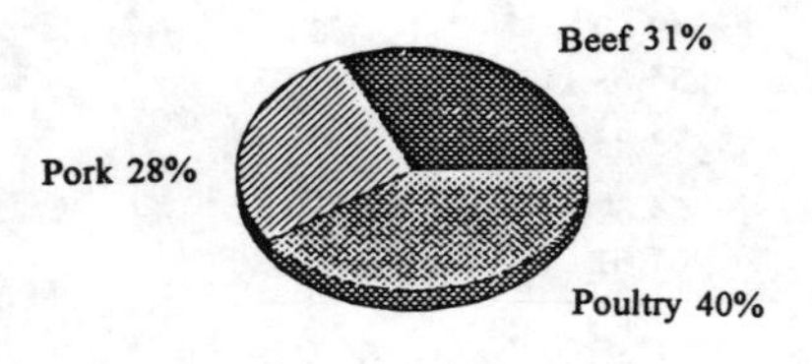

Market Share - 2000

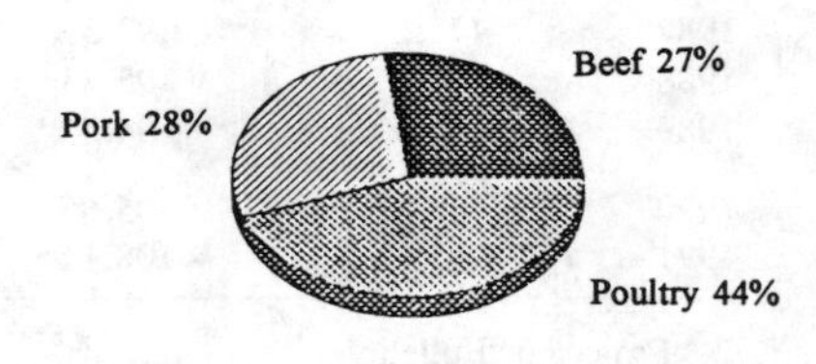

Source: USDA; The Heming Group; Beef Industry Council; National Live Stock and Meat Board.

Figure 1.12. Per capita meat consumption, market share.

Table 1.7. Per Capita Consumption of Red Meat, Poultry, and Fish. Retail Weight Basis, 1960–1991

Year	Red Meat	Poultry	Fish	Total Red Meat, Poultry, and Fish
1960	131.7	34.1	10.3	176.1
1965	134.1	41.2	10.8	186.1
1970	145.9	48.7	11.8	206.4
1975	136.3	47.7	12.2	196.2
1980	136.8	59.1	12.5	208.4
1985	133.8	65.7	15.1	214.6
1990	120.1	80.6	15.0	215.7
1991	120.2	83.7	14.9	218.8

Source: AMI 1992 Meat & Poultry Facts.

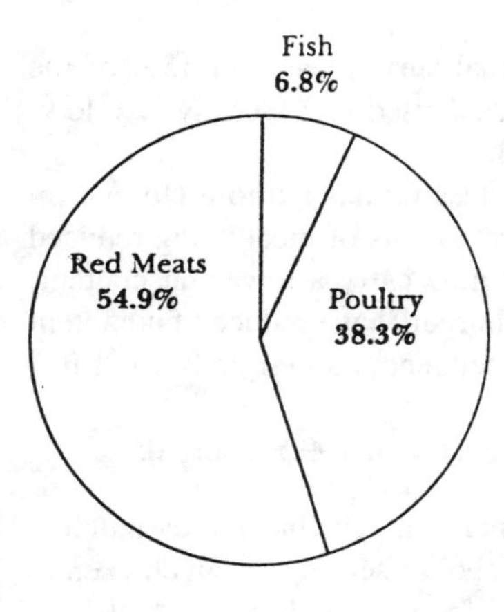

Figure 1.13. Meat consumption shares.

large settlements of a single ethnic group develop and customs are preserved. Religious traditions have been continued for centuries by some sects and are adhered to through the requirements for ritualistic slaughter and processing procedures by Jewish Kosher and Mohammadan Halal laws. These are followed closely on a daily basis by the orthodox observers and most especially on religious high holidays by all of those faiths.

Additionally, seasonal or temperature variables impact on eating patterns on a short-term basis. Prices of muscle foods influence buying decisions between competing meats as well as with other foods.

The American consumer's food buying decisions today are driven by (1) variety of choices represented by new products, many of ethnic origin, with Spanish and Southeast Asian influences most evident; (2) convenience, with both parents working, microwaveable and fast foods are in greater demand; (3) health and diet considerations, in 1991 more than 600 new products were introduced featuring reduced salt, sugar, cholesterol, and fat content; and (4) price, especially for the lower economic third of the population.

Recent Trends in Eating

With concerns about ingestion of fat, saturated fatty acids and cholesterol and their contributions to atherosclerosis, arteriosclerosis, and heart attack, there has been considerable deliberation about the relationship of diet and health. The USDA has recommended that no more than 30% of calories be derived from fat and 10% or less from saturated fat. The American consumer has embraced "Lite" foods. Such items are generally sugar free, low calorie, and low fat. In fact, by definition the term "Lite" must assure a reduction of at least 25% fat, from the norm. The primary reasons for using "Lite" foods have been (1) better health, (2) reduction of cholesterol, (3) maintenance of weight, and (4) appearance.

Figure 1.14 depicts the results of a 1991 national survey wherein 45% of the populace was using low-cal and low-fat foods; 22% used low-fat only, 9% low-cal only and 24% were nonusers of "Lite" products.

This demand on the part of the consumer for less fat has led to a closer trim of fresh meats with the maximum fat cover on retail cuts of meat being reduced from ½ in. to ¼ in. to ⅛ in. Processed meats now carry a lower fat content, usually under 25% and some much lower. Beef burgers have reduced fat content from 30% (legal limits) to as low as 9%. The tendency is definitely to "Lite" foods, meat included.

Balzer, writing for the *Meat and Poultry Magazine* in 1992 reported:

- Fat and cholesterol continue to be consumer concerns but not as much as just a year ago. The percent of households serving fried chicken regularly—not a low-fat, low-cholesterol food, though it is chicken based—has changed little over time.
- Pork, fresh and processed combined, has suffered the most from the increased popularity of chicken and turkey.
- The trend now is to eat more meals at home. More restaurant meals are takeout.

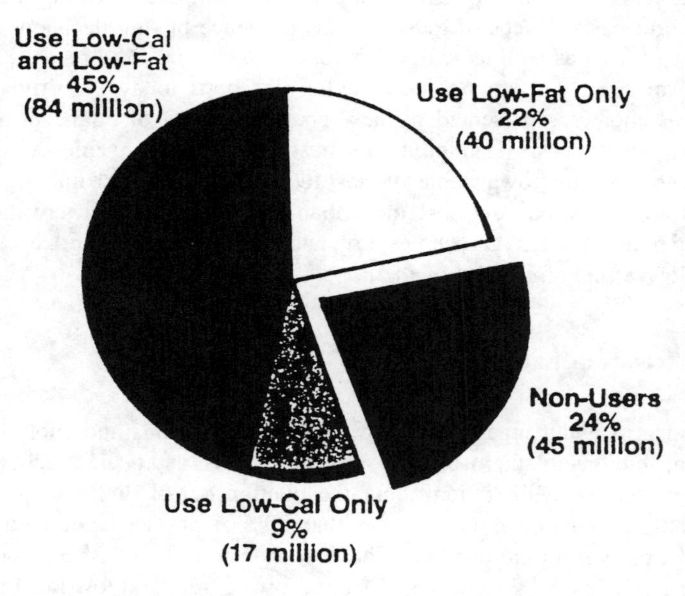

Source: Calorie Control Council 1991 National Survey

Figure 1.14. One hundred forty-one million consumers see the "Lite."

- Brown-bagging is increasing, especially over the last 2 years, but no one seems to be capitalizing on the trend.
- BBQ grilling is growing, perhaps because more women are working and more men are cooking as a result.
- Three-fourths of all individuals eat chicken at least once in an average 2 weeks, but the average consumer eats chicken once a week.
- More children are eating chicken. The increase jumped 19% from 1982 to 1990, compared with increases of 5% for adult males and 7% for adult females.
- The most skipped meal is lunch—not breakfast, as most people in the industry seem to think.
- Red meat is still the most preferred food eaten at home.
- What we eat in restaurants is mostly beef burgers, but note the realignment of the top five:

1977 Preference	*1990 Preference*
1. Burgers	1. Burgers
2. Fish	2. Pizza
3. Beef	3. Chicken
4. Chicken	4. Fish
5. Pizza	5. Mexican (mostly beef)

A 1992 survey of consumer attitudes about meat, conducted by the National Live Stock and Meat Board (Table 1.8) provides some interesting insight into the rationale of today's food shoppers.

The 21st Century

What lies ahead in the 21st century pertaining to the role of muscle foods in the human diet? There is no question but that muscle foods will continue to be the food of choice around which meals and diets will revolve. It cannot be replaced

Table 1.8. Consumer Attitudes About Meat

	Consumers
Main meal must include meat	24%
Meat is best part of meal	25%
Meat is healthier than other foods	12%
Meat can fit into reduced fat diet	54%
I am trying to avoid cholesterol	48%
Would buy more meat if it were less expensive	13%

Source: National Live Stock and Meat Board (1992).

as the single most naturally occurring nutrient-dense foodstuff. Even if a suitable substitute should be concocted, the worldwide availability and distribution of "erzartz meat" could not be accomplished. On a global basis, people consume that which is most readily available as foodstuff. Meat is of universal and primary concern as a source of essential nutrients.

There may be fluctuations in the popularity of various muscle foods as has been observed during the past several decades. Diet and health concerns as well as price and availability will dictate many of these decisions. As facts become known and research demonstrates the efficacy of new knowledge and its application to humans, more and more food consumption decisions will be based on that information. Eating habits then will tend to focus on those areas that give evidence of greatest health benefits.

With continued trends of more two-parent wage earners or single-parent or childless homes, there will be greater need and demand for more table-ready and microwaveable foods that require a minimum of preparation time. In any event, the principal nutrient item of menus will be muscle foods. These will be in a boneless, relatively fat-free form that will maintain an attractive, flavorful condition that will be highly nutritious and have a high satiety value.

Selected References

USDA, *Agricultural Statistics Board Reports*. 1992. United States Department of Agriculture, Washington, DC.

American Meat Institute. 1993. *Meat Facts,* 1993. American Meat Institute, Washington, DC.

Basu, J.M. and R.C. Cassens. 1980. *A Brief History of the U.S. Meat and Livestock Industry*. American Meat Science Association in Conjunction with Meat Industry Magazine and National Live Stock and Meat Board, Chicago, IL.

Breidenstein, B.C. and J.C. Williams. 1989. *Contribution of Red Meat to the U.S. Diet*. National Live Stock and Meat Board, Chicago, IL.

Economic Research Service. 1992, *National Agricultural Statistical Service Reports. 1992*. Economic Research Science, Rockville, MD.

Hinman, R.B. and R.B. Harris. 1947. *The Story of Meat*. Swift and Co., Chicago, IL,

Kiplinger, Agriculture Letter. 1992. *Farm and Food Facts,* Kiplinger, Washington, DC.

The National Provisioner. 1981. *Meat for the Multitudes*. Staff of National Provisioner Magazine Vols. 1 and 2. The National Provisioner, Chicago, IL.

J.J. Putnam and J.E. Allshouse. 1992. *Food Consumption, Prices and Expenditures,* Statistical Bulletin Number 840, 1970–1990, United States Department of Agriculture, Washington, DC.

USDA-FSIS, 1992. *Meat and Poultry Inspection*. Report of the Secretary of Agriculture to the U.S. Congress, United States Department of Agriculture, Washington, DC.

2

Structure and Properties of Tissues

Richard J. McCormick

Introduction

Muscle tissue, which becomes a highly prized food, is also a specialized apparatus that allows the living animal movement. Properties associated with postmortem muscle, or meat, are entirely dependent on the unique architecture of muscle, how muscle functions, and how function is regulated. Muscle, whether skeletal, cardiac, or smooth, is a living system that is able to convert chemical energy into mechanical work. Intracellular contractile protein structures generate force through their coordinated shortening. A second and often overlooked aspect of muscle movement and work involves the transmission of force within the muscle, to the skeleton or to other muscles. The role of force transmission is usually ascribed to the extracellular connective tissue network of muscle. Knowledge of the characteristics of intracellular elements, collectively known as the muscle cytoskeleton, has added new dimensions to the understanding of force transmission in muscle. Understanding both facets of muscle function, force generation and force transmission, and the specialized structures associated with each are essential to understanding the properties of meat.

Muscle is not only an organ uniquely suited to the generation and transmission of force, it is also a remarkably flexible tissue which adapts to changing demands. Thus, a recurring theme of this chapter is the flexibility, or plasticity, of the muscular system. Plasticity results from mechanisms that operate at the molecular level and extend to cellular (contractile and metabolic) and extracellular (connective tissue) elements of muscle. The ability to promote desirable changes in muscle properties, and thus improve meat quality, depends on the capacity of muscle to adapt to varied management practices.

Nearly 40 years have passed since A.F. Huxley and his co-workers published their initial observations on the structural changes that occur when muscle contracts. These findings, which resulted in the sliding filament theory of muscle contraction, also formed the underpinnings of meat science. Today's knowledge and understanding of meat properties and factors that influence its characteristics have made remarkable strides in the past 30 years. For example, the well-accepted concept that toughness of cooked meat is directly related to sarcomere length or degree of overlap of actin and myosin filaments of muscle is grounded in the knowledge of how living muscle functions. Control of deleterious conditions such as cold shortening has relied on an understanding of the excitation–contraction coupling process of muscle. Likewise, the current understanding of the basis of water-holding capacity in muscle owes much to the detailed description of muscle ultrastructure. Future advances in meat science will depend on the understanding of the most current knowledge of muscle structure and function and the ability to apply this knowledge to the challenges of converting muscle to meat. Meat that is unacceptably tough still remains a leading consumer complaint. A more complete understanding of muscle developmental processes, as well as postmortem changes, related to both intracellular and extracellular components of muscle is necessary if fresh meat products are to compete successfully in the marketplace.

Muscle Anatomy and Morphology

Vertebrate Skeletal Muscle

Vertebrate skeletal muscles including the muscles of meat-producing animals can be characterized by their shape. Skeletal muscles are classified anatomically as strap, fusiform (cigar-shaped), pennate, or deltoid. Muscle shape is largely a function of muscle fiber arrangement within the muscle, and point or points of tendinous attachment to bones or other muscles. Muscles with fibers that run roughly parallel to the long axis of the muscle with tendinous inserts at either end are called strap or fusiform. Strap muscles are more nearly cylindrical along their length than the cigar- or spindle-shaped fusiform muscle. *Semitendinosus* is a muscle of the strap type and of commercial importance in meat-producing animals. Fusiform muscles, which are typical of the shank or flexor muscles of the limb, are characterized by prominent tendinous insertions, high concentrations of connective tissue, and are of lesser commercial significance because they are tough. Muscles in which the fibers are arranged in a featherlike pattern and angled relative to the long axis of the muscle are described as pennate (Fig. 2.1). Pennate muscles may be high in connective tissue (limb muscles) and tough or low in connective tissue content (spinatus muscles of shoulder) and relatively tender.

Muscle nomenclature often includes terms that describe their position relative to the skeleton (*biceps femoris*), number of tendinous attachments (biceps, triceps), or muscle fiber direction (oblique, lateral). A thorough description includ-

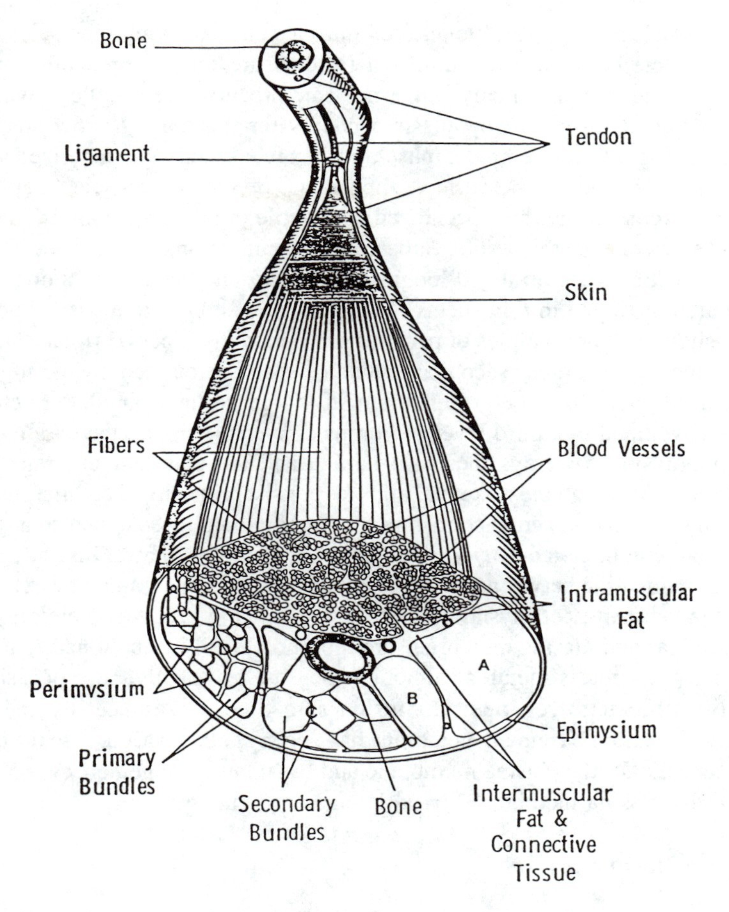

Figure 2.1. Diagrammatic representation of the gross structure of a muscle of the leg (R. Rowe, Meat Research Laboratory, Division of Food Research, CSIRO, Brisbane, Australia).

ing location and nomenclature of the major muscles of the carcass of meat producing animals is presented by Swatland (1984).

Muscles, regardless of shape or functional properties, contain a continuum of connective tissues consisting predominantly of the fibrillar collagens which encompass and invest cellular and extracellular elements. Entire muscles are surrounded by a heavy sheath of connective tissue called the epimysium. Epimysium thickens as it becomes contiguous with tendon, which forms the point or points of attachment of muscle to bone or to other muscle. The predominant cells in muscle are long, multinucleated muscle cells called myofibers, or simply fibers. Muscle cells are syncytial; that is, they are formed prenatally in meat animals by the fusion of many single-nucleate precursor cells called myoblasts. The end-to-end fusion of myoblasts accounts for the long, narrow shape and multiple nuclei of muscle fibers; muscle fibers are correctly characterized as cells which are also fibers. The striated appearance of skeletal muscle fibers arises from the sarcomeres within specialized contractile organelles (myofibrils) within each muscle cell (see the section Subcellular Organization of the Skeletal Muscle Fiber). Myofibers are arranged longitudinally into bundles of fibers or fasciculi which are enveloped in thin sheets, or trabecula, of connective tissue known as perimysium. Smaller bundles of perimysium-bound myofibers (primary bundles) are grouped into larger, secondary bundles also surrounded by perimysium. Secondary bundles may be grouped into still larger tertiary bundles. Each myofiber is enveloped in a third level of fibrous connective tissue, the endomysium. The endomysium overlays the basal lamina and the plasma membrane of the muscle cell. Although most skeletal muscles consist primarily of cellular material, that is, muscle fibers, actual contact between individual muscle cells at any point along their length, called a myomyous junction, is relatively rare. Instead, contact and communication between and among muscle cells is maintained by extracellular endomysial and perimysial connective tissues (Fig. 2.1). The endomysium, basal lamina, and plasma membrane are morphologically, functionally, and biochemically discrete structures, although they are physically, and occasionally visually, difficult to separate. Confusion sometimes arises because the terms sarcolemma and/or basement membrane have been applied variously to the plasma membrane alone, the plasma membrane and basal lamina together, or, occasionally, to the plasma membrane, basal lamina, and endomysium.

Plasma Membrane

Sarcolemma is the cell or plasma membrane; the basement membrane represents the extracellular connective structure in opposition to the muscle cell, but it is distinct from the cell and its plama membrane. The plasma membrane is the muscle cell membrane and completely surrounds each fiber. It defines the limits of the muscle cell and is similar in characteristics to the membranes of other eukaryotic cells. The plasma membrane is often called the plasmalemma or the sarcolemma. It consists of a trilaminate structure with dense inner and outer borders about 2.5 nm thick separated by a less dense middle region about 3 nm thick. The thinness of the plasma membrane (8 nm) makes its resolution by light microscopy difficult. Both the inner and outer walls of the membrane consist of phospholipid molecules oriented such that the polar groups form the outer surface and the hydrophobic fatty acid chains comprise the inner core of the plasma membrane. Membrane proteins are embedded in the lipid bilayers

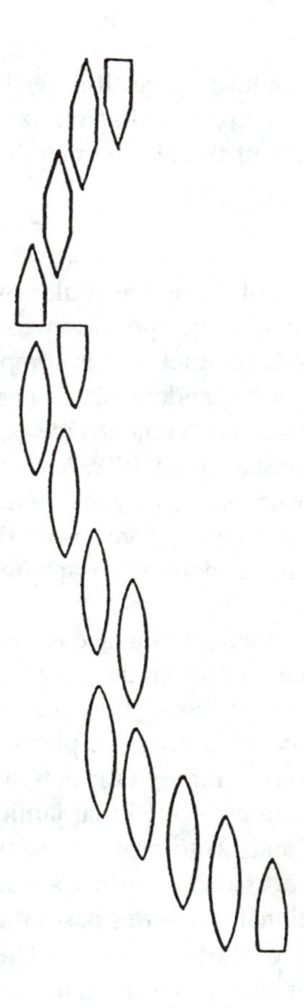

Figure 2.2. Diagrammatic representation of entire semitendinosus muscle of the goat demonstrating fiber overlap. The proximal compartment is modeled with four fibers in series, showing approximately 35% overlap at each end. The distal compartment is modeled with 10 fibers in series showing close to 50% overlap. (Gans, C. and F. de Vree, 1987. Functional bases of fiber length and angulation in muscle. *J. Morphol*. 192: 63–85 1989.)

also with polar portions facing or in contact with the inside or ouside of the muscle cell and hydrophobic regions associated with the lipid of the membrane. The exposed surface of the plasma membrane reveals a series of pores or caneolae, some of which in skeletal msucle are the opening of the T-tubule system to the surface of the fiber (see the section Force Generation). The pattern of cross-striations arising from aligned sarcomeres within the myofibrils are clearly visible through the membrane. Along and around the length of the muscle fiber the plasma membrane forms a relatively smooth surface. However, at the tapered ends of the fiber the surface is characterized by extensive invaginations and clefts. The plasma membrane of the muscle fiber has several functions: (1) it forms a selective barrier that regulates transport of ions and molecules in and out of the cell, (2) it has the ability to transmit an action potential (an electrical impulse) generated by a nerve (see the section Force Generation), and (3) it is involved in the transmission of force produced by the contractile apparatus within the muscle cell. In meat it forms a barrier against leakage of the soluble components of the sarcoplasm into the extracellular space. It is easily perforated by freezing and thawing, and frozen-thawed muscle is characterized by a large exudate that issues from the cell after rupture of the plasma membrane.

Basal Lamina

The basal lamina represents the beginning of the extracellular matrix which surrounds and envelopes all muscles cells; it is in apposition to the plasma membrane. The basal lamina is usually 30–70 nm thick and is composed of two discrete regions. The external layer which borders endomysial connective tissue is called the reticular lamina or lamina densa. The cellular side is the laminal rara or lamina lucida. The basal lamina is described as a feltlike surface composed of fine reticular fibers including some elastin and collagen microfibrils. The lamina densa forms a substratum for the overlaying connective tissue of the endomysium, although collagen fibrils of the endomysium apparently do not penetrate the lamina densa.

The lamina lucida, bordering the plasma membrane, is a lucent zone composed of a very fine meshwork of collagen (Type IV) and the proteoglycans and glycoproteins comprising the ground substance (see the section Connective Tissue in Muscle). The basal lamina is virtually inseparable from the plasma membrane by mechanical means although it can be removed through the action of bacterial collagenase and/or acid conditions. The functions of the basal lamina are (1) to regulate ion and molecular diffusions rates and (2) to provide the link between intracellular components and the connective tissue of the extracellular matrix. In meat, its contribution to textural or functional properties has not been widely elucidated. Cooking meat causes shrinking of collagen fibers. The associated force development may cause expression of fluid from the muscle cells which is in part due to heat-induced changes in the basal lamina (see the section Connective Tissue in Muscle). The effect of postmortem aging, potential proteolytic alterations in the basal lamina, and their influence on the properties of meat likewise await investigation.

Muscle Fiber Shape and Orientation

Skeletal muscle cells are notable for their remarkable length relative to their width. Muscle fibers are 10–100 μm in diameter at their widest point but typically may be millimeters to centimeters in length. Average muscle fiber diameters in meat-producing animals range from 40 to 60 μm. Muscle fibers are described as cylindrical; however, fibers in cross sections are often polygonal or irregular in shape because pressure exerted by the surrounding muscle mass or perimysium distorts fiber shape. Furthermore, fibers taper at both ends and the tapered portion can extend up to 50% of the fiber's length (25% at both ends) depending on the total length of the fiber. The angle of taper at either end has been estimated in one muscle to be about 1°. Assessment of muscle fiber cross-sectional area (diameter) is complicated by the presence of tapered fiber ends. In some muscles, fibers extend the length of the fasciculus (or muscle) and attach to connective tissue (tendon) at either end. Such attachments are called muscle–tendon junctions. In many muscles, particularly strap muscles, fibers are shorter than the muscle or the fasciculus and terminate by tapering to an end within the fasciculus (intrafascicularly terminating fibers). In muscles with intrafascicularly terminating fibers, transverse cross sections may include sections through full diameter fibers as well as sections of fibers cut at some point along their tapering length. In meat-producing animals, the number of fibers in a muscle does not increase significantly after birth. Thus, postnatal muscle growth results from the enlargement or hypertrophy of individual muscle fibers. Muscle fiber diameter is important to meat quality. Muscle texture, fine versus coarse, and concentration of connective tissue are functions of muscle fiber diameter. Shear force of meat, as reflected by the biting and chewing process of eating, will be influenced by muscle fiber size, the relative amounts and contractile state of myofibrillar and connective tissue components, and degree of myofibril fragmentation occurring during the postmortem period.

Muscle Fiber Orientation

In long, nonpennate muscles, fibers usually do not extend the length of the muscle (or fasciculus) from one point of tendon insertion to the other. Instead, they are arranged in end-to-end series with their tapering ends significantly overlapping adjacent fibers (Fig. 2.2). Such an orientation of fibers within a fasciculus results in compartmentalization of groups or populations of fibers with discrete innervation (see the section Force Transmission). The in-series arrangement of short fibers which terminate within a fasciculus suggests that physiological limitations to muscle fiber length exist. If the onset of contraction in response to a nerve impulse (action potential) begins in a portion of a very long fiber, then force generation in the fiber would occur before the action potential activated distal regions of the fiber. Thus, a contracting region of the fiber would exert force in relaxed portions. Series-fibered muscles with distinct innervation obviates this possibility (Gans and de Vree 1987).

As noted, the process of consuming meat involves shearing action and the subsequent fracture of larger pieces of meat into smaller ones. The orientation of fibers longitudinally or at right angles to the direction of shear will affect perceived tenderness. For example, meat is usually sliced at right angles rather than parallel to the length of the muscle fibers.

Fiber–Tendon and Fiber–Fiber Junctions

A physical link between muscle fibers and connective tissue is essential for the transmission of force generated within the muscle cell to other muscles and the skeleton. There are three forms of junctions which can occur in muscle tissues: (1) myomyous junctions or direct mechanical junctions between neighboring muscle fiber basal laminae, (2) muscle–tendon (myotendinous) junctions or direct attachments of muscle fibers to tendon collagen fibers, and (3) associations of in-series fibers which are maintained via the connective tissue of the endomysium. Myomyous junctions are apparently an insignificant means of fiber-to-fiber associations in most vertebrate skeletal muscle. Muscle–tendon junctions are characterized by clefts and invaginations of the muscle fiber end which forms extensions of the fiber into the connective tissue tendon of the junction. Between these muscle fiber extensions are complementary regions of connective tissue which extend into the fiber (Fig. 2.3). The effect of this three-dimensional, interdigitating array of muscle fiber and connective tissue is to greatly amplify the surface area of muscle cell–connective tissue adhesion. Thus, the muscle–tendon junction provides a large surface area over which force can be transmitted. Morphologically, filamentous structures (connecting domain) have been described that physically link intracellular (internal lamina) with external (lamina densa of basal lamina) domains.

In the case of series-fibered muscles, in which fibers end within the fasciculus, associations and force transmission between adjacent fibers are maintained by the surrounding endomysial connective tissue. Along the tapered ends of the muscle fiber, and perhaps along the nontapered portion as well, force and tension are apparently transmitted from the cell interior through the plasma membrane to

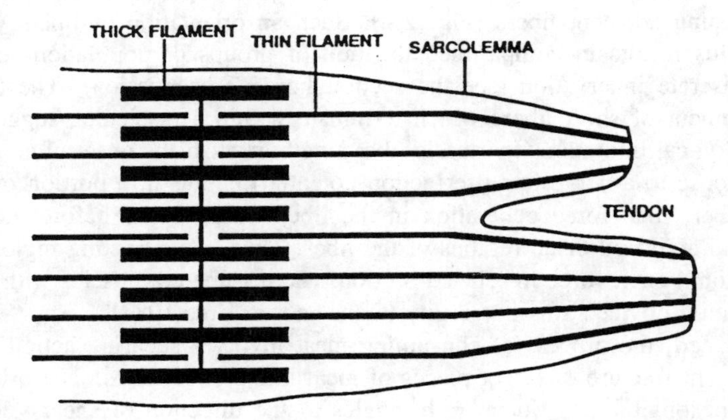

Figure 2.3. Diagrammatic representation of myotendinous junction showing tapered, invaginated end of a muscle fiber forming a junction with tendon. Note extension of thin filaments of sarcomere and their insertion at an angle into the bifurcated end of muscle fiber. (After M. Greaser, 1991, Proc. Forty-fourth Recip. Meat Conf., p. 2.)

the overlying endomysium and then to adjacent cells (Trotter 1990). Anatomical adaptations facilitating force transmission in muscles are described in the section Force Transmission.

Vertebrate Cardiac Muscle

The muscle of the heart (myocardium) possesses a contractile apparatus and, therefore, a mode of force generation similar to skeletal muscle. Like skeletal muscle, it has a striated appearance due to the organization of myofibrils, and the myocardium has an analogous distribution of connective tissues. Unlike skeletal muscle fibers, the muscle cells of the heart (cardiocytes) are much smaller. The diameter of cardiocytes is about 15 μm and length about 150 μm. Cardiocytes are branched and, thus, anastomose with adjacent cardiac fibers (Fig. 2.4). A distinguishing feature of cardiac muscle fibers is that they are arranged in series and link end to end to other fibers by connecting structures known as intercalated disks (Fig. 2.4). The intercalated disk is located transverse to the long axis of cardiac fibers and consists of a series of clefts and projections that interdigitate with corresponding extensions and invaginations of the adjacent cell.

The dense band which identifies intercalated disk in electron micrographs is called fascia adheres. The opposing plasma membrane of adjacent cardiac cells can approach very closely (within 2 nm) along regions of the disk. In areas of the intercalated disk, the plasma membrane possesses specialized adaptations (denser regions) which anchor intracellular filaments (actin) to the disk. The junction formed by intercalated disks represents a direct mechanical link between cardiac muscle fibers (myomyous junction) and, unlike skeletal muscle, is the predominant method of force transmission in cardiac muscle.

Cardiac muscle fibers are mononucleate or binucleate, with the nucleus centrally located in the muscle cell. Per unit volume, cardiac muscle cells are richer in mitochondria and cytoplasm (sarcoplasm) than skeletal muscle fibers. Cardiac muscle tissue also contains specialized tissues which conduct electrical impulses which consist of modified, large-diameter, cardiac muscle fibers, called Purkinje fibers. The primary function of the sinoatrial node, atrioventricular node, atrioventricular, bundle and Purkinje fibers is to coordinate contraction among and between cardiac fibers and regions of the heart, a form of regulation that does not exist in skeletal muscle.

Unlike skeletal muscle, heart muscle is under involuntary control. Contractions are initiated from within the muscle fibers themselves rather than by the neurons and axons originating with the central nervous system. The heart, of course, functions as a pump. In terms of function, it differs from skeletal muscle because it neither pulls upon other muscles or skeleton nor does myocardium possess connective tissue attachments to muscle or bone.

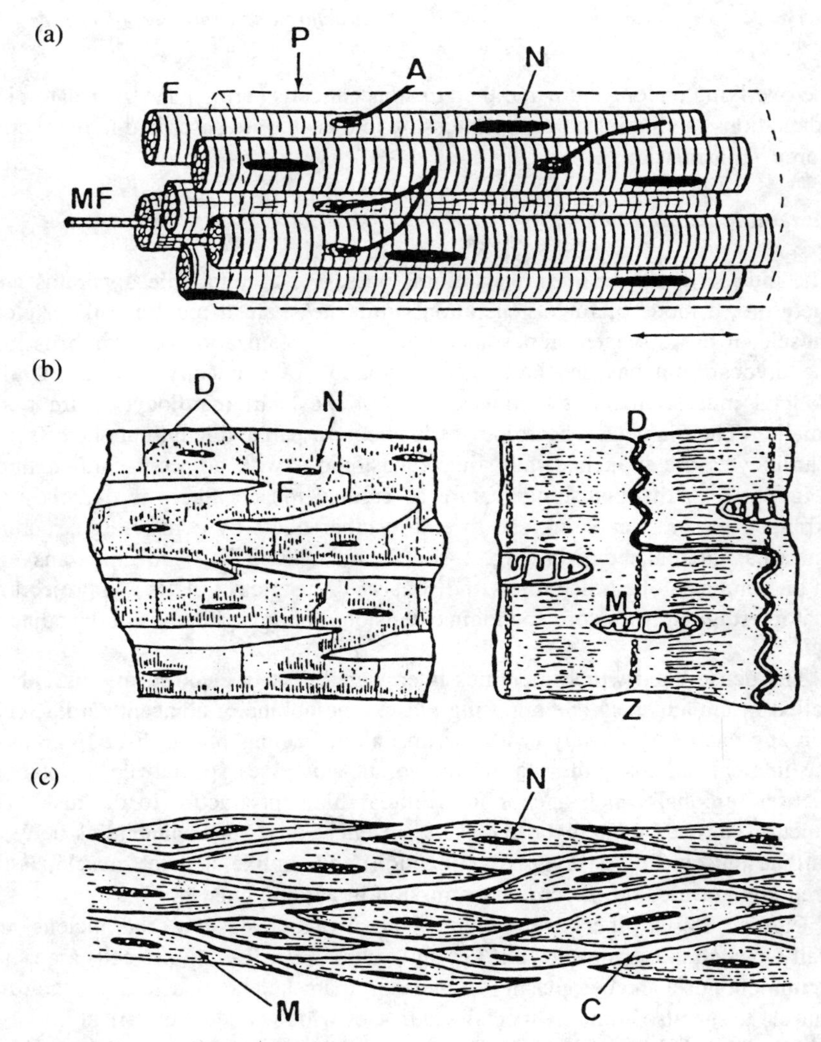

Figure 2.4. Schematic representations of the different cell structures in (a) vertebrate skeletal muscle, (b) vertebrate cardiac muscle, and (c) vertebrate smooth muscle. (a) Long, straight fibers (F) each containing many nuclei (N) and connected to motor nerves (A). The fibers are composed of myofibrils (MF) and several fibers for a fascicle surrounded by a connective tissue layer, the perimysium (P). (b) At left shows the arrangement of the branched fibers in cardiac muscle, which are linked through intercalated disks (D). The single nuclei (N) are centrally located. The enlargement (right) shows the dense granular structure on each side of the intercellular boundary (z: Z-band; M: mitochondrion). (c) Shows the arrangement of the spindle-shaped cells in vertebrate smooth muscles. The nuclei (N) are centrally placed, region M contains the contractile material, and there is much collagenous connective tissue (C) in the intracellular space. These cells are often very much longer and thinner than shown. (After J. Squire, *Muscle Design, Diversity and Disease,* Benjamin/Cummings Co., New York, 1986, p. 37.)

Smooth Muscle

Smooth muscle, as the name implies, lacks the striated pattern of skeletal or cardiac muscle. Smooth muscle cells are located primarily in the gastrointestinal tract (visceral smooth muscle), vasculature (vascular smooth muscle), and uterus and, like cardiac muscle, are considered to be under involuntary control.

Smooth muscle fibers are usually described as elongated spindle or fusiform-shaped cells with a single, centrally located nucleus. Actually, smooth muscle fiber shapes vary considerably depending on tissue of origin (Fig. 2.4). Smooth muscle fibers also vary greatly in size, with the smallest cells in small blood vessels only 20 μm long. The largest smooth muscle cells which occur in the pregnant human uterus may be 0.5 mm in length and 10–20 μm in diameter. Smooth muscle cells possess myofibrils; however, they lack the alternating transverse bands that characterize skeletal and cardiac (striated) muscles. Smooth muscle fibers are surrounded by a thick coat, which corresponds to the basal lamina of skeletal muscle fibers. Fine collagen and elastin fibers, fibroblasts as well as capillaries, and nerves form a loose network between bundles of smooth muscle cells. Due to the thick basal lamina region, adjacent smooth muscle cells are separated by a distance of 40–80 nm. Force transmission between cells can occur via the basal lamina/connective tissue matrix surrounding individual cells. Dense regions on the cell surface that correspond to points of intracellular filament attachment may also represent sites of cell-to-cell association, suggesting direct mechanical linkage and direct force transmission between cells.

Subcellular Organization of the Skeletal Muscle Fiber

In mature muscle, fiber nuclei, about 35 per millimeter of muscle fiber length, are located peripherally along the interior surface of the plasma membrane. Most of the interior volume of the cell, about 75–85%, is occupied by organelles unique to muscle cells called myofibrils. Myofibrils, which are of uniform diameter and similar width (about 0.5–1.0 μm) regardless of species muscle or fiber type run the length of the muscle fiber parallel to its long axis. Myofibrils are long, slender, tubelike filaments consisting of repeating units joined in series called sarcomeres. Myofibril numbers vary from cell to cell such that muscle cell size (diameter) is a function of myofibril number. All myofibrils and their constituent sarcomeres within a muscle fiber are arranged in register such that bisecting a muscle fiber perpendicular to its long axis will section equivalent regions of sarcomeres and myofibrils, the exception being the tapered ends of muscle fibers which terminate within a fasciculus.

Mitochondria, termed sarcosomeres in skeletal muscle, are the site of cellular respiratory processes and are located both between myofibrils in the interior of the muscle cell (core mitochondria) and peripherally. Mitochondria occupy 2–

10% of the muscle cell volume depending on muscle fiber type (see the Muscle Fiber Types) and its capacity for aerobic (respiratory) metabolism. Mitochondria and their associated enzyme systems consume oxygen. Thus, the oxygen consumption rate of muscle during the postmortem period depends largely on mitochondria number and their continued enzymic activity, which is pH dependent (see Chapter 3). Like the muscle cell plasma membrane, mitochondria are subject to rupture during the freeze–thaw process. Liberation of enzymes, or enzyme isoforms unique to the mitochondria, into the sarcoplasm and their detection in the muscle exudate is the basis of biochemical tests for differentiation of fresh meat from that which has been previously frozen.

A second membrane system consists of transverse tubules (T-tubules) and the sarcoplasmic reticulum (SR). The intracellular membrane systems consisting of SR and T-tubules, together with plasma membrane, form the apparatus that conveys a neural impulse or action potential to the interior of each muscle cell and facilitates excitation–contraction coupling. The T-tubule system consists of a network of tubules which pass through the plasma membrane as tubular lumen. The T-tubule system invades the interior of the cell and surrounds each myofibril. The T-tubule system terminates in the sarcoplasmic reticulum (SR) to which it is coupled at the SR–T-tubule junction. The SR store, releases, and sequesters intracellular calcium and also surrounds each myofibril.

The organelles of muscle cells are bathed in a cytosolic fluid known as the sarcoplasm. The sarcoplasm is a protein-rich, aqueous solution that occurs primarily in the interfilament spaces within each myofibril. Quantitatively, the major protein constituents of the sarcoplasm include the enzymes of glycolysis, creatine kinase, and myoglobin. The composition of sarcoplasm is far more complex than these predominant proteins would indicate, however. Over 200 individual components have been identified, including proteases such as the calpains, their inhibitors, nucleotides, fatty acids, carbohydrates, and minerals. The proteins as well as nonprotein constituents are water soluble but are most efficiently extracted from muscle in low-ionic strength buffer at neutral pH.

Intracellular proteases are either compartmentalized in organelles known as lysosomes or distributed throughout the sarcoplasm. The lysosomal-bound enzymes or cathepsins are acidic proteases with optimal activity between pH 4.0 and 7.0. Cathepetic activity toward muscle proteins is regulated by compartmentalization of the enzymes in lysosomes (which have an acidic pH) and the neutral pH of the sarcoplasm. Cathepsins A, B, C, D, E, and L have been identified, but only cathepsins B, D, and L possess substrate specificities that would permit degradation of muscle proteins. Cathepsins B and D hydrolyze myosin and actin in vitro. Cathepsin L is specific for myosin, actin, α-actin, troponin, and tropomyosin. Although these cathepsins have the potential to degrade the predominant proteins of the sarcomere, evidence suggests that relatively little degradation of myosin or actin actually occurs in situ during the postmortem period. Certainly, the amounts of endogenous cathepsins in muscle are sufficient to degrade most of the actin and myosin in muscle during normal aging periods if they achieved optimal activity. Probably because the pH of muscle never becomes acidic enough to fully activate the cathepsins and their lysosomal compartmentilization prevents enzyme–substrate association, the degree of proteolysis due to catheptic activity remains minimal. Alkaline proteases which exist in the sarcoplasm as well as in mast cells are active at a pH considerably more basic (8–10.5) than that of intracellular muscle. The alkaline proteases also have as their substrates muscle proteins; however, their role in specific proteolytic processes and postmortem tenderization are not well defined.

Another class of muscle proteases active at neutral pH are referred to as calcium-activated factor (CAF), calcium-activated neutral protease (CANP), calcium-dependent protease (CDP), or more recently calpains. As the names imply, calpains are calcium-activated proteases. Regulation of calpain activity in muscle in vivo is not well understood. However, inhibitors in the sarcoplasm, autoproteolysis, and differential calcium requirements are possible regulatory mechanisms under physiological conditions. There are two well-characterized forms of calpain in skeletal muscle, one form that is active at millimolar of calcium concentrations (m-calpain) and a second which is active at micromolar calcium concentrations (μ-calpain). Regulation of calpain activity during postmortem aging is likewise complex and not yet completely understood. However, evidence suggests that it plays an important role in postmortem tenderization in beef and lamb. The μ-calpain form of the protease is probably the only form active under postmortem conditions because millimolar levels of calcium, necessary for activation of m-calpain, are never achieved. An endogenous inhibitor of calpains in muscle tissue is calpastatin. Elevated levels of calpastatin have been attributed to reduced proteolysis postmortem in meat, but regulation in vivo and in vitro is not well understood. In the presence of calcium, calpains also undergo autoproteolysis, that is, one molecule of calpain will hydrolyze another. Although the protease activity of calpains will eventually be destroyed by autoproteolysis, it is yet unclear if initial autoproteolysis is actually necessary for activation of calpains. Postmortem conditions of pH, temperature, intracellular calcium concentration, and the fresh or frozen–thawed state of meat all play an interactive role in regulating the degree of calpain-induced proteolysis. Much attention has been focused on calpains as the proteases primarily responsible for tenderization in the early postmortem aging period, particularly because the Z-line is degraded by calpain activity. This early postmortem alteration of the cytoskeletal structure of the sarcomere, which is histologically verifiable, also coincides with increased tenderness. To date, most evidence suggests that postmortem tenderness improvements that are attributable to alterations in myofibrillar proteins result from specific degradation of a limited number of sarcomeric proteins. This "nicking" or proteolytic scission of one structure of the myofibril, the sarcomere, is sufficient to produce an effect on texture. The major proteases of skeletal muscle, their characteristics, and protein substrates are summarized in Table 2.1.

Table 2.1. Proteases of Muscle

Protease	Localization	MW	Optimal pH	Substrate
Alkaline protease	Mast cells?		8.5–9.0	Muscle proteins
Muscle alkaline protease MAP		22,000	9.5–10.5	Tyr-leu, phe-phe, peptide bonds
Serine protease	Mast cells	22,000–24,000	8.0–9.0	Actin, myosin, tropomyosin, troponin T, I
Myosin-cleaving enzyme (serine protease)	Mast cells	26,000–27,000	8.4–9.0	
Calpains	Cytoplasm, myofibril	80,000 30,000 + 30,000	7.0–7.5	C protein, desmin, filamin, tropomyosin, troponin T, I, Z-disks
Cathepsin A	Lysosome		5–5.4	
Cathepsin B	Lysosome	24,000–27,000	5.2	
Cathepsin C	Lysosome		5–7	
Cathepsin D	Lysosome	42,000–45,000	4.0	
Cathepsin L	Lysosome	24,000	4.1 (for myosin) 7.0 (for troponin–tropomyosin?)	

Source: Modified from Bandman, *The Science of Meat and Meat Products.* [Price and Schweigert (1987)]

Sarcomere Organization

The Sarcomere

The repeating contractile unit of the myofibril is the sarcomere. Viewed two dimensionally (Fig. 2.5) it consists of two sets of thin filaments each anchored at one end in a structure: the Z-disk and interdigitating thick filaments. The Z-disk is oriented perpendicular to the thin filaments and the long axis of the myofibril. The sarcomere is, thus, that portion of the myofibril between two adjacent Z-disks. The thin filaments are isotropic—they do not refract birefringent light—and are called I-bands. The thick filaments are anisotropic—they refract birefringent light—and are called A-bands. A-bands are bipolar and possess projections along either end of the filament. In the center of the A-band, a bare zone free of these projections exists. The bare zone is bisected by a structure, the M-line, that occurs perpendicular to the axis of the sarcomere. The bare regions are referred to as H-zones. That portion of the bare region not overlapped by the I-band is known as the pseudo-H-zone. Faint lines, the N-lines, oriented parallel to the Z-disk within the I-band may also be visible. The letter nomenclature of Z-disk, M-line, H-zone, and N-lines derive from the original German description of the structures: Zwishenscheibe—the disk between; Mittellinie or middle line; heller for bright or clear region; and neben for adjacent lines.

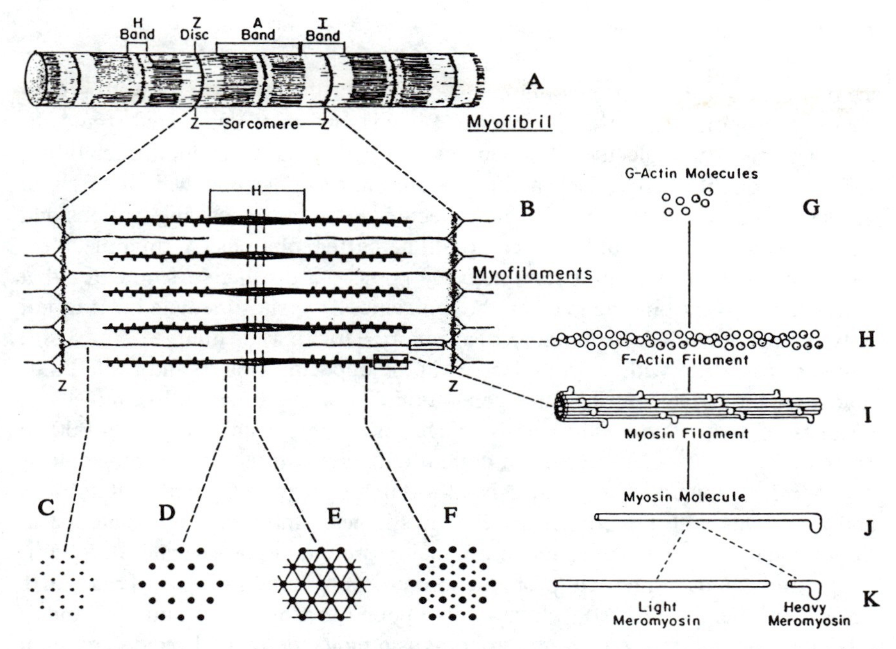

Figure 2.5. Diagram of the organization of skeletal muscle. A represents a myofibril showing characteristic striated appearance; B the sarcomere; C, D, E, F the arrangement of filaments in cross section at the levels indicated; G, H, I, J, K the thin and thick filaments and their constituent proteins. (After D. Fawcett, *A Textbook of Histology*, W. B. Saunders Co., Philadelphia, 1986, p. 282.)

Visualized in cross section, a sarcomere will reveal various patterns of actin filament, myosin filaments, or actin and myosin filaments. Bisected in a non-overlapping region of the A-band or I-band, only myosin or actin filaments will be seen. However, a section through the overlapping region of a sarcomere will reveal myosin filaments, each surround by six actin filaments (Fig. 2.5).

The distance between the Z-disks varies as muscles contract or lengthen, thereby changing the sarcomere length. The width of the H-zone, located in the center of the A-band, remains constant irrespective of sarcomere length. However, the dimensions of pseudo-H-zone vary with degree of I-band overlap and sarcomere length. Thus, the basis of muscle contraction resides in back-and-forth movement of thick and thin filaments relative to each. As rigor mortis ensues, postmortem, sarcomeres shorten and assume a fixed length. Excessively shortened sarcomeres are associated with decreased tenderness. Final sarcomere length in postmortem muscle is a major determinant of meat tenderness (see Chapter 3).

The Thin Filament

Thin filaments are composed of three major myofibrillar proteins: actin, tropomyosin, and troponin. The thin filament or I-band consists of two coiled strands of polymerized actin molecules (filamentous or F-actin) that wrap about each other, forming a double helix. Each actin molecule has a molecular weight of about 40,000 and associates with adjacent molecules via noncovalent interactions and tends to be sparingly soluble in water or dilute buffer solutions. Actin molecules, generally visualized as globular, actually possess a distinctive symmetry that corresponds to six binding sites on the molecule. Two sites mediate head-to-tail polymerization with adjacent molecules within a strand, a site binds tropomyosin, two sites associate with actin molecules in the opposing strand, and a site binds myosin in the A-band during force generation. Tropomyosin, a rodlike molecule, consists of two polypeptide chains, each with a molecular weight 34,000 to 36,000, which associate to form a coiled helix. Each tropomyosin molecule is about 385 Å long and associates head-to-tail to form a filament that follows and associates with the coil of the F-actin filament. Intimately associated with tropomyosin and F-actin is the third major protein of the I-band, troponin. Troponin is an asymmetrical protein and consists of three subunits. Troponin T (molecular weight 37,000), which is also bound to troponin subunits C and I, links the troponin molecule to the tropomyosin molecule in the I-band. Troponin C (molecular weight 18,000) binds Ca^{2+} and confers Ca^{2+} sensitivity to the troponin–tropomyosin–actin complex. Troponin I (molecular weight 23,000), the inhibitory subunit, binds tightly to troponin C and actin and only slightly, if at all, to tropomyosin or troponin T. When intracellular Ca^{2+} concentration rises from 10^{-7} M ("off"-state) to 10^{-5} M ("on"-state) troponin C undergoes a conformational change that, in turn, alters the linkages between troponin T and tropomyosin, and troponin I and actin. When the troponin–tropomyosin–actin complex is in the off-state, the interaction of actin molecules with myosin heads in an overlapping myosin filament is prevented by the position of tropomyosin along the F-actin filament that covers the myosin binding site on each actin molecule. When the regulatory complex is in the on-state, Ca^{2+}-induced conformational changes in troponin allow movement of the tropomyosin molecule along the F-actin filament, exposing the myosin binding sites on each actin monomer (Fig. 2.6) see the subsection (Regulation of Crossbridge Formation.).

Z-Disk

Each F-actin filament is anchored in a Z-disk which consists predominantly of the protein actin as well. Viewed in cross section, Z-disks present a lattice structure in which F-actin filaments enter, zigzag, and exit into the next sarcomere, now staggered relative to the opposite sarcomere. Depending on muscle

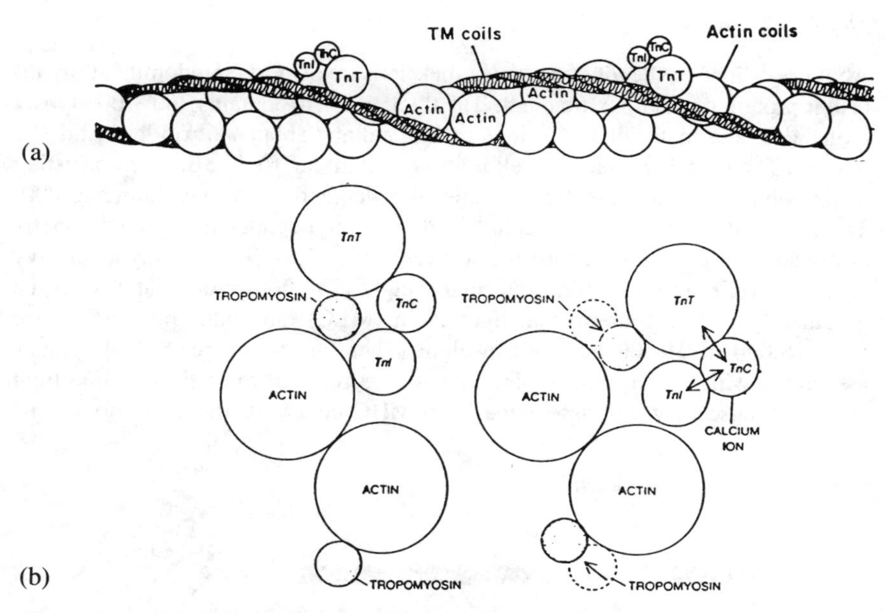

Figure 2.6. (a) Coiled coil—formed by the filament of tropomysin molecules (coiled coils) wound in each of the two grooves of the actin helix in a thin filament of muscle. The actin molecules are polar; they all point in the same direction in a double helical array. The tropomyosin filaments consist of polar tropomyosin molecules, bonded head to tail, that lie near the grooves, each molecule spanning seven actin monomers. A troponin complex is about a third of the way from one end of each tropomyosin molecule. In this schematic diagram only the troponin complexes on one side of the actin helix are shown. (b) Proposed model for configuration of actin, tropomyosin and troponin (Tn) subunits. TnT = troponin–tropomyosin subunit, TnI = troponin-inhibitory subunit, and TnC = troponin–calcium-binding subunit. The diagram on the left shows the resting state on which the TnT subunit binds tropomyosin and the TnI subunit binds to actin. The calcium concentration is low and the links between the troponin subunits are relatively loose. In the drawing on the right, the calcium concentration is high enough to interact with the TnC, which tightens the linkages between the troponin subunits and causes tropomyosin to move deeper into the groove exposing the binding site for the myosin-actin interaction. [After C. Cohen, *Scientific American* **233:** 36 (1975).]

type and species, Z-disks differ in complexity. For example, the Z-disks of fish are narrower (38 nm) than the Z-disks of mammals. In addition, alpha actin and desmin are also major constituents of the Z-disk. Associated with this complex structure are other proteins occurring in lesser quantities including Eu-actin, filamin, synemin, vimentin, vinculin, zeugmatin, and a 55,000 molecular weight protein.

The Thick Filament

About one-third of the total protein in muscle is myosin, the predominant myofibrillar protein of the thick filament. The thick filament contains 300–400 myosin molecules, and in skeletal muscle is cylindrical and about 1.6 μm long and 150 Å wide. The myosin molecule itself is an asymmetrical, rodlike protein consisting of six polypeptide subunits. Native molecular weight of myosin is about 500,000. The long tail of the molecule consists of two polypeptides in a coiled α-helix terminating in two globular heads at one end (Fig. 2.7). The two myosin heavy chains (MHC) are similar in molecular weight (200,000). Associated with the globular heads are four myosin light-chain (MLC) molecules ranging in size from 16,000 to 21,000 molecular weight. There are two types of light chains associated with the myosin heads. One type is referred to as the alkaline light chains because they are dissociated from MHC under alkaline conditions. The

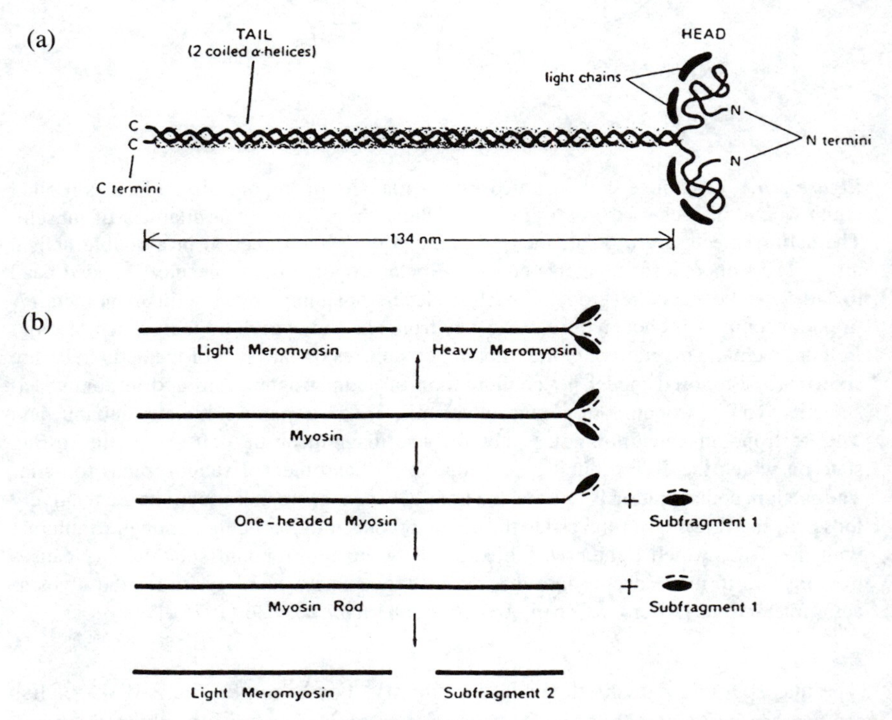

Figure 2.7. Diagrammatic representation of myosin molecule depicting its six subunits (two heavy chains forming the tail and head regions and four light chains associated with the head). (b) Diagrammatic illustration of the way a myosin molecule can be cleaved into various subfragments by the action of proteolytic enzymes. [After G. Offer, *Companion to Biochemistry* (A. T. Bull et al., eds.) Longmans, Harlow, Essex U.K. 1974.

second type is known as DTNB light chains because they are dissociated by 5,5′-dithiobis 2-nitrobenzoic acid.

In addition to being a structural protein and the primary constituent of the thick filament, myosin is an enzyme possessing ATPase activity. The globular head regions of myosin bind and hydrolyze ATP to ADP. Myosin ATPase activity is dependent on the association of the alkali light chains with the head region of MHC. Removal of the alkali light chain results in the loss of myosin ATPase activity, although in the absence of MHC the alkali light chains possess no ATPase activity. The DTNB light chains appear to have a regulatory function mediating the level of ATPase activity.

The intact myosin molecule is susceptible to proteolytic cleavage at several points. Digestion with trypsin cleaves the long tail of the molecule producing two fragments: light meromyosin and heavy meromyosin, the latter contains the globular head region of the molecule. Further treatment with papain will cleave one or both globular heads known as subfragment(s). Treatment with just papain produces a long taillike fragment known as the myosin rod, whereas trypsin cleavage of the myosin rod produces light meromyosin and subfragment 2 (Fig. 2.7). Intact myosin molecules pack tail to tail in a staggered array to form bipolar myosin filaments with globular heads at either pole and a bare zone in the center of the filament. Thus, the myosin filament, which is roughly cylindrical in shape, possesses a regular spacing of protruding heads around the periphery. In the central bare zone of the myosin filament, a transverse element positioned perpendicular to the A-band is apparent. This region, known as the M-band, forms a series of bridgelike structures that are the only elements known to interconnect the myosin molecules with the filament. Proteins associated with the M-band include myomesin, M-protein, creatine kinase (the enzyme which catalyzes the rephosphorylation of ADP to ATP), and two enzymes of glycogen metabolism, phosphorylase, and glucosidase.

Other proteins associated with the A-band include C-protein, H-protein, I-protein, and F-protein, so named because of their relative electrophoretic migration. Of these proteins, C-protein is best characterized. It forms a series of barrel stavelike bands that encircle the tail regions of myosin molecules in the middle third of each half of the bare zone. The function these proteins may be both structural, supplying stability to the myosin filament, and regulatory, mediating the assembly of myosin molecules into the thick filament.

Myosin and the other proteins of the thick filament are insoluble in water or low-ionic-strength buffers but are soluble in salt solutions of ionic strength 0.3–0.7.

Cytoskeletal Structure

A third group of intracellular muscle proteins, the cytoskeletal proteins, are distinct from sarcoplasmic or myofibrillar proteins and serve to support and stabilize the contractile apparatus of the muscle cell, both laterally and longitudinally. The most abundant cytoskeletal protein is titin. It constitutes about 8% of the myofibrillar protein, second only to actin and myosin. Titin is also the largest of the muscle proteins, with a molecular weight of about 3,000,000, a length of 900–1000 nm, and a diameter of 4–5 nm. It is one of the larger naturally occurring proteins. Titin occurs in thin elastic filaments that run parallel to the thick and thin filaments of the sarcomere. Titin may be the major constituent of "gap filaments," thin, elastic filaments that apparently form a bridge across the sarcomere and link Z-disk to Z-disk. The function of titin and/or gap filaments is apparently to provide structural support to the contractile elements of the sarcomere and to maintain the thick and thin filaments in register. Titin and gap filaments may be involved in contraction and play a role in meat texture. Nebulin is also a large (600,000–800,000 molecular weight) protein approaching 1 μm in length. Nebulin is thought to run parallel to and be in close association with the thin filaments. Desmin, a small cytoskeletal protein (55,000 molecular weight), occurs within and between Z-disks of adjacent sarcomeres. Its function is to maintain lateral association between sarcomeres.

The influences of structural and regulatory proteins of the sarcomere, myofibril, and the cytoskeleton on the physical properties of meat are significant. The sliding filaments of the sarcomere, the basic contractile unit, will determine sarcomere length in meat post-rigor, a major determinant of tenderness. The proteins of the contractile apparatus and the structures they comprise vary in their susceptibility to proteolytic degradation postmortem, likewise contributing to the degree of meat tenderness. Finally, the interfilament space between the thin and thick filaments is bathed in aqueous solution (sarcoplasm) and is the primary water depot in muscle. Sarcomere length (degree of thick and thin filament overlap), combined with solubilization of myosin, will play key roles in the functionality of muscle in processed meat products.

Force Generation

Contraction: Actin–Myosin Interaction

The sliding filament model of muscle contraction is easily visualized by considering the structure of the sarcomere described in the section Sarcomere Organization. The thin filaments, anchored in Z-disks, move toward each other, as do the Z-disks of the ever-shortening sarcomere as thin filaments and thick filaments move past each other in rachetlike fashion. The basis of muscle contraction lies in the unique filament structure of the sarcomere, the transient affinity of actin for myosin under conditions existing in muscle, and the conformational changes the myosin molecule and the troponin–tropomyosin–actin complex undergo. Actin–myosin interaction resulting in movement of the thin relative to thick filament involves the cyclic attachment–detachment–attachment of the S-1 head region of myosin to the myosin binding site on actin molecules. Crossbridge cycling and shortening can be briefly described stepwise:

1. Assuming that myosin and actin have formed a crossbridge complex, one molecule of ATP binds to each myosin head or S-1 fragment.
2. Myosin dissociates itself from actin and the S-1 head of myosin undergoes a conformational change from 45° to 90° relative to the axis of the myosin molecules in filaments.
3. ATP associated with the S-1 head is hydrolyzed to ADP + inorganic phosphate, both of which remain associated with S-1.
4. Myosin binds actin if the actin binding sites are available.
5. The S-1 head region undergoes a conformational change, swiveling the S-1 head from 90° to 45° while it is bound to actin-producing movement of actin filament relative to myosin the length of one crossbridge.
6. Inorganic phosphate and ADP are released from S-1 and replaced by a new molecule of ATP which initiates a new cycle.

Contraction resulting from the sliding movement of thin relative to thick filament is no longer in the realm of theory and is the accepted model for force generation in muscle. The most convincing evidence for the roles actin and myosin play in contraction derives from the direct relationship demonstrated between a force or tension and sarcomere length. When sarcomere length is just less than 2 μm to about 2.5 μm, overlap of actin and myosin filaments (and the number of interactions they can undergo) is maximized. Tension and force generation is also greatest in this range (Fig. 2.8). This is the situation in vivo. However, in terms of the textural properties of meat, an analogous relationship will exist postmortem between actin and myosin filament overlap and increasing shear force.

Although aspects of the sliding filament model of muscle contraction have been elucidated, many details of the biochemical and molecular events that occur during contraction are still the subject of investigation. For example, a persistent problem with the crossbridge model has been an inability to demonstrate a conformational change in the myosin molecule of sufficient magnitude to account for observed movement during contraction. Recent crystallization and X-ray diffraction studies of myosin S-1 head (Rayment et al. 1993) suggest that structural changes that could produce the necessary shift in orientation of S-1 head during crossbridge cycling may occur.

Alternative models to the myosin crossbridge theory of contraction have also been proposed. A mechanism whereby conformational changes in the actin filament, rather than in myosin heads, accounts for movement of one filament relative to the other has been described (Schutt and Lindberg 1993). Actin and myosin interaction, crossbridge cycling, and contraction in the absence of ATP

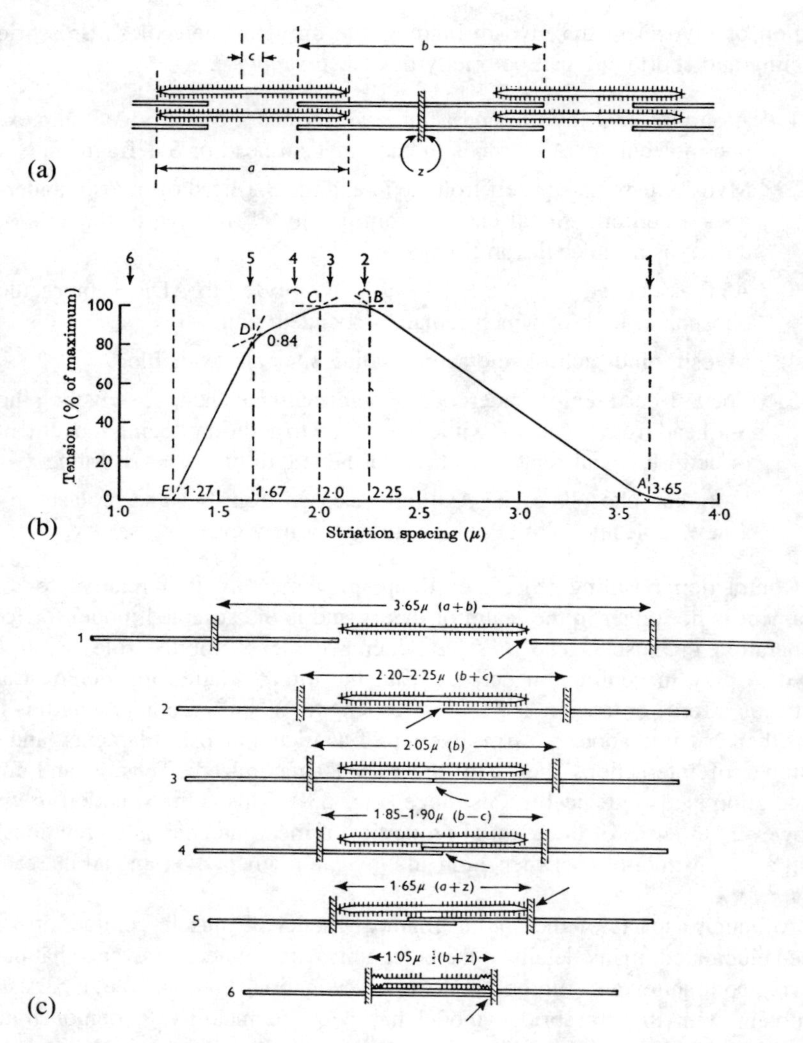

Figure 2.8. The alteration of tension generated by the contractile apparatus as a function of sarcomere length (b) and its relationship to the varying overlap of thick and thin filaments (a and c). Maximum tension is developed (b, points 2, 3, 4) when sarcomere length is just less than 2 μm to about 3 μm, which corresponds to maximum filament overlap and maximum crossbridge formation. As the sarcomere shortens, further actin–myosin interactions are apparently disrupted. An analogous situation develops in post-rigor meat where shear force is likewise dependent on degree of actin–myosin filament overlap and sarcomere length. [After Gordon et al., *J. Physiol. (Lond.)* **184,** 170–182 (1966).]

hydrolysis, a theory that may explain current inability to account for work done during contraction with limited heat produced, has been proposed (Brenner 1991).

Regulation of Crossbridge Formation

The structure and composition of the thin filament confer on the contractile apparatus of the sarcomere (and of the muscle fiber) the means of regulating the interaction between actin and myosin and, consequently, contraction or shortening. The myosin-binding site on each actin molecule is not always exposed and available for binding the S-1 head of myosin. When the contractile apparatus is in the "off" position (not contracting), the tropomyosin molecules in head-to-tail alignment parallel thin-filament actin following the pitch of helix and covering the myosin-binding sites. Tropomyosin is held in the "off"-state by troponin. The conformation of troponin is determined by intracellular free calcium levels that in the "off"-state are about 10^{-7} M. As a contraction response is initiated intracellular calcium levels rise to 10^{-5} M and the calcium-sensitive troponin subunit undergoes a conformation change that, in turn, shifts tropomyosin molecules deeper into the groove formed by the coiled F-actin filaments and away from the actin–myosin-binding sites. Crossbridge interaction is then free to proceed until calcium is sequestered and intracellular concentrations approach 10^{-7} M. This sequence of events represents a simplified description of a steric blocking mechanism for regulation of contraction.

Excitation–Contraction Coupling

The series of events that links a neural impulse from the central nervous system via an electrical impulse to the steric blocking mechanism within each muscle that regulates thick and thin filament interaction is excitation–contraction coupling. Components include the motor nerve fiber endings terminating in the myoneural junction at the surface of the muscle fiber and the electrical signals carried along nerves and muscle fiber to the sarcoplasmic reticulum via action potentials.

Innervation

Innervation of skeletal muscles is complex and a thorough review of the subject is presented in *The Structure and Development of Meat Animals* (Swatland 1984). Skeletal muscle is voluntary and all neural impulses which control shortening and contraction originate from motor neurons or cell bodies linked to the spinal cord. Motor axons lead out from ventral root of the motor neuron linking each motor neuron to the group of muscle fibers it will innervate. The anterior horn cell, its axon, and all branches associated with neuromuscular junctions and the muscle fibers innervated comprise the motor unit. All muscle fibers in the motor unit contract in synchrony. However, synchrony of muscle contraction results from incremental recruitment of different motor units which vary in speed and force of contraction. When a peripheral nerve enters a muscle, via the neurovascular hilum, it branches into smaller segments following the perimysial connective tissue and further separates into terminal axons which innervate a myofiber of its motor unit. The terminal axon joins the muscle fiber at a neuromuscular junction. At the neuromuscular junction, the terminal axon branches and contacts the basement membrane sheathing the muscle cell via grooves in the myofiber called synaptic clefts.

Action Potentials

Signals that originate in the central nervous systems are conveyed along nerves (axons) to individual muscle fibers and then along the muscle fiber by electrical activity on the surface of the cells (nerve and muscle) known as action potentials. Action potentials are created by the reversal of the membrane potential (difference in charge) which always exists between the inside and outside of cells. In resting nerve and muscle cells, this potential is 80–85 mV. The charge differential between the inside and outside of the cells is due to an excess of anions that accumulate in the intracellular fluids and a concurrent excess of cations occurring in the extracellular space. The separation of net opposite charges across the cell membrane, positive outside and negative inside, results in the establishment of a membrane potential. Membrane potentials in nerve and muscle cells result from active transport of ions through the membranes, selective permeability of the membrane to diffusion of ions, and the ionic composition of intracellular and extracellular fluids. Extracellular fluids contain high concentrations of sodium (Na^+) and chloride (Cl^-) and low concentrations of potassium (K^+) and nondiffusible anions. In contrast, the intracellular fluid contains high concentrations of K^+ and nondiffusible anions. Concentration gradients of Na^+ and K^+ across the cell membrane are maintained by active transport of Na^+ out of the cell and K^+ into the cell. Active transport is accomplished by an ATP-dependent "Na^+-K^+ pump" located within the cell membrane which moves three Na^+ ions out for every two K^+ ions moved into the cell. The permeability of the cell membrane to diffusion of K^+ is 50–100 times greater than its permeability to Na^+ diffusion. Thus, K^+ compared to Na^+ passes easily through the membrane. Diffusion of positive charges (i.e., K^+) does not continue indefinitely; once the membrane potential is achieved, it impedes further flow of K^+ out of the cell, and equilibrium is established. An action potential is initiated by a sudden 100-fold to 1000-fold increase in Na^+ permeability of the cell membrane. The change in cell membrane permeability is the result of the release of the neurotransmitter acetylcholine from vesicles located in the membrane of nerve cells, or within the synaptic cleft in the basement membrane of the muscle fiber, in response to a nerve impulse or action potential.

Acetylcholine acts on the acetylcholine receptor protein, a protein which spans the cell membrane and is responsible for the membrane's rapid shift in ion diffusibility. As soon as acetylcholine has bound to its receptor and the reversal of membrane potential has occurred as a result of rapid influx of Na^+ into the cell, acetylcholine is destroyed by the enzyme acetylcholinesterase. The excitatory response to the muscle fiber is terminated and membrane potential is reestablished by the Na^+-K^+ pump. Production of an electrical impulse in the form of an action potential is similar in nerve and muscle cells with the exception that the impulse is transmitted much more quickly along the nerve than the muscle cell.

Sarcoplasmic Reticulum and Calcium Balance

The end result of a nerve impulse to skeletal muscle is the release of calcium from the terminal cisternae of the sarcoplasmic reticulum (SR) into the sarcoplasm resulting in an increase in intracellular Ca^{2+} concentration from 10^{-7} to 10^{-5} M. The increase in Ca^{2+} concentration promotes a conformational change in Ca^{2+}-sensitive troponin, an unblocking of the myosin interaction sites on F-actin, and a contractile response. The action potential, which reaches the muscle cell as a wave of depolarization, then moves left and right from the motor end plate along the surface of the fiber and is carried to the interior of the cell via the transverse tubule system (T-tubules). The T-tubules form a network of tubes which enter the sarcolemmal surface at approximately right angles to the muscle fiber and extend to the center of the cell. In skeletal muscle, the T-tubules are superimposed on the myofibrils at the Z-disks. Extending over the thin-filament regions in both directions are the terminal cisternae of the SR. Longitudinal vesicles of the SR emanating from the terminal cisternae meet in the center of the sarcomere and form the fenestrated collar of the SR. Collectively, these SR structures are called the triad (Fig. 2.9).

Force Transmission

The preceding sections described how force in muscle is generated and how these events are regulated. Ultimately, muscles transmit force to the skeleton via tendons or the epimysial connective tissue of muscles directly attached to bones. Movement is accomplished by the coordinated shortening and lengthening of groups of muscles that, in turn, manipulate the skeleton in leverlike fashion. Although the concept of transmission force generated within a muscle to other muscles or the skeleton is fairly obvious, the transmission of force from fiber to tendon or from fiber to fiber involves several specialized anatomical arrangements among myofibrils, plasmalemma, and the surrounding connective tissues.

Myotendinous Junctions

The junction of a muscle and tendon is a site of force transmission for skeletal muscle fibers. The tapered ends of individual muscle fibers form longitudinal

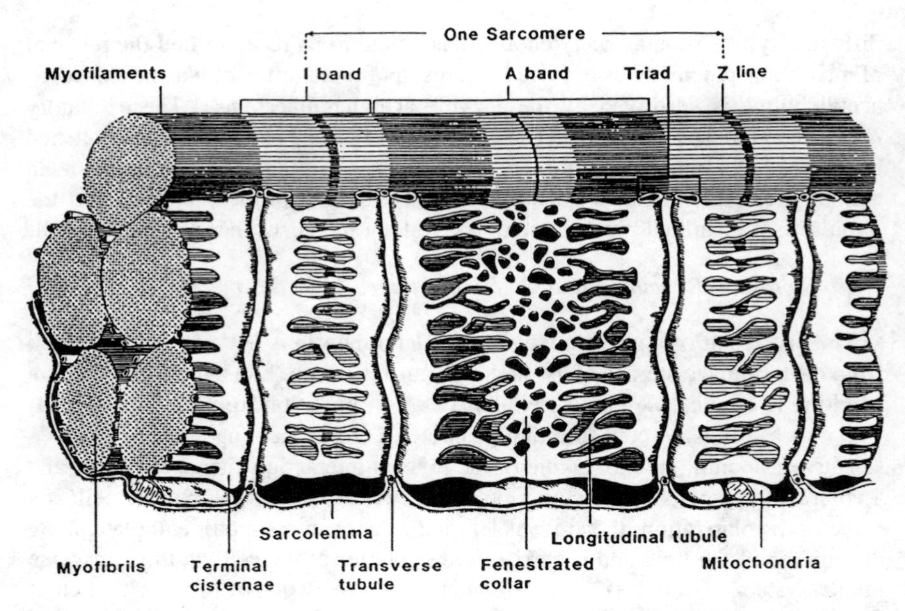

Figure 2.9. Diagram of T-tubules and sarcoplasmic reticulum in longitudinal view of skeletal muscle. [After L. Peachey, *J. Cell Biol.* **25,** 209–231 (1965).]

projections that interdigitate with extensions of tendinous connective tissue. The interfacial surface of the muscle fiber is, thus, greatly increased. In the region of the myotendinous junction, actin filaments extend beyond the terminal sarcomeres of the myofibril and insert at an angle into the fingerlike extensions of the muscle fiber. Transmembrane linkage of actin filaments within the longitudinal extension of the muscle fiber end to surrounding connective tissue via proteins is apparently a feature of the myotendinous junction. The myotendinous junction, which greatly amplifies the contact between muscle fiber and tendon, significantly reduces the load on the muscle fiber at the muscle–tendon junction (Fig. 2.3).

Myofibril to Plasma Membrane Associations

The occurrence of intrafascicularly terminating fibers within many muscles (described in the section Muscle Anatomy and Morphology) requires a second mechanism of force transmission that can transmit force from fiber to fiber. Myofibril to plasma membrane attachments, which span the membrane and create a junction with the connective tissue of the endomysium, transmit tension developed in the muscle fiber to adjacent fibers (Trotter 1990). The association of myofibril to overlying connective tissue via the plasma membrane is maintained by a complex of several proteins. Actin filaments are the myofibrillar components that anchor the contractile apparatus of the cell (myofibrils) to the plasma membrane. The proteins talin and vinculin at once bind actin and the transmembrane protein integrin. Integrin spans the plasmalemma and links the extracellular matrix endomysium to the myofibrils through binding to talin on the intracellular side of the muscle fiber. A transmembrane glycoprotein complex which binds the protein dystrophin, which, in turn, associates with actin filaments, forms another membrane–myofibril attachment. Morphologically, sites of plasma membrane–myofibril attachments, called costameres, have been described as a riblike, lattice structure occurring regularly in register with I-bands and at right angles to the length of the muscle fiber. Evidence that plasma membrane–myofibril attachments occur along the body of the muscle fiber, not just at the ends, was provided by experiments in which all myofibrils in the center of a muscle fiber were severed. Such muscle fibers, lacking continuity of the myofibrils, were capable of generating nearly as much tension as intact fibers, indicating that lateral transmission of force of myofibrils to the plasma membrane must occur.

Muscle Fiber Types

Classification

The muscle fibers that compose mammalian skeletal muscle are not identical but can be classified into types on the basis of contractile properties, energy supply, structure, and color. Thus, individual muscle fibers can be considered fast or slow twitch, oxidative or glycolytic, and red or white. Fast contracting muscle fibers are predisposed to glycolytic metabolism and are well endowed with the enzymes of glycolysis and glycogenolysis but lack an abundance of mitochondria or myoglobin, hence their whiter appearance. They also fatigue more rapidly than slow contracting fibers but exhibit differential fatigue resistance. Slow contracting fibers possess a more oxidative metabolism, more mitochondria, and more myoglobin for O_2 transport to support respiration. They are generally fatigue resistant. Whereas some muscle fibers can certainly be typed as fast or slow, oxidative or glycolytic, red or white, and so on, the classifications of muscle fibers is much more complicated because there exists a continium of these properties as well as considerable overlap between the three major classes. Furthermore, there is overlap in the use of the two energy sources: the complement of mitochondria and associated enzyme systems, myoglobin, glycolytic enzymes, and glycogen in muscle fibers. Such fibers, which possess some characteristics of both fast and slow contracting fibers, are called intermediate. There have been numerous attempts to classify muscle fibers on the basis of their characteristics and some of the systems are summarized in Table 2.2. A detailed summary of the structural and biochemical properties of red and white muscles and fibers is presented in *Muscle and Meat Biochemistry* (Pearson and Young 1989).

As the multiplicity of fiber classifications suggests, choosing the criteria by which to type fibers is somewhat arbitrary. Most widely employed methods to

Table 2.2. Terminologies for Describing Fibers of Mammalian Muscle

Method	Fiber Spectrum			Reference[a]
Histology and physiology	Slow red	Fast white	Fast white	Ranvier
Oxidative enzymes, phosphorylase	I	II	II	Dubowitz and Pearse Engel
Mitochondrial distribution, ATPase	B	C	A	Stein and Padykula
Histochemical profile	III	II	I	Romanul
Mitochondrial distribution	Intermediate	Red	White	Padykula and Gauthier
Z-line width	Red	Intermediate	White	Gauthier
Oxidative enzymes, ATPase	I	IIA	IIB	Brooke and Kaiser
	β	αβ	α	Yellin and Guth
	β-Red	α-Red	α-White	Ashmore and Doerr
Motor unit physiology and histochemistry	S	FR	FF	Burke et al.
Homogenous muscle: physiology histochemistry, biochemistry	Slow twitch, oxidative	Fast twitch, oxidative, glycolytic	Fast twitch, glycolytic	Peter et al.

[a] Table and references therein from Eisenberg (1983).

type fibers of skeletal muscle and meat are those based on the oxidative enzymes and ATPase activity. Using these criteria, different nomenclatures have been adopted (Table 2.2). The first, proposed by Brooke and Kaiser, describes slow twitch, red oxidative fibers as Type I, intermediate fibers as Type IIA, and fast twitch white glycolytic fibers as Type IIB. The Ashmore and Doerr system utilizes the same criteria but names slow, red fibers, β-red; intermediate, α-red and fast, white fibers α-white. In these systems fibers are typed or classified by the nature of the myosin ATPase activity and intensity of their reaction to histochemical stains of oxidative enzyme activity, usually, NHDH–tetrazolium–reductase (NADH-TR) and/or succinate dehydrogenase (SDH). Type I oxidative fibers exhibit low Ca^{2+}-activated ATPase activity and stain intensely for NADH-TR or SDH, typical of slow contracting fibers with predominantly oxidative metabolism. Type IIB glycolytic fibers have high Ca^{2+}-activity ATPase activity but neglible NADH-TR or SDH activity traits which characterize fast contracting fibers with a glycolytic metabolism. Intermediate fibers tend to exhibit features of both Type I and Type IIB fibers. Burke and co-workers added fatigue resistance to histochemical and histological characterization of muscle fiber types.

Several differences between red and white fiber types are of particular interest in terms of postmortem muscle characteristics and meat quality. The more highly developed SR–T-tubule system of white muscle fibers compared to red influences the response of muscle to changes in temperature and pH early postmortem, making muscles composed of white fibers more resistant to prerigor shortening than red. The higher lipid and pro-oxidant iron content of red compared to white fibers (and muscles composed of them) predisposes red muscle to greater oxidative changes. White muscles, because of their predominantly glycolytic nature, exhibit much more rapid lactate accumulation early postmortem and the conditions associated with rapid postmortem glycolysis.

The Molecular Basis of Muscle Fiber Types

It is obvious that, although classifying a fiber based on one or two traits may serve to place it in a particular group of fibers, the diversity of muscle fiber is underestimated by any classification system because differences among a population of fibers represent a continuum rather than discrete groups of fibers, each with identical characteristics.

The diversity among fiber types is due to the wide range of metabolic activities and the multiplicity of protein isoforms which occur in muscle fibers. At least 10 different myosin heavy-chain genes have been identified across several species. Expression of different myosin genes results in multiple myosin heavy-chain isoforms that can occur simultaneously in the same muscle. The variable ATPase activity of fast and slow twitch fibers reflects the expression of specific myosin isoforms. Myosin light chain, actin, tropomyosin, and troponin, in addition to myosin heavy chain, may exist as multiple isoforms in muscle. Expression of specific myofibrillar isoforms occurs in response to growth and developmental transitions, hormonal influences, and other environmental factors such as use and exercise.

In addition to myofibrillar proteins, the metabolic enzymes of muscle as well as connective tissue proteins can also occur as isoforms. Thus, the number of possible combinations of isoforms within a single muscle, or even a single fiber, is immense. Differential protein expression is the mechanism by which muscle adapts to a wide range of conditions and requirements.

Connective Tissue in Muscle

Extracellular Matrix: Definition

Each muscle is surrounded by a connective tissue sheath called the endomysium. Bundles of muscle fibers are enveloped in a heavier layer of connective tissue termed the perimysium, and whole muscles are covered with another connective tissue layer, the epimysium. All connective tissues in muscle are contiguous, yet each possesses distinctive structural and chemical properties. The connective tissues of muscle all occur outside the muscle cells and provide a scaffold or matrix at each level of organization to support individual muscle fibers, groups of fibers, or the muscle itself. Collectively, connective tissues are called the extracellular matrix (ECM).

Connective Tissue Components

The predominant connective tissue protein in muscle is collagen. Associated with collagen are relatively small amounts of proteoglycans and other proteins including elastin.

Collagen

Collagen represents about 2–6% or more of the dry weight of muscle, depending on muscle type and animal age. There are three morphologically distinct collagen depots in muscle: the epimysium, the connective tissue sheath surrounding individual muscles; the perimysium, a three-dimensional collagen network surrounding large and small bundles of muscle fibers, and which contains lipid deposits (marbling) and muscle vasculature; and the endomysium, a layer of connective tissue encircling individual muscle fibers. The epimysium is generally easily separated from the body of the muscle. Perimysium and endomysium, connective tissues not practically separated from meat, comprise the intramuscular connective tissue (IMC). Perimysium comprises the vast bulk of IMC, about 90%, and is generally held to be the main contributor to variations in connective-tissue-related meat quality. The role endomysium plays in meat texture is less well understood.

Collagen Biosynthesis

Collagen, like all proteins, is synthesized intracellularly; however, newly synthesized collagen molecules are secreted from the cells in which they were produced and function extracellularly. Collagen biosynthesis is, thus, divided into events that occur intracellularly and those which happen outside the cell. Sequentially, major intracellular events include (a) stimulation of discrete collagen genes followed by synthesis of specific mRNA for the different procollagen alpha chains, (b) message translation and subsequent enzymatic hydroxylation of selected proline and lysine residues and glycosylation of specific hydroxylysine residues, (c) molecule folding and triple helix formation, and (d) proteolytic excision of the terminal propeptides as the processed collagen molecule is transported through the cell membrane to the extracellular space. Each alpha chain twists into a left-handed polyproline helix with three residues per turn. The three alpha chains are wound into a right-handed superhelix which forms a molecule about 1.4 nm wide and 300 nm long (Fig. 2.10). The triple helical molecule is flanked by short non-helical regions called telopeptides. The helical alpha chains contain a GLY-X-Y sequence repeated 340 + 2 times per molecule, where X or Y are often proline or hydroxyproline. Fibrillar collagens are, thus, about one-third glycine and one-quarter proline and hydroxyproline with a molecular weight of about 300,000. Once in the extracellular space, collagen molecules align themselves into microfibrils, in quarter stagger array; cross-linking is initi-

A Primary sequence

GLY–PRO–Y–GLY–X–Y–GLY–X–HYP–GLY–

B Triple helix

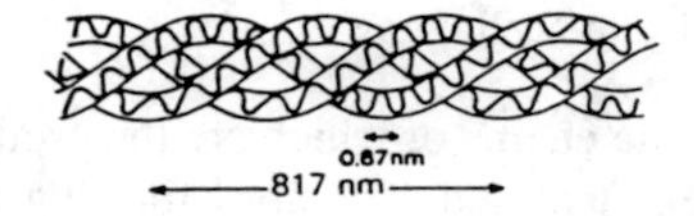

C Collagen fibril

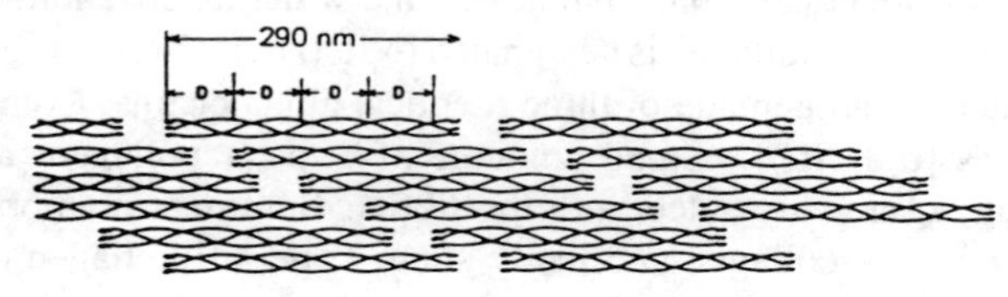

D Collagen fibre

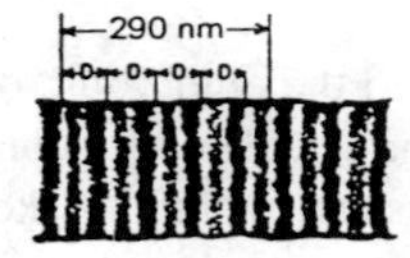

E

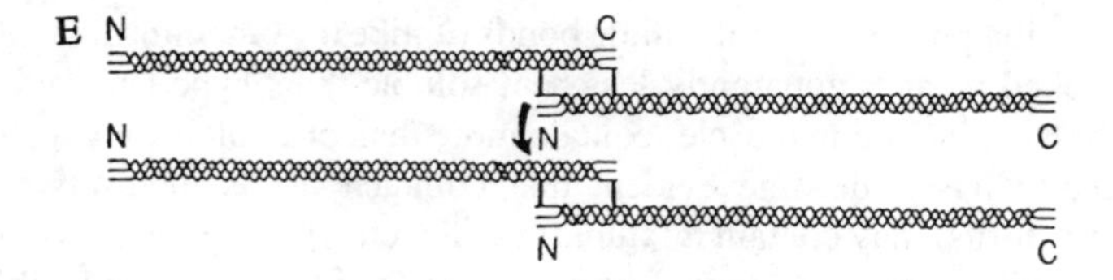

F

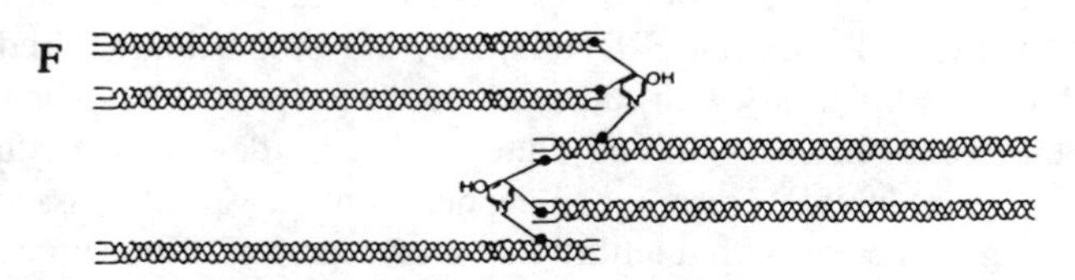

Figure 2.10. Diagram of collagen amino acid sequence (a), molecule (b), filament (c), and fiber (d) structure. [After D. Etherington and A. J. Bailey, *Metabolism of Collagen and Elastin in Comprehensive Biochemistry* (M. Florkin, eds), Elsevier Scientific Publishing Co., New York, 1980, p. 310.] Molecular packing and cross-linking of collagen molecules (head-to-tail sites) by divalent cross-links (e). Formation of a trivalent cross-link (hydroxylysylpyridinoline) (f) could produce molecules cross-linked in typical quarter stagger array as in (e) with the addition of collagen [molecule(s)] in register (no stagger). [After D. Eyre, Collagen stability through covalent crosslinking, in *Advances in Meat Research, Vol. 4, Collagen as a Food* (A. M. Pearson et al., eds.), Van Nostrand Reinhold, New York, 1986, p. 74.]

ated and larger diameter fibrils are formed either by the addition of microfibrils or by association with other fibrils (Fig. 2.10). Covalent cross-linking continues as fibers grow and age. Collagen biosynthesis is a complicated process that entails complex posttranslational processing of molecules.

Phenotypic Composition of Muscle Collagens

Fourteen collagen types, the alpha chains of which are the products of at least 25 discrete genes, have so far been identified. The fibril-forming collagen Types I and III comprise the vast bulk of IMC collagen. Type I collagen is a heterotrimer consisting of two identical alpha chains and one which is different. To describe its composition, Type I collagen is designated $[\alpha\ 1\ (I)]_2\ \alpha\ 2\ (I)$. Type III collagen is a homotrimer which is made of three identical alpha chains. Compositionally it is denoted as $[\alpha\ 1\ (III)]_3$. Small amounts of Type V collagen, also a fibril-forming collagen, are associated with the fibrils. Basement membranes consist primarily of Type IV collagen and likely some Type VII collagen occurring as "anchoring fibrils" connecting the basement membrane to the overlying endomysium. The role that the quantitatively "minor collagens," that is, Types IV, V, and VII, play in meat texture is unknown.

Type III collagen is generally considered the embryonic or precursor form in fibrillar collagens consisting of Types I and III. Fetal or neonatal tissues are rich sources of Type III collagen. In most tissues, including skeletal muscle, there is a general shift with chronological age to increased proportions of Type I collagen. Type III collagen fibrils are likewise smaller in diameter than Type I collagen fibrils, possess a few disulfide bonds (unlike the other fibrillar collagens), and in cooked meat is apparently less heat soluble than Type I collagen.

Most tissues, including muscle, contain more than one collagen type; however, only recently has it become evident that collagen fibrils themselves (termed heterotypic fibrils) may contain mixtures of collagen types. Molecules of different collagen types associate in mixed fibrils via reducible and nonreducible lysine aldehyde-derived cross-links. For example, in cartilage, a tissue containing fibrils composed of Types II, IX and XI collagens, the proportions of reducible to nonreducible cross-links vary with collagen molecule type and cross-link location. Evidence that collagen cross-linking pattern, that is, degree of maturation of reducible to nonreducible cross-links, depends on the specific association of different collagen species within a fibril has been reported. Biochemical, as well as immunofluorescence, studies indicate that Type I and III collagens also occur together in the same fibril. Lysine aldehyde-derived covalent cross-links linking Types I and III molecules in bovine IMC have been documented.

Cross-link Biosynthesis

The well-known textural changes that occur in meat as animals grow and mature are most directly correlated to the progressive maturation of muscle collagen. Mature collagen cross-link concentrations increase with age for all species.

Structural, biochemical, and physiological aspects of collagen cross-linking are detailed in several comprehensive reviews (Reiser et al. 1992). Cross-linking is initiated by the oxidative deamination via the enzyme lysyl oxidase of specific lysines or hydroxylysines which produces peptidyl aldehydes, termed allysine or hydroxyallysine, respectively. The head-to-tail lateral alignment of collagen molecules in a quarter-stagger array allows the aldehyde functions to react with other peptidyl aldehydes or unmodified lysine or hydroxylysine residues on adjacent alpha chains. The initial condensation products form reducible cross-links that are divalent. Reducible cross-links contain Schiff base double bonds which can be reductively labeled and can join only two collagen molecules. There are two major pathways by which cross-links form: the first, the allysine pathway which is based on lysine aldehydes and produces aldimine crosslinks; the second, the hydroxyallysine pathway, produces cross-links arising from hydroxylysine aldehydes. Amadori rearrangement of the initial aldimine crosslinks formed between lysine and hydroxylysine aldehydes can produce ketoamine derivatives. The reducible cross-links vary in their stability, with ketoamine cross-links being heat stable and aldimine cross-links heat labile. Cross-linking of collagen is a progressive process, and the reducible cross-links undergo further condensation reactions and are replaced with mature nonreducible cross-links. In the hydroxyallysine pathway a mature, nonreducible, trivalent cross-link which has been identified is hydroxylsylpyridinium (HP) (Fig. 2.10).

Initial studies by Bailey and co-workers demonstrated the presence of reducible cross-links in IMC, and a positive relationship of toughening in cooked muscle to the concentration of heat-resistant ketoamine cross-links. HP residues apparently arise from the condensation of two ketoamine cross-links, a mechanism of formation which is confirmed by the stoichiometric relationship between the disappearance of ketoamine molecules and the appearance of HP. Further, because the reducible cross-links are transient, their concentration in tissue diminishes as collagen ages or matures. Thus, an inverse relationship can, and often does, exist between degree of mature cross-linking and the measured concentration of reducible cross-links. The progression of cross-link formation along the hydroxyallysine pathway in skeletal muscle is rapid; the nonreducible cross-link concentration of IMC from meat animals less than 1 year old is often already 50% or more of values obtained for animals 5–7 years old.

The progressive nature of collagen cross-link biosynthesis does not mean, however, that in every muscle there is a steady, irreversible progression of lysine aldehyde-derived cross-links from less mature to mature forms. Although there is a good correlation between maturation of muscle collagen cross-links and chronological age, it is also apparent that rate of cross-link formation and directional shifts in the concentration of mature cross-links, irrespective of age, can be altered.

Age, Growth, and Adaptation

There is remarkably little variation in the collagen concentration of a skeletal muscle with growth and maturation. In general, variations in skeletal muscle collagen concentrations are relatively slight, indicating that synthesis, accretion, and turnover of cellular and extracellular proteins in muscles remain in equilibrium over much of the life span of the animal. Some exceptions are the elevated collagen concentrations in the muscles of very young animals compared to larger, more mature animals, and diminished collagen concentration in the muscles of double-muscled cattle. Collagen concentration is often slightly increased in the muscles of intact males compared to castrates.

The steady increase in mature collagen cross-linking is due to progressive and ongoing cross-linking reactions that occur within fibrillar collagen and with the slowing of collagen synthesis rates as animals reach maturity. Less collagen synthesis and turnover provide existing fibrillar collagen time to progressively cross-link or mature. Variations in animal growth rate, which influences muscle growth and collagen synthesis and turnover, have the potential to either reverse or accelerate typical maturation profiles. A muscle effect on cross-link concentration is also apparent. In general, muscles with considerably higher concentrations of collagen such as *biceps femoris* or *soleus,* which also tend to be slow twitch, possess higher reducible and nonreducible cross-link concentrations than muscles of lower collagen concentration. These observations emphasize the difficulties in attempting to determine the contribution that muscle collagen concentration makes to meat texture, particularly when different muscles are compared.

Although a steady increase in cross-link concentration in muscles with aging is typical, the properties of extracellular collagen of muscle, including cross-linking profile, are extremely variable.

Proteoglycans

Proteoglycans, formerly called mucoproteins or mucopolysaccharides, are glycoproteins with a high carbohydrate content, up to 80%. They consist of a core protein to which glycosaminoglycan chains (GAGs) are covalently attached. GAGs are polymers of repeating disaccharides which result in macromolecules varying in molecular weight from 30,000 to several million depending on the size of the core protein and its constituent GAGs. GAGs are made up of chondroitin and dermatan sulfates, keratin sulphates, heparan sulphates, and hyaluronic acid. GAGs are generally negatively charged due to the presence of sulfate and carboxylate groups, thus they repel each other, occupy a large volume for their molecular weight, and bind large amounts of water. In the extracellular matrix of muscle, proteoglycans are a quantitatively minor component. Their possible role in determining meat quality is not known. Proteoglycans are important in maintaining interaction of extracellular matrix components, and in water distribution and retention. Their association with collagen indicates they may play a role in determining collagen fibril size and collagen fibril stabilization, factors which suggest a role for proteoglycans in muscle growth and development and meat quality.

Glycoproteins

Glycoproteins are proteins to which one or more saccharides are covalently bonded. They differ from proteoglycans in that the carbohydrate fraction of the glycoprotein is generally much smaller. Fibronectin and laminin are two of the better characterized glycoproteins of the extracellular matrix. Fibronectin apparently mediates the binding of cells to ECM components such as collagen fibrils and proteoglycans. It is closely associated with, but not part of, the basement membrane of the muscle fiber. Laminin is an integral component of the basement membrane closely associated with both the collagen network of the basement membrane and the underlying cell. Little is known of the role or function of these noncollagen constituents of the ECM in muscle growth, development, or in meat quality.

Elastin

Elastin is a rubberlike connective tissue protein which, as the name implies, confers elasticity and stretchability to tissues subject to deformation such as skin, arteries, and muscle. It is heavily cross-linked with desmosine and isodesmosine cross-links which arise from lysine residues. Elastin is noted for its extreme insolubility and resistance to physical disruption by heat. It occurs only in small quantities in skeletal muscle; *semitendinosus* muscle possesses more elastin than other muscle of the carcass.

Adipose Tissue and Bone

Intimately associated with muscle are two additional components of the carcass, the specialized connective tissues adipose and bone. Adipose tissue and its primary constituent lipid are important factors in meat quality and nutritional value, carcass yield, and market value. Bone, while generally not consumed, often comprises a fraction of retail meat cuts and plays important roles in muscle growth, carcass yield, and estimation of carcass maturity.

Adipose Tissue

Adipose distribution within a carcass varies by anatomical sites (specific muscles or regions of the carcass) as well as by systemic locations (intramuscular, intermuscular, subcutaneous, and visceral adipose depots). Intracellular fat droplets within the muscle fiber are not, strictly speaking, adipose tissue depots.

Adipose may consist of small clusters of cells, called adipocytes, such as those that exist between bundles of muscle fiber, within the perimysium and that constitute "marbling fat." Alternatively, extremely large deposits of adipocytes including intermuscular, subcutaneous, and visceral fat depots are also adipose tissue and represent the major fat storage areas of the body.

The mature adipocyte has an average diameter of 80–120 μm and is filled with neutral lipid, primarily triacylglycerol. Phospholipid and sterols also contribute to lipid composition of adipocytes, but these lipids are associated most with the adipocyte cell membrane.

Species differences exist for depot site as well as fatty acid composition. In pig carcasses, the bulk of adipose tissue is located in the subcutaneous depot, whereas in cattle and sheep carcasses, subcutaneous and intermuscular depots contain most of the fat tissue. The fatty acid composition of adipose tissue lipids in meat animals is mainly dependent on factors regulating de novo fatty acid biosynthesis. Because oleic, stearic, and palmitic acids are the major end products of fatty acid synthesis, these fatty acids comprise the majority of adipocyte fatty acids. Dietary fat contributes to the fatty acid pool in adipocytes but to a much lesser extent in ruminants than in pigs because of rumen biohydrogenatation of unsaturated fatty acids. Pigs can deposit dietary unsaturated fatty acids. However, soft, oily pork products, as well as the potential for rancidity, has discouraged production of pork with fat that is high in unsaturated fatty acids. Regardless of species, the more internal depots contain higher proportions of saturated fatty acids than external depots. The reasons for this are not clear, but it may be related to temperature variations between internal and external depots. Current theory suggest differences in fatty acid composition may be inherent to the respective depots, and adipocyte size may also be a factor; these latter two hypotheses have not yet been tested.

When meat animals receive diets that provide excessive energy, fat deposition is inevitable. In meat products, this fat contributes both positive and negative attributes. On the positive side, eating quality of meat is improved because of fat, largely because of juiciness and flavor enhancement. The negative factors primarily involve increased serum cholesterol in many individuals who consume animal fat in their diets. Lean red meat, however, is not necessarily hypercholesterolemic; several current studies have shown the opposite to be true. Current consumer attitudes and marketing trends discourage the production of fat in meat animals.

Bone

The carcass skeleton, which accounts for 12–25% of its weight, consists of endochondral bone. Endochondral bone formation occurs by ossification within an existing cartilage model. Ossification begins when chondrocytes of the cartilage matrix, the precursor to bone, hypertrophy and are invaded by capillaries. New bone is formed when the extracellular matrix becomes calcified (ossified) with the deposition of hydroxyapatite crystals. The formation of endochondral bone involves not only mineralization of a connective tissue matrix but also the transition of collagen types comprising the matrix. The cartilage model, prior to ossification, consists of Types II, IX, and XI collagen (see the section Connective Tissue in Muscle). These typical cartilage collagens are replaced by Type I collagen, with small amounts of Type V also present, as endochondral bone is formed. Mature bone is about 20–25% collagen and the balance mineral. The collagen of bone becomes progressively more cross-linked with maturation; the major mature cross-links in bone are HP and its dehydroxy analogue lysylpyridinoline (LP). The ratio of HP/LP in serum and urine can be used to estimate rate of bone growth or resorption. HP/LP ratios can further serve to differentiate serum or urine hydroxyproline derived from the skeleton from muscle or other organ hydroxyproline because LP occurs only in significant quantity in bone.

With growth and development, the composition of the skeleton shifts from cartilage to calcified cartilage to cancellous bone to cortical (dense) bone. The long bones of the carcass consist of a shaft, the central portion or diaphysis, and the enlarged ends, called the epiphyses. The junction of epiphysis and diaphysis is formed by a thin disk of cartilage called the epiphyseal plate. In animals whose skeletal growth is not yet complete, bone lengthens as the face of the growing epiphyseal plate moves through the transition from cartilage to calcified cartilage to bone, and newly formed bone is added to the diaphysis.

The maturity of meat carcasses is assessed by visually estimating the degree of ossification of specific bones of the carcass. In cattle, the extent of ossification of the spinous processes of the thoracic vertebrae, exposed when the carcass is split longitudinally, is used to estimate animal age. In both sheep and cattle, redness and absence of flatness of the ribs are indicators of youthfulness. Lamb carcasses are differentiated from mutton by the exposure of the distal epiphyseal plate of the metacarpal bone ("break joint"). Breed and gender influence the rate of bone maturation; thus, physiological maturity as estimated by bone condition may not be an accurate indicator of chronological age across a species.

Bone growth and skeleton size are also determinants of muscle growth, ultimate muscle mass, and meat quality. Large-framed meat animals produce more muscle than animals with small bones and skeletons. Larger breeds also mature more slowly than smaller breeds which results in leaner meat because lipid accretion is associated with maturation.

Selected References

Bailey, A.J. and N.D. Light. 1989. *Connective Tissue in Meat and Meat Products.* Elsevier Applied Science, New York.

Brenner, B. 1991. Rapid dissociation and reassociation of acto-myosin crossbridges during force generation. A newly observed facet of cross-bridge action in muscle. *Proc. Natl. Acad. Sci.* 88:10490–10494.

Croall, D.E. and G.N. DeMartino. 1991. Calcium-activated neutral protease (calpain) system: Structure, function, and regulation. *Physiol. Rev.* 71:813–847.

Eisenberg, B.R. 1983. Quantitative ultrastructure of mammalian skeletal muscle. In: *Handbook of Physiology, Section 10 Skeletal Muscle*. L.D. Peachey, (ed.). Pages 73–112. American Physiological Society, Bethsda, MD.

Gans, C. and F. de Vree. 1987. Functional bases of fiber length and angulation in muscle. *J. Morphol.* 192:63–85.

Nimni, M.E. and R.D. Harkness. 1989. Molecular structures and functions of collagen. In: *Collagen, Volume I Biochemistry*. M.E. Nimni, ed. Pages 1–77. CRC Press, Boca Raton, FL.

Pearson, A. and R. Young. 1989. *Muscle and Meat Biochemistry*. Academic Press, San Diego, CA.

Price, J.F. and B.S. Schweigert (eds.). 1987. *The Science of Meat and Meat Products*. Food and Nutrition Press, Inc. Westport, CT.

Rayment, I., H.M. Holden, M. Whittaker, C.B. Yohn, M. Lorenz, K.C. Holmes, and R.A. Milligan. 1993. Structure of the actin-myosin complex and its implications for muscle contraction. *Science* 261:58–65.

Reiser, K., R.J. McCormick, R.B. Rucker. 1992. Enzymatic and nonenzymatic cross-linking of collagen and elastin. *FASEB* J. 6:2439–2449.

Schutt, C.E. and U. Lindberg. 1993. A new perspective on muscle contraction. *FEBS Lett.* 325:59–62.

Swatland, H.J. 1984. *The Structure and Development of Meat Animals*. Prentice-Hall, Englewood Cliffs, N.J.

Trotter, J.A. 1990. Interfiber tension transmission in series-fibered muscles of the cat hindlimb. *J. Morphol.* 206:351–361.

3

Postmortem Changes in Muscle Foods

Cameron Faustman

Introduction

The biology of living skeletal muscle and the processes which occur during its conversion to meat are critical to an understanding of quality in muscle-based foods. There is a variety of animal species from which muscle foods are obtained. Although some species-specific differences exist, the changes associated with the conversion of muscle to meat are essentially the same. Slaughter of livestock generally involves stunning, exsanguination, removal of hide, hair, or feathers, evisceration, and washing. In commercial plants, these procedures are accomplished in less than 1 h. Onset of rigor mortis varies from 3 to 6 h in smaller animals such as poultry, and 24–36 h in large animals like cattle. Commercial harvesting of fish may be accomplished in a variety of ways. In large operations, fish are caught, skinned, scaled or shelled, eviscerated, filleted, and stored frozen within 1 h of harvest. In smaller operations, boats may harvest fish and store these on ice until further processing when the boat reaches port.

Hultin (1985) has noted that differences in eating quality of fish versus that of traditional livestock meats are related to the different physical environments in which each of these live. For instance, fish live in a buoyant environment and do not have the amount of connective tissue for support that is required by land animals. Thus, textural problems associated with connective tissues are less of a concern in fish than in meats from traditional livestock species.

Regardless of the species being harvested, death is accompanied by an inability to deliver oxygen within the body, and subsequent anoxia. When normal life processes are halted, many of the biochemical reactions present in the living state retain some degree of activity in the nonliving state. These reactions are responsible for profound quality changes during storage. For considerations of the conversion of muscle to meat, the nonliving or postmortem period can be divided into two stages: postmortem pre-rigor and postmortem post-rigor.

Postmortem, Pre-rigor Conditions in Muscle

Myofibers—The Cells of "Living" Muscle

The myofiber or skeletal–muscle cell is relatively large and is formed by fusion of individual myoblasts. The myofiber is unusual in that it contains many nuclei and has an elaborate cytoskeletal apparatus for contraction. It contains many of the cell components found in other tissue cell types and, in general, these organelles perform the same functions as they do in other cell types. For instance, mitochondria utilize oxygen and other substrates to produce energy in the form of adenosine triphosphate (ATP). However, some of these organelles perform specialized functions within myofibers. The endoplasmic reticulum is highly specialized for regulating calcium concentrations within the myofiber and, as a result, exerts a strong influence on control of muscle contraction. Within muscle, this organelle is referred to as the sarcoplasmic reticulum.

When ATP is split into adenosine diphosphate (ADP) and inorganic phosphate (Pi), energy is released and is harnessed for contractile activity. In the living state, ATP for muscle function may be obtained by aerobic or anaerobic metabolism (Fig. 3.1). Aerobic metabolism includes the processes of glycolysis and respiration (citric acid cycle and electron transport). This process utilizes oxygen and is a more efficient means of ATP production than anaerobic metabolism. The presence of sufficient oxygen and substrate, primarily glucose, allows continued production of ATP by aerobic processes. In the myofiber, glucose is stored as the polysaccharide, glycogen.

Creatine phosphate (CP) is a chemical compound which serves as a storage for the high-energy phosphate necessary to regenerate ATP from ADP. Creatine phosphate availability is especially crucial during times of high-energy demand by the muscle (i.e., intense exercise). Creatine kinase is an enzyme that catalyzes a reversible reaction whereby ATP may be regenerated from ADP utilizing CP:

$$\text{ADP} + \text{CP} \xrightarrow[\text{Creatine Kinase}]{} \text{ATP} + \text{Creatine}$$

During intense muscle activity when direct ATP production by mitochondria is limiting, creatine phosphate and creatine kinase provide a rapid, short-term supply of ATP. The reversible nature of the above reaction is important in that it allows regeneration of creatine phosphate from creatine and ATP during periods of rest.

AEROBIC (Glycolysis, citric acid cycle and electron transport)

$$\text{1 Glucose unit} + \text{36 Pi} + \text{36 ADP} + 6\ O_2 \longrightarrow \text{36 ATP} + 6\ CO2 + 42\ H_2O$$

ANAEROBIC (Glycolysis)

$$\text{1 Glucose unit} + \text{3 ADP} + \text{3 Pi} \longrightarrow \text{3 ATP} + \text{2 lactate} + 2\ H^+ + 3\ H_2O$$

Figure 3.1. Yield of ATP in aerobic and anaerobic metabolism.

Anaerobic metabolism includes only the process of glycolysis and results in the accumulation of lactate. In the living state, anaerobic metabolism is adopted when the muscle experiences an oxygen debt such as in situations of intense exercise; lactate is transported to the liver and converted to glucose. Although anaerobic glycolysis yields only three ATP per glucose unit (from glycogen), it does permit muscle function to continue despite anoxic conditions. The process of anaerobic metabolism also occurs following the slaughter/harvest of meat-producing animals. At death, the animal's ability to obtain and deliver oxygen to tissues is lost, but some aerobic metabolism can still occur at carcass tissue surfaces where oxygen is readily available. Although the capacity to produce ATP through aerobic processes in deeper locations within the tissue has been compromised, all biochemical components necessary for anaerobic metabolism in cells are present and functional, and glycolysis proceeds. As a result, changes occur in the concentrations of glycolytic substrates and reaction products until a point is reached at which some reaction in the glycolytic process is arrested and metabolism ceases. When this occurs, the ability to produce ATP is completely lost and rigor mortis is established. Adenosine triphosphate is necessary for contraction and subsequent release of myosin from actin during a contraction cycle. The loss of ATP over time following death results in a concomitant increase in fixed actomyosin crossbridges. Thus, as ATP concentration decreases and approaches zero, a greater percentage of myosin heads remains fixed to actin and the extensibility of muscles decreases.

During the postmortem pre-rigor process, several changes occur in the concentrations of glycolytic substrates and products. The changes in concentrations of hydrogen ion (H^+) (expressed as pH), acid-labile phosphorus (ATP), CP, and extensibility from the point of death until the onset of rigor mortis are presented in Fig. 3.2.

These compounds can be measured, and their concentrations at a given time postmortem are related to the rate and extent of glycolysis. It is important to note that the concentration of ATP does not begin to decrease immediately postmortem but rather remains at physiological levels for a brief period before declining. This is due to the regeneration of ATP from CP and continued anaerobic glycolysis.

Conditions Affecting Onset of Rigor

Rigor mortis is a temporal process occurring during the time course of postmortem glycolysis and is characterized by progressive stiffening of the muscle. The loss of ATP during postmortem anaerobic glycolysis is the event causing the onset of rigor. As long as there is a pool of ATP present, crossbridge cycling between myosin and actin molecules can occur. When ATP is exhausted, the myosin and actin molecules remain locked together and yield the stiff nature of muscle in rigor.

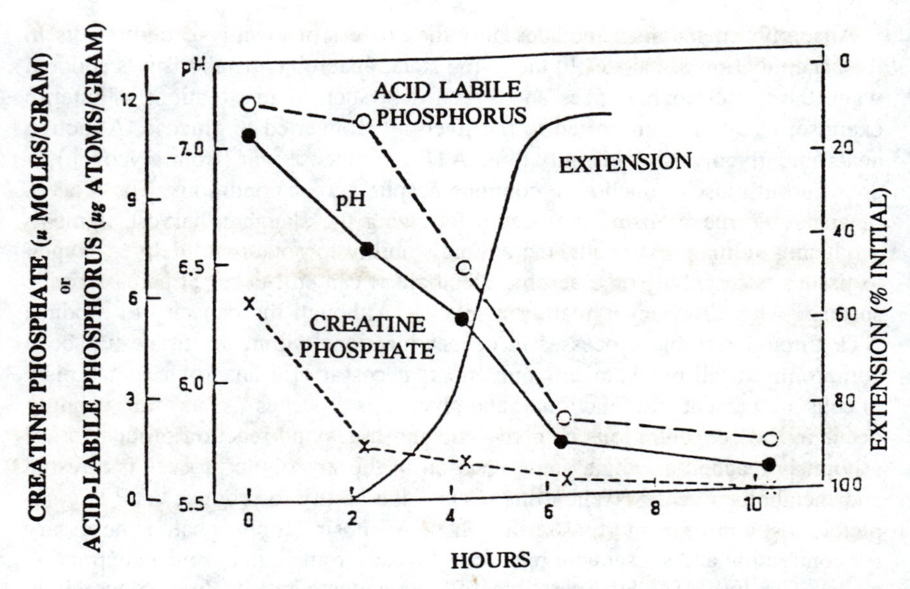

Figure 3.2. The changes in biochemical metabolites during the onset of rigor mortis. (From Newbold 1966.)

The rate of postmortem glycolysis and the extent to which it occurs have implications for muscle food quality. As anaerobic glycolysis proceeds from the point of slaughter to rigor mortis, changes occur in the muscle. The production of H^+ (Fig. 3.1) leads to a more acidic condition which, in turn, is measured as a decrease in muscle tissue pH.

The rate of biochemical reactions is strongly affected by temperature. A rule of thumb for chemical reactions in general is that an increase of 10°C results in a doubling of reaction rate. At death, livestock (not including fish) muscle temperature is at the normal physiological level (≈38–40°C). Once the animal carcass has been processed, it is placed into a cooler at 4°C (40°F) or less. The rate of carcass cooling will influence the degree to which glycolytic reactions are slowed and the time course of rigor onset.

Fat cover and/or presence of the hide will act as insulators and can significantly affect the rate of postmortem temperature decline in the carcass. "Hog-dressed" lamb and veal are processed with their pelt/skin attached to prevent excess moisture loss; pig carcasses typically have the skin left on. In these cases, the rate of cooling would be expected to be slower than for species with the hide removed. In turn, a faster rate of postmortem glycolysis may be expected because of the higher temperature of the carcass. Fat cover also affects the rate of carcass cooling. Carcasses with greater external fat cover will cool more slowly than those with less fat cover. It is important to note that, within a given carcass, various muscles will display different cooling rates based on their location. The more exterior muscles will cool more rapidly than deep-seated muscles, allowing for a slower rate of glycolysis.

Cold Shortening, Thaw Shortening, and Electrical Stimulation

Cold shortening is a phenomenon that occurs in pre-rigor muscle and results in less tender meat. The "cold" refers to the rapid cooling which must occur in order to observe the effect (Fig. 3.3). "Shortening" refers to short sarcomere lengths characteristic of highly contracted muscle. It is generally believed that rapid chilling compromises the ability of sarcoplasmic reticulum and mitochondria to retain calcium. Calcium is released into the sarcoplasm in an uncontrolled manner; this causes substantial contraction to occur in the presence of ATP. The contractile activity results in shortening of muscle fibers which, in turn, corresponds to a decrease in meat tenderness. There has been some concern that the selection of livestock for production of carcasses with reduced fat cover (and thus more rapid cooling) may predispose the muscles of these carcasses to cold shortening.

The process of electrical stimulation has been adopted by the meat industry as a means of countering cold shortening in beef and lamb. Electrical stimulation is the postslaughter application of an electrical current to the animal carcass.

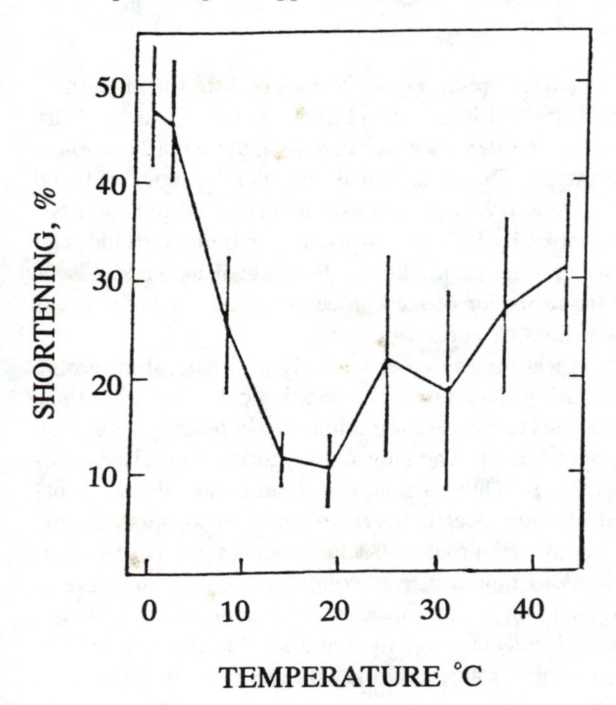

Figure 3.3. Relationship between muscle shortening and temperature. (From Locker and Hagyard (1963).]

Pearson and Dutson (1985) have provided an excellent review of electrical stimulation. In the living state, the signal to contract is provided by a membrane depolarization from the nervous system. The physical application of an electric current stimulation mimics this system and results in accelerated muscle metabolism, and serves to hasten the onset of rigor mortis. Thus, for a given time, postmortem, anaerobic glycolysis will have progressed to a more advanced "stage" in carcasses that have experienced electrical stimulation than in those which have not. This allows rigor onset to occur before the musculature reaches low temperatures at which cold shortening takes effect.

If meat is frozen prior to rigor onset and subsequently thawed, it will shorten dramatically and be extremely tough. This phenomenon is referred to as "thaw shortening." The process of pre-rigor freezing halts the various anaerobic metabolic reactions occurring in the muscle. The process of freezing can damage the sarcoplasmic reticulum and destroy its ability to regulate calcium concentrations within the myofiber. During thawing, all components necessary for muscle contraction are still present, but control of the reactions has been lost. As a result, anaerobic metabolism proceeds at a very rapid rate and is concomitant with severe contraction.

Stress and the Postmortem Pre-Rigor Condition

Pale, Soft, and Exudative Meat

Animals which are exposed to stressful conditions prior to slaughter may produce meat that is pale, soft, and exudative (PSE), or dark, firm, and dry (DFD). The PSE condition is most often observed in pork and is a major problem for the pork industry worldwide. The condition is prevalent in hogs afflicted with the Porcine Stress Syndrome (PSS) but does occur in normal pigs as well. PSE meat is accurately described by its name. Muscles are quite soft and tend to sag; meat surfaces are watery and light colored. When cooked, PSE meat is very dry and unpalatable. Its utility for cooked processed meats manufacture is limited due to its poor water-holding capacity.

Animals which yield PSE meat display an abnormally high rate of anaerobic glycolysis immediately following slaughter. The rapid biochemical reactions produce heat so quickly that muscle temperature immediately postslaughter may exceed the normal physiological level. The elevated glycolytic rate also means that muscle pH decreases rapidly. Thus, a relatively high-temperature/low-pH condition occurs. Elevated physiological temperatures combined with low pH are sufficient to disrupt some muscle protein structure (denaturation). As such, the ability of muscle proteins to bind water is compromised and moisture is expressed on cut surfaces, yielding a wet appearance. In addition, the surface wetness results in an increased reflectance of light and a paler appearance.

At 45 min postmortem, the pH of normal pork will generally be >6.4–6.5, whereas the pH of PSE pork will be <6.0. The ultimate pH of meat from PSE carcasses is generally a bit lower (~ 5.2–5.4) than that of pork from normal carcasses (~ pH 5.6–5.8). Thus, in the PSE condition, postmortem anaerobic metabolism occurs at a faster rate and to a slightly greater extent than in carcasses from normal animals. It is important to note that the PSE condition is not an "all or nothing" phenomenon and that the extent to which meat quality is compromised can vary over a wide range.

Dark Firm and Dry Meat

The dark, firm, and dry (DFD) condition is a meat quality defect observed primarily in beef and pork. The DFD phenomenon is typified by meat which has a darker than normal appearance. Meat surfaces appear dry and the pH of the meat is higher than normal. The pH range of normal meat is 5.3–5.8. DFD meat will generally exceed pH 6.0–6.2 and may reach as high as pH 6.8. Conditions of stress are also associated with the DFD condition, although the mechanism by which this stress-associated meat quality defect occurs is different from that of the PSE phenomenon. DFD meat is obtained from animals which have been exposed to long-term preslaughter stress. In some cases, the simple act of penning strange animals together may induce sufficient stress. In response to the stress, affected animals will utilize a significant portion of their muscle glycogen stores. When slaughter occurs, the glycogen pool (which has been substantially depleted during exposure to stress) has less glucose available for glycolysis, and the extent to which anaerobic metabolism can occur is abbreviated. This, in turn, results in decreased formation of H^+ and lactate, and the ultimate pH of the meat obtained is higher than normal. In the DFD situation, the extent (and not the rate) of anaerobic metabolism is shortened, causing rapid rigor onset. It should be noted that muscle of high ultimate pH has a greater water-binding ability than meat of normal pH. Thus, although DFD beef is undesirable for the retail meat case, it may have important value to meat processors.

Postmortem Changes: Post-rigor

The exact point at which conversion of muscle to meat is complete is difficult to assess, although the establishment of rigor mortis is generally accepted to be this point. However, while the functional role of skeletal muscle is lost and rigor has been established, metabolic activity of the tissue has not ceased. Various biochemical processes may still occur and some of these have significant implications for the quality of postmortem muscle as food.

Post-Rigor Changes Affecting Texture and Tenderness

Historically, meat has been aged to improve tenderness. Aging is necessary as meat is often unacceptably tough immediately following rigor onset. The amount of time required for aging is species dependent with approximately 12 days needed for beef, 3–5 days for pork, and 1–2 days for chicken. The aging process may be expedited by a process called "high-temperature conditioning." High-temperature conditioning refers to the holding of carcasses at temperatures of 59°F (15°C) or greater, as opposed to the normal cooler storage at 4°C (Pearson and Dutson 1985). This conditioning can be applied in the pre- or post-rigor state and has been found to be very effective in improving meat tenderness.

It is believed that, during the aging process, tenderization occurs as a result of protein degradation or proteolysis. These proteolytic processes originate within the myofiber and are responsible for degradation of cellular constituents. Although it may appear strange for the cell to possess the means for its own destruction, these agents are necessary in life for breakdown of older "worn-out" protein so that new "fresh" protein may be utilized to maintain cell function. The proteolytic enzymes in meat that have been most studied are the cathepsins and calpains.

Cathepsins are located within lysosomes and operate best at acidic pH values (generally $pH<5.2$). When myofibrillar proteins are incubated with various cathepsins in vitro, they are degraded. Some investigators believe that catheptic enzymes are able to act at the pH of meat to degrade myofibrillar proteins (primarily myosin and actin), thus producing a more tender product. The relative role of cathepsins in meat tenderness has been a hotly debated issue. Cathepsins are effective proteolytic agents and have been identified in meat.

Calpains, also known as calcium-activated factors, are proteases that require calcium ions (Ca^{2+}) for activity. There are two major types of calpains: one which requires a high concentration of free calcium ($\approx 300\ \mu M$) for activation, and the other which requires much less free calcium ($\approx 5\ \mu M$); both calpains are found in the sarcoplasm of myofibers. The amount of calcium available in normal muscle cells precludes the high-calcium-requiring calpain as a major contributor to meat tenderization. Both types of calpains require a high pH, ≈pH 6.6–6.8, for optimal activity. This value is substantially higher than pH 5.4–5.8 of normal meat and maximal activity of calpains would most likely occur during the early postmortem (pre-rigor) condition.

The subject of proteolytic enzymes in postmortem muscle has received significant attention. Enzymes require specific conditions (e.g., temperature, pH) for optimal activity, and if these can be determined and maximized in meat, then a means for improving meat tenderness might be realized. In addition, it is not unrealistic to consider the possibility that animals could be bred for high-proteolytic-enzyme activities, or that genetic engineering might somehow play a role in achieving more tender meat.

Fish Texture

Trimethylamine oxide (TMAO) is a naturally occurring chemical compound in marine fish. It is not found in any appreciable amount in freshwater fish but is plentiful in marine gadoid species (e.g., pollock, haddock, cod). TMAO appears to assist in maintaining tissues of the live fish in osmotic balance with its environment. TMAO may be readily converted along one of two pathways to yield trimethylamine, or dimethylamine and formaldehyde.

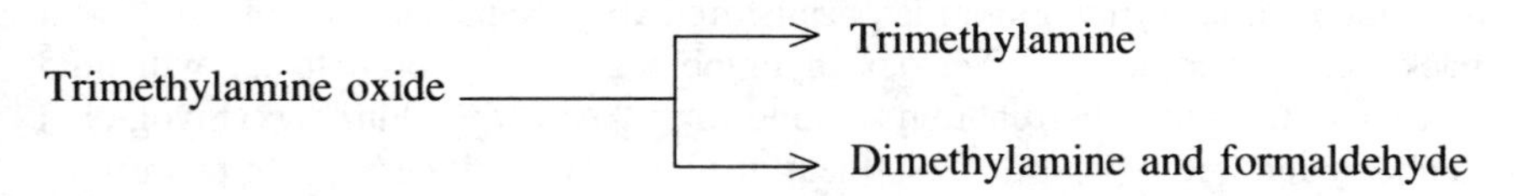

The production of trimethylamine from TMAO is mediated by bacteria; trimethylamine contributes to typical "fishy" odor. TMAO may also be converted by enzymes endogenous to the fish tissue to dimethylamine and formaldehyde. Formaldehyde is chemically reactive and is believed to act as a cross-link between various muscle proteins. This is undesirable, as it results in toughening of the fish tissue. Textural deterioration is sometimes observed following freezing of fish muscle. Because freezing of foods does not generally result in complete ice formation of the water present, solutes dissolved in the water phase of foods can become concentrated as a result of the freezing process. Thus, enzymes and substrates involved in fish toughening may be brought together and effect the observed changes. In part, the occurrence of fish toughening is attributed to protein cross-linking caused by formaldehyde formation from TMAO.

Post-Rigor Changes Affecting Color

The red color of meat is due to the presence of the heme protein, myoglobin. Some residual blood may also be present in meat, but it is generally minimal and is of little practical concern in considerations of meat color. The degree of meat pigmentation is directly related to myoglobin content. In general, myoglobin concentration within a given muscle will differ with species or age (myoglobin concentration is increased with animal age) and is dependent on muscle fiber distribution (Lawrie 1985). Muscles comprised predominantly of red fibers contain more myoglobin than muscles with a high white fiber content (i.e., turkey thigh versus breast muscles).

The heme group within myoglobin is a planar chemical structure that contains a centrally located iron atom. The iron atom has six coordination sites available for chemical bonds. Four of these bonds anchor the iron atom within the heme structure. A fifth bond connects the iron atom to the amino acid chain of the globin protein. The sixth coordination site is available for binding a variety of chemical groups. The chemical group bound at the sixth site and the oxidation state of heme iron are two major determinants of meat color. The oxidation state determines which molecule may be bound at the sixth site of heme iron. Iron is a transition metal capable of existing in an oxidized ferric (+3) form or reduced ferrous (+2) form. For considerations of fresh meat color, there are three oxidation-state/ligand combinations of note, each of which has its own distinctive color (Fig. 3.4). It is important to emphasize that these all occur within the small heme portion of the larger myoglobin protein. When heme iron is in the ferrous form and lacks a ligand at the sixth position, it is referred to as "deoxymyoglobin." The color of deoxymyoglobin is purplish-red and is characteristic of fresh meat packaged under vacuum. Ferrous myoglobin that is exposed to air will bind oxygen at the sixth coordination site and form "oxymyoglobin." Oxymyoglobin is cherry red and typical of fresh meat displayed in retail cases. The process by which deoxymyoglobin binds oxygen and is converted to oxymyoglobin is called "oxygenation." It is important that this not be confused with "oxidation." Oxidation means the loss of an electron by any chemical species. The process of oxidation occurs in myoglobin when ferrous (+2) iron (deoxymyoglobin or oxymyoglobin) is converted to ferric (+3) iron and leads to the third form of myoglobin found in fresh meat, metmyoglobin. Metmyoglobin is brownish-red in color and is characterized by ferric iron with a water molecule bound at the

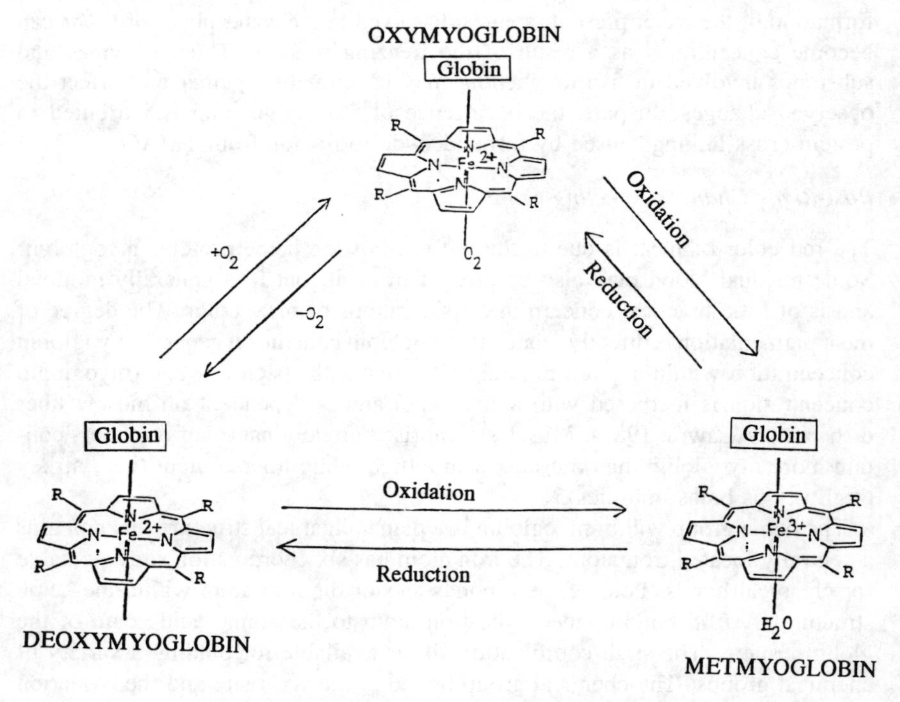

Figure 3.4. The relationship between myoglobin oxidation state and meat color.

sixth position. The oxidation of deoxymyoglobin or oxymyoglobin leads to the formation of metmyoglobin; this process occurs gradually over the surface of meat cuts during storage and/or display.

The myoglobin form which predominates in the surface of meat determines the perceived color. A typical scenario for color expression of the various myoglobin forms can be observed during the cutting of meat. The deep portion of a fresh piece of meat is anoxic, and when sliced will reveal an interior that is purplish-red in color. Following exposure to air for 20 to 30 min, the deoxymyoglobin will oxygenate to form cherry-red oxymyoglobin. As display time increases, oxymyoglobin will oxidize to metmyoglobin and the portion of meat displaying undesirable brownish discoloration will increase. The rate of metmyoglobin formation is dependent on several factors including the specific muscle, display temperature, type and intensity of lighting, and bacterial load.

Meat color is an extremely important sensory characteristic by which consumers make judgments of meat quality. Color that deviates from the accepted cherry red for beef is discriminated against. This has several interesting consequences. One of these is that meat color determines the shelf-life of meat. Cuts of meat with too much metmyoglobin will be viewed as old and undesirable for consumption.

The process of vacuum-packaging of fresh meat has several benefits, including significant shelf-life extension over that of its aerobically packaged counterpart. The lack of oxygen in the packaging atmosphere results in a purplish-red appearance. Although this is readily explained in terms of myoglobin chemistry, the nontraditional color is held suspect by a majority of consumers. This has been a primary stumbling block to the meat industry's attempts to market retail-ready, vacuum-packaged fresh meat cuts. Until the meat industry can effectively educate consumers about meat color, maintenance of myoglobin in the oxymyoglobin form remains an industry priority. There are two basic strategies for accomplishing this goal: the quenching of oxidative processes and enhancement of the biochemical reduction of metmyoglobin.

The delaying of myoglobin oxidation is accomplished in a variety of ways. These include storage and display of meat under refrigerated conditions, hygienic preparation of meat cuts, and selective use of lighting. In addition, the application of antioxidants such as ascorbic acid (vitamin C), citric acid, or α-tocopherol (vitamin E) may extend color shelf-life.

Biochemical reduction is a process in which a chemical compound gains an electron; it works in opposition to oxidation. Metmyoglobin reduction refers to the conversion of ferric (+3) heme iron to ferrous (+2) heme iron. When this occurs, metmyoglobin concentration decreases and that of deoxymyoglobin or oxymyoglobin increases. A source of reducing equivalents (electron donors) must be present. Metmyoglobin cannot bind oxygen and is physiologically inactive. The reduction of metmyoglobin in the living state is an extremely important process for maintaining an adequate pool of ferrous myoglobin for oxygen binding. The process of reduction in living tissue is catalyzed by an enzyme called 'metmyoglobin reductase'. The activity of this enzyme in postmortem post-rigor muscle is only beginning to be understood. Enhancement of metmyoglobin reduction in meat is a logical approach for prolonging acceptable fresh meat color, but is presently a difficult concept to apply in commercial practice.

Influence of Mitochondria

Mitochondria utilize oxygen in performing their normal role within cells. They have a high affinity for oxygen and may remain active in meat stored at normal refrigeration temperatures. It appears that meat cuts that possess high mitochondrial activity have a shorter color shelf-life than those cuts with lower mitochondrial activity.

Pre-rigor muscle contains a greater proportion of actively respiring mitochondria than post-rigor muscle. The color of pre-rigor muscle is more purplish-red than that of post-rigor muscle. The consumption of oxygen by mitochondria makes the oxygen unavailable for binding by myoglobin; the deoxymyoglobin form predominates and yields the observed purplish-red color. One drawback to the predominance of deoxymyoglobin in either pre- or post-rigor meat is its lower stability relative to its ferrous oxymyoglobin counterpart. O'Keefe and Hood (1982) have hypothesized that post-rigor meat with a higher concentration of mitochondria will have a greater proportion of its ferrous myoglobin in the deoxymyoglobin state and that it will be converted to metmyoglobin faster than if oxymyoglobin had been allowed to form.

The appearance of dark, firm, and dry (DFD) meat is partly related to mitochondrial activity. Mitochondria survive and function better at the high-pH conditions of DFD meat than at the pH of normal meat. As such, the relative oxygen-consumption activity of DFD meat is greater, deoxymyoglobin predominates, and subsequent oxidation to metmyoglobin is enhanced. The dark appearance of DFD meat may also be related to the hydration state of myofibrillar proteins. The elevated pH which is characteristic of DFD meat allows for greater hydration of myofibrillar proteins in DFD meat than in normal, lower-pH meat. Moisture is bound more tightly by these proteins and, thus, less moisture is expressed on cut meat surfaces. This results in less reflected light from the meat surface, yielding the dry, dark appearance.

Post-Rigor Changes Affecting Flavor

Lipid Oxidation

The lipid in foods contributes significantly to perceived flavor. The goal of this section is to illustrate the processes involved with development of rancid flavors.

The three major classes of lipids in meat are triacylglycerols, phospholipids, and cholesterol. Although cholesterol has been an issue of nutritional concern, it has no real impact on flavor and will not be discussed further in this chapter. Phospholipids are located in membranes of cells and subcellular organelles, whereas triacylglycerols predominate in lipid droplets and fat depots. The fat which is evident on the external portion of meat cuts or as intramuscular marbling is triacylglycerol. However, phospholipids appear to be the more important lipid class in considerations of off-flavor development. The basic structures of phospholipid and triacylglycerol are shown in Fig. 3.5. Glycerol serves as a backbone for each of these two lipid classes. The first and second positions of these lipid molecules may be occupied by any of a variety of fatty acids. Fatty acids are chains of carbon atoms with a carboxylic acid group at one end. They vary in length according to the number of carbon atoms which comprise their backbone and may be saturated or unsaturated. Saturated fatty acids are more solid at room temperature and contain no double bonds between carbon atoms. Unsaturated fatty acids may contain one (monounsaturated) or several (polyunsaturated) double bonds between the carbon atoms and are generally liquid at room temperature. The third position of triacylglycerol is occupied by another fatty acid, whereas that of phospholipid contains phosphate and usually a nitrogen-based chemical group. The phosphate portion of phospholipids is important for the function of the lipid within membranes. It should be noted that the proportions of saturated, monounsaturated, and polyunsaturated fatty acids (PUFA) found in animal tissues are species dependent (Table 3.1), and, in monogastric species, may be influenced by diet. In general, fish muscle contains the greatest concentration of polyunsaturated fatty acids followed by poultry and pork.

The double bonds located within PUFAs are sites of chemical reactivity. Oxygen is a necessary ingredient for lipid oxidation and may react with these sites to form peroxides, which lead to rancidity. Polyunsaturated fatty acids are especially susceptible to oxidative rancidity because of their high number of reactive double bonds. The formation of lipid breakdown products leads to

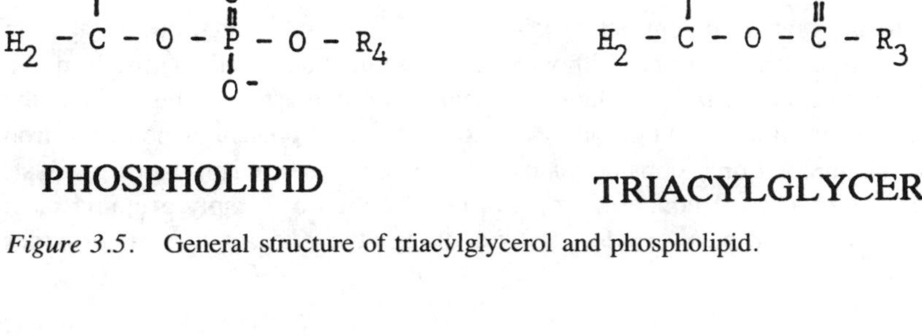

$H_2 - C - O - C(=O) - R_1$
$H - C - O - C(=O) - R_2$
$H_2 - C - O - P(=O)(O^-) - O - R_4$

PHOSPHOLIPID

$H_2 - C - O - C(=O) - R_1$
$H - C - O - C(=O) - R_2$
$H_2 - C - O - C(=O) - R_3$

TRIACYLGLYCEROL

Figure 3.5. General structure of triacylglycerol and phospholipid.

Table 3.1. Fattty Acid Content of Muscle Foods

Species	% Saturated	% Monenoic	% Polyenoic
Beef	40–71	41–53	0–6
Pork	39–49	43–70	3–18
Mutton	46–64	36–47	3–5
Poultry	28–33	39–51	14–23
Cod (lean fish)	30	22	48
Mackerel (fatty fish)	30	44	26

Source: Hultin (1985).

development of undesirable flavors and odors. Those muscle foods with high concentrations of PUFAs (e.g., fish) typically develop rancid flavors and odors faster than foods with less PUFA. The oxidative rancidity process cannot occur in the absence of oxygen. Thus, vacuum-packaging of meat products provides longer shelf-life by excluding oxygen from the packaging environment. The interaction of oxygen with PUFA to cause rancidity is a nonenzymatic process. Lipid oxidation and rancidity may also be caused by enzymatic processes occurring within the muscle food.

Enzymic-based lipid oxidation occurs in muscle foods and has also been termed "microsomal lipid oxidation." "Microsomes" do not constitute a specific cellular organelle but refer to membrane fractions of these (e.g., sarcoplasmic reticulum). This process requires certain biochemical cofactors for activity including reduced forms of nicotine adenine dinucleotide phosphate (NADPH) or nicotine adenine dinucleotide (NADH), adenosine diphosphate (ADP), and iron ions. The enzymatic nature of the process implies involvement of membrane-bound proteins. Cooking of meat provides sufficient heat to denature enzymes, and thus enzymic/microsomal lipid oxidation will not occur in cooked meats. During normal physiological functioning, the enzymes found in these subcellular organelle membranes produce chemically reactive substances known as radicals. These are a necessary part of normal cell functioning, and in the "living state" the cell has a variety of mechanisms for protecting itself against the undesirable actions of radicals. In postmortem muscle tissue, many of these protections are lost and radicals may hasten lipid oxidation and cause rancidity.

Metal Ions and Warmed-Over Flavor

All metal ions are potent catalysts of nonenzymic lipid oxidation. Meat is an excellent source of iron and, although this is a nutritional benefit, the iron can also serve to enhance lipid oxidation in meats. Iron in meat is bound in the heme (heme iron) portion of myoglobin or hemoglobin, or is present as nonheme iron (NHI). Nonheme iron is considered to be the more potent lipid oxidation catalyst. The concentrations of nonheme iron can be increased by simply grinding meat through a cast-iron meat grinder, or by cooking. The cooking process provides sufficient heat to denature myoglobin, allowing iron which is bound within the heme molecule to be liberated.

Warmed-over flavor (WOF) is a flavor defect that occurs in reheated meat products. The initial cooking of meat increases the NHI concentration within meat. During the time period between initial cooking and reheating, the iron acts to catalyze lipid oxidation. This occurs in cooked meat that is stored under refrigerated conditions (i.e., leftovers); the warm temperatures of the reheating process may also accelerate lipid oxidation. The outcome is that rancid flavors develop, resulting in WOF. The degree to which WOF is detected is dependent on the individual consumer. Warmed-over flavor is a special concern for manufacturers of precooked meat products.

Post-Rigor Changes for Predicting Freshness: K *Value and Nucleotide Catabolism*

During early postmortem metabolism, the supply of ATP within muscle is kept at a high level by regeneration via creatine phosphate, and by glycolysis. As the postmortem condition progresses, creatine phosphate reserves become depleted and, as glycolysis slows, ATP production is reduced. The muscle responds physiologically by using an enzyme, myokinase, to convert 2 moles of ADP to ATP plus AMP (Fig. 3.6). AMP then enters a catabolic (breakdown) pathway in which it is sequentially converted to other compounds including inosine and hypoxanthine. The catabolism continues and both inosine and hypoxanthine accumulate during storage of muscle foods. The concentrations of inosine and

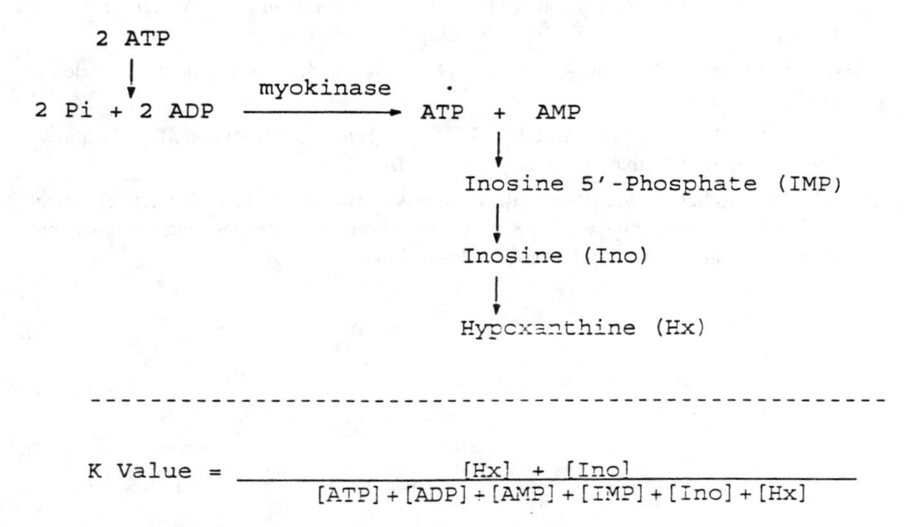

Figure 3.6. Breakdown of ATP and calculation of the K-value. [Adapted from Uda (1990).]

hypoxanthine, expressed as a percentage of total ATP-related compounds present, can be used as an indicator of freshness in muscle foods. The longer a piece of meat has been stored, the greater the relative concentration of hypoxanthine and inosine. This concept has been most extensively applied in fish where a *K* value is calculated as shown in Fig. 3.6. Longer storage times are correlated with higher *K* values and reduced freshness. The *K* value at which fish are considered no longer fresh is species dependent. *K* values have been less thoroughly studied in traditional red meats but theoretically should be just as applicable as freshness indicators.

Selected References

Greaser, M. L. 1986. Conversion of muscle to meat. In: *Muscle as Food*. P. J. Bechtel, ed. Academic Press, New York.

Hultin, H. O. 1985. Characteristics of muscle tissue. In: *Food Chemistry,* 2nd ed. Owen R. Fennema, ed. Marcel Dekker, New York.

Lawrie, R. A. 1985. *Meat Science*. Pergammon Press, New York.

Locker, R. H. and C. J. Hagyard. 1963. A cold shortening effect in beef muscle. *J. Sci. Food Agric*. 14:787.

Newbold, R. P. 1966. Changes associated with rigor mortis. In: *The Physiology and Biochemistry of Muscle as Food*. E. J. Briskey, R. G. Cassens, and J. C. Trautmann, eds. The University of Wisconsin Press, Madison.

O'Keefe, M. and D. E. Hood. 1982. Biochemical factors influencing metmyoglobin formation in beef from muscles of differing colour stability. *Meat Sci*. 7:209.

Pearson, A. M. and T. R. Dutson, (eds.). 1985. *Advances in Meat Research, vol. 1 Electrical Stimulation*. The AVI Publishing Co., Westport, CT.

Pearson, A. M. and R. B. Young, (eds.). 1989 *Muscle and Meat Biochemistry*. Academic Press, New York.

Price, J. F. and B. S. Schweigert, (eds.). 1971. *The Science of Meat and Meat Products,* 2nd ed. W. H. Freeman and Co., San Francisco, CA.

Uda, F. 1990. Studies on evaluation of the freshness of fish and chicken meat. Development of simple methods for measuring *K* value, an index of fish freshness and establishment of *K* value–storage time relationship. *Mem. Tokyo Univ. Agric*. 31:1.

4

Edible By-products from the Production and Processing of Muscle Foods

Robert E. Campbell and P. Brett Kenney

Introduction

The production of by-products from food animals is important to the economics of muscle food production. In U.S. muscle food industries, profits to packers are less than the value of "by-products." For ocean fish, surimi may be the major product manufactured, improving the economics of that industry.

By-product processing and production impact on the environment. The recovery of by-products from muscle food industries reduces pollution and produces materials for a variety of uses.

Edible by-products contribute to enjoyment of food, with unique flavors, textures, and aromas. Finally, by-products contribute millions of pounds of high-quality protein and energy to a world that is often undernourished.

By-Product Classification

There are three major categories of by-products from food animals. These are **edible, inedible,** and **pharmaceutical.** Each class has subcategories, and categories overlap. For example, organ meats considered edible in one part of the world are not in other places. This chapter discusses only edible by-products; however, other by-products are important to the economics of food animal production and human health and comfort (see Table 4.1).

Edible by-products are defined legally in most countries. The label "edible" or "inedible" is often because of ethnic and cultural considerations. Here "edible by-products" are defined as animal products, other than skeletal muscle, that are eaten. Discussion of by-products inedible in the United States will be limited, and pet foods will not be discussed unless they impact "edible" by-products.

Table 4.1. Partial List of Inedible and Pharmaceutical Co-Products of the Muscle Foods Industries

By-product	End Products
Beef:	
Horns	Ornaments, buttons
Pituatary glands	Hormone products
Thyroid glands	Hormone products
Slunks	Fetal blood products / very expensive leather
Blood plasma	Biological products (microbial growth, etc.)
Red blood cells	Blood agar and other blood products
Whole blood	Blood meal animal feeds and fertilizer
Lungs	Pet food and heparin / fibrinogens
Bones	Gelatin for film, dicalcium phosphate, animal feed
Small intestines	Raquet gut / suture gut, mucosa → heparin
Hides	Leather, gelatin
Gall stones	Aphrodisiacs, folk medicine
Penis	Ornamental canes and other ornaments
Inedible trimmings	Greases and animal feeds, specialty lubricants, soaps, explosives (glycerine), animal feed
Hair	Artist brushes, animal feed (after hydrolization)
Pancreas	Insulin, other peptide and hormone products
Pork:	
Skin	Burn dressing, technical gelatin, specialty leather
Pituatary glands	Hormone products
Other glands	Hormone products
Pancreas	Insulin, other peptide and hormone products
Hearts	Valves for human heart repair
Inedible trimmings	Grease for animal feeds, specialty lubricants, protein for animal feed, fatty acids
Intestines	Sutures, heparin
Other inedible organs	Animal feed / pet feed
Hair	Artist brushes
Sheep:	
Hide	Leather goods, vellum
Inedible organs	Animal feed[a]
Inedible trimmings	Animal feed,[a] specialty lubricants, fatty acids
Blood	Blood agar, fertilizer
Intestines	Sutures, animal feed[a]
Poultry:	
Inedible trimmings	Animal feed, specialty lubricants, fatty acids
Feathers	Ornamental uses, feather meal for feed
Feet	Ornamental uses, chicken meal for feed
Fish:	
Inedible trimmings	Animal feed, specialty lubricants, fatty acid
Skin	Some species only, specialty leathers
Bones and other inedible material	Fish meal for feed
Swim bladders	Selected species, specialty gelatin

[a] Since the scrapie problem, use of sheep by-products as animal feeds have been restricted.

Some by-products can be processed as edible or inedible products; in these cases, only edible products are characterized.

Two major categories within edible by-products are **Fresh** and **Further Processed.** Fresh by-products are consumed or sold basically as removed from the animal. Fresh product may be chilled, frozen, or ground. Even though by-products are chilled or frozen, they are still "fresh" coproducts; liver is recognized as liver frozen or chilled and is consumed as liver. Additionally, fresh products may be cooked prior to consumption.

Further processed by-products are significantly changed by thermal, chemical, or mechanical processes. Gelatin, for example, looks and performs very differently from skin and bones its raw materials.

Further processed by-products are classified by method of processing. Major categories are thermal, mechanical, chemical/washing, and combination processes. A section on other processes is included in this chapter.

Fresh By-products

Fresh by-products are those sold in the same form as harvested from animals. These products are often further processed by other manufacturers into edible end products for human consumption. The following descriptions provide by-products' common use and processing notes. More facts are presented in tables; Table 4.2 has weight ranges of fresh coproducts; Table 4.3 has analytical values, calories, and some cooking methods for fresh by-products, and Table 4.4 describes packaging, distribution, and shelf-life for coproducts in this section.

Individual Fresh By-products

Brains

Beef brains are a low-cost by-product in the United States due to a European ban on imports of cattle raised with growth hormones. Formerly, the European market created a demand for brains, supporting higher price levels. Now, this product is saved only when the pharmaceutical demand for pituitary and pineal glands is sufficient to pay for labor to remove brains. Once removed, brains are washed to remove blood clots and packed in bulk or individually for consumers. Brains are sold to the end consumer because there is not a further processed market.

Pork brains are rarely saved in the United States because there is no market for them. Labor costs to remove brains is more than pork brains are worth, so pork brains are rendered.

Sheep brains may be vehicles for the causative agent of scrapie and bovine spongiform encephalopathy (BSE); so, the heads of sheep are landfilled, as renderers will no longer accept them.

Table 4.2. Weights of Selected Meat Production Co-products in Pounds per Head (kg)

Co-product	Beef	Pork	Ovine	Poultry
Brains	0.61–0.77[a] (0.28–0.35)	Not applicable	0.18–0.25[a] (0.08–0.12)	Not applicable
Cheek meat	3.5–4.5[b] (1.6–2.0)	0.62–0.78[b] (0.28–0.35)	Not applicable	Not applicable
Ears	Not applicable	0.38–0.50[c] (0.17–0.23)	Not applicable	Not applicable
Esophagus	0.56–0.58[b] (0.25–0.26)	0.14–0.18[b] (0.06–0.08)	0.05–0.10[b] (0.02–0.05)	Not applicable
Feet	8.0–12[c] (3.6–5.4)	2.0–4.0[c] (0.91–1.8)	Not applicable	Not applicable
Giblets	Not applicable	Not applicable	Not applicable	0.14–0.20[d] (0.06–0.09)
Gizzards	Not applicable	Not applicable	Not applicable	0.07–0.10[d] (0.03–0.05)
Gullet Meat	0.10–0.20[b] (0.05–0.09)	Not applicable	Not applicable	Not applicable
Head Meat	1.9–2.0[c] (0.86–0.91)	0.52–0.65[c] (0.24–0.30)	Not applicable	Not applicable
Heart	2.6–4.5[a] (1.2–2.0)	0.40–0.54[a] (0.18–0.25)	0.26–0.39[a] (0.12–0.18)	0.02–0.04[d] (0.01–0.02)
Kidneys	1.3–2.6[a] (0.60–1.2)	0.29–0.48[a] (0.13–0.22)	0.22–0.30[a] (0.10–0.13)	Not applicable
Lips	1.8–2.0[c] (0.82–0.91)	0.22–0.28[c] (0.10–0.13)	Not applicable	Not applicable
Livers	8.0–14.0[b] (3.5–6.2)	2.5–3.7[a] (1.2–1.7)	1.0–1.4[c] (0.50–0.63)	0.05–0.10[d] (0.02–0.05)
Necks	Not applicable	Not applicable	Not applicable	0.09–0.13[d] (0.04–0.06)
Skin	Not applicable	0–20[c] (0–9.1)	Not applicable	Not applicable
Spleens	1.2–2.4[a] (0.55–1.0)	0.19–0.30[a] (0.10–0.14)	0.10–0.15[a] (0.05–0.07)	Not applicable
Tails	1.5–2.5[b] (0.68–1.1)	0.20–0.25[b] (0.09–0.11)	Not applicable	Not applicable
Tendons	0.50–1.0[c] (0.23–0.45)	Not applicable	Not applicable	Not applicable
Testes	0.75–1.2[a] (0.34–0.56)	Not applicable	Not applicable	Not applicable
Tongue	3.1–3.9[a] (1.4–1.8)	0.33–0.46[a] (0.15–0.21)	0.15–0.22[a] (0.07–0.10)	Not applicable
Tongue trim	1.5–2.5[b] (0.68–1.1)	Not applicable	Not applicable	Not applicable

[a] From W. F. Spooncer (1988).

[b] From Boren and Weiss (1988).

[c] From the author's files.

[d] From USDA (1979).

Table 4.3. Moisture, Fat, Protein, Caloric Values and Typical Cookery Methods for Selected Fresh By-products[a]

Co-product[b]	Moisture[c]	Fat[c]	Protein[c]	Calories[c] per 100 g	Typical Cookery
Brains	78.3	9.3	9.8	126	Pan fry
Cheek Meat	66.5	15.5	16.5	205	Fry as ground product
Pork ears	61.3	15.1	22.5	233	Simmer
Esophagus	68.5	8.8	21.5	165	Fry as ground beef
Feet	58.3	18.8	22.1	264	Simmer
Giblets	74.8	4.5	17.9	124	Pan fry
Gizzards	76.2	4.2	18.2	118	Pan fry
Gullet meat	62.5	20.0	16.5	246	Fry as ground beef
Head meat	63.0	18.0	17.5	232	Fry as ground product
Heart[d]	76.0	17.0	5.0	120	Simmer
Kidneys[d]	78.0	3.0	16.4	105	Braise
Lips	62.5	17.5	19.2	235	Fry as ground product
Livers[d]	70.0	5.0	21.0	140	Pan fry
Chicken necks	60.0	15.5	17.1	208	Fry
Pork skin	61.8	15.1	22.5	235	Simmer
Spleens	74.2	6.0	17.6	125	Braise
Tail (ox)	68.6	10.1	20.0	177	Stew
Tendons	65.3	5.3	27.7	159	Stir fry
Testes	71.5	4.9	21.5	132	Breaded/fried
Tongue	65.5	17.0	16.0	225	Simmer

[a] Values in this table come from the authors' files, *Nutrient Value of Muscle Foods* (National Live Stock and Meat Board, 1988), Anderson (1988), USDA (1979), and Bittle (1981).

[b] Organs of different species are often similar in composition and except where species is noted, values represent averages of reported beef, pork, and lamb values.

[c] Percentages and calories based on raw (uncooked) product.

[d] These products have more than 1% carbodydrate in them.

Cheek Meat

Cheek meat comes from the masseter muscles. These muscles are high in connective tissue; however, they are lean (10–15% fat) and valuable as a lean source in ground beef and bologna products. Care must be taken when using cheek meat. In emulsion products, overuse can lead to jelly pockets, and in ground beef, gristle can have a negative impact on acceptability (Cross et al. 1976; Wiley et al. 1979).

Pork cheek meat is similar to beef cheek meat in its uses and its muscle source. Similar cautions apply because of connective tissue. Sheep cheek meat, when saved, is used in ground lamb products. Low demand and volume reduce its value.

Table 4.4. Typical Packaging, Distribution State, and Suggested Shelf-Life of Some Co-products

			Master Carton	Shelf-life (years)		
Co-product	Packaging	Preservation	Size (lbs)	Beef	Pork	Poultry
Brains	indiv.[a]	fz.[b]	10–40	0.5	0.5	n.a.[c]
Cheek meat	bulk pack	fz.	40–80	1.0	0.75	n.a.
Pork ears	bulk pack	fz.	40–60	n.a.	0.5	n.a.
Esophagus	bulk pack	fz.	40–80	0.75	0.5	n.a.
Feet	bulk and/or indiv.	fr.[d]/fz.	10–20 (pork) 20–80	0.5	0.5	n.a.
Giblets	bulk (1 lb)	fr./fz.	10–30	n.a.	n.a.	0.5
Gizzards	bulk (1 lb)	fr./fz.	10–30	n.a.	n.a.	0.5
Gullet meat	bulk	fz.	40–80	0.75	n.a.	n.a.
Head meat	bulk	fz.	40–80	0.75	0.5	n.a.
Heart	bulk	fz.	40–80	1.0	0.5	0.5
Kidneys	bulk	fr./fz.	40–80	0.5	0.5	n.a.
Lips	bulk	fz.	40–80	1.0	0.5	n.a.
Livers[e]	indiv./usu. 2 per carton	fz.	20–40	0.75	0.5	0.5
Chicken necks	bulk made into MSP[f]	fz.	40–80	n.a.	n.a.	0.5
Pork skin	bulk/combos[g]	fr.	2000–2500	n.a.	< 7 days	n.a.
Spleens	bulk	fz.	40–80	0.5	0.5	n.a.
Tail (ox)	bulk	fz.	10–60	1.0	n.a.	n.a.
Tendons	bulk	fz.	20	0.5	n.a.	n.a.
Testes	bulk	fz.	10–20	0.5	n.a.	n.a.
Tongue	indiv. and bulk	fz.	10–40	1.0	1.0	n.a.

[a] indiv. = individually wrapped prior to bulk packaging or master carton.

[b] fz. = frozen.

[c] n.a. = not applicable.

[d] fr. = fresh.

[e] MSP = mechanically separated poultry.

[f] Combos are large bulk containers containing up to 2500 lbs.

Esophagus

This smooth muscle is the stripped esophagus, also called weasand. It is very lean (<10% fat) and is used like cheek meat. Connective tissue content is acceptable however; this material produces ground beef with soft texture if used as a major portion of the formulation. Weasand is generally frozen in blocks. Pork weasand is similar to beef weasand in usage and texture.

Gullet Meat

In beef, this tissue is removed from the larynx and consists of muscles that move the larynx and the base of the tongue. Extreme care must be taken to avoid getting thyroid gland in this material. As little as 0.5 g of thyroid gland can cause thyrotoxicosis with symptoms like hot flashes, heart palpitations, and high blood pressure. Use of gullet meat has been implicated in thyrotoxicosis outbreaks. Gullet meat is 25–30% fat; this and problems with the thyroid gland keep some packers from recovering gullet meat. Pork gullet meat and weasand meat are synonymous. To our knowledge, pork thyroid glands have not been implicated in thyrotoxicosis.

Head Meat

Head is removed from beef skulls near the poll and includes ear muscles. It contains nearly 20% fat and has less connective tissue than cheek meat. It is used mostly in ground beef and batter-type formulations.

Pork head meat is used as a sausage raw material. It is fatter (25–30%) than beef head meat and lower priced. Head meat contains more collagen than most skeletal muscle, so usage in batter products is limited.

Heart

Beef heart is generally consumed in batter-type products or hamburger with appropriate labeling. Heart demand is small for home consumption. Hearts are split for inspection, then valves and cartilage are removed prior to packing. Heart meat is material left after the heart cap (auricles) and bone have been removed (deHoll and deHoll 1983).

Pork and beef hearts are used as raw materials in sausages. Heart meat has lower solubility (Saffle and Galbreath 1964) than skeletal muscles. This lower function is reflected in heart prices. Research with acetylations (Eisle and Brekke 1981) and enzymatic modifications (Smith and Brekke 1984) has improved heart function. Improved function is associated with improved solubility.

Lamb hearts are sold to consumers in specialty markets but are often rendered due to low demand.

Kidneys

Beef kidneys are generally removed after the carcass has been chilled. The peritoneal membrane is removed and the kidneys are packed, frozen, and shipped. There is no domestic market for kidneys, except as pet food. Asia provides a small export market for vacuum-packaged beef kidneys. Kidneys are removed at slaughter, vacuum-packaged, blast chilled, and shipped to Asia.

Pork kidneys are considered edible in the United States; however, most pork kidneys are sold as pet food. Low human demand keeps the value of kidneys attractive to pet food producers.

Lamb kidneys are sometimes sold to consumers; however, many of them are rendered for the same reasons as lamb hearts.

Lips

Beef lips are used as a raw material for various sausage products, including coarse and batter products. Their high collagen content limits lips in batter-type products. Pork lips are similar to beef lips in use and composition.

Livers

Beef livers are peeled from the surface membrane prior to packaging. Additionally, the hepatic artery, associated fat, and connective tissue is removed. Livers are usually wrapped in plastic and packaged two per box. Most livers in the United States are sold frozen. Packaging requirements for export markets can differ widely. Prior to the ban on product from cattle produced with growth hormones, Europe was a major importer of livers from the United States. Some beef livers are used in sausage; however, most are sold to slicing and batter/breading operations. Lamb livers are used and handled similarly to beef livers.

Pork livers for human consumption are further processed, usually into liverwurst. Pork livers have more connective tissue than beef livers so their texture deters fresh consumption.

Spleens

The U. S. government considers spleens edible, but few are processed for human consumption. Spleens are used mostly for pet food. Some research with spleens as "color enhancers" in restructured beef (Marriott et al. 1985) has not increased usage. Pork and lamb spleens are called melts, and lamb carcasses are often shipped with the spleen in the carcass. This practice adds weight to the lamb carcass. The major drawback to the use of spleens as human food in the United States is the requirement that they be labeled when used in cooked sausage. Spleens may not be used in frankfurters or bologna (deHoll and deHoll 1983).

Tails

Beef tails or oxtails are usually sold frozen to soup manufacturers. A small market for vacuum-packaged fresh tails exists; these oxtails are sold in supermarkets for home use. Tails are mostly sold as material for use in soup and other products where the tail is a flavoring agent.

Pork tails are also used in soup. Final products are soup stocks and meat-flavored broths. Tails are packed frozen in bulk containers. Tail meat might have a larger market if labeling for mechanically separated beef and pork was similar to mechanically separated poultry.

Tongue

Beef tongues are generally used on the east and west coasts. Domestic demand is from further processors who cure the tongues for sandwich meat. There is also an export market for tongue; prior to the European ban, this was a larger market. Tongues are graded on color and defects. Defects include cuts on the membrane, warts, and physical abnormalities. The most desirable tongues are white and free of defects. Most tongues are sold individually wrapped and frozen. Packaging varies, depending on customer and shipping requirements.

Pork tongues are used in products such as tongue loaf and head cheese. Additionally, tongue trim is used in sausage. Some tongues are sold frozen; others are cooked and skinned.

Fresh By-Products Specific to Beef

Sweet Breads

This is the common name for the thymus gland. Although more commonly consumed in Europe, the sweet bread is a delicacy and not generally consumed at home due to a lack of knowledge on preparation. Sweet breads are removed from the neck area of young animals, trimmed of extraneous meat and fat, and washed before packaging. Sweet breads are individually wrapped prior to packing, and frozen. Typically, the sweet bread is battered, breaded and fried or sauteed before eating.

Tendons

Although not allowed as edible in the United States, tendons are sold to Asian markets, improving the value of beef animals by $1.50 per head. Tendons are removed from the posterior border of the metacarpus between the distal end of the radius/ulna and the foot. The product is trimmed to uniform lengths, washed and trimmed of all red meat, and packed into layers separated with plastic sheets and frozen. Tendons are chopped finely and used in stir-fried foods as a flavor and texture component.

Testes

Beef testes are not produced in great quantity in the United States because most "table beef" is from steers and heifers. Testes are available from smaller packers that slaughter bulls; they are often marketed as "Rocky Mountain Oysters." Typical cooked testes are sliced, battered, breaded, and deep fat fried.

Tongue Trim

This is sold as lean trim trimmed off the tongue body to meet specifications. This meat is fairly lean (<20% fat) and can be used as manufacturing meat.

Fresh By-Products Specific to Pork

Ears

Pork ears are used as a raw material because of their high collagen content. Ears are the source of gelatin for products like souse, tongue loaf, and head cheese.

Feet

Pig feet are used in products from soup stock to direct human consumption. High proportions of bone and collagen make pig feet relatively inexpensive. Pig feet are simmered prior to eating to break down collagen into gelatin.

Salivary Glands

Pork salivary glands are used as "filler" meat in sausage. They provide inexpensive protein. Salivary glands are generally sold frozen in bulk boxes.

Skin

Pork skin has two major uses. Most pork skin goes to gelatin manufacturers to make Jello and marshmallows. The other market is "pork rinds." Skins are generally sold fresh, in combo bins, to gelatin manufacturers, because freezing causes gelatin quality and yield to deteriorate (Hinterwaldner 1977a).

Snouts

Pork snouts are "filler" meat in sausage. They are high in collagen and low priced. Snouts consist of the upper lip, nose, and underlying muscle. They are packed and bulk frozen. They are also used in head cheese or souse.

Fresh Poultry By-Products

Feet

Chicken feet are available in some markets catering to populations of Caribbean and Asian origins. They are sold as soup raw material and not widely distributed in the United States.

Giblets

Poultry giblets are the neck, liver, heart, and gizzard. When sold as giblets, they remain with the bird, as an insert. Giblets are used for stuffing or soup ingredients. With value-added poultry, more giblets are sold separately or further processed.

Gizzards

Gizzards are sold in refrigerated containers. They are also used for flavoring in soups and stuffing mixes.

Hearts

Poultry hearts are sold at retail in small containers. Hearts are also used in soup and stuffing mixes.

Livers

Livers are handled similarly to hearts.

Necks

Necks from poultry are rarely sold at retail, except when a retailer has broken whole birds into parts. Necks are used for mechanically deboned poultry meat, a further processed by-product. Necks are also used as soup stock.

Poultry Skin

Poultry skin is a sausage ingredient. It is usually packed in frozen containers. Generally it is a filler ingredient in poultry or mixed-species batter sausages. In some products it may be listed as "poultry by-products"; in others, skin cannot be added in higher proportion than occurs naturally.

Fishery By-Products

By-products represent a significant portion of income generated by the fish industry. Ongoing concerns on diet/health issues and a need for economical, high-protein, low-fat food has opened an expanding market for surimi.

Surimi is a mechanically processed product from underutilized species of fish, like Alaskan Pollock (*Theragra charcogramma*), that are not suitable for filleting. Fish mince from the deboning process may contain undesirable materials. To minimize the adverse effects of these on color and aroma, fish mince is washed. Washing removes soluble components, and lipid will float and be removed. After washing, it is centrifuged to remove water. Freezing reduces the utility of fish mince, so it is necessary to add cryoprotectants. Surimi typically contains 0.3% phosphate (0.15% sodium tripolyphosphate and 0.15% tetrasodium pyrophosphate) and 4% each sucrose and sorbitol. Starch and egg white may be added to enhance gel-forming abilities. Wheat starch is commonly used because it has better freeze–thaw stability and is less prone to retrogradation than other plant starches.

Surimi-based products are classified as molded, fiberized, composite-molded, and emulsified items. Fish surimi has been used successfully in processed meat products as a binding adjunct. A problem with using surimi in a processed meat product is fishy aroma. A rule of thumb for use in meat products is not more than 5%. Control of temperature during formulation is critical for optimal protein–water interactions; function may be adversely affected because fish proteins are more susceptible to temperature denaturation than mammalian proteins as fish live at lower temperatures than mammals. During surimi manufacture and use as a raw material, temperature should not exceed 10°C (50°F).

Fish mince is another coproduct of fish, similar to surimi, except no cryoprotectants are added and mince is not washed. Fish mince is used frozen fish sticks. It is mixed with small pieces to form fish blocks. Minced fish is the product of mechanically deboned fish skeletons, whole fish from species less desirable for fillets, and fish that are too small to fillet. After filleting, approximately 40% of a fish's weight is head and skeleton. Deboning machines allow the fisheries industry to recover highly functional and nutritious muscle from this portion. Deboning machine drums have 3–5-mm holes. Three-millimeter drums are most often used because they do not destroy the integrity of the mince to the degree that smaller holes would; larger holes may allow too many bone particles and other materials to pass through. Shellfish not suitable for direct sale due to size or defects, such as shrimp or crab, may also be deboned to produce flavoring for analogues made from surimi.

Kamaboko is an elastic fish cake that is produced from washed fish material; it is popular in Japan. During grinding, salt, starch, and sugar are added. The ground mixture is formed into a desired shape and heated.

Concentrates and isolates may be prepared from minced fish for use in snack items, baked products, and beverages. Lipid components are removed from these concentrates using solvents to minimize the off-flavor and odor problems; however, these solvents cause reduced protein solubility. These changes can be minimized when fish protein isolates are codried with sucrose or sorbitol. Acylation with either acetic or succinic anhydride can improve the solubility and functionality of these proteins. Additional uses of fish mince consist of codrying the muscle with cereal. The muscle also may be drum dried. Dried whiting is particularly useful because it can be rehydrated and used as an extender.

Fish roe is an expensive by-product of the fish industry. Processed (salted) roe or eggs is commonly called caviar, especially sturgeon eggs. Fish gelatin prepared from swim bladders of selected fish (sturgeon, cod, catfish, carp) is used as a clarifying agent for beer and wine. Livers from several fish species can be processed to extract oils high in vitamins A and D as well as B vitamins. Oil content will vary by specie.

Processed By-products

Many by-products require significant further processing before use as food. Heat is the most common form of processing. Therefore, heat processing of by-products will be discussed first. Other processes such as centrifugation will be discussed in conjunction with heat processes when they are applicable. Inedible by-products will not be addressed except where they impact on the processing of edible by-products.

Heat Processing

Rendering

Rendering is a heat process applied to edible fat and bone from food animals to separate fat, moisture, and solids. The process is similar for all food animals. Products produced by rendering include tallow from beef and sheep, lard from pork, chicken and/or turkey fat, and fish oil.

Edible rendering systems (Fig. 4.1) start with collection bins for bone and fat scraps from meat-plant processing areas. Most slaughter fat is considered inedible; however, fat lining the intestines (caul fat) is edible, provided it has not been contaminated. Bin material is ground into 1-cm particles to maximize yield of

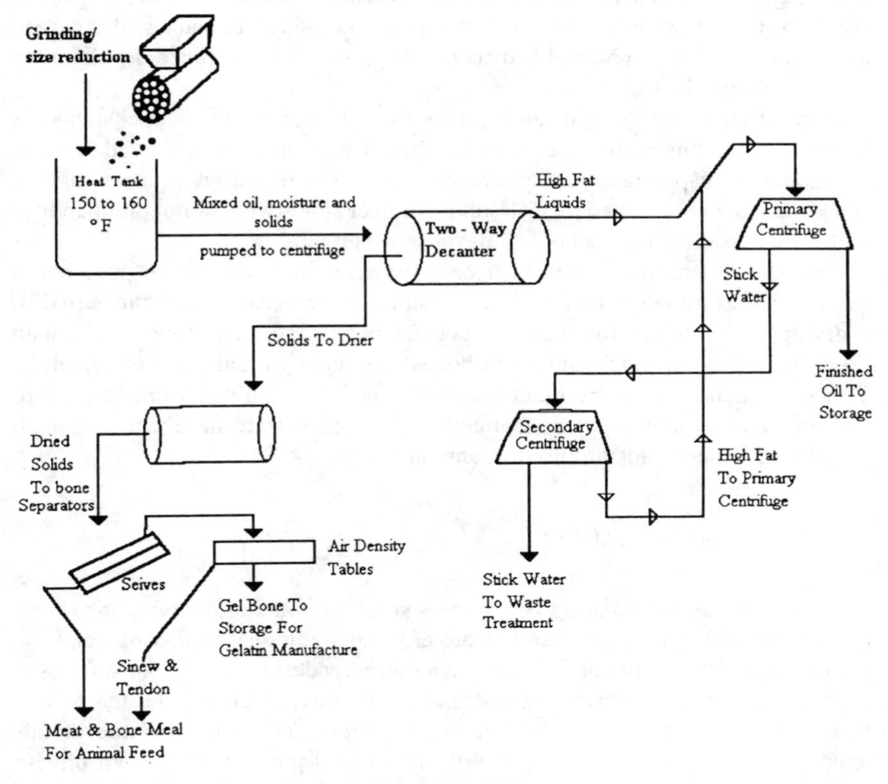

Figure 4.1. Flow diagram of an edible rendering system.

edible fat. Ground material is fed to a heat tank and heated to a temperature sufficient to melt fat and to disrupt fat cells. Temperature will vary depending on the species being rendered. In beef, the heat tank will generally run 66–71°C (150–160°F), with pork and poultry 6–9°C (10–15°F) cooler, and sheep sightly hotter.

The heated slurry of bones and fat is fed into a two-phase decanter to separate liquids (fat and water) and solids. The moisture/fat (liquid) material is 10–30% moisture at this point and requires two more steps. This material is heated to 96°C (205–207°F) and put through a "polishing" step to make finished oil. The "polishers" are high-speed liquid centrifuges which separate virtually all the water from tallow. The most efficient way to work this process is to make "clean oil" on the first pass, leaving "dirty stick water." The second stage makes "wet oil" and "clean stick water," reducing the discharge of fat to′the environment. Oil from the second stage is recycled back through the polishing stage to make clean oil. Finished product (tallow, lard, poultry fat, etc.) should contain less than 0.2% moisture.

Another product from rendering is "stick water" coming from the secondary centrifuges. This material contains water-soluble proteins and fine suspended solids the centrifuge does not remove. Until recently, this material was discharged as wastewater. Environmental and economic pressures have forced development of a salable product from "edible stick water."

Stick water is 3–5% solids as it comes from the centrifuge. The challenge is to concentrate this material economically and prevent protein degradation. A solution is to evaporate excess water by using a vacuum, falling film evaporator. Stick water is concentrated to 35% solids, (higher solids would limit pumpability) and sold as beef, pork, or poultry flavor concentrate.

Gel bone is produced from beef only because sulfur content in pork bones results in lesser quality gelatin. Solids coming off a decanter are transferred to a drying drum, where the moisture content is reduced from 50% to less than 10%. This dried material consists of bones, insoluble proteins, and other solids. Bones are denser than the other materials and are separated from them with vibrating density tables. Once separated, bones are sieved to size them for gelatin producers. Other solids are used in animal feed.

Low-Temperature Rendering

Low-temperature rendering is a process similar to traditional rendering, however, it is designed to recover the protein that traditional rendering sends to animal food. Raw material for low-temperature rendering is *only* the soft tissue from a beef or pork fabrication operation. The raw material is heated to less than 49°C (120°F), to melt fat but avoid protein denaturation. Time at this temperature is critical to the quality of the finished product, from both protein functionality and microbiological standpoints. This temperature is too low to pasteurize product effectively and may promote the growth of microorganisms.

After heating, the melted material is centrifuged to remove fat, and a slurry with 70% moisture, 20% protein, and 10% fat remains. It is critical to product quality to reduce temperature from 49°C (120°F) to below freezing as rapidly as possible. Chilling methods include drum freezing, blast freezing, and immersion freezing. Fat stream from this operation is returned to traditional rendering. Products produced with low-temperature rendering include partially defatted beef fatty tissue, partially defatted pork fatty tissue, "chopped beef," "chopped pork," and, just recently approved. (Breedlove and Brown 1991), "fat reduced" beef and pork. These products have regulatory definitions based on raw materials and processing parameters that are several pages long, limiting discussion here. These products are a lower-cost source of lean raw material. The largest users of this material are pizza manufacturers who use them in beef and pork toppings. Labeling regulations limit use in some products.

Drying

Drying is used less frequently than other heat processing procedures. Dried edible by-products include blood meal and blood plasma, which are used in sausage. This material is always spray dried. Any other method degrades protein.

Cooking and/or Frying

Two common coproducts are cooked or fried pork rinds and cracklings. Pork rinds (skins) are the more popular of the two items and several major snack food manufacturers make them. Pork rinds are deep fat fried at temperatures (350–400°F/177–204°C) that cause the skin to puff due to steam produced internally. The product is then salted, spiced, and packaged for consumer sale.

Edible cracklings are produced when edible fat is rendered. Cracklings are proteins left over from the rendering process. To make cracklings, fat is drained from the rendering tank and cracklings are pressed. Pressing removes much of the residual fat trapped in the protein matrix. To dry cracklings further, they are deep fat fried, salted and/or spiced, and packaged. Pork skins and cracklings are processed and packaged to be shelf stable. Low water content and high salt (4–6%) levels prevent microbial spoilage, and packaging in barrier packages with a nitrogen atmosphere slows oxidation.

Beef small intestines, although not considered edible in this country, are cooked and packaged for export, particularly to Japan. Intestines are cleaned of internal contents and split open. The open intestine is then cut into lengths as specified by the customer (usually 50 cm or 18 in.) and water cooked at 90°C (194°F) for 15 min. The strips are then layered into boxes with separators between layers and frozen before shipping.

Mechanical/Combination Processing of By-products

Many muscle food coproducts are mechanically manipulated. These processes include mechanical deboning, extrusion, pressing, centrifugation, and filtration.

Mechanical Deboning

Mechanical deboning is a process used to recover edible skeletal muscle from bones of food animals that is not economically recoverable otherwise. Most mechanically deboned meat (MDM) comes from poultry. The major reason that less beef or pork MDM is produced is labeling restrictions that processors feel are prohibitive. MDM for beef, pork, and lamb must be labeled as mechanically separated (species) and must meet several requirements for protein, protein efficiency ratio, fat, and calcium content. Further, the processor must have an approved quality control program. Rather than deal with labeling and processing restrictions, most processors do not manufacture or use MDM from beef or pork.

Mechanical deboners force ground bones against a screen, and meat on the bones is forced through the screen. The most common mechanical deboners use a metal augur to force meat through a screen. The material going through the screen is very fine, almost batterlike in its consistency. Therefore, it is only suitable for sausage production. Because of the severe treatment, this material has to be handled carefully, as it is more susceptible to microbial spoilage than whole muscle. Most mechanically deboned poultry is frozen immediately after processing and used as a raw material in sausage, usually poultry or mixed-species frankfurters and bologna.

Another type of deboner, common to the fish processing industry, is a belt press device made by Baader™ that presses the fish bones between a belt and a screen, forcing the fish paste through the screen. This type of device works well with fish because fish bones are flexible.

The Baader™ is also used to produce **beef connective tissue.** This material is used in sausages. The raw material is shank meat. Shank is processed for ground beef and separated connective tissue is packed and frozen for further processing. Research shows that this process can add $1–3 per head of beef (Hunt and Campbell 1992). Mawson (1992) has indicated that the Baader™ may have applications for defatting and desinewing mutton.

Centrifugation

Centrifuges are used in several processes for recovering edible by-products. Blood from beef and pork slaughter can be separated into plasma and red cell fractions. Plasma is usually spray dried and used as a protein additive. Red blood cells are dried for fertilizer; however, some research indicates red blood cells can be used as a color ingredient for sausage.

The major use of the centrifuge in muscle foods processing is as a separator of fat and water. Because fat and oil from muscle foods are combined with water and solids, it is necessary to separate the components.

Two types of centrifuges are used in rendering. First a solid–liquid separation device (see Fig. 1) separates animal processing by-products into solids and liquid (60–80% fat). The solids are further processed. Liquids go to the "primary" fat/moisture separators, and oil is produced. Oil should have a moisture content of less than 0.2%.

Stick water from the primary centrifuges is 2–5% oil and this goes to secondary centrifuges. Secondary centrifuges recover most of the oil in stick water. Outputs are a high-fat stream (60–95% fat), reprocessed by the primary centrifuges, and stick water that is sent to the sewer. Stick water is usually less than 0.5% oil. Thus, a properly functioning rendering system recovers valuable coproducts of muscle foods production and reduces the waste material released to the environment.

Centrifuges also recover tissue from low-temperature rendering. Partially defatted tissue is a valuable coproduct and is used extensively in sausage. Newer raw materials (Breedlove and Brown 1991) using a variation of low-temperature rendering are fat-reduced beef and fat-reduced pork.

Washing Processes

Virtually all food animal coproducts are washed in their processing. Several by-products undergo washing procedures that are more than simple cleaning. Washing processes include scalding, chemical treatment, and depilatory agents.

Beef Feet

Beef feet are scalded at 60°C (140°F) for 5–10 min, often with bleaching agents. Beef feet are scraped to remove hair and the outer layer of skin. The keratin portion of the hoof is also removed. Feet are then packed and frozen.

Beef Tripe

Beef tripe is washed under severe conditions. The treatment includes 45 min of agitation at 60°C (140°F). For 20 min of this treatment, alkali (NaOH) is added. This loosens the inner layer of the tripe and bleaches tripe from dark green or black to white/yellow. Two types of tripe are processed; the first, made from the rumen, is "tripe." The second is made from the reticulum and is called honeycomb tripe. Honeycomb is named from the raised hexagonal patterns on the inside surface. Honeycomb tripe is more valuable than tripe.

After tripe is washed, it is chilled with ice water prior to packaging. Tripe is packaged depending on customer requirements and frozen. The domestic market for tripe is in areas with an Hispanic population. Tripe is exported to Mexico and Asia.

Pork Tongues

Pork tongues can be scalded prior to sale. Scalding occurs at 71°C (160°F) for 20–30 min. Then tongues are stripped of their outer layer. Scalded tongues are frozen and sold for use in head cheese and sausage products.

Pork Small Intestines

Processed pork small intestine is sold as chittlings. The intestine is flushed of all contents with warm water. If the wash water is too hot, the mucosa will come loose from the wall of the intestine and casings will be produced rather than chits. Small intestines are placed over a nozzle and contents are flushed from the lumen. After washing, the color of chittlings is white to pale yellow without brown or green spots. Chits are then cut to length, depending on market demand, and packed in tubs. Tubs are marketed fresh to retail customers. Major markets for chitterlings are in large cities and the southeastern United States. African-Americans are traditional consumers of chittlings. Chittlings are often fried or boiled with vegetables.

Pork Stomachs

Pork stomachs are scalded to remove the inner lining and to bleach color, as with beef tripe. This product is called a hog maw. Demand for maws is smaller than for tripe. A traditional dish made with maws or sheep stomachs is haggis, an oatmeal-based sausage product stuffed in the stomach.

Collagen Manufacture and Processing

Several sources of collagen and its end product, gelatin, are from food animals. Collagen and/or gelatin can be used in many products; several of which are edible.

Beef Bones

Beef bones are a major source of gelatin. Beef bone gelatin is generally used for photographic film; however, some, not suitable for film, becomes edible product. Pork bones are unsuitable for film so they are not processed for gelatin.

Making beef bones into gelatin is a lengthy process. Bones are dried and sized in rendering. Then bones are treated with 3–10% mineral acid (H_2SO_4 or HNO_3) for 3–7 days. Acid removes the minerals in bone as dicalcium phosphate, a food additive. The material left after acid washing is ossein. Ossein is treated with alkali (saturated calcium hydroxide) for 30–60 days depending on the bone source. This treatment breaks collagen cross-links in the ossein (Ward and Courts 1977).

Ossein is then subjected to a series of hot-water treatments at successively higher temperatures. Heating starts at about 54°C (130°F) and progresses in 6–9°C (10–15°F) intervals to boiling. These "boils" extract gelatin of varying quality, with the higher-quality gelatin coming from the lower-temperature "boils." The "boil" process takes 1–3 days.

Liquor from the gelatin boil is concentrated from 30–50% solids in a vacuum concentrator, depending on viscosity. Concentrated liquor is then extruded. Extrusion can be as "spaghetti" that is dried in a hot air belt drier, or as a sheet, dried on a drum drier. Dry gelatin is ground, tested for quality, and blended to produce gelatin with the desired gel strength, color, and water-binding capacity. It is interesting to note that there is no way to predict finished product quality from raw materials (bones), and the end product quality varies considerably from one batch to the next. Seventy to 80% of the gelatin will be used in film. The rest will end up in foods including Jello, marshmallows, juices, wines, beer, and jellied meat products. Other products include gelatin capsules for medicinal use and protein supplements. This is one of the few processes where "lower"-quality product goes to food use.

Pork Skin Processing for Gelatin

Gelatin is also produced from pork skin. Gelatin manufacturing for pork skin is different than for beef bones. Pork skin for gelatin is produced when scalded (skin on) hogs are cut. Skin from the belly, ham, and backfat is made into gelatin. Skin is processed quickly, because freezing or spoilage degrades gelatin quality (Hinterwaldner 1977b).

Collagen in pork skin is less highly cross-linked (Hinterwaldner 1977a) than collagen in beef bones, therefore initial processing to break cross-links is less stringent. Skins are soaked in a 5–10% solution of mineral acid for 1–5 days. This removes fat and breaks cross-links prior to heat processing. After acid treatment, skins go through a "boil" stage of manufacture, similar to ossein. Pork skin gelatin is dried similarly, ground, and packaged for manufacture into foods.

Collagen Production

Collagen as a meat processing additive has been studied (Kenney et al. 1986). Collagen for these studies was produced from the corium layer of beef hide. Corium collagen is also used to manufacture sausage casings.

Manufactured Casings

Some casings are made from collagen, rather than using "natural" casings from internal organs. Collagen casings are made from hide chorium collagen.

Corium collagen manufacture starts with a hide dehairing alkali treatment.

This swells collagen and allows the hide to be split into corium (bottom) and grain (top grain hide) layers. Top grain hide goes to leather manufacture, and the corium layer is processed into edible collagen.

Corium is treated with acid to decalcify it, ground into small particles (<2 cm), and treated with acid to swell the fibers and create a "dough" of collagen fibers. The dough is approximately 4.5% solids. When the dough reaches the required consistency, it is extruded (~ 100 psi) through a mold, to make the casing tube. The tube is passed through a "fixing" chamber. Chemicals used to fix the casing include ammonia, ammonia hydroxide, and concentrated salt solutions. Casings are washed to remove the ammonia salts and manufacturing adjuncts. Washed casing is plasticized with glycerin and sometimes dextrose. Casings are dried to 12–15% moisture and folded (shirred) into compact tubes that are easily placed on a sausage-stuffing horn. Shirred tubes are 17–31 cm (7–12 in.) long, and extend to 150–230 cm (60–90 in.) (Hood 1987).

Natural Casings

Natural sausage casings are produced from a variety of animal sources. Natural casings are the collagenous outer linings from food animals' gastrointestinal tracts (GIT) and other organs. To produce casings from intestines, the intestine is washed. The mucosa is removed with heat and pressure to loosen and squeeze it out of the casing. This is called sliming (Rust 1988). Mucosa can be used to make heparin, a pharmaceutical product. Casings are flushed to remove remaining mucosa and contents. Then they are packed in salt. Natural casing units are called hanks and should be 110 yards (100 m) long. Casings can also be preflushed. These casings are rinsed free of salt and packed in brine for freshness. Flushed casings are easier to handle and more expensive than salt-packed casings.

Other organs used for casings include large intestines, bungs, caecums, stomachs, and bladders. Prior to use, the inner lining is removed, and usually a salting process preserves the casing. Additional processing includes sewing pieces together for uniform sizing, and presleeving, which is packing the casing on a sleeve so that it can be slid on a stuffing horn quickly and easily.

Other Processes for Edible Coproduct Production

Refining

Tallow and lard are sometimes refined after extraction and polishing. Refining removes fatty acids, producing oil with longer shelf-life than unrefined tallow or lard. Refined oil is clarified with diatomaceous earth and clay filters to remove particles. Refined tallow is white, as opposed to slightly yellow for unrefined tallow. Lard is also whiter (less gray) after refining.

Hydrogenation

Lard melts at lower temperatures than tallow. To increase lard's melting point, hardness, and improve frying characteristics, lard is hydrogenated. Hydrogenation increases the saturation of the fatty acids in lard. Hydrogenation takes place at elevated temperatures and pressures in the presence of a catalyst, such as nickel. Hydrogenation also reduces susceptibility to oxidative rancidity.

Supercritical Extraction

Supercritical extraction is an experimental process for muscle foods. This process is used to extract caffeine from coffee and cholesterol from butter. Experimental work has used this process to extract cholesterol from tallow. Possibly this process could separate fat from lean (McHugh and Krukonis 1986).

Supercritical extraction uses high-pressure (300 atm) carbon dioxide to extract fatty substances. This process is now used to extract fat from meat samples for analysis and may have application in food.

Packaging, Distribution, and Storage

The packaging, distribution, and storage of edible coproducts is as varied as the products. Packaging (Table 4.2) fits the needs of buyers of a given by-product. Packaging varies from a tank car that holds 83,990 kg (185,000 lbs) of tallow to a 500-g (1-lb) retail container of chicken livers.

Coproduct packaging protects the product during shipment and storage. In most cases this protection is from spoilage agents like bacteria and oxygen. Packaging also prevents drying and provides protection from damage during shipping. Packaging must deliver acceptable product, at affordable costs, or the coproduct will not be sold. In several cases, the cost of packaging is crucial to the decision to produce a coproduct.

Packaging considerations can affect the cost of the final product. A larger package is generally less expensive on a unit (i.e., pound) basis than a smaller package. A major consideration is customer acceptance of the final product. Purchasers can vary considerably, from customers who take tallow by the tanker load to others who buy 2.5-kg (5-lb) tubs of lard at the supermarket. Another consideration is ergonomics; most people cannot comfortably pick up a box that weighs more than 36 kg (80 lbs). Thus, the typical by-product package weighs 22.7–36 kg (50–80 lbs).

Packaging for coproducts can be classified into four types. The largest is bulk shipment. Here, the package is reusable, like a rail car or tank truck. The next size package is combo bins, with about 1000 kg (2200 lbs) of product per bin. Combo bins are generally used for nonfrozen shipments for use in further processed products. An example is chilled beef fat, shipped from a producer to a low-temperature renderer who will make partially defatted beef fatty tissue. The next category of packaging is for items that are generally frozen in cardboard containers (sometimes with plastic liners) weighing 22.7–36.3 kg (50–80 lbs). Many by-products are packaged this way for two reasons; people cannot easily lift larger containers, and these containers freeze quickly enough to prevent spoilage. The final type of packaging is for retail ready product. Retail packages are usually 5 kg (10 lb) or smaller. Common examples are 0.5–1-kg (1–2-lb) containers of poultry giblets, 2.5–4.5-kg (5–10-lb) containers of lard, and 1–2.5-kg (2–5-lb) packages of frozen sliced liver. Virtually all retail packages leave the point of packaging in some sort of master carton (primary package), so the retail package is a secondary or tertiary package.

By-product distribution is another production factor. Bulk shipment of oil is often by rail due to the economies of rail shipping. Export of animal fats and oils is done by either container or tanker ship. Bulk tallow, lard, and chicken fat shipments go directly to customers equipped to handle bulk product. Frozen product is generally distributed by truck through the frozen food retail system for retail sale, and through frozen storage warehouses, for further processing. Retail products are usually shipped to distribution points, and then smaller units are redistributed to individual retail stores.

Storage of animal coproducts is similar to other perishable food items. Virtually all edible by-products will spoil if stored improperly. Improper storage will cause economic loss. Fresh by-products generally have a shelf-life of 45 days or less. Frozen by-products have shelf-lives that range from 6 months to 2 years. Anything in frozen storage longer than 1 year will have deterioration. Tallow and lard can be stored for 6 months or more if antioxidants have been added and storage conditions are optimal. Other items, such as salt-packed casings or dried gelatin, deteriorate very slowly; however, improper conditions will cause any edible coproduct to deteriorate.

By-product Value

Various sources have estimated the value of by-products at 5–15% of the value of a food animal (Goldstrand 1988). The total value of by-products depends on the market at any given time. Furthermore, there are considerable differences between species. However, without any argument, one can state that by-product value is very important to animal foods processing industries.

By-product value impacts all phases of meat animal economics. The first area affected is the price a processor can bid for live animals. The value of coproducts is calculated into this bid. In the beef and pork industries, profit margins have traditionally hovered near 1% of sales. All items with value are considered in the value equation. If coproducts are worth between 5% and 15% of the total value of the animal, then the value of coproducts is more than packer profits.

Therefore, it is critical to packer profitability to make good decisions about which by-products to manufacture.

The third area where by-product value affects the muscle foods economy is consumer purchase of meat and muscle foods. By-products spread costs of producing muscle foods over more end products. This reduces the cost of muscle food, compared to what it would be without coproducts.

Value Decision Points in Edible By-Product Production

It is critical to make good decisions about what by-products to produce, particularly where the value of an edible end-product is close to the value of an alternative finished product. A good processor knows the costs of producing a by-product. Some of these costs are energy, yield, equipment usage costs, labor, packaging, refrigeration/freezing, and transportation/storage. For edible by-products like spleens, there are only two end products: edible spleen or raw material for animal feed. Other products, like large intestines in pork, have three or more end products (chitterlings, casings, heparin, and sutures).

Using beef spleens as an example, a calculation of the value of spleens as a rendered product or as whole spleens is shown. To estimate rendered value, it is necessary to know the moisture, fat, protein, and ash content of spleen. Published values (Bittel et al. 1981) for spleens are

Moisture—74.2%	Fat—6.0%
Protein—17.6%	Ash—1.3%

Meat scraps (rendered meat solid by-product) generally contain 9% fat and 5% moisture. Therefore, to obtain the meat scrap yield divide the nonfat solids of the spleen (protein and ash) by 0.86 (the amount of nonfat solids in meat scraps). In this example

$$(17.6 + 1.3)/0.86 = 26.6\% = \text{Meat scrap yield of raw spleens}$$

To determine how much tallow is derived from this process, multiply total meat scrap yield by 0.09 (percentage of fat in typical meat scraps), and then subtract total fat in spleens from the fat in meat scraps.

$$(26.6)(0.094) = 2.5 = \text{percentage of total spleen as fat in meat scraps}$$
$$6.0\ (\%\ \text{fat in spleen}) - 2.5 = 4.0\% = \text{Tallow yield from spleens}$$

Finally, as meat scraps are sold on a 50% protein basis, one needs to know the protein content of meat scraps from spleens to calculate the incremental value of the meat scraps. This is done by taking the original amount of protein in spleens and dividing it by the meat scrap yield:

17.6/26.6 = 66.16% protein in the meat scraps from spleens

Thus, from 100 lbs of spleens that are rendered, there is

26.6 lbs of 66.16% protein meat scraps
4.0 lbs of edible tallow

One final mathematical manipulation is necessary to find the rendered value of spleens. Because meat scraps are sold on a 50% protein basis, the 26.6 lbs of meat scraps at 66.6% protein needs to be on a 50% protein basis. To do this, multiply the meat scrap yield by the protein percent and then multiply this figure by 2, to obtain a 50% protein equivalent weight:

$$(26.6)(0.6616)(2) = 35.2$$

Thus, from 100 lbs of spleens that are rendered, one realizes

35.2 lbs of 50% protein meat scraps
4.0 lbs tallow

Now, to calculate the rendered value of the spleens, determine the value of the meat scraps and tallow. To do this, using prices (*National Provisioner,* September 1992) for tallow at $0.16/lb, and 50% meat and bone meal of 0.1125/lb:

$$(35.2)(0.1125) = \$3.96$$
$$(4.0)(0.16) = \$0.64$$

Therefore, the gross return for rendering 100 lbs of spleens is $4.60.

At $.07/lb for a 50-lb box of frozen spleens, one must still look at the processing costs associated with packaging and freezing spleens before the decision is made to sell edible spleens. If the box costs $1.00 ($0.02/lb) and freezing costs are $0.01 per pound, and transportation costs another $0.01 per pound, it is more economical to render the spleens than to sell them as edible. To ensure maximum profitability, this kind of analysis has to be performed on each edible coproduct.

As discussed in the Introduction, muscle food coproducts can be categorized as edible, inedible, and pharmaceutical. The inedible category is vast; covering everything from leather, to animal feed, to photographic film (Table 4.1). Pharmaceutical uses of coproducts enhance the quality of human life by treating some diseases. In the production of by-products, often the question is asked, "What will be produced from this raw material?" This chapter has focused on edible coproducts; however, the value of other potential by-products must always be considered when making the decision to produce an edible product. Therefore, even though we have only discussed edible products, all by-products have an impact on the cost of muscle food production.

Research Needs

Research addressing uses of food animal coproducts must look at enhancing by-product value. Enhancing the value of by-products will improve economics for all involved in animal food production. There is substantial room for improvement in the use of these products. From an edible product standpoint, many of the by-products are high in minerals and protein, and low in fat (Table 4.3); so they are underutilized compared to muscle foods with the same characteristics. Problems to overcome include strong flavors and unfamiliar textures with some by-products; however, a major hurdle is the perception that coproducts are somehow less wholesome than other products. Research and lobbying for label reform may also be fruitful, especially for mechanically deboned pork and beef. Research into new processing techniques is another way to enhance the value of coproducts. Finally, research to improve processing of all coproducts will enhance the value of muscle foods. An example would be the development of a method to predict optimum processing times for bones to make gelatin.

Nutritional Aspects of Muscle Food By-products

Generally, muscle food coproducts, particularly liver, heart, brains, and kidney, contain higher levels of protein, vitamins, and minerals than the skeletal muscle of the animal from which the coproduct was harvested. These same organs generally contain less fat but more cholesterol than skeletal muscle. This is a general statement; the exact content of a given nutrient depends on many variables. Table 4.3 shows the proximate analysis and caloric content of some edible coproducts.

Although many muscle food by-products are nutritious, consumer perception of by-products is poor, so relatively few people take advantage of nutrients available in coproducts. Improved utilization of coproducts in the diet is an important area of nutritional research.

Environmental Impact of Coproduct Utilization

Skeletal muscle content of muscle food animals ranges from 32% live animal weight for lamb to 52% for pork (Goldstrand 1988). This leads to the conclusion that 48% or more of the food animal is by-products. If by-products were not valuable, and had to be disposed of as waste, there would not be a muscle foods industry. The costs of waste disposal would outweigh any possible advantage. Fortunately, recovery and use of these products provides substantial value and reduces waste.

Animal production and feeding nutritional by-products from other industries improve environmental quality. Examples of industries that benefit from food animals converting material unfit for human consumption to wholesome nutritious food include the following: the citrus industry—citrus pulp and skins are fed to livestock; the brewing and distilling industries—brewers and/or distillers grains are fed to many animal species; potato processing—waste peels and scraps are fed to cattle; corn milling and ethanol production—their waste products are feeds for various species; oilseed industries—the seed hulls and oilseed processing waste are fed to various species. There are many other industries that feed their coproducts to muscle food animals, thus reducing the environmental impact of these industries and reducing the cost producing their primary products.

Another area in which muscle food industries impact environmental concerns is the large portion of muscle food by-products that are directly recycled back into food animals. These coproducts are not acceptable for human consumption, but they reduce the cost and environmental effects of muscle foods.

Selected References

Anderson, B. A. 1988. Composition and nutritional value of edible meat by-products. In: *Edible Meat By-Products,* Advances in Meat Research, Vol. 5, Pages 15–42. Elsevier Science Publishing Co., New York.

Anonymous. 1988. *Nutrient Values of Muscle Foods.* The National Live Stock and Meat Board, Chicago.

Bittle, R. J., P. P. Graham, and K. P. Bovard, 1981. Mechanically separated spleen: Its composition and protein efficiency ratio. *J. Food Sci.* 46:336

Breedlove, A. L. and W. L. Brown. 1991. Fat reduced beef. In: *Proceedings of the 1991 Meat Industry Research Conference.* American Meat Institute, Washington, DC.

Booren, A. M. and G. M. Weiss, 1988. Lean skeletal trimmings incidental to slaughter, in *Edible Meat By-Products,* Advances in Meat Research, Vol. 5, Pages 219–230. Elsevier Science Publishing Co., New York.

Carpenter, J. A. and R. L. Saffle, 1964. A simple method of estimating the emulsifying capacity of various sausage meats. *J. Food Sci.* 29:774.

Cross, H. R., C. E. Green, M. S. Stanfield, and W. J. Franks Jr., 1976. Effect of quality grade and cut formulation on the palatability of ground beef patties. *J. Food Sci.* 41:9.

deHoll J. C and J. F. deHoll. 1983. *Encyclopedia of Labeling Meat and Poultry Products,* 6th ed., Meat Plant Magazine. St. Louis, Mo.

Eisle T. A. and C. J. Brekke. 1981. Chemical modification and functional properties of acetylated beef heart myofibrillar proteins. *J. Food Sci.* 46:1095.

Goldstrand, R. E., 1988. Edible meat products: Their production and importance to the meat industry. In: *Edible Meat By-Products,* Advances in Meat Research, Vol. 5, Pages 1–12. Elsevier Science Publishing Co., New York,

Hinterwaldner, R. 1977a. Raw materials. In: *The Science and Technology of Gelatin.* A. G. Ward and A. Courts, eds. Academic Press, London.

Hinterwaldner, R., 1977b. Technology of gelatin manufacture. In: *The Science and Technology of Gelatin* G. Ward and A. Courts, eds. Pages 315–365. Academic Press, London.

Hood, L. L., 1987. Collagen in sausage casings, In: *Collagen as a Food,* Advances in Meat Research, Volume 4, Van Nostrand Reinhold, New York.

Hunt, M. C. and R. E. Campbell. 1992. Lean, fat and connective tissue from beef shanks, processed with a Baader™ Desinewer. In: *The 38th International Congress of Meat Science and Technology, Proceedings, Volume 6,* Clermont-Ferrand, France. Pages 1219–1222. 1992.

Kenney, P. B., R. O. Hendrickson, P. L. Claypool, and B. R. Rao, 1986. Influence of temperature, time and solvent on the solubility of corium collagen and meat proteins. *J. Food Sci.* 51(2):277–280, 287.

Marriott, N. G., P. P. Graham, C. K. Shaffer, and J. W. Boling. 1985. Effect of spleen in restructured beef. *J. Food Qual.* 8:237–245.

Mawson, R. F. 1992. *Mechanical Meat/Fat/Sinew Separation*. MIRINZ, Meat Research, Hamilton, New Zealand.

McHugh, M. A. and V. J. Krukonis. 1986. *Supercritical Fluid Extraction: Principals and Practice*. Butterworths, Boston,

Rust, R. E. 1988. Production of edible casings. In:*Edible Meat By-Products,* Advances in Meat Research, Volume 5. Pages 261–279. Elsevier Science Publishing Co., New York.

Saffle, R. L. and J. W. Galbreath. 1964. Quantative determination of salt-soluble protein in various types of meat. *Food Technology*. 18(12):1943.

Smith, D. M. and C. J. Brekke. 1984. Functional properties of enzymatically modified beef heart protein. *J. Food Sci.* **49**:1525–1528.

Spooncer, W. F. 1988. Organs and glands as human food. in *Edible Meat By-Products,* Advances in Meat Research, Volume 5. Pages 197–218. Elsevier Science Publishing Co., New York.

USDA. 1979. *Agricultural Handbook 8–5*. U.S. Government Printing Office. Washington, DC.

Wiley, E. L., J. O. Reagan, J. A. Carpenter, and D. R. Campion. 1979. Connective tissue profiles of various raw sausage materials, *J. Food Sci.* 44:918.

Ward, A. G. and A. Courts (eds.). 1977. *The Science and Technology of Gelatin*. Academic Press, London.

5

Processed Meats/Poultry/Seafood

James R. Claus, Jhung-Won Colby, and George J. Flick

Processed Muscle Foods

Overview of Processed Meats, Poultry, and Seafood

Diversity is one of many very desirable characteristics of processed muscle foods. Processed muscle foods are convenient, versatile, and wholesome, and contribute positively to the diet by providing an excellent source of high-quality digestible protein (amount and proportion of essential amino acids), water-soluble vitamins (B vitamins), fat-soluble vitamins (A, D, E, K), minerals (very bioavailable heme iron, zinc), and essential fatty acids.

Most of the commercial fish species are overharvested and there is a need to increase fisheries production through increased processing efficiency, the utilization of nontraditional or underutilized species, and aquaculture. The development of processed seafood products has enabled the commercialization of catches that were previously unmarketable due to low consumer appeal or unfamiliarity, high processing costs, and limited shelf-life.

The sophistication of equipment and techniques has resulted in an explosion of convenient seafood products in the market. Fish can be processed into appealing ready-to-eat products such as fish sticks or portions and a wide variety of surimi-based seafood analogues. Descriptions of the processing of minced products, surimi-based products, battered and breaded products, dried, smoked, and cured products, and other further processed seafood products will be discussed later in this chapter.

Processed meats consist of sausages, cured and smoked, noncomminuted meats (ham and bacon), restructured products, and canned products. Processed meats can be successfully manufactured from beef, pork, lamb, venison, chicken, turkey, and seafood.

Sausages

Sausages can be classified by various methods including curing, particle size or degree of chopping, composition, fermentation, addition of smoke, and thermal processing. However, the USDA system for classifying sausages is probably the most widely accepted, general system that is based primarily on six traits that include whether the product is fresh, cured, cooked, smoked, fermented, and dried.

Fresh sausages are manufactured from fresh or frozen, uncured meat. Fresh sausages may contain meat by-products which, if used, must be properly declared on the label. Many consumers are not fully aware that any sausage with an ingredient label indicating only the species name as one of the ingredients contains only skeletal muscle indicative of that species. Fresh sausages must be refrigerated and cooked prior to consumption. Fresh sausages are typically formulated with pork and often beef obtained from lean and fat trimmings removed during the fabrication of primal cuts. Fresh sausages are uncured, ground, seasoned, and often stuffed into edible casings. These sausages can be cooked by grilling and panfrying. Although the incidence of *Trichinella spiralis* in pork is very low in the United States, those sausages containing pork should be cooked to at least 60°C so that the larva, if present, is inactivated. Depending on the specific product's standard of identity, fresh sausages may contain up to 30% fat (fresh beef sausage) or 50% fat trimmings (fresh pork sausage). Examples of fresh sausages include fresh pork sausage, bratwurst, breakfast sausage, and whole hog sausage.

Uncooked, smoked sausages are essentially the same as fresh sausages except that natural hardwood or liquid smoke is applied to the surface sufficient to develop a desirable color and flavor. These sausages include kielbasa, mettwurst, and smoked, country-style pork sausage.

Cooked sausages include cured and uncured products that are made from comminuted, seasoned meats. Liver sausage and braunschweiger are often water cooked. **Cooked, smoked sausages** are considered a class of sausage that include frankfurters, bologna, and cotto salami. Cooked, smoked sausages can be made from any combination of red meats and poultry, with some restrictions as to the amount of poultry, if primarily a red meats product. Cooked sausages also can be manufactured to contain only poultry. Most cooked sausages are served cold; however, one of the more popular cooked sausages, frankfurters, are served hot either by boiling or grilling. Another classification of cooked sausages includes luncheon meats, loaves, and jellied products.

Dry and semidry sausages are products made from fresh meats that are cured and fermented typically by the addition of a bacterial starter culture that metabolizes an added simple sugar to produce a mild to very acidic tangy flavor. Fermentation results in lowering the pH (production of lactic acid) of the meat product which aids in preservation. Dried sausages are usually not cooked as the combination of fermentation and drying produces a product that is shelf-stable. Dry and semidry sausages include cervelat, chorizos, Lebanon bologna, mortadella, pepperoni, salami, and summer sausage.

Cured and Smoked, Noncomminuted Meats

These processed meats are cured and smoked sometimes as bone-in primal cuts (ham, loin, picnic shoulder) or boneless products. To produce boneless products, the primal cut can be muscle boned, sectioned, and formed by recombining the same muscles as in the original primal or by combining only certain muscles from a given primal with the same muscles from other animals of the same species. Although not smoked, boiled ham which is cooked in water or steam is a sectioned and formed product. Boneless products include Canadian-style bacon, boneless ham, turkey ham, buffet ham, and honey glazed ham. Bacon is a cured, smoked product produced from a pork belly. Beef jerky is a cured and smoked, noncomminuted or restructured product. It is a thinly sliced, marinated, dried product that unlike other cured and smoked, noncomminuted meats does not require refrigeration.

Restructured Products

Restructured products typically refer to uncured items that have been produced from raw meat materials that have been coarsely ground, flaked, or sectioned in some manner and then recombined to form a product of very uniform composition, shape, and sensory characteristics. Sectioned and formed products may consist of an individually selected whole muscle or similar muscles cut into smaller chunks, but not ground, that can be recombined to form the final product. Restructured products can be made from any muscle food and include restructured roast beef, turkey roll, roast turkey breast, flaked and formed steaks and chops, and nuggets that are often used in fast-food chains. In addition, nuggets are usually covered with batter and breaded.

Canned Products

Ground as well as sectioned and formed meat products have been successfully canned. Some products are heat-treated such that refrigeration is not required (shelf-stable), whereas others must be refrigerated and have a considerably shorter shelf-life. Some products heated to the point of being considered as commercially sterile have a distinctly different flavor profile than similar products heat processed only to achieve pasteurization. Canned products include Vienna sausages (small frankfurterlike sausages that are cut to a uniform length), cured boneless ham, meat stews, corned beef hash, luncheon meats, and seafood.

Raw Materials and Nonmeat Ingredients

Meat Ingredients

Meat raw materials can vary in a variety of attributes and as such must be carefully selected to produce a consistent product. Meat raw materials vary in composition (moisture, fat, protein, collagen) and cost. In addition, the functional or physicochemical properties of the fat (melting point, color, flavor stability), protein (bind, texture formation, gelation), and color contribution can vary between different raw materials.

The predominant meat raw material used to manufacture processed meats is skeletal muscle. However, other raw materials are used in processed meats in order to reduce formulation costs while increasing the value of the raw material. In addition to skeletal muscle, raw materials include the variety of organ meats (tongues, snouts, livers, hearts), partially defatted tissue, fat, and blood.

The binding ability of a raw material is one of the most important characteristics. Bind can encompass the ability to create adequate protein-to-protein interactions necessary to hold meat pieces together and produce the desired texture, stabilize fat, and chemically bond water. Generally, skeletal muscle offers the most desirable combination of these properties based on the amount of extractable actin and myosin. Leaner skeletal muscle yields a higher percentage of these proteins. This is true of support-type muscle and the major exercise-type muscles. However, muscles in the extremities of the appendages and areas within a given muscle near the origin and insertion are higher in connective tissue (collagen). The amount of high connective tissue containing meats should be closely monitored and restricted because of the reduced overall functional properties or functionality such as protein binding, fat stabilization, and water binding. The physiological state of the muscle used can have a dramatic effect on the extractability and functionality of the raw material. Pre-rigor meat is known to have greater functionality than post-rigor lean. In pre-rigor lean, actin and myosin are more extractable and have a greater role in protein functionality compared to post-rigor meat where actomyosin is more prevalent.

Color is another characteristic that varies considerably with the source of raw material. Color is predominantly affected by the myoglobin content. Significant differences in myoglobin content occur due to species, animal age, and muscle function. Lamb and beef typically have the most myoglobin followed by pork and chicken white meat. Bovine muscle can vary from 1 to 20 mg/g of myoglobin on a wet tissue basis, with veal having the least, followed by beef (5–10 mg/g). Exercise-type muscles (i.e., *M. rectus femoris*) typically have more myoglobin than support-type muscles (i.e., *M. longissimus thoracic et lumborum*).

Nonmeat Ingredients (Adjuncts)

There are numerous nonmeat ingredients that can be used in processed meats. The most commonly used ingredients include water, nitrite, cure color accelerators, salt, phosphate, and sweeteners.

Water, one of the most common nonmeat ingredients, has many vital functions in processed meats including assisting in ingredient distribution, solubilization of meat proteins, temperature control, machinability, improving various sensory traits, and cost reduction. The addition of some water allows the added salt to solubilize faster and, therefore, improves the extraction of salt-soluble proteins (actin and myosin). Grinding or mincing high-fat meats is improved by the addition of water. Water decreases the frictional forces between meat particles and decreases the viscosity of finely comminuted batters. A decrease in viscosity may be advantageous where uncooked, minced meats must be pumped as a means of transport from one part of the processing plant to another. Water also aids in temperature control during grinding, mixing, and mincing. Sometimes processors will add ice as the source of moisture to avoid having the fat render out or excessive protein denaturation as a result of elevated temperatures during grinding and chopping. Fat reduction in processed meats is a major trend in the meat industry today. Simply reducing the fat content of processed meats results in products that are more rubbery, firmer, and often less juicy. Because proteins will bind water, those that are involved in protein-to-water interactions will not be available to bind other proteins and, as such, result in a reduction in protein-to-protein interactions and a softer product.

Nitrite is the most important ingredient incorporated in the manufacture of cured meats without which most of these products would not exist as they are today. Nitrite serves as a vital bacteriostatic control over the outgrowth of spores produced from *Clostridium botulinum*. These bacteria can grow under anaerobic conditions such as those created by canning or vacuum-packaging. The growth of vegetative cells can be controlled using normal smokehouse heat processing. However, spores produced from these bacteria can survive normal heat processing. Although canning will destroy the spores, it is not practical for all processed meats. Without the addition of nitrite, if the spores are provided the correct conditions, the spores may produce vegetative cells that are responsible for producing a toxin that is extremely lethal. As heat processing is lethal to vegetative cells, survival of spores in meat without nitrite would mean that growth of vegetable cells from the spores would not be restricted by other competing microflora. Without the presence of normal spoilage bacteria, consumers would not be able to determine if the product was unfit for consumption because the toxin produced by *Clostridium botulinum* is odorless. If a can of thermally processed meat appears swollen, microbial growth is the potential cause and the can should be discarded. Legal limits for nitrite are 120 ppm for bacon, 156 ppm for cooked sausages, and 200 ppm for cured hams, loins, and shoulders. These levels are well above that necessary to provide protection against the outgrowth of *Clostridium botulinum* spores.

Nitrite is involved in cured color development and flavor protection. Nitrite reacts with myoglobin and, upon heat processing, forms a heat-stable pink cured pigment. Nitrite contributes to flavor stability [prevention of warmed-over flavor

(WOF)] by complexing with heme iron, which, if free, could act as a potent catalyst in lipid oxidation.

Cure color accelerators (e.g., sodium ascorbate, sodium citrate, and sodium erythorbate) are ingredients added to promote the formation of nitrosyl hemochrome. Cure color accelerators allow manufacturers to heat process cured meats sooner by minimizing the time necessary for the nitrite to react with myoglobin. Sodium erythorbate can be added up to 550 ppm. However, half of the sodium erythorbate can be replaced with sodium citrate in a formulation. Sodium ascorbate and sodium citrate also have antioxidant properties that help to maintain the color and flavor of the cooked product. Sodium citrate is known to bind metal ions (copper, iron, chromium) that can promote the development of oxidative rancidity.

Salt has three basic functions: (1) add or enhance *flavor;* (2) *solubilize or extract proteins* that are essential for improving moisture retention and in forming the necessary bind and texture in the finished product; (3) *extend the shelf-life* of the product. The most common salt used is sodium chloride (NaCl). Typically meat products are formulated to contain 2–3% salt but may range from 1.5% to 5%. As with several other ingredients, salt is considered self-limiting. The desired degree of saltiness in some products may be different than other products and can be modified by other ingredients. The most common ingredient used to counteract the astringent taste of products with high salt is sugar. This information can be beneficial to the processor experiencing product failures (low bind, slicing defects, dry products). If the processor has some latitude in elevating the salt content to improve protein extraction and consequently bind, then a concurrent increase in sugar would be beneficial in order to maintain a product similar in flavor. However, increasing the sugar content in order to mask higher levels of salt is not desirable because consumers wish to avoid elevated sodium intake. Processors can elect to substitute potassium chloride (KCl) for NaCl. A major disadvantage of using KCl is its bitter taste. Therefore, its use is usually limited to 0.5–0.7% on a finished weight basis.

Processed meats that are formulated with a lower level of salt are usually the premium-type products that are not highly extended with various nonmeat ingredients including water. Lowering the salt content can greatly reduce the amount of salt-soluble myofibrillar proteins that are extracted and are essential for desirable water-holding ability and bind. Lower salt content requires even closer attention to using high-quality, fresh raw meat, adequate physical manipulation (mixing, massaging, tumbling), and very tightly controlled thermal processing.

Salt can be both beneficial and detrimental relative to preservation. Salt can retard microbial growth and subsequent spoilage in fresh and cooked processed meats. Yet, salt has undesirable oxidative effects on meat products, especially on fresh sausages that are frozen and stored. Processed meats frozen for extended periods of time may become rancid and unacceptable from a flavor standpoint.

Often a binder, manufactured from lower-quality minced meats such as shank meat that is high in connective tissue, is used to increase the bind between larger meat chunks and fill in void spaces. Because the binder meat is very finely minced, the opportunity for spoilage is increased. Fresh minced shank meat may remain acceptable for 1–2 days prior to use, whereas salted, minced shank meat may be stable for twice as long. Salt used as a cover pickle can extend the shelf-life of meat for further processing.

Phosphates have many functions in meat products; however, one of the greatest benefits is the improvement in water-holding capacity. Processing yields are significantly improved and purge accumulation in the package during storage is reduced. Products formulated with phosphates retain more natural juices and added water during heat processing and subsequent reheating. Phosphates can increase the pH and ionic strength of the comminuted meat which contributes to the water-holding capacity. The further the pH is from the isoelectric point of a protein, the higher the water-holding capacity. The isoelectric point of a protein represents that pH in which the protein has a net charge of zero (equal negative and positive charges). Because water is a polar compound, it can orientate and form hydrogen bonds with the charged amino acid side groups of the protein (Fig. 5.1). This association is dramatically increased if the net charge on the protein is either more negative (alkaline) or more positive (acidic). A protein that has a net charge of zero will have the lowest affinity for water and will be the least soluble. Elevating the pH is one of the functions of alkaline phosphates.

Phosphates also work *synergistically* with salt to extract myofibrillar proteins and improve water-holding capacity. Phosphates are involved in the dissociation of actomyosin which enhances salt's ability to solubilize myosin. Phosphates have the ability to chelate (sequester) certain metal ions. This function may be of benefit as traces of iron and copper can act as catalysts in lipid oxidation. As such, phosphates can have a role in flavor preservation in addition to maintaining product appearance and color. Use of phosphates in some products has been reported to develop a soapy flavor. During the manufacture of some products such as bacon, phosphates migrate and crystalize on the surface of the unsliced slab.

Use of acidic, alkaline, and blends of phosphates are restricted to 0.5% in the finished product. Alkaline phosphates are the most frequently used phosphates. Some of the approved phosphates include sodium tripolyphosphate, sodium hexametaphosphate, sodium pyrophosphate, and monosodium phosphate. Besides some flavorings that can be used in cured meats, phosphates are the most difficult ingredient to dissolve.

Sweeteners contribute to the sweet taste of the product. In addition, sweeteners have the ability to attract water and develop the surface color of some products through browning reactions. Browning is a result of the sugar reacting with the amino group of the proteins during heat processing. Sweeteners vary in the

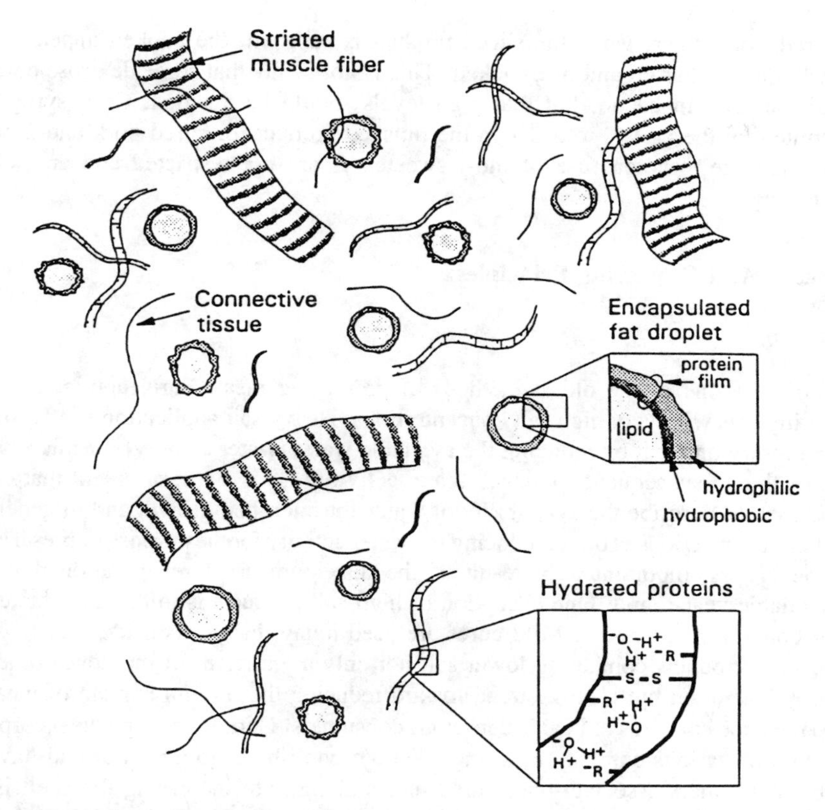

Figure 5.1. Sausage batter.

degree of sweeteness and are usually compared to sucrose (cane sugar) which is given a value of 100. Other sweeteners include maltose (33–45), corn syrup (40–70), dextrose (70), molasses (74), and honey (100–170). Knowledge of the different degrees of relative sweetness is useful when substituting one sweetener for another. Some of these sweeteners can contribute other flavor components that make them unique, such as the honey flavor of honey along with a sweet taste to the product. Processors that manufacture high-solids processed meat usually prefer dextrose over sucrose because more dextrose (1.4 lbs dextrose versus 1 lb sucrose) is required to achieve an equivalent degree of sweetness. Substitution of one sweetener for other, particularly across product types, should be made only with an understanding of how that sweetener will perform in that product. Suppose a processor has been successfully using dextrose in a cured, smoked ham and wants to substitute the sucrose used in a bacon pickle with dextrose. Because the cured bellies are not heat processed to very high temperatures the processor will probably not notice any visual changes. As dextrose is a reducing sugar, when the sliced product is panfried, the cooked appearance will appear charred and overcooked. This is not to say that some dextrose could not be used in bacon, but that high levels should be avoided. Corn syrup is limited by the USDA to 2.0% in the finished product for cured pork and 2.5% for sausage, whereas use of most sweeteners are not restricted and are self-limiting.

Basic Meat Processing Principles

Curing

Curing is one of the oldest methods of preserving meat. Early application of curing was with very high salt concentrations. Heavy salt application is effective in preserving meat by reducing the available water (water activity) for microbial growth and subsequent spoilage. Water activity (a_w) is a measurement that can be used to describe the availability of water for microbial growth and in general chemical reactions. Today, reducing the water activity for the purpose of preservation is less important as a result of the development of refrigeration that is available year-round. One exception of high-salt products is in the manufacture of country-cured meats. Mild cures are used today that are considerably lower in salt. Products containing lower salt not only are a result of the development of refrigeration but also the trend toward reducing the sodium content of meats to meet the concerns of health-conscious consumers. Curing, through the incorporation of various cure adjuncts (ingredients), contributes to the color and flavor that consumers associate with cured meats along with increasing the shelf-life and food safety of these products. The two main ingredients that make up the cure are **salt** and **nitrite.**

Basic Curing Reactions

Sodium nitrite is the primary ingredient used to produce the *heat-stable,* pink cured pigment, nitrosyl hemochrome. Sodium nitrite, through a series of chemical reactions, reacts with the meat pigment, myoglobin. Physiologically, the role of myoglobin is to sequester oxygen transported through the blood by hemoglobin. Knowledge of the differences in myoglobin content is important for predicting the extent of cured color development and can be used to produce a restructured product that is uniform in appearance and color.

The **heme ring** of myoglobin is the **nonprotein portion** that is directly involved in the cured color formation. The heme ring contains the 6th coordination position associated with the iron atom (Fe) which is the site where oxygen, water, and other reactive ligands can bind. Myoglobin (Mb) can exist in three chemical states: deoxymyoglobin (DeoxyMb; Fe^{2+}, ferrous iron), oxymyoglobin (MbO_2; Fe^{2+}, ferrous iron), and metmyoglobin (MMb; Fe^{3+}, ferric iron). MMb is the oxidized form of myoglobin, which usually corresponds to product that is older or abused in some manner (elevated storage temperatures, microbial growth). When approximately 40 to 60% of Mb is in this oxidized state, the meat will appear discolored and brown. To facilitate cured color formation, sufficient *reducing equivalents* either native to the meat (nicotinamide adenine dinucleotide, NADH; nicotinamide adenine dinucleotide phosphate, NADPH) or added as a curing adjuncts (cure color accelerators; sodium erythorbate or ascorbic acid) must be present to provide a source of electrons as follows:

$$\text{NADH} + \text{H}^+ \xrightarrow{\text{oxidation}} \text{NAD}^+ + 2\text{H}^+ + 2e^-$$
$$\text{MMb (Fe}^{3+}) \xrightarrow{\text{reduction add } e^-} \text{DeoxyMb (Fe}^{2+})$$
$$\text{Nitrite} \xrightarrow{\text{reduction}} \text{Nitric oxide}$$
$$\text{DeoxyMb (Fe}^{2+}) + \text{nitric oxide} \longrightarrow \text{nitric oxide myoglobin}$$
$$\text{Nitric oxide myoglobin} \xrightarrow{\text{heat}} \text{nitrosyl hemochrome}$$

Therefore, lean that is older or more brown in appearance as a result of metmyoglobin formation will require a higher level of reducing equivalents than fresh meat. Somewhat confusing is the fact that nitrite is a *very strong oxidizer* and, therefore, would contribute to an oxidized state of myoglobin. This is often seen in uncooked cured sausage batters after the addition of sodium nitrite. Nitrite when added to meat is reduced to nitric oxide (NO). Nitric oxide is a colorless, odorless gas which reacts with myoglobin to form nitric oxide myoglobin (red color) in the case of reacting with DeoxyMb or nitric oxide metmyoglobin (brown) in the case of MMb. In order to form the heat-stable cured pigment, iron must be in the reduced state. For nitric oxide metmyoglobin, the reducing equivalents associated with the meat or that added to the meat are important. Upon heating, sulfhydryl groups of some proteins become available and provide some reducing equivalents. Heating also results in the denaturation of the globin (protein) portion of myoglobin. The resulting heat-stable pigment is nitrosyl hemochrome (nitroso hemochromogen). This pigment is quite stable if light and oxygen are excluded and is stable to heat. Light can cause the release of the bound nitric oxide, and oxygen oxidizes the heme ring, resulting in a brown, discolored cured meat. To minimize problems with oxygen, it is important that mixing, massaging, and stuffing be performed under vacuum. In addition, the cooked product should be vacuum-packaged. To reduce the effect of light, some package films today provide a barrier to ultraviolet light. In addition, some cured meat manufacturers place a relatively large pressure-sensitive label on the outside of the clear vacuum package to minimize the product's exposure to light.

Curing Methods

Curing methods can be classified as ***dry, immersion,*** or ***injection cure*** and can be used to some extent in combination. ***Dry cure*** represents the application of salt and sodium nitrate and possibly some sodium nitrite directly to the surface of the meat (primarily pork). With sufficient quantities and time (dependent on temperature), these ingredients will be absorbed into the meat and can penetrate the entire muscle. Examples include parma and prosciutti ham that can be consumed without cooking. Dry curing for thicker cuts (shoulder, ham) requires 2–3 days per pound. Dry cured ham is a uniquely flavored product that does not require refrigeration. This is one of the few meat products in which a certain level of rancidity is actually desired to develop a characteristic flavor sought after by consumers.

Immersion curing requires proper refrigeration and can be used for thick and thin cuts of meat. Although, immersion curing is more rapid than dry curing, it is slower than injection curing. In immersion curing, enough ***pickle*** is prepared to completely cover or submerge the meat that is placed in stainless-steel or selective approved plastic tubs. A pickle can be prepared by dissolving salt, sugar (sucrose, dextrose), seasonings, and sodium nitrite in water. In some cases a prepared brine may be used as a source of salt. A ***brine*** is simply a saturated salt (26.4% salt at 15.5°C) solution. This means that for every 100 lbs (45.5 kg) of saturated brine, there is approximately 26 lbs (11.8 kg) of salt. The salinity of the brine can be checked using a salometer, which is a device that measures the density of the solution on a scale of 0 to 100°S, where 100°S represents a salometer reading equivalent to a saturated salt solution at a given temperature. The salometer is a weighted, enclosed blown-glass device that floats at different heights depending on the concentration of the salt. Therefore, the greater the salt concentration, the higher the salometer floats. A salometer reading of 100°S would represent the most buoyant solution and result in the salometer the farthest out of the solution.

Most curing operations that formulate large batches of pickle (5000–6000 lbs) do not have the capability to weigh the water, brine, or the completed pickle to verify the final amount. Instead, this is usually done volumetrically. Water has a weight of approximately 8.34 lbs per gallon and a typical pickle may weigh 9.25–9.5 lbs per gallon. Commercially, a brine can be prepared using an above-ground silo-type brinemaker. The major advantage in using a brine is that associated with the time savings normally required for the salt to become dissolved each time a pickle is prepared if added dry. Some potential disadvantages with using a brine prepared with some brinemakers are that the brine weight, volume, and, to a lesser extent, concentration can vary relative to changes in temperature and, as such, requires close monitoring by a quality control technician. The **purity of the water** and salt should be monitored particularly for metals (copper, iron, chromium).

Injection curing is the most widely used method in the industry today. This general procedure facilitates the distribution of ingredients and greatly reduces the cure time. Injection can be accomplished by artery pumping, a hand-held stitch device, or multineedle injection. ***Artery pumping*** involves locating the femoral artery of the ham and inserting a stainless-steel needle in the open end. Pickle is mechanically injected by the use of a pump that draws up the pickle from a storage vessel and then infuses the pickle throughout the ham via the vascular network. For artery pumping to be effective, the vascular system must be intact. Artery pumping usually requires a lower percentage of injection compared to multineedle injection to achieve the same composition in the finished products. Hams are usually placed on a scale so that the operator can determine when the correct percentage of pickle has been injected based on the starting weight.

Hand-held stitch devices having one to four needles are more commonly used by small curing operations in which relatively low daily volumes are produced. However, some curing operations supplement the artery curing method with a hand-held stitch device. The forecushion of the ham is often injected to assure that this area also receives sufficient cure, as artery pumping alone may not be adequate.

Multineedle injection machines are currently the most widely used system for incorporating pickle. These machines may have 100–250 stainless-steel needles. Each injection needle will have multiple orifices to facilitate cure injection. The meat is transported by conveyor to the injection site. Product advances on the machine's conveyor upon each reciprocating cycle of the injection needles. As the needles come in contact and enter the product, valves open, and pickle is injected. Both bone-in and boneless products can be injected.

Curing time varies greatly depending on the type of product being produced and what ingredients are incorporated. Fresh pork bellies can be pumped with a multineedle injection system and processed in a smokehouse within 1 h. Large hams may have to be stored under refrigeration overnight before thermal processing. In order to produce sectioned and formed hams, the pumped meat may have to be massaged 16–18 h for the purpose of extracting sufficient salt-soluble myofibrillar proteins necessary to bind the chunks together upon thermal processing and to permit the uniform distribution of cure ingredients as well as sufficient time for cure color development to occur.

Batter Formation

Batter formation involves finely comminuting lean meat in the presence of salt, some water, and possibly phosphate prior to incorporating the fat meats in order to form a meat batter that will properly stabilize the lipid fraction, optimize water retention during heat processing, and provide a product with desirable textural traits. The lean meat is first comminuted in the presence of salt as this fraction contains the highest concentration of myofibrillar proteins (actin, myosin, and actomyosin) that must be extracted by the salt in order to form a stable batter. As the fat component is added and minced, the fat particles are encapsulated by the extracted protein. The extracted proteins have emulsification properties which facilitate surrounding and interacting with the fat particles or droplets. An emulsifier has a hydrophobic region and a hydrophilic region. Because fat is hydrophobic, the hydrophobic region of the emulsifier readily associates with the fat while the hydrophilic region can interact with other hydrophilic regions of other proteins and water to stabilize the batter matrix. A meat batter is not considered a true emulsion because it does not consist of simply two immiscible liquids in which one is dispersed (dispersed phase) in another liquid (continuous phase). A meat batter is actually a very complex mixture consisting of suspended particles (collagen fibers, myofibrils, cellular organelles), immobilized and free water, lipid droplets and particles, and hydrated, solubilized and nonsolubilized myofibrillar proteins (Fig. 5.1). In a meat batter, the continuous phase consists of water, dissolved salt, and the myofibrillar proteins existing at various states of hydration and solubilization. The discontinuous phase consists of suspended particles and fat droplets. Extraction and hydration of the proteins assist in stabilizing the batter by increasing the viscosity. Increased viscosity decreases lipid droplet aggregation, flocculation, and subsequent coalescence. If, upon comminuting the batter, the temperature becomes too high (> 15.6°C, 60°F) and the batter is overchopped, the viscosity may decrease, and the fat may become too fluid. Fat droplets could then coalescence into droplets that are too large to be properly stabilized by the protein matrix, whereupon during thermal processing the fat would render out and form fat caps or pockets. A certain degree of softening of the fat is desirable so that the hydrophobic portions of the extracted protein can associate more thoroughly to stabilize the batter.

Fermentation

Fermentation is the result of controlled bacterial growth of selected microorganisms that produce lactic acid as a by-product of their metabolism of a sugar. Lactic acid accumulation results in a decline in pH and is responsible for developing the desirable tangy flavor and texture, preservation, increased rate of drying, and wholesomeness of the sausage.

The primary principle behind texture (bite, mouthfeel, firmness) formation in fermented sausages is different than that known in cooked sausages. In fermented sausages, the texture, in part, is a result of acid coagulation and dehydration of the myofibrillar proteins, whereas in cooked sausages the texture is developed by heat denaturation of the salt-extracted myofibrillar proteins. In fermented sausages, extensive extraction of salt-soluble proteins is not necessarily desirable because too much could retard the drying process. During manufacture of fermented sausages the salt is usually added near the end of the comminution process to minimize extraction.

A drop in pH to less than 5.3 inhibits the growth of spoilage organisms and potential pathogens such as *Staphylococcus,* particularly with the application of natural smoke on the surface of the product. Since the myofibrillar proteins have an isoelectric point at or near 5.1, reducing the pH to near or at the isoelectric point results in the proteins having the lowest water-holding capacity, which facilitates the drying process.

Cold temperatures during comminution are vital in order to provide a clean cut and, in particular, to avoid fat smearing. Fat smearing coats the surface of the lean which interferes with the ability of the lean particles to bind together during fermentation. Desirable temperatures of the lean meat during particle size reduction are approximately −3.3 to −1.1°C (28°F) and slightly lower for the fat meats.

Fermentation can be accomplished using starter cultures, backslopping, or by natural fermentation. Use of starter cultures is by far the most common method used because of the greater control and more consistent results, reduction in fermentation time, decreased opportunity for spoilage, and improved food safety. Frozen concentrates or lyophilized starter cultures can be purchased that consist of selected species of *Lactobacillus* (*L. plantarum*), *Leuconostoc, Micrococcus* (*M. aurantiacus*), and/or *Pediococcus* (*P. cerevisiae, P. acidilactici*).

Backslopping entails saving a portion of a previous fermented batch that has viable lactic acid bacteria for incorporation into a new batch of meat on the following day. Another practice used in the industry similar to backslopping is to prepare a small batch of inoculated ground meat (25–30% fat) to serve as seeding for subsequent production. Fresh meat is ground and mixed with the appropriate starter culture and sugar. This mix is stored for 2–3 days (depending on storage temperature) prior to use to verify the viability of the starter by tracking changes in pH. Once progress of fermentation has been verified, a portion of this batch can be incorporated as a seeding for regular production. Natural fermentation involves utilizing the natural microflora that exist in the raw meat. Very inconsistent results are obtained using this technique as it is somewhat by chance that the appropriate inherent organisms will effectively outcompete the growth of spoilage organisms.

Regardless of the method selected, a source of easily metabolizable carbohydrate such as dextrose is added to permit the production of sufficient levels of lactic acid. Inoculated, comminuted product is stuffed into casings and placed in "greening rooms" or smokehouses where the temperature of 23.9–37.8°C (75–100°F) and humidity (80–90% relative humidity) are controlled according to the specific needs of that product to facilitate fermentation. Increasing the incubation temperature can greatly increase the rate of fermentation. The end point of fermentation (desired pH) can be controlled by either limiting the amount of substrate or via heat processing. Lactic acid bacteria are inactivated at temperatures above 48.9°C (120°F) and some at lower temperatures depending on the culture. If the end point is controlled by heat processing, it is possible to maintain a certain level of nonfermented sugar to provide some sweetness.

Although not considered a fermentation, glucono delta lactone (GDL) can be added to produce the tangy flavor similar to lactic acid associated with microbially fermented sausages. In the presence of moisture, GDL is converted to gluconic acid upon heating. GDL is permitted in fermented sausages at levels of 0.5–1%, depending on the specific product. One potential disadvantage of using chemical acidulation is that the drop in pH may be too rapid and result in the coagulation of the proteins before the fat is stabilized, thereby resulting in the product fatting out during thermal processing. Encapsulation of GDL with a food-grade material with different responses to acid conditions and heat has been shown to overcome this problem.

Some products are cold smoked before entering drying rooms. Cold smoke dries the surface somewhat and, very importantly, some of the smoke volatiles that are deposited on the surface have bacteriostatic and bacteriocidal properties. Drying should be gradual and uniform. Ideally, the drying rate at the surface should be slightly greater than the rate of moisture migration from the interior of the product to the surface. If the rate of water migration is faster than the drying rate, then the surface of the product will be too wet and result in mold and possibly bacterial growth. Mold growth is the most common problem in the production of fermented and dried sausages. To inhibit mold growth the sausage can be dipped in a solution of 2.5% potassium sorbate. Fermented and dried sausages do not need to be heat processed, whereas semidry sausages must be heated to a minimum of 58.3°C (137°F), particularly if noncertified trichinae-free pork is used. The extent of drying (30–40% moisture for dry sausages and 50% for semidry sausages) and the time (10–150 days) necessary to reach the required dryness end point varies according to the specific sausage. The end point is specified as the moisture/protein (M/P) ratio for that particular product. Dry salami must be dried to an M/P of 1.9, whereas pepperoni must achieve 1.6. The low moisture content of dry sausage in combination with the low pH makes these products shelf-stable (not requiring refrigeration).

Process and Product Optimization

Raw Material Analysis

To achieve the highest degree of consistency relative to manufacturing one batch of the same product to the next, a *representative* sample of each of the raw materials that are used to formulate the product must be collected. This is one of the most critical steps in formulating a further processed meat item. Consider a processor that is preparing to manufacture a 2000-lb batch of frankfurters. Suppose that the processor only has two raw meat materials (4000 lbs of lean pork trim and 1000 lbs of fat trim) to manufacture the product. From each of these two meat sources a representative sample must be collected and analyzed chemically. Often the meat sources are not previously ground and mixed, but rather pieces and strips of meat with obviously different proportions of lean to fat. Ideally, the meat should be coarse ground and appropriately mixed before a sample is taken. Regardless of what preparatory steps have been taken, the chemical analysis amounts to utilizing a 2–3-g sample (preferably in triplicate) to predict the moisture content of the corresponding meat source. Suppose the moisture was performed in triplicate (3×3 g, 0.02 lb) for the lean pork trim, then the samples analyzed represents only 5 ppm ($0.02 \div 4000$) of the 4000 lbs.

Many muscle food processors determine chemically the moisture, fat, and protein content of their raw materials. However, not all processors have the instrumentation or technical expertise to perform all three. In addition, time may be a limiting factor. Fortunately, there are ways to estimate some of the chemical traits given certain information. A given is that the chemical composition of muscle tissue consists of moisture, fat, protein, and ash. Furthermore, the total of these four equals 100%. Depending on the raw material, ash typically represents less than 1% (0.4–0.6%). Therefore, if the processor has the capabilities to determine two of the components besides ash, then the remaining component can be estimated by the difference. Suppose the moisture content of the lean pork trim was 69% and the fat content was 11%, estimate for the protein content.

Given: moisture (M) + fat (F) + protein (P) + ash = 100%

$$69\% + 11\% + P + 0.5\% = 100\%$$

$$80.5\% + P = 100\%$$

then the protein content equals 19.5%

Another very important guideline is the M/P ratio. With either sufficient documentation of the chemical analysis of typical raw materials utilized within a given processing facility or through the use of table values, it is possible to estimate the chemical composition of a raw material if either the moisture or protein content is known. For instance, a typical M/P ratio for lean beef trim is 3.5. Therefore, if the moisture content was determined to be 67%, then the protein content could be determined as follows:

M/P ratio for lean beef = 3.5

$$P = M \div 3.5$$

$$P = 67\% \div 3.5 = 19.1\% \text{ protein}$$

Substituting these values into $M + F + P + \text{Ash} = 100$

$$F = 100 - 67 - 19.1 - 0.5 = 13.4\% \text{ fat}$$

Estimating the composition as described previously should not be considered an equivalent substitution for complete chemical analysis. However, these two guidelines (total of M, F, P, and ash equals 100) and the M/P ratio are very useful tools in monitoring the results from a complete chemical analysis.

Postmortem age and quality (temperature and microbial status) affect the functionality of meat proteins. Meat trim and pieces that have been temperature abused or have been refrigerated for an extended period of time will not provide the same functional attributes as fresh meat. Such meat will contribute less bind and water-holding capacity than fresh meat. Pale, soft, exudative (PSE) pork is inferior for both curing and sausage manufacture. PSE results from a rapid decline in pH (accelerated anaerobic glycolysis) early postmortem when the carcass temperature is still elevated (Fig. 5.2). The combination of low pH and elevated temperatures causes considerable protein denaturation. This differs from the normal decline, which is more gradual. In some cases the pH may exhibit only a slight decline and maintain a high ultimate pH. This latter case results in beef that is described as dark, firm, and dry (DFD). DFD meat actually has superior water-holding capacity as the high-alkaline conditions protect the proteins from denaturation and the high ultimate pH is further away from the isoelectric point of the salt-soluble proteins than in the normal tissue pH.

The classical description of a PSE ham is one that has a very pale, two-toned appearance of the surface where the ham is separated from the loin; soft as a result of a low water-holding ability and protein denaturation; and exudative to the point that water exudes from the cut surface. This description is more indicative of an extreme case of PSE. PSE actually occurs in various degrees and in some cases is visually difficult to separate from a normal ham. A higher percentage of PSE carcasses is seen with extreme changes in environmental conditions.

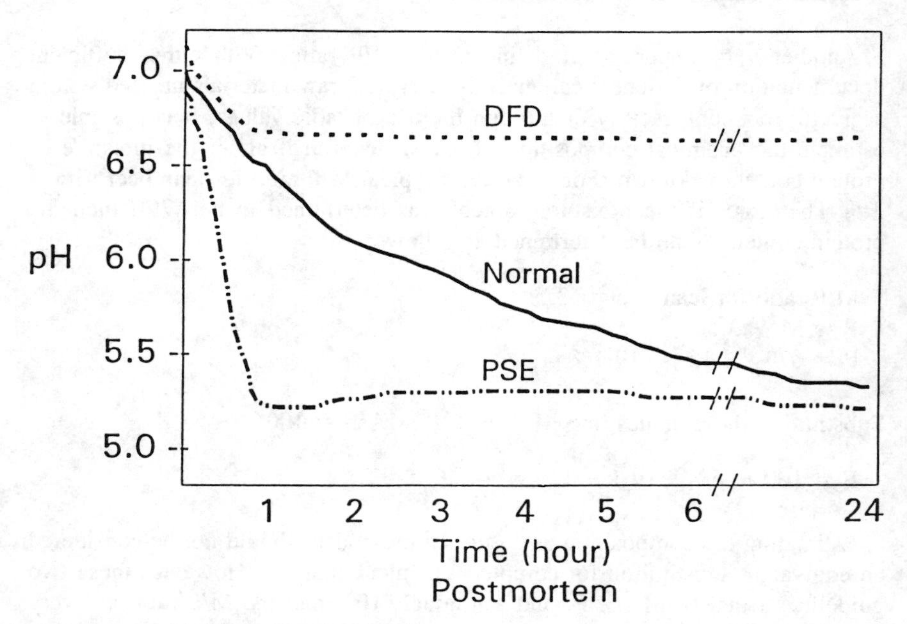

Figure 5.2. Various rates of pH decline.

During the winter months, pigs become acclimated to the cold temperatures. If these pigs were shipped to the abattoir in the spring during the first warm spell, they would be exposed to considerable stress. Because such environmental changes are inevitable, from a processing standpoint, if possible it may be advantageous to alter the formulation used to manufacture certain products to avoid product failure. An increase in sodium phosphate and salt may be beneficial. Suppose a processor has a commodity-type product that has been formulated to achieve the highest yield and lowest cost. This type of product cannot tolerate much variability in raw material attributes (protein function), ingredient incorporation, thermal processing, and storage characteristics. Under ideal conditions, manufacture of the product may be very acceptable. However, if a major climatic change occurs prior to the animals having enough time to adjust to these changes, even the "normal" pigs may produce meat that is clearly inferior in functionality. The first indication that a loss in function has occurred may be reduced processing yields. In addition, actual product defects may be observed. Boiled hams, which represent cured pieces of meat that are recombined, may have been formulated to contain a relatively low protein content and, therefore, may exhibit extreme purge accumulation. If this defect is adequately severe, the boiled ham can be completely surrounded by liquid and appear to float in the package. This defect may further be accentuated if the film used does not have the appropriate characteristics. The film must have an appropriate degree of shrinkability and formability and maintain a certain degree of adherence to the product after heat processing. If the film lacks sufficient adherence ability, the voids created will promote fluid accumulation. On the other hand, if the film adheres too tightly to the exterior of the product, excess ham will be peeled away when the product is removed from the package.

Formulation

Product yield information can be a useful management tool to determine whether or not production goals and product consistency have been achieved. When formulating a product, two of the processing components that greatly affect the final composition of the finished product are the cooking and chilling losses. If the finished composition is not tightly controlled, the fat and the texture of that product may not meet expectations and the product may be in violation relative to the fat content and USDA added water.

Product yield is important to production management because the cost to manufacture the product is directly associated with yield. Product yield can be used as a guide to determine if the product has met the specifications for that product. If the yield is higher or lower than normal, this could have a dramatic impact on the composition of the finished product. A low yield could imply that inferior quality raw meat ingredients were used, the amount of nonmeat ingredients added was incorrect (low pump), transfer problems existed, product was overcooked, and/or delayed chilling occurred prior to packaging. A low yield could mean that the product may be drier than normal, and if the loss were related to overcooking, then most of the loss would be associated with moisture loss. Therefore, a concentrating effect on the remaining ingredients would occur. The salt concentration may be increased to a level of saltiness that is distinctly different from the normal product and, therefore, render this batch unacceptable.

If the overall yield is not within an acceptable range, the next step would be to review the individual manufacturing steps for that production run to determine where the variation may have occurred so that future corrective measures can be made. A typical example of production records could be as follows for a processor that manufactures a cured and smoked boneless ham:

	Pounds	
1. Tunnel boned hams	1000	
(green weight)		(100%)
2. Pumped weight	1300	
(pump)		(30%)
3. Tumbled weight	1295	
(tumble loss)		(0.4%)
4. Stuffed weight	1278	
(stuffing loss)		(1.3%)
5. Cooked weight	1120	
(cooking loss)		(12.4%)
6. Chilled weight	1101	
(chill loss)		(1.7%)
7. Overall yield	1101	
(green to finish)		(110.1%)

Relative to formulating a new product, some of the key elements that an inspector will want to verify include (1) the percentage pump to be used, (2) percentage drain if applicable, and (3) the usage level of the restricted ingredients (phosphate, nitrite, erythorbate). Sometimes, processors will want to modify an existing formulation but in a manner that does not require a label change relative to the order of predominance of ingredients used in a product. It may simply be in response to marketing requesting that the sodium content be reduced and/or the sugar content increased. Any changes in the formulation can affect the percentage of the other ingredients and may result in a violation if not closely monitored.

The following problems illustrate several mathematical calculations that given a minimum of information a processor can determine the (1) percentages of various ingredients in the finished product on a finished weight (FW) basis with provisions included to adjust for smokehouse and chill losses, (2) meat block (MB, meat components only), which corresponds to the percentage limit of many restricted ingredients, and (3) pickle, which allows a pickle maker to determine the amount of each ingredient necessary to prepare the pickle.

Assume that a processor wants to formulate an injected, cured product to contain various ingredients based on the finished weight and some based on the meat block weight in the case of certain restricted ingredients. In addition, similar products are known to have a 12% cooking loss and 3% chilling loss.

1. Formulate

Ingredient	Percentage Pickle	Percentage Meat Block (% MB)	Percentage Finished Wt (%FW)	Smoke loss (%)	Chill Loss (%)
Water			7.0[a]	12[a]	3[a]
Salt			2.0[a]		
Sugar			1.5[a]		
Phosphate		0.500[a]			
Nitrite		0.015[a]			
Erythorbate		0.055[a]			
Totals	100%				

[a] Given.

2. (a) Unknowns: total %FW of all ingredients, yield, %pump.
   (b) Assumption: all loss is moisture only.
3. Calculations
   (1) Determine the total percentage of ingredients in the finished product.

$$\text{pFW} = \text{partial sum of ingredients listed under \%FW} = 7.0 + 2.0 + 1.5 = 10.5$$

$$\text{pMB} = \text{partial sum of ingredients listed under \%MB} = 0.50 + 0.015 + 0.055 = 0.57$$

$$\text{Total \%FW} = \text{pFW} + X$$

$$X = \frac{100 - \text{pFW}}{100 \div \text{pMB} + 1}$$

$$X = (100 - 10.5)(100/0.57 + 1)^{-1} = 0.507$$

$$= 10.5 + 0.507 = 11.007$$

   (2)

$$\text{Yield} = \frac{100}{100 - \text{total \% FW}} \times 100 = 112.37$$

   (3) Once the yield is known the %MB of the ingredients (except water) can be calculated from the %FW.

%MB salt	= %FW salt × yield/100	= 2.2474
%MB sugar	= %FW sugar × yield/100	= 1.6856
%MB PO_4		0.5000
%MB nitrite		0.0156
%MB erythorbate		0.0550
		4.5036

(4) To determine %MB water, first calculate the %Pump:

$$\%\text{Pump} = \frac{\left(\dfrac{\text{yield}}{1 - \text{chill loss\%} \div 100}\right)}{1 - \text{smoke loss\%} \div 100} - 100$$

$$\%\text{MB water} = \%\text{Pump} - \text{all }\%\text{MB} = 31.64 - 4.504 = 27.136\%$$

(5) Once all %MBs are known, calculate % in pickle by (%MB ingredient) $(\%\text{Pump}/100)^{-1}$:

%water in pickle	27.136/0.3164 =	85.765
%salt in pickle	2.2474/0.3164 =	7.103
%sugar in pickle	1.6856/0.3164 =	5.327
$\%PO_4$ in pickle	0.5000/0.3164 =	1.580
%nitrite in pickle	0.0156/0.3164 =	0.049
%erythorbate in pickle	0.0550/0.3164 =	0.174
		≈ 100

Depending on the requirements of the finished product, a cured pork product could be pumped 10–40%. USDA regulations are now in place that specify a minimum Protein Fat Free (PFF) value that corresponds to a particular finished product composition and consequently the amount of ingredients that can be added (USDA 1984). Examples of specific PFF values for cured pork products include cooked ham (20.5), cooked ham with natural juices (18.5), and cooked ham water added (17.0). PFF values also are specified for various types of ham patties, chopped ham, and uncooked cured products. PFF is defined as

$$\text{PFF} = \frac{\%\text{ Meat protein}}{100 - \%\text{ Fat}} \times 100$$

To achieve lower PFF values, more nonmeat ingredients must be incorporated into the product. In effect, when more nonmeat ingredients are added, the protein is diluted to a greater extent. Cured pork products with PFF values lower than 17.0 (cooked ham) or 16.5 (cooked shoulder) can be produced. These products must be labeled with the common and usual name (cooked ham or cooked shoulder) followed by the qualifying statement "and water product—*X*% of weight is added ingredients." In order to produce this type of product, processors must have at least a partial quality control program approved by the USDA. This program details which personnel are responsible for preparing and monitoring the program, what the product is to contain, how the product is manufactured, what critical processing points (e.g., pump percentage, end-point temperature) are to be used to assure that the product meets manufacturing specifications, and, in the event of noncompliance, what corrective action is to be taken. This type of product is regulated based on the PFF value and on a yield basis once the lower PFF limit has been exceeded. The processing plant first establishes what formulation and process are sufficient to achieve the lower limit (e.g., < 17.0 for ham and water product—*X%* of weight is added ingredients). Each time the product is produced, the manufacturing information, particularly the starting raw meat weight and the finished packaged weight of the product, must be documented to determine the compliance with the approved program and the stated *X%* of the weight that represents added ingredients. Assume that a processor pumped 5000 lbs of boneless ham with a pickle to add 30% weight and then massaged the pumped product in combination with 2.8% dextrose based on the pumped weight. After the product was cooked and chilled the final weight of finished product was determined to be 6667 lbs. The *X%* of the weight that the added ingredients represents is determined by subtracting the weight of the meat from the finished product (6667 − 5000 = 1667 lbs of added ingredient) and dividing this number by the weight of the finished product (1667/6667 × 100 = 25%). The 25 represents the *X* that must be declared on the label. This example assumes that minimal transfer or processing losses have occurred. Such a product could have been water cooked in a moisture-impermeable film using cook-in-a-bag technology. Products cooked in fibrous casings would lose some moisture during heat processing, and, as such, this loss would have to be accounted for in order to achieve the targeted percentage of added ingredients.

In 1988, the USDA began regulating the amount of added substance that products such as cooked sausages could contain, wherein the combination of added water (AW) and fat (limited to 30%) could not exceed 40% in the finished product (Fig. 5.3). Although many combinations of fat and AW could be formulated, there are some raw material and technical limitations. Typically, a mini-

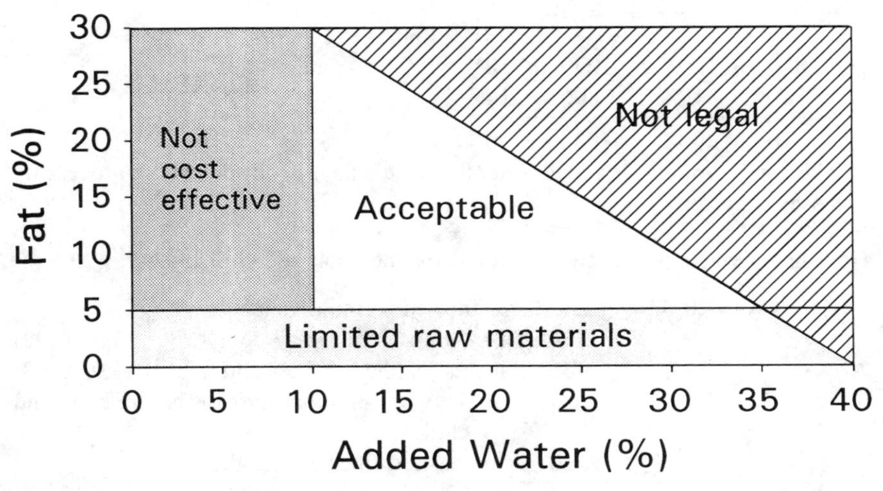

Figure 5.3. Forty percent sausage rule. (Adapted from J. R. Claus, *Proceedings 44th Reciprocal Meat Conference*. American Meat Science Association; Chicago, IL., 1991, p. 94).

mum of 10% water is added to the formulation to facilitate batter formation and ingredient distribution. In addition, there are limited sources of lean meats with less than 10% fat and such meats would be very costly. In general, substituting water for fat in the formulation can reduce the rubberiness normally associated with fat reduction alone. However, significant reductions in the fat content while maximizing the AW may lead to problems with firmness.

AW is defined as % moisture − (4 × % protein). Therefore, a cooked frankfurter containing 62% moisture and 12% protein would have an AW of 14% (62 − 4 × 12) and as such should not contain more than 26% fat (26 + 14 = 40%). A question that can be asked is how much water (X) is required to obtain a 10% added water in the final product given

1. 100-lb meat block
   17% protein (by analysis)
   60% moisture (by analysis)
2. 3 lb dry cure and spices, etc.
3. 8% cook loss (assume only water)
4. 2% chill loss (assume only water)

USDA Regulation: Added water = % moisture − (4 × % protein)

$$\frac{60 + X}{(100 + 3 + X)} \times 100 - 4 \times \frac{17}{(100 + 3 + X)} \times 100 = 10$$

$$\frac{6000 + 100X}{103 + X} - \frac{6800}{103 + X} = 10$$

$$\frac{6000 + 100X - 6800}{103 + X} = 10$$

$$\begin{array}{rcl} -800 + 100X &=& 1030 + 10X \\ 800 - 10X & & 800 - 10X \\ \hline 0 \quad 90X &=& 1830 \quad 0 \end{array}$$

$20.33 = X$, amount of added water that would have to be added if no processing losses occurred.

How much water is needed to account for the cook and chill losses?

100 + 3 + 20.33 = 123.33 lbs finished formula weight
134.05 lbs (adjustment for 8% cook loss, 123.33/0.92)
136.79 lbs (adjustment for 2% chill loss, 134.05/0.98)
This weight represents lbs before cook and chill.
−123.33 lbs desired finished weight

13.46 lbs water (to account for cook and chill losses)

Total water added to formulation:
20.33 + 13.46 = 33.79 lbs water

The **Pearson square method** is widely used in the meat industry to determine the quantities of various raw materials that are necessary to blend two or more raw materials together in order to achieve a particular composition of a given component. Often processors need to determine how many pounds of two or more materials are necessary to achieve a particular fat target. For instance, a beef processor wants to produce a sausage that consists of 95% meat (meat block) and 5% nonmeat ingredients (e.g., soy protein, flavoring, salt) and achieve a fat content of 30%. Assume that this processor has two meat sources; lean meat (15% fat) and a fat meat (60% fat). To determine the amount of each necessary, the target % fat when the two meat sources are blended together needs to be determined. This is done by dividing the finished fat percent by the percent the meat block (lean meat plus fat meat) represents as follows:

$$\frac{30}{95} \times 100 = 31.6\% \text{ fat target of meat block only}$$

Using the Pearson square (Fig. 5.4), the percentage of the lean and fat are positioned at the corners of the square with the meat block fat target in the center. Next, the absolute difference between the meat block fat target (31.6) and each of the two meat sources is determined and positioned on the opposite, diagonal corner. The sum of the differences is determined (45.0) and the percentage that each represents is listed to the right (note, the sum of these should equal 100%). Finally, these percentages are used to determine the pounds of each meat necessary to achieve the fat target (59.95 lbs of 15% lean and 35.05 lbs of 60% fat meat).

A second approach utilizes **matrix algebra.** This approach can easily be used to solve more comprehensive formulation problems. In fact, matrix algebra serves as the basis for linear programming used in prepared computer programs such as Least Cost Formulator™ (Least Cost Formulations Ltd. Inc., Virginia Beach, VA). These programs permit the formulation of products with numerous requirements (fat content, bind, color, salt content, added water, species, cost, etc.)

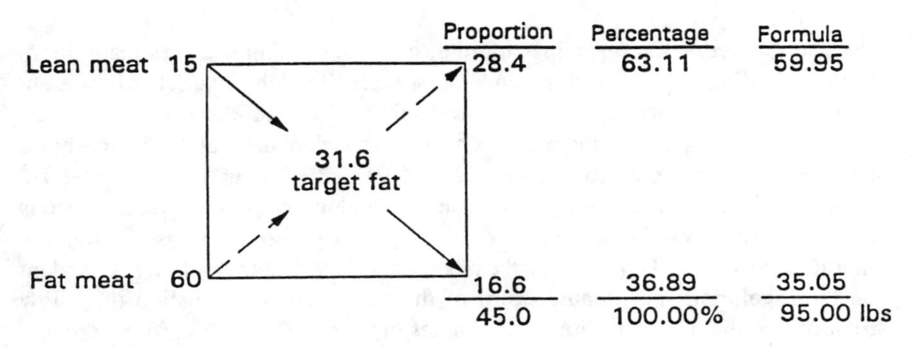

Figure 5.4. Pearson square.

made from raw material having several attributes (moisture, fat, protein, bind, cost/lb).

Matrix approach to solving the formulation problem (L = lean meat and F = fat meat):

$$
\begin{array}{lrl}
\text{Equation 1} & L + F & = 95.00 \\
\text{Equation 2} & 0.15L + 0.60F & = 30.00 \\
\text{Eq. 1} \times -0.15 & -0.15L - 0.15F & -14.25 \\
\hline
\text{Sum of two above} & 0.00 + 0.45F & = 15.75 \\
\text{Solve for F} & F & = 15.75 \div 0.45 = 35.00 \\
\text{Solve for L} & L + 35.00 & = 95.00 \\
 & L & = 60.00
\end{array}
$$

Verification: Substitute values for L and F into Equation 21.

$$0.15 \times 60 + 0.60 \times 35 =$$
$$9 + 21 = 30 \text{ lbs fat}$$

The **Li method** developed by Xingchu Li (Olson 1990) is another procedure that can be used to solve the same problem and avoids the rounding errors inherent with the Pearson square method. The Li method uses shortcut procedures based on matrix algebra.

Most of the hand-held electronic calculators or computers with sufficient memory can be programmed to solve simple blending problems. However, as indicated previously, linear programming offers the capability to formulate products with multiple requirements from a tremendous number of meat and nonmeat ingredients. These programs enable the processor to specify many requirements or constraints on the finished product. Table 5.1 lists several examples of such specifications.

Particle Size Manipulation

Grinding

Particle size reduction (grinding, milling, chopping, flaking) serves many functions, including improvement in tenderness, ability to blend a variety of different ingredients and meats together, and production of new products.

Grinding involves forcing portions of meat though a metal plate that has holes bored to a specific diameter (e.g., 3.2, 6.4, 9.5, 12.7 mm; 1/8, 1/4, 3/8, 1/2 in. respectively). Grinding is a particle size reduction step in which meat is continuously forced through the holes of the plate between the arms of a rotating multibladed knife. The knife cuts the pieces to a given length, partially dependent on the revolutions per minute (rpm) of the knife. The combination of a slow rotation and the relatively large size of holes in the plate produces ground product

Table 5.1. Examples of Various Requirements for Further Processed Meats

Item	Requirements	Basis[a]
Fat content	>15% and <30%	FW
Protein content	≥11%	FW
Collagen	≤4.0%	FW
Salt	2.5%	FW
Bind[b]	>150	FW
Color[c]	>375	FW
Added water (AW)[d]	>15%	FW
40% Rule (fat + AW)	≤40%	FW
Sodium nitrite (cure)	156 ppm	BW
Sodium erythorbate	550 ppm	BW
Species[e]	B>P	BW
	C<10%	BW

[a] Basis: FW = finished weight (cooked product); BW = combined weight of all meat (lean and fat) items.

[b] Bind: reported as Georgia bind units.

[c] Color: units based on myoglobin content.

[d] Added water: AW = %moisture − (4 × %protein).

[e] Species: B = beef, P = pork, C = chicken.

that has distinct particles. Grinding is frequently the only particle size reduction process used to manufacture fresh breakfast sausage.

Low temperature and even partially frozen meat will result in a cleaner cut and more distinct particles that are less likely to have fat smearing. Other factors that influence how well the product is ground include the sharpness of the knife, flatness of the machined surface of the plate, how evenly these two pieces fit together, and pressure of the screw.

Milling and Chopping

Milling is similar to grinding in that a plate and knife are used. Milling equipment (emulsion mill) typically uses much smaller holes in the plate and operates at very high rpm. Milling equipment is suitable for preparing finely comminuted batters used to produce bologna and frankfurters.

Bowl choppers decrease particle size by a series of rotating knives that closely conform to the contour of the metal bowl that holds the meat and nonmeat ingredients. Bowl choppers can both mix and cut as the bowl rotates. Some bowl choppers are fitted with vacuum capabilities to minimize air incorporation and the ability to control product temperature. Bowl choppers can be used to manufacture both coarse and finely comminuted products.

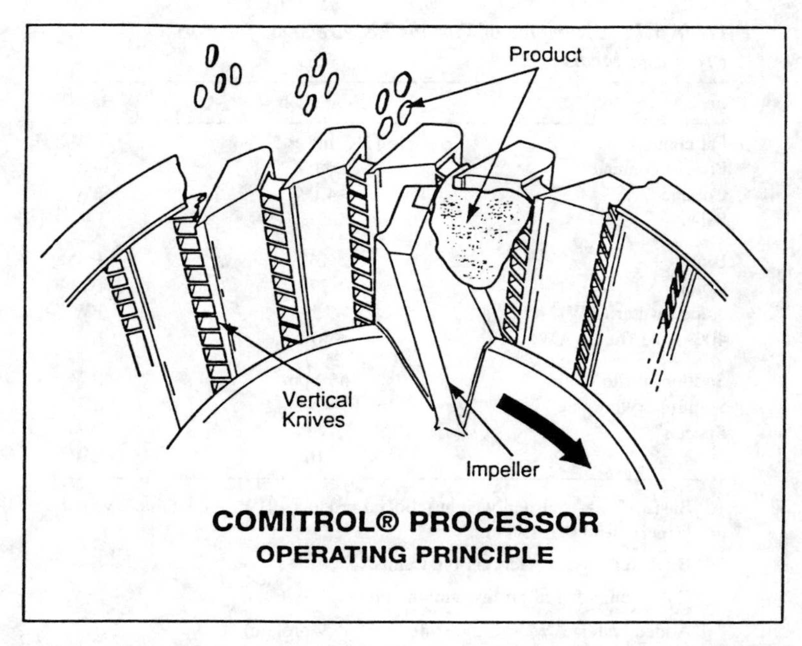

Figure 5.5. Flaker operating principle. (Courtesy of Urschel Laboratories, Inc.)

Flaking

Flaking is a process in which frozen and tempered meat pieces are cut into flat slices (flakes) as a result of the force generated by a rotating impeller (Fig. 5.5) at high speeds (3000–5000 rpm). The flakes can be mixed with salt and other ingredients prior to being stuffed into a casing to form a log. After a period of frozen storage the shape of the log can be modified by being processed in a hydraulic press that creates a particular shape. The log can then be cleaved into chops or steaks of various thickness.

Mechanical Separation of Meat

Mechanical separation of lean tissue from various nonedible tissues has been successfully utilized for a variety of muscle foods including fish, chicken, and red meats. The separation principle of some equipment relies on a perforated filter to screen flesh from nonflesh components. A wide range of machinery for fish separation is available. Fish processed through a deboner can yield from 62% to 93% of original weight as edible meat. There are two types of meat–bone separators. The principle of one type is that the fish is pushed against the outside of a rotating perforated cylinder. The pressure on the product forces meat through the perforations, leaving bones, skins, and scales on the outside. The second type, also used for other muscle foods, forces the product, inside the perforated cylinder, through the perforations under pressure. The soft flesh is pushed through the perforations with the undesirable materials sent to the end of the cylinder to be removed. Excessive pressure exertion on fish during the mincing process will permit bones and kidney to pass through the perforated cylinder and to mix with the minced meat.

The design of the machines has been improved to solve some of the problems with mincing. The dead spaces from the flesh-carrying areas have been removed to decrease the spoilage rates of minces. Ferric contamination of mince, which accelerates oxidative fat degradation, has been eliminated by replacing the parts with stainless-steel and nonmetallic materials. Oxidation of the mince product can further be reduced by separating under water or under washing solutions.

Tumbling and Massaging

Purpose

Tumbling and massaging are methods used to accelerate the curing process, facilitate extraction of salt-soluble proteins, improve texture, bind, water-holding capacity, and product yield. Without the advent of massagers or tumblers the higher processing yields and lower production costs enjoyed today in many sectioned and formed meats could not be achieved.

Equipment and Process

Tumblers are designed so that meat chunks are continuously elevated and dropped to provide physical impact on the falling chunks. Some degree of massaging also occurs as the chunks slide on each other as the tumbler rotates. The metal units that actually contain the meat of some tumblers are rotated end over end for a period of time to create the necessary tumbling action. A **massager** can essentially be a mixer motor mounted above the meat container attached to a vertical drive shaft that has several large horizontally mounted, rounded, paddlelike arms located at various depths in the container. Some massagers are cylindrical-shaped drums that have helical flights which gradually move, mix, and massage the product under vacuum as the entire drum revolves (Fig. 5.6). Some equipment has been designed to combine both actions of massaging and tumbling in addition to vacuum capabilities. Depending on the design of the equipment, some can be used to physically manipulate cured pork bellies, bone-in products, boneless chunks of meat, and in some cases coarsely ground meats. However, tumblers typically provide somewhat more destructive action because of the impact force generated from tumbling and, therefore, cannot be used on all products. Pumped and tumbled bellies produce a more uniform product as the pickle uptake can be more tightly controlled and pickle pockets can be

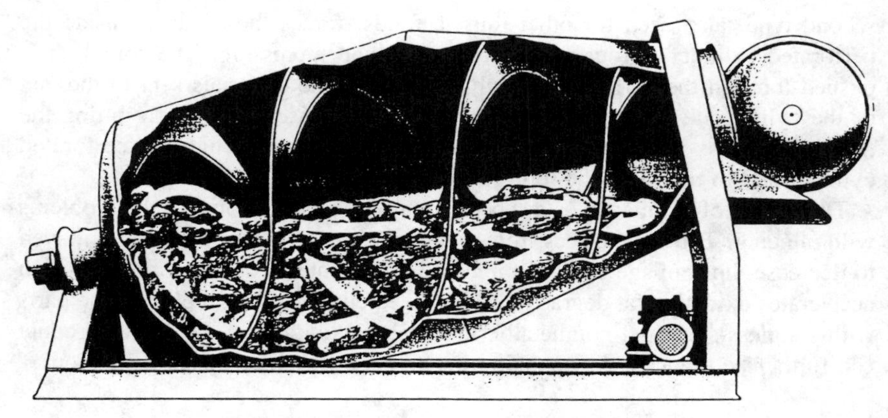

Figure 5.6. Challenge massager. (Courtesy of Challenge-RMF Inc.)

reduced. The tighter control of pickle uptake results from the ability to pump the bellies at or somewhat below the target pump and then adjusting the product to the exact percent pump by adding pickle directly to the tumbler. In contrast, nontumbled bellies are pumped to meet the target percent pump and then immediately combed and hung on smokehouse trees during which time some of the pickle is lost prior to heat processing regardless of whether or not the product was underpumped or overpumped. Massaging and tumbling both offer this excellent control over pickle uptake and consequently more control over product yield and finished composition. Tumbling or massaging time depends on the particular product being produced and the design of the equipment. Highly extended (high pickle injection), sectioned, and formed products that are to be water cooked in a moisture impermeable film may require 8–12 h of massaging. Restructured roast beef may only require 30–45 min. Semiboneless products that could possibly benefit from extended massaging times cannot be massaged too long because the semi-intact nature of that type of product could be destroyed. Often, the cycles used with this equipment are not continuous but rather cycled on for 10, 15, or 20 min and then off for a 30-, 15-, or 10-min rest period, respectively. The rest periods in noncontinuous cycles are used to provide time for the pickle to equilibrate within the meat.

Thermal Processing and Smoking

Thermal Processing

Smokehouse Processing

Smokehouse (processing ovens) processing provides more flexibility than water cooking as to the types of products that can be cooked. In addition, smokehouses are designed to control humidity, heating rate, conditions necessary for appropriate smoke deposition, surface appearance, and product composition.

Relative humidity (RH) is the relationship between the dry bulb (DB) and wet bulb (WB) temperature. All smokehouses have two thermocouples (DB and WB) positioned in the interior. The WB has a sock (cloth material) covering the sensing end. The sock is long enough to hang down into a water reservoir so that moisture can osmotically travel to the surface of the thermocouple. As heat is increased in the smokehouse, a temperature rise is recorded by both the DB and WB. However, at the point where the moisture on the sock begins to evaporate, the WB temperature will be lower than the DB temperature. *The greater the difference between the DB and WB temperature, the lower the RH.* When the DB and WB temperatures are equal, the RH is 100%. Controlling the RH can be successfully used to alter the amount of smoke deposition and color of a smoked product. Processors usually associate high RH values with faster cooking rates. However, the surface temperature of the product is a better indicator of the rate of heating. To illustrate this with all other factors the same, if two boneless products were cooked [76.7°C DB (170°F) and 62.8°C WB (145°F)] to the same end-point temperature of 64.4°C (148°F), a product cooked in a moisture impermeable casing would take less time to reach the same end-point temperature as another in a moisture-permeable casing. The reason for the difference is the product in the moisture-impermeable casing would have a higher surface temperature because moisture could not migrate to the surface and provided evaporative cooling.

Canning and Pasteurization

Canning is a very effective means of preservation. To produce shelf-stable products, sufficient heat processing must be utilized to destroy spoilage and pathogenic microorganisms. Heat processing is only effective if the product is in an hermetically sealed container that maintains its integrity through processing, distribution, and ultimate utilization by the consumer. Potential microorganisms include bacteria, molds, and yeasts. Vegetative cells are those that can most easily be destroyed by heat. Some microorganisms have the ability to produce heat resistant spores. Subsequent outgrowth of the spores leads to the proliferation of vegetative cells that may result in product spoilage and potentially food-borne illnesses. In order to destroy certain spores, the product must be heat processed to at least 115.6°C (240°F) for extended periods of time. Factors that affect the survival of microorganisms during heat processing include salt, pH, moisture and fat content, nitrite, and level of microbial contamination. The potential danger and impact of botulism is a major concern in the manufacture of shelf-stable products. Therefore, manufacturers must take every precaution to assure that approved procedures are stringently followed. Specific procedures for canned meats, including corned beef hash, beef stew, chili con carne, vienna sausages, and meat balls with gravy, are well documented by Pearson and Tauber (1984).

The general steps in canning involve the selection of raw materials, precanning preparation, filling, vacuumization, can closure, retorting, and chilling. Prior to canning, the raw materials may be cured, seasoned, chunked, ground, and/or precooked. Cans or other containers should be clean prior to filling. Once the containers have been properly filled they are sealed under mechanical vacuum or through the utilization of steam in the container to create a vacuum once the product is cooled. After closure, the exterior of the container should be cleaned and the product immediately heat processed and then chilled with potable water.

The seafood canning industry primarily involves salmons, sardines, shads, tuna, herring, mackerels, cods, haddocks, roe, clams, and oysters. The order of importance of canned product is salmon, tuna, sardines, shrimp, clam products, mackerel, and oysters in descending order. These seven products account for 96% of the value of all canned fishery products.

Fish used for canning are eviscerated and trimmed, and sectioned to portions. They are placed in brine to draw the blood from the tissues. This gives the flesh the desirable firmness and flavor. The length of the brining period depends on the size and the fat content of the fish. Most fish are dried or cooked prior to canning; one exception is salmon. Sardines are either dried and then fried, or they are steamed and dried prior to the heating process. The packaged fish are passed through a steam chamber from 5–15 min prior to sealing. This exhausting step ensures the formation of a proper vacuum, as well as shortening the processing time. The canned fish are sterilized in retorts under pressure at 110°C (230°F) to 116°C (240°F). The exact processing time varies for each specific product. For example, 0.5 kg (1 lb) of salmon is processed from 80–90 min at 116°C (240°F) to 118°C (245°F); while 0.5 kg (1 lb) of tuna is sterilized at 116°C (240°F) for 70 min.

Pasteurization is heat processing at a lower temperature (usually below 100°C) than the sterilization process involved in canning. Pasteurization is defined in terms of lethality and not by a simple time–temperature process. Although there is less damage to the texture, color, flavor, and nutritional content of food, the shelf-life of the product is shorter than those processed for commercial sterilization. Pasteurization can be accomplished by heat processing in water. Water cooking can be used for products packaged in moisture-impermeable containers or films. Water cooking does not produce products that are shelf-stable but require refrigeration.

Boiled ham is a sectioned and formed product that can be packaged in a moisture-impermeable film that is tested to withstand the normal water temperatures of 71.1–79.4°C (160–175°F) during the cooking process. The U.S. Department of Agriculture requires approval of such film to ensure that under specified water-cook temperatures the film does not react or deposit undesirable components on the meat. Boiled ham is frequently cooked to an internal temperature of 65.6°C (150°F).

Shellfish meat from shrimp, oysters, crawfish, and lobsters undergo thermal processing which extends its shelf stability under refrigeration by preventing the growth of *Clostridium botulinum* spores. Products are cooked by submersion in vats or tanks of water for appropriate time and temperature specific for each product. Although the meat is cooked, freshness is still maintained. Excessive heat may cause meat discoloration. The product closest to the container has the greatest exposure to high heat and will show the highest degree of discoloration.

Shellfish meat is placed in containers (rigid or flexible films), sealed, and pasteurized to an internal temperature of 66–99°C (150–210°F), preferably 82–91°C (180–195°F) for 1–4 min at the geometric center of the container. It is recommended that a minimal thermal process having an *f*-value (85°C reference temperature) equal to 32 min (with a *Z*-value of 16) be applied. It may be possible to carry out pasteurization with exposure to microwave energy (frequency of 2450 MHz; power of 10–30 kW) with a process time of 3–5 min at 80°C and a holding time of 7 min at 80°C (Giese 1992). The pasteurized product is cooled to 13°C (55°F) within 3 h, and placed in refrigeration to further reduce to 2°C (36°F) or below within 18 h. Thermotolerant psychrotrophic obligate anaerobes (nonpathogenic *Clostridium* spp.) have been found to survive traditional pasteurization processes and grow at temperatures as low as 2°C (36°F). These microorganisms originate in the harvest waters and appear to be prevalent after severe hydrographic (as hurricanes) conditions. Although the spoilage from these microorganisms occurs only sporadically, processors should consider their possible presence and economic significance when developing pasteurization schedules.

Oysters are given a mild thermal process so the product can be sold as a fresh, rather than cooked product. Oysters are heated so the internal temperature reaches a maximum of 40°C (140°F), but the process usually does not exceed 49°C (120°F). Oyster meats maintain the firmness and fresh flavor while bacterial deterioration is retarded.

Smoking

Smoking gives the product a desirable color, aroma, and flavor. Smoking of meat products entails depositing various volatile substances on the surface as a result of burning hardwood shavings from fruitwoods, hickory, maple, or oak. There are hundreds of chemical compounds present in smoke and the type of chemicals produced in the smoke depends on the wood, water content of the wood material, and the temperature and method of heating. Two compounds reported in smoke, polycyclic aromatic hydrocarbons and nitrosamine, are considered carcinogenic. The presence of these compounds can be reduced by decreasing the temperature, using an electrostatic filter, or liquid smoke.

Smoking has a role in preservation via reducing oxidative rancidity and decreasing microbial populations on the surface of the meat. Many different phenols that are deposited on the surface have bacteriostatic and antioxidant properties. In addition, some of the compounds have bacteriocidal activity.

Smoking hardwood contributes to the desirable reddish-brown appearance of the surface as the carbonyls from the smoke react with amino acids to form furfural compounds. The phenols are largely responsible for the smoky flavor. Organic acids are involved in surface coagulation of the proteins and proper skin formation which is vital to casing peelability.

Cold and Hot

Dampened hardwood shavings can be augered onto the surface of a heated plate enclosed in a chamber connected to a smokehouse in order to generate smoke. Smoke generators are designed so that the amount of oxygen can be adjusted to control the rate at which the wood burns. Smoke can be deposited on the surface of meat during thermal processing (hot) or as a cold application on the surface of uncooked products such as fermented and dried sausages.

Humidity in the smokehouse affects smoke deposition and color. Higher humidities during the smoking stage will increase smoke deposition but may yield an undesirable muddy brown appearance. If the surface is too wet, a streaked appearance can occur on the surface, whereas a very dry surface as a result of too low humidity during smoking can result in poor smoke adherence, a pale color, and high shrinkage. A desirable range for appropriate smoke deposition and surface color is 30–38% relative humidity.

Smoke is generally applied to fish either during cold- or hot-smoking. During cold-smoking the temperature is maintained below 43°C (110°F). Heat and smoke are applied indirectly because an increase in temperature could damage the muscle texture. Heat and smoke are produced from different sources and supplied to the product by a fan or through a common outlet. Proper application of cold-smoke takes 18–24 h. Salmon, black cod (sablefish), and herring are commonly cold-smoked products. Drying is an important step in achieving a product with high quality. It must be slow and the humidity carefully controlled to produce the preferred external hardening.

A majority of fish products are hot-smoked. The temperature may reach up to 82°C (180°F), resulting in a fully cooked product. Sablefish is commonly smoked in an electrically heated smokehouse for 7 h, with the temperature at the last hour of processing reaching 100°C with an internal flesh temperature of 75°C. Cod fillets are usually placed in 70–80% brine for 4–10 min and smoked for 2–5 h at 27°C (80°F).

Liquid smoke is an alternative to natural smoke. Liquid smoke is manufactured from burning wood under commercial conditions such that selected volatile constituents can be recovered from the smoke and concentrated in a liquid form.

Liquid smoke is added to brine, applied as a dip postbrining, or sprayed on the product. Spraying liquid smoke on muscle foods is a more controlled application, imparting a more even distribution of the smoked flavor. For fish, the best results are obtained if the air velocity is low, air temperature is raised 5.6–7.2°C (10–15°F), and liquid smoke is applied in a fine mist. The major advantages of using liquid smoke include:

1. Ease of eliminating known carcinogenic compounds that can be found in smoke
2. Environmentally more compatible for processing facilities located near populated areas
3. Can be incorporated into the product, whereas natural smoke only has limited surface penetration (1–3 mm) in part because of the skin formation
4. Application (spraying on surface) is faster

Liquid smoke does have some limitations such as potential differences in the flavor and color of the finished product and the higher cost of the concentrate compared to sawdust. Also, processors have to purchase and install the equipment necessary to uniformly apply the liquid to the surface of the product.

Smoke-flavored additives may also be added to impart the smoked flavor. There are three kinds of additives: synthetic smoke flavors, natural smoke flavors, and nonsmoke related substances, such as yeast components with smoke flavor and aroma characteristics.

Processed Meat Manufacturing

Fresh Ground Sausage

Fresh ground pork sausage is one of the simplest types of processed meats to manufacture. Frozen lean is processed though a flaker and ground with various chilled (2–4°C) pork trimmings (60–70% lean) and 2% ice (optional; for temperature control to facilitate grinding and avoiding smearing) based on the meat weight through a 2.54-cm (1-in) plate and then transferred to a mixer grinder. The ground meat is mixed (3 min) with 1.7% salt, 0.35% sugar, 0.4% white pepper, 0.10% mace, and 0.10% sage based on the meat weight prior to being ground through a 0.64-cm (0.25-in.) plate. The ground mix is stuffed into various fibrous cellulose or collagen casings and made into links. Casings can range in diameter from 19 to 30 mm. Although linking machines can be adjusted to produce a certain length, some variation can occur relative to the delay time that the stuffed casings are stored prior to the linking operation. The variation in length is due to a softening effect moisture has on some casing materials which then affects how the linker processes the stuffed casing. Therefore, for consistency, the time between stuffing and linking should be regulated.

Because fresh sausage does not include nitrite, care should be taken to avoid possible cross-contamination with nitrite-containing processed meats and should either be manufactured before those containing nitrite or after all equipment has been thoroughly cleaned. Fresh sausage must be cooked prior to being consumed.

Cured, Noncomminuted Meats

Cured, non-comminuted meats include ham, turkey ham, bacon, corned beef, and beef jerky. Because ham and bacon represent a sizable contribution to the total U.S. tonnage of processed meats, examples of the manufacturing procedures for these products will be discussed.

Cured and smoked bacon is produced from pork bellies. Cured and smoked bellies are referred to as slabs and can be sliced or cut into smaller chunks for the retail market. Sliced product can be sliced so that the individual slices mostly overlap one another (shingled) and then vacuum-packaged to extend the shelf-life. Slabs also can be sliced onto sheets of food grade approved paper so that none of the slices overlap. This is accomplished by slicing the slab onto a continuous dispensing roll of paper positioned directly at and under the slicer blade. The distance between the slices can be regulated by the speed of the conveyor that carries the paper. This produces what is referred to as layout bacon for the hotel, restaurant, and institutional (HRI) trade. Typically, there are 8–10 slices per sheet and the sheets are stacked one on top of the other until the desired boxed weight has been achieved. This product can be gas flushed and pillow-packed with a combination of CO_2 (60%) and nitrogen (40%) to provide extended refrigerated storage (65–70 days). Pillow-pack means that unlike a vacuum-packaged product where the objective is to have the packaging film skin tight to the product, a gas mixture is used to flush the package of air and then seal in a certain volume of this mixture such that the film is not in tight contact with the product. Layout bacon can be microwaved right on the paper, which reduces the food service preparation time.

Various trimming specifications are used depending on the type of product being produced. Derinded bellies should be rectangular and have perpendicular cuts. Several reference points are used to specify how to properly trim the belly (Fig. 5.7). The scribe line is produced when the costal bones are cut to remove the wholesale loin from the belly. Some processors may leave 2.54–5.08 cm (1–2 in.) of belly dorsal to the scribe line; others may remove trim at this line. A second reference point used is the distance from where the first rib was removed to the shoulder cut. A final reference point is relative to the largest gland located in the flank region. Bellies with deep lean cuts, usually done with round rotating mechanical knives that expose underlying adipose tissue (called snowballs), decrease the slicing yield of the high-quality slices.

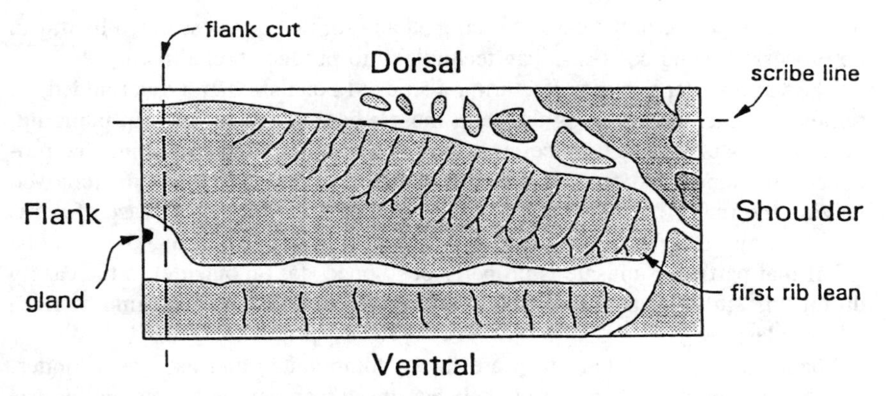

Figure 5.7. Fresh pork belly for bacon.

Fresh bellies trimmed and weighing 9–15 lbs can be pumped 16% with the following pickle: water (71.39%), salt (15.4%), sugar (10.0%), phosphate (2.8%), sodium erythorbate (0.34%, 544 ppm), and sodium nitrite (0.075%, 120 ppm). Pumped bellies are hung on smokehouse trees from the flank end using bacon combs. Proper combing facilitates producing a slab that requires less trimming because of irregular shaped sides.

Pumped bellies can be immediately smoked and heat processed in a smokehouse programmed to provide 1 h at each dry bulb setting (100, 110, 120, 125) and then 140°F dry bulb until an internal temperature of 53.3°C (128°F) is achieved. The slabs should have approximately a 12% cooking loss and a 2% chill loss. Therefore, a 12% cooking loss on 100 lbs of bellies pumped 16% (116 lbs of pumped product) results in 102.1 lbs of heat processed slabs [13.9 lb cook loss; 116 − (0.12 × 116)] and a 2% chill loss yields approximately 100 lbs of chilled product [2.1 lb loss; 102.1 − (0.02 × 102.1)]. After processing, the weight of the heat processed, chilled slab must not exceed the before pump weight for regulatory purposes.

Chilling and, particularly, tempering are critical to the pressing and slicing operations. Slabs can be rapidly chilled using a brine chiller and then while still hung on the smokehouse trees are refrigerated in coolers for sufficient time to dry the surface and to allow the slabs to temperature equilibrate to −4.4 to −3.3°C (24–27°F) prior to pressing. Proper tempering a slab near freezing temperatures when pressed results in a thicker, more uniformly shaped slab that retains the new shape to facilitate slicing and improve slicing yields.

Trimmings from cured, smoked bacon can be used as a sausage material for flavoring or processed into bacon bits. Bacon is not a fully cooked product and must be cooked prior to consumption to assure destruction of trichinae.

Various types of hams can be produced, including bone-in, semiboneless, and boneless. These cured products can be stuffed into fibrous casings and smokehouse processed or in the case of sectioned and formed boneless cured ham, can be processed using cook-in-a-bag technology to produce boiled ham.

The various steps in manufacturing a smoked boneless ham (water added) or turkey ham include trimming the ham, muscle boning and separation, pumping, massaging, stuffing, heat processing, smoking, chilling, and packaging. To produce a sectioned and formed boneless ham, the skin and bones are removed leaving the major muscles largely intact. The external fat, in the case of pork, may be completely removed or in some cases a 0.64-cm (0.25-in.) trim may be left if that particular muscle is properly positioned (fat on outside) in the casing during the stuffing operation. Seam fat and glandular tissue are trimmed from all muscles.

Shank meat or lean ham trim are finely comminuted and used as a binder. The binder helps to fill the voids between the larger muscle pieces and assists in binding these pieces together. In addition, the binder assists in absorbing any residual-free pickle. Binder is prepared by comminuting the shank meat or lean ham trim in a bowl chopper and mixing in water, salt (2.5%), sodium phosphate (0.3–0.5%), and sodium nitrite (150–200 ppm).

Boneless muscles are pumped 32% with a pickle consisting of water (85.4%), salt (7.3%), sugar (5.6%), sodium phosphate (1.5%), sodium erythorbate (0.15%), and sodium nitrite (0.05%). Pumped product is massaged for 6–8 h with the addition of 5–10% (based on the pumped weight of the hams) of prepared binder.

One cushion (*semimembranosus*) and two flats (*biceps femoris* and *semitendinosus*) can be selected and recombined to make the boneless pork product. Careful control of which and how many of these muscles are recombined should be done because these muscles may differ in chemical composition of the unprocessed muscle and, if pumped at the same time, may not have picked up and retained the same amount of pickle. Massaged product is then tightly stuffed into fibrous cellulose casings and heat processed to an internal temperature of 66.7°C (152°F). The smokehouse schedule is programmed to provide 3 h at a dry bulb (DB) setting of 71.1°C (160°F) and wet bulb (WB) turned off prior to finishing the cycle with a DB of 85°C (185°F) and WB of 73.9°C (165°F) yielding a relative humidity of 55%.

Chemical analysis of the finished product should be done to verify compliance with PFF regulations for an added water ham as many factors can affect the final composition including composition and quality of the raw boneless meat, and differences in massaging, heat processing, and equipment used.

Uncured, Comminuted, and Noncomminuted Meats

Restructured Products

Uncured, comminuted, and noncomminuted meats include restructured steaks, roasts, blocks, and portion and sticks. Restructured steaks often involve taking lower-value raw materials and altering the composition (fat, connective tissue), structure and shape in order to form a value-added processed meat. Advantages of restructuring include production of products that are less expensive, more uniform, and provide the opportunity for greater portion control than whole muscle cuts. One major disadvantage is that because these products do not contain nitrite and many utilize salt as a binder, autoxidation of the lipid fraction can be a problem.

Lower-value meats may be higher in connective tissue and less tender. Tenderness can be improved by trimming the silver-skin (connective tissue), tendinous areas and utilizing blade tenderization. Pieces of meat are usually bound together by salt (often in conjunction with phosphates) extraction of myofibrillar proteins but can also be bound in the uncooked state by the addition of calcium alginate.

Particle size reduction is accomplished by chunking or flaking. Chunking basically means the muscle is ground through a grinder plate to sufficiently reduce the size of the meat pieces to facilitate restructuring. The intent of restructuring is to produce a product that closely simulates the palatability and appearance of intact, whole muscle cuts such as steaks, chops, and roasts. Therefore, the meat pieces should not be ground to the point that the integrity of the muscle structure is completely destroyed. Flaking produces thin slices or flakes from frozen and tempered −3.9°C (25°F) meat. Regardless of which type of particle reduction is selected, the chunks or flakes are mixed for 4–10 min with salt (0.50–0.75%, based on meat weight) and sodium tripolyphosphate (0.20–0.30%, based on meat weight) to extract protein and to contribute to flavor.

Mixed product is tightly stuffed into a suitable casing or polyethylene bag, sealed, and rapidly blast frozen (−28.9 to −34.4°C) and tempered (−5.6 to −2.8°C) prior to being pressed and shaped into the desired configuration for slicing. As with tempering of bacon slabs for slicing, too cold a temperature may cause the product to shatter, whereas too warm may prevent a clean, uniform cut that will retain the desired shape.

Roast beef and cooked turkey breast can be produced with sectioning and forming technology. The approach is basically the same as that used to produce a boneless sectioned and formed ham, except that the meat is not cured.

Blocks, Portion, and Sticks

Fish fillet blocks are prepared by placing boneless, and often skinless fillets, or fish minces, or a combination of both in a fiberboard container. The fillets are placed parallel or perpendicular to the length of the container. All depressions are filled with fillets. Fish blocks are usually 7.5 kg (13.5 lbs) or 8.4 kg (18.5 lbs). The blocks are placed in a multiplate compression freezer with spacers and frozen. The spacers are smaller than the width of the block so that a slight compression can be formed which will facilitate cutting. The blocks are cut with a band saw or gang saw (series of circular saw blades on a single axis) or guillotined on a shearing machine. The size of the fish portions are between 43 and 142 g (1.5 and 5 oz).

Cooked Sausages

Cooked sausages include frankfurters, wieners, bologna, luncheon meats, loaves, liver sausage, and braunschweiger. Most cooked sausages are stuffed into cellulose or animal-based casings. The manufacturing procedures for frankfurters and bologna are essentially the same with slight changes in seasoning. In addition, because of the differences in the size of the casings that are used, differences in the smokehouse schedules exist. Smokehouse schedules for frankfurters are relatively short compared to bologna. Loaf items usually represent coarsely ground seasoned meats that are filled into rectangular open-topped pans. Loaves may be glazed or garnished (pimentos, pickles, and pepper) and contain pickles, pimentos, and olives within the loaf. Liver sausages are manufactured from the livers and skeletal muscle of various species. Liver sausage must contain at least 30% liver. Braunschweiger is basically a liver sausage that can only be manufactured from beef, pork, or veal (livers and muscle).

The first step in manufacturing a cooked sausage is to establish the requirements of the finished product desired. The basic and common requirements are species (beef, pork, chicken), bind, color, fat content, added water, amount of protein, and addition of various nonmeat ingredients. Ingredients include salt content (1.75–2.3%), sugar (0.75–1.0%), and seasoning based on the finished weight, and alkaline phosphate (0.3–0.5%), sodium erythorbate (550 ppm), and sodium nitrite (156 ppm) based on the meat weight. Next, the analysis of the raw meat materials needs to be determined to formulate how much of each of these must be used to achieve the specific requirements of the finished product.

In some commercial operations, the lean meats are ground and mixed with water, salt, and nitrite to produce a *preblend* that represents about 75–80% of the gross batch weight. After thorough mixing, a representative sample is collected and finely chopped for subsequent moisture, fat, and protein analysis. This information is necessary to determine the appropriate quantities of lean corrector, fat corrector, and water to more accurately achieve the formulation targets. Lean correctors consist of lean ground meats that have less fat than the target for the product being produced and the fat corrector has a fat content higher than the target. The preblend, correctors, water, and remaining ingredients are combined and mixed. The next step is critical in forming a stable batter. A variety of different pieces of equipment can be used to achieve this goal. The mix can be processed in a vacuum bowl chopper to an end-point temperature not to exceed 15.6°C (60°F). Bowl choppers offer the flexibility of additional mixing and controlling the particle definition by regulating the speed and duration of the rotating knives along with when each of the meat components is added to the bowl. An emulsion mill also can produce a finely comminuted batter by grinding the meat mix through a very small diameter plate (1.4 mm or 1.7 mm) in which a knife turns at very high revolutions per minute (3000–4000 rpm). The resulting batter is vacuum stuffed into the appropriate casing, smoked, and heat processed. With the advent of coextruders a variety of new products can be produced that have a meat batter outer layer and are filled with numerous prepared foods such as vegetables, cheese, and other meat mixes.

Fermented and Dried Meats

Pepperoni is an example of a fermented and dried sausage. It is produced by grinding fresh meat through a 0.95 cm (0.38 in) plate to assist in maintaining particle definition by reducing mixing and chopping time. Fresh meat can include pork and beef trimmings of various fat contents. Frozen meat (20–30% of the batch weight) is ground frozen through a 2.54–3.81 cm (1–1.5% in) diameter plate and used to help maintain a cold temperature during chopping. Frozen meats could include beef cheeks and pork head meat which are high in connective tissue and, therefore, are ground more effectively in the frozen state. All pork used in the formulation should be certified as trichinae-free by approved storage times at specific freezer temperatures. Ground meats are mixed in a bowl chopper for 90 s with salt (3%), dextrose, and spices. The starter culture is added and the mix is chopped until the fat particles are slightly less than 0.63 cm (0.25 in.). Final chopping temperature should not exceed −3.3 to −1.1°C (26–30°F). The product is stuffed into a fibrous casing (40–50 mm in diameter), and, depending on the culture used, is placed in a smokehouse set to provide a dry bulb temperature of 32.2–43.3°C (90–110°F) and a relative humidity of 80–90%. The pH should be monitored regularly to determine when the appropriate end point has been achieved. For this type of product, between 10 and 25 h are required. Fermentation end point can be controlled by the amount of added metabolizable carbohydrate (usually 6–12 oz per 100 lbs of meat) or heat. Because temperatures above 48.9°C (120°F) will inactivate lactic acid starter cultures, temperature control during fermentation is vital, particularly if the product is smoked as the heat generated during smoking could elevate the temperature above this critical temperature.

The American Meat Institute published guidelines (1982) for the production of fermented dry and semidry sausages that recommend the utilization of time–temperature controls necessary to achieve a pH of 5.3 called degree/hours during the fermentation process. Degree/hours are calculated as the product of the hours at a particular temperature times the temperature difference above the critical temperature for staphylococcal growth 15.6°C (60°F). Recommended degree/hours vary according to the maximum fermentation temperature used [< 1200 degree/hours, 32.2°C (90°F) < 1000 degree/hours, 32.2–37.8°C (90–100°F); < 900 degree/hours, > 37.8°C (100°F)].

With the proper pH end point, tang, and color achieved, the product is hung in drying rooms maintained at 12.8–15.6°C (55–60°F) and a relative humidity of 65–75%. Drying times for a 50-mm casing could range from 20 to 35 days. The exact drying end point is determined when the moisture/protein ratio is 1.6 or less.

Pickled, Spiced, and Marinated

Marinated fish are baked, fried, or salted in salt brine or brine containing vinegar, prior to pickling in vinegar. In southern Europe, marinated fish are fried in hot oil and packed in vinegar with spices. Marinated fish production is popular in continental Europe, primarily in Germany. Herring is the principal raw material used in Germany, although fish species such as haddock and anchovies are used to a lesser extent.

Pickled seafood is preserved with vinegar or vinegar and spices. The acetic acid of vinegar is responsible for imparting its preservative effect with 15% acetic acid inhibiting bacterial growth. Because commercial vinegar grades contain an average of 6% acetic acid, marinated fish products are subject to spoilage. Products in 5% or more acetic acid will retard spoilage for weeks to months when the product is properly stored in a temperature-controlled environment.

Water, vinegar, salt, sugar, spices, herbs, and other ingredients affect the flavor, texture, color, and, to some degree, the shelf-life of the product. Vinegar should have a clear consistency with no uncharacteristic odors and flavors. The use of distilled vinegar with a confirmed acetic acid content is generally recommended. Salt should be free of calcium and magnesium impurities, which will impart a bitter flavor to the product.

Pickled herring is the most popular fish product. There are several varieties of this product with the content of the sauce as the principle difference between them. Some common products are Bismarck herring (boned herring with sides joined together, and packed in vinegar and spices), Kaiser Friedrich herring (Bismarck herring in mustard sauce), and rollmops (Bismarck herring rolled around a piece of dill pickle), and cut spiced herring (sliced herring with spices).

In the United States, herring is cured during the fishing season, held in storage, and made into pickled products as demanded by the market. A typical curing method starts with cutting and dressing herring and removing the kidney. Fish are rinsed with fresh water and covered with 80–90°S brine, containing 120-grain distilled vinegar with 2.5% acidity. Fish are removed after salt is allowed to completely penetrate the flesh and before the skin starts to wrinkle and lose color. The curing time depends on the processing temperature, the size, and the freshness of the fish. A 5-day cure is generally practiced in the industry, although there is a variation of 3–7 days.

After the curing step, herring is packed in metal drums lined with plastic bags. The drums are then headed and filled with 70°S salt–vinegar solution and shipped to marketing centers for final manufacture. Herring is cut into fillets or left whole with the removal of the backbone. They are repacked in kegs filled with diluted vinegar–salt solution with 3% acidity and testing 35°S and stored at 1.1°C (34°F) until needed for further manufacture. The final process involves soaking the fish in cold water for 8–10 h, draining, and replacing the water with a salt, vinegar, and water solution for 72 h. The solution consists of 1 lb (0.5 kg) of salt, 1 gal (3.8 L) of 6% white distilled vinegar, and 1 gal (0.5 kg) of water. The pickled herring is then made into rollmops, cut spiced herring, or Bismarck herring with the addition of various spices.

A popular variation to spiced herring in Europe is fried marinated fish. Young herring, anchovies, sardines, or other small fish are cleaned, washed, and dried in open air for 1 h. They are then fried in hot oil, cooled, and drained. They are placed in barrels, kegs, or other containers for marinating in hot spiced vinegar–salt solution.

Mollusks, such as oysters, mussels, and clams (to a lesser extent), are popular pickled products. Cockles, a bivalve mollusk of the family Cardiidai, are a prevalent French product. They are cured in a 3% salt brine, followed by 3 days in solution containing vinegar and salt, each at 3%. This product is further processed by draining the brine and packaging in spiced vinegar.

Minced Fish and Surimi

Production of minced fish started in the 1970s with the development of a meat–bone separator which recovered the fish flesh from whole fish, fillets, and by-products. The process allowed greater use of nontraditional species and provided for a greater tissue recovery from all harvested fish. Minced fish is the component produced from the separation of the flesh from the skin, bones scales, and fins into a comminuted form. The technology allows fish that are too small for filleting or contain too many bones to be utilized. The process can be used to recover flesh from waste products such as fish frames.

Sources of Raw Material

Minces are prepared predominately from gadoid fish which include cod, hake, haddock, pollock, and croaker. Other sources of minced products are the flatfish (soles and flounders).

The large stocks of southern blue whiting is a potential source for mince production. Much research has been focused on the development of mince products from the American mullet. Other underutilized species studied by American and Canadian workers include sea trouts, croaker, ribbonfish, argentine (smelt), cusk, turbot, grey cod, thorneyhead, red hake, and menhaden.

Much of the published work on mince production focuses on the by-catch from the shrimp fisheries of the Caribbean and the Gulfs of Mexico and California. The world mean by-catch fish length is 10 cm (4 in.) with a low proportion of marketable species. However, the small mean size makes evisceration a major problem. But the high consistency of the catches makes it a likely source for mince production. Eight species constitute 74% of the by-catch. There is a low incidence of fatty pelagic fish in the Caribbean, the Arabian Gulf, and elsewhere.

Although the world's catch of small pelagic species exceeds 20 million tons annually, half of the landings are used for fish meal. Mince technology may offer the optimum potential for increased exploitation of these fish. However, the small pelagic fish seem to be the most difficult for mince production out of all the materials studied because of the difficulty in filleting.

Major species studied for mince production include the mackerels, horse mackerels, herring, sardines, sardinellas, spruts, anchovies, and menhadens. Another source of mince is the freshwater species, which contribute more than 10 million tons to the annual world fish catch with a potential for an additional several millions tons. The bone structures of most freshwater fish are delicate, making filleting difficult; therefore, mincing can significantly increase the recovery of the flesh.

The fat contents of the freshwater species are low, with the exception of the freshwater pelagic, primarily the shads. The bland taste of these minces can be used in nonfish products and as meat extenders. Some applications, contents, and labeling on the use of minces as extenders are controlled by federal food regulatory agencies. However, the presence of highly unsaturated fatty acids requires particular attention to the storage conditions.

Physical and Chemical Changes During Mincing

Fish mince quality depends on the raw material and on the separation process. During the separation procedure the raw material is fractionated into distinct components. The most apparent part is the bone fraction, which can make up 15% of whole or gutted fish, or more than 30% of filleting wastes. The skull and the body of the vertebrae are more calcified than spines, ribs, or pin bones. Therefore, mincing heads and frames can give a higher proportion of hard, brittle fragments. Quality specifications for minced fish limit the bone content based on weight, number, or size.

A variety of mince contaminants can damage texture and texture stability. One feature that distinguishes gadoid fish from other marine and freshwater fish is the use of trimethylamine oxide (TMAO) as the principal osmo-regulatory agent. The enzymatic degradation of TMAO to dimethylamine (DMA) and formaldehyde can cause severe textural damage when formaldehyde cross-links with the flesh proteins. This enzymatic reaction, although limited to certain species, occurs predominately in frozen storage. The kidney is the primary source of this enzyme. Other sources are pyloric caeca, blood and blood clots, dark brown, lateral muscle, and, perhaps, skin. During the mincing of these species, the formaldehyde-mediated denaturation is accelerated.

Minces produced from demersal and pelagic species are susceptible to fat degradation because of the presence of unstable, polyunsaturated lipids in the skin, in the subcutaneous and dark lateral tissue, in the viscera, and in the brain and nerve tissues. Minces containing these materials may result in undesirable flavor changes and oxidative rancidity. The mincing process degrades and disperses enzymes that degrade fat. These enzymes are contained in the viscera and the dark muscle. Mincing also accelerates nonenzymatic oxidation through increased surface area and dispersing organic and inorganic oxidative catalysts. The hemoproteins (hemoglobin and myoglobin) catalyze nonenzymatic degradation.

Contamination by melanoid skin pigments darkens the minced product. No problems have been reported from bacterial spoilage or from heavy metal and pesticide contaminants. However, parasite contamination may arise in the minced product. Worms and larvae in the flesh can survive the mincing process and become transferred unaltered.

A potential danger may be the formation of nitrosamine in nitrite-treated DMA-producing species. Hemoglobin, which acts as a catalyst, can accelerate these reactions when dispersed during mincing. Washing and antioxidant treatment can reduce the effect, but DMA-producing species should not be used for nitrite-cured mince products. Toxins may be present in minces contaminated with high levels of sphingomyelin and other complex lipids. Sources of toxins are nerve tissue, brain, and the reaction products from oxidative fat degradation. Certain tropical species may contain potent neurotoxins if the mince is contaminated with viscera. Potential contaminants may be removed from the minced product when the raw material is sorted carefully to remove the undesirable fractions.

The mincing process itself increases the microbial flora. There may be an increase in bacterial population in the mechanical deboner throughout the mincing process. Also, the increase in friction between the fish and the perforated head of the mincer generates heat which also promotes bacterial growth. In whole fish, the highest bacterial counts are found in the viscera, on the skin, and on the gills. However, the bacterial count of mince from whole (gut-in) fish is similar to counts in mince from degutted fish. Generally frames produce higher counts than mince from fillets, V-cuts, or frame trimmings.

Textural Properties of Mince

Separated fish minces are amorphous granular slurries. Coarse minces contain fibers and fiber bundles, whereas fine minces and minces from soft textured species have a homogeneous pasty consistency. Some forming or structuring must be incorporated to achieve a higher textural integrity.

Reforming techniques can impart flakiness to the mince products. Alginate gels are used to set the mince into a sheet structure, followed by layering and compacting the sheets. A fibrous characteristic can be achieved by incorporating spun vegetable protein fibers, extrusion-textured vegetable protein, precooked fish muscle, or alkali/acid-precipitated mince protein fibers.

Individual portions are constructed from the mince products by several methods. Band saws and circular saws cut regular portions from frozen mince blocks. Another method of portioning the blocks is through frozen extrusion forming. Although the yield loss from sawdust is eliminated, shear damage to the muscle proteins may result. A wide range of machinery is available for the low-pressure extrusion forming of fresh minces. A variety of shapes and sizes can be produced such as fillets, shrimp tails, balls, as well as regular geometric portions. Adding salt, phosphate, soy protein, and gums to the mince produce optimal characteristics for extrusion and better control of the texture. Colors, flavors, and seasonings can also be used to enhance the product acceptability.

Products Formed with Mince Meat

Some of the products made from minced sources are fish blocks and dried fish cakes. Other applications include frozen, canned and smoked mince, dried mince flour, and silage from mincing waste. The low levels of fatty pelagic species and DMA/formaldehyde-producing gadoid species permit the production of frozen mince with good storage stability.

In the United States, mince has been suggested for use as an extender in meat-based sausages, patties, and burgers. The color of frame mince and pelagic mince is effectively masked in the production of frankfurters. Highly comminuted, soft textured pastes are used in spreads and dips. Coarsely chopped products are used in fishburgers and loaves. Mince are used in fish patties, which are structured by using functional additives such as salt, phosphate, and alginate. Mince utilization includes other formed products, such as fried or extrusion-expanded, starch-based snack products, sliced salmon or saithe (pollock) analogues, and filled products.

Another use of mince is as an ingredient in composite products. Mince is bound with cereal flours or starches to form fish cakes, rissoles, and croquettes. Higher levels of mince are used in traditional products such as fish balls (Southeast Asia, Scandinavia) and Gefilte fish (Israel, Europe, United States). A range of high-quality products is manufactured using the extrusion-forming techniques. The firm, elastic textures are good qualities for shellfish and mollusk analogues. Shrimp incorporated into mince portion can improve the acceptability of the textural properties of the product.

There is extensive use of mince in surimi-based products. This industry is dominated by Japanese "kamaboko" (described below) production. Most mince products outside of Japan are fine-textured, heat-set emulsion products rather than heat-set protein gels.

Surimi

The surimi process involves further processing of minced meat to manufacture surimi and kamaboko. The traditional methods for making surimi-based products were revolutionized with the discovery of producing a stable frozen surimi in 1960. Large-scale on-shore production of pollock surimi was established in 1964. Kamaboko manufacturers no longer were dependent on unstable local fish catches and raw surimi. Instead, they had a year-round availability of frozen surimi.

Sources of Raw Material

General statements can be made regarding the raw material for surimi or kneaded products. Freshwater fish species as well as marine species inhabiting warm waters are generally not suitable. The most important desirable property for raw materials is gel-forming capability. In general, freshwater fish and dark-fleshed fish have lower gel strength than saltwater fish and white-fleshed fish. Dark meat fish are not good sources for surimi-based products. Sources of raw material for good-quality product are cold-water-inhabiting fish species, white meat fish, and demersal fish. There are exceptions to these rules, as meat from different species can be combined together. Almost all fish species can be utilized with the addition of squid, which imparts the necessary elastic property.

Most surimi is produced from Alaska pollock because it is an inexpensive source of white flesh with good texture after freezing. For kneaded products, better sources would include sea bream, croaker, sharp-toothed eels, lizardfish, and rays.

Production of Surimi

The development of the screw press and the rotary washing screen or sieve enable the automation of the process. The screw press is an efficient dewatering machine. It reduces the water content of the washed mince, maximizing the concentration of protein in the surimi. Prior to the development of the screw press, water was removed from surimi with a basket-type centrifuge, by feeding washed minced meat batch by batch. The efficiency of the centrifuge is limited, processing only about 0.5 metric tons of minced meat a day. A single screw press, on the other hand, has a 20 metric ton capacity. The new system can be integrated into a continuous operation with the washing procedure.

The rotary screen is inserted between the washing tank and screw press. It is a device that combines the functions of washing and preliminary dewatering, improving the efficiency of both washing and dewatering procedures.

A machine that improved the straining procedure is the refiner. A refiner strains wet slurry to remove fragments of bones, ligaments, and scales prior to dewatering in a screw press. The removal of membranes, bones, and tendons

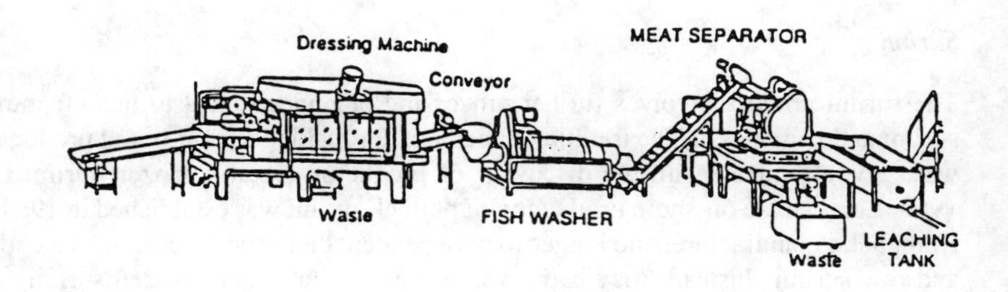

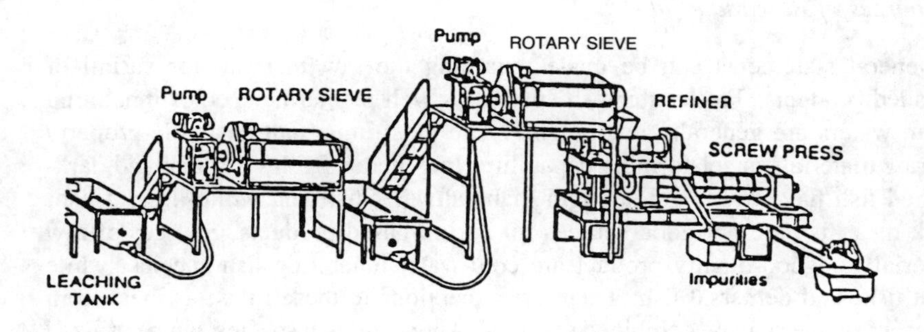

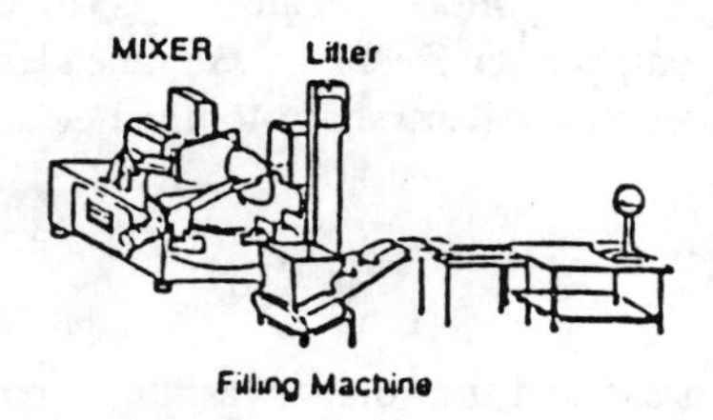

Figure 5.8. Organization of surimi production equipment. (From R. Martin and G. Flick, *The Seafood Industry,* Van Nostrand Reinhold, New York, 1990, p. 152.)

was previously applied after the dewatering step. The straining of the samples was slow, and the quality of the surimi was affected by the heat generated from the mechanical pressure. The refiner is more efficient in removing impurities and there is no heat generated. The refiner works directly on the washed meat and the high water content helps to buffer the adverse effects of temperature on the surimi sample. Figure 5.8 illustrates the organization of equipment involved in surimi production.

Further Processing of Surimi-based Products

Surimi is a Japanese term defined as "minced meat." It is an intermediate raw material for the traditional Japanese kneaded foods called "kamaboko." Imitation shrimp, scallop, and crab meat products are other end products.

Surimi is further processed to have two major distinguishing features. It has a gel-forming capacity, which allows it to assume almost any desired texture. It maintains a long-term stability during frozen storage with the addition of cryoprotectants (sugars).

Minced meat becomes raw or unfrozen surimi after the washing step, which removes fat and water-soluble constituents (Fig. 5.9). The flavor components are removed during this leaching process. The washing step also isolates the myofibrillar proteins, which are insoluble in fresh water. These proteins possess the essential gel-forming capacity.

When raw surimi is mixed with antidenaturants and frozen, it becomes frozen surimi. The antidenaturant additives which are primarily sugar compounds (sucrose, sorbitol) make surimi resistant to denaturation during freezing. This denaturation step is an irreversible change in the protein. The denaturation of muscle proteins, particularly actomyosin, during frozen storage are prevented with the addition of cryoprotectants. Actomyosin becomes denatured as a result of aggregation caused by intermolecular cross-linkage formation. Cryoprotectants prevent denaturation of actomyosin by increasing the surface tension of water, as well as the amount of bound water. This will prevent the growth of ice crystals and

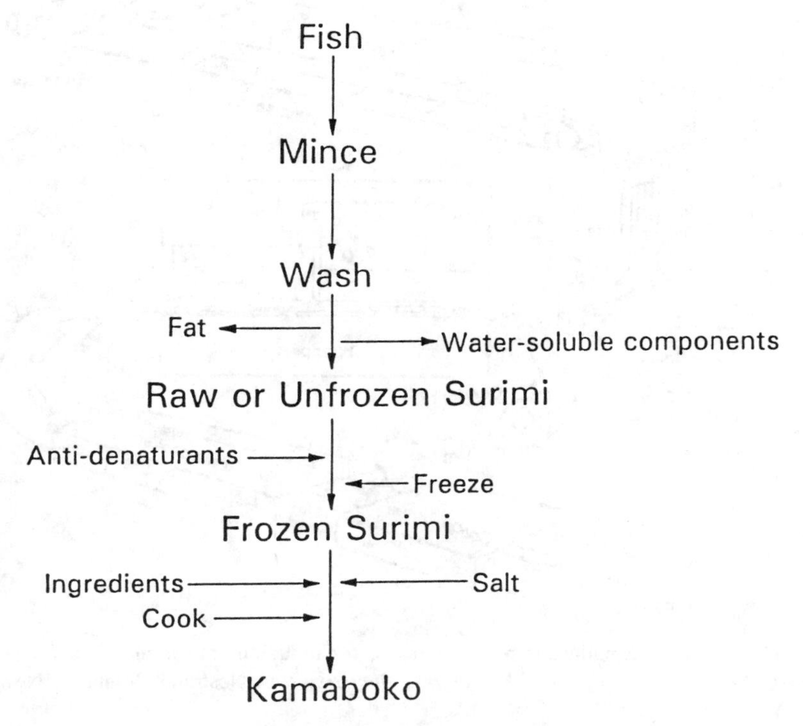

Figure 5.9. Schematic diagram of surimi production.

the migration of water molecules from the protein, thereby the native form of protein is stabilized during frozen storage. As a result, there is a reduction in the gel strength. Protein denatures without the addition of these cryoprotectants, resulting in the loss of the gel-forming capacity of surimi. Because approximately 95% of all surimi is frozen with the addition of cryoprotectants, the term surimi generally implies frozen surimi.

To make imitation crab and other traditional kamaboko products, surimi is partially thawed, mixed with a small amount of salt, blended with other ingredients (starch and egg protein) and flavors, kneaded, and formed to create the desired texture and shape. This mixture is sometimes referred to as a surimi paste. The product is then cooked by steaming, broiling, or frying. Figure 5.10 shows a small-scale production line for imitation crab product.

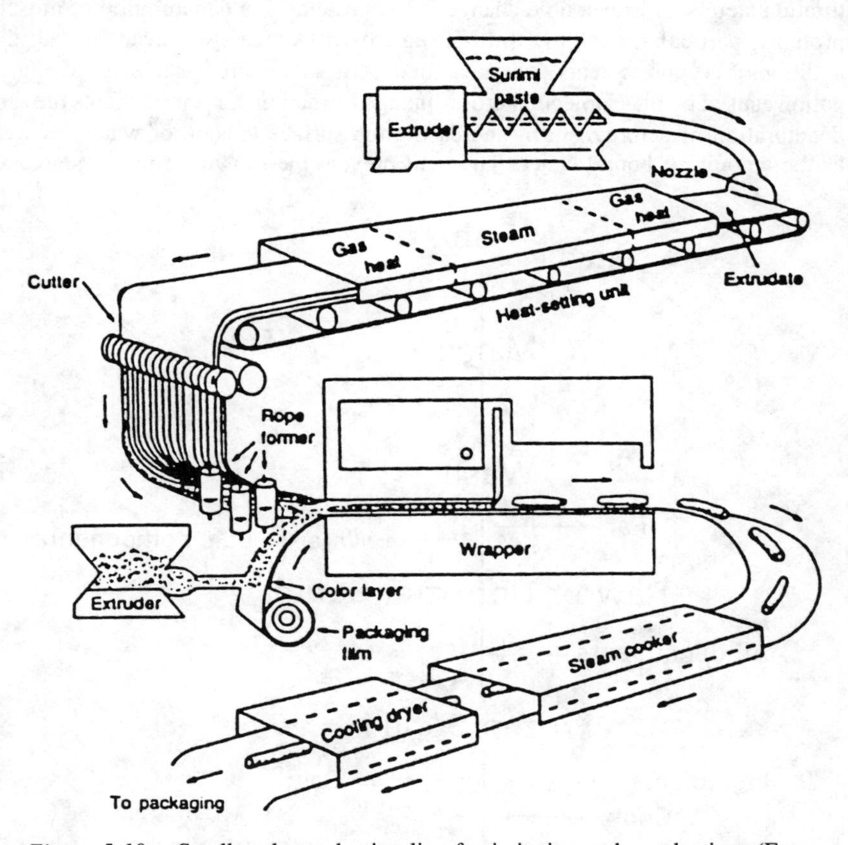

Figure 5.10. Small-scale production line for imitation crab production. (From R. Martin and G. Flick, *The Seafood Industry,* Van Nostrand Reinhold, New York, 1990, p. 143.)

Kamaboko products are classified into three major categories: steamed, broiled, and fried. The typical steamed kamaboko is called "itatsuki," which is mounted on a thin board. This group also includes imitation seafood, "naruto" and "hapen," spongy marshmallowlike products which contain entrapped air. "Chikuwa" is a typical broiled kamaboko, which is tubular and is shaped like a hollow bamboo stem. The typical fried kamaboko (age-kamaboko) products are "satsuma age" and tempura. "Age" kamaboko are shaped like balls, bar, or squares. There are other names for kamaboko depending on product shapes, such as "susa" (bamboo-leaf shaped), "soba" (needle shaped), "date-maki" (whirled or rolled), and "kezuri" (chipped).

A homogenous gel of ground fish muscle is the main ingredient of kamaboko. Other ingredients are sugar, starch, sweet sake, sodium glutamate, and egg whites. Typical ingredients in "Odawara" kamaboko are 76–84% surimi, 11.9–19.5% sugar, 4.8–6.5% sweet sake, 4.2–5.3% salt, 1.2–2.0% sodium glutamate, 0–6.5% potato starch, and a small amount of egg white. Surimi-based imitation crab products contain 55% surimi, 25% water, 8% egg white, 5% starch, 2.5% crab essence, 2.3% seasoning, 1.5% salt, 1% sweet sake, 0.6% sugar, and 0.1% natural coloring.

Surimi paste is extruded into various shapes to resemble the natural product. Surimi-based products are in the shapes of shellfish meat, such as king or snow crab legs, crab claws, lobster tails, scallops, and shrimp. The closer the analogue resembles the natural product, the greater the sophistication of the extrusion process. Surimi-based products are categorized into four groups according to the methods of fabrication and structural features: molded, fiberized, composition molded, and emulsified.

Surimi-based Imitation Seafood Products

Crab analogue form of kamaboko was first marketed in 1978 as imitation crab legs. Now the market offers lump and chunk forms. The first surimi products were imported to the United States from Japan in 1978. The market doubled between 1978 and 1984. U.S. sales of surimi-based analogues increased from approximately 2 million in 1980 to around 300 million in 1985. The U.S. production of surimi from Alaska Pollock for 1992 was 131,497 metric tons. Presently, imitation scallops and shrimps, as well as other varieties, are manufactured by U.S. producers, employing the basic technology of surimi production.

End products called "neri-seihin" (surimi-based products) are manufactured from surimi, the intermediate raw material. Approximately 90% of surimi-based products are various types of fish cake called "kamaboko." Fish sausage, fish ham, and fishburgers represent less than 10% of surimi-based products. The imitation crab and other surimi-based shellfish analogues are included as kamaboko.

Smoked Seafood

Seafood is cold- and hot-smoked under controlled conditions. Products generally undergo curing and drying processes prior to smoking. Products may be cured and dried without further processing with smoke. Salt is the primary curing ingredient. Additional agents such as sugar, spices, and nitrite (in some cases) are used. The addition of salt contributes to the shelf-life, safety, and flavor of the end product. But smoked products must be properly refrigerated for preservation.

Salting

Salt is added to the product mainly to improve the flavor. Salt will aid in the delay of spoilage and help in the inhibition of *Clostridium botulinum* Type E organisms. Dry salt or salt in brine is applied to fresh or thawed fish before hot- or cold-smoking.

Lean fish, such as cod, are normally dry salted. The fish is headed, eviscerated, and split open to the tail with the backbone removed. Fish are washed and any remaining viscera is removed. They are salted flesh side up in a large barrel or piled on a kench (solid floor), stacked in layers, and each layer is alternated with a layer of salt. The top layer of fish is placed with the skin side up. Light salted fish contain 4% salt after 2–8 days of curing. Heavy salted fish contain 20% salt after 21–30 days curing. A heavy salted cod may be dried to about 40% moisture content after the end of the curing period.

The next step involves drying fish to reduce the water content by 25–38%. Fish may be dried naturally in the sun or artificially in an indoor drying chamber. The temperature of the chamber is usually 23.9°C (75°F), with a range of 15.6–26.7°C (60–80°F). The relative humidity is 50–55%. The final salt content may be one-third the weight of the end product. The product is packaged in small wooden boxes or plastic films. Lightly cured products must be stored under refrigerated temperatures. However, heavily cured fish can be stored at ambient temperatures for years without any evidence of bacterial growth.

A technology was developed at the Gloucester Laboratory at Northeast Fisheries Center that decreases the overall processing time necessary to produce a salted dried product. Fish are ground to increase the surface area and then covered immediately with brine and additional added salt to maintain saturation. Drying rate is increased by the small particle size of the product. The dehydrated fish can be ground into a meat flour or used in reconstituted form. Approximately three times its weight in water is added and allowed to soak for 15–30 min. Although a large quantity of water is absorbed during the rehydration process, the dehydrated fish never approaches the water content of fresh fish. This is caused by the changes in physical structure and chemical nature of fish during drying.

Another method of curing is using a salt brine. Processors may use brine curing or dry salting or a combination of the two. The brining step may aid in hardening the fish texture, providing flavor, and helping to preserve the product. Brines are liquid salt solutions. One gallon of saturated brine at 15.5°C (60°F) contains 2.64 lbs of salt and is 100°S (100° Salometer). The concentration of the solution must be measured within 5°S to predict the correct amount of time to cure the product.

A typical brine has a salometer reading from 30–50°S. During the brining process, water migrates out of the fish tissues by osmotic pressure and salt is absorbed. As a result, the water content of the fish tissues decreases over time. Keeping fish in the brine longer than the allotted period will make the product unpalatable due to the high salt content.

There are several factors affecting the rate of salt absorption by the fish. Salt penetrates skinless and thin areas faster than thick areas with skin. Areas with low fat absorb more salt than flesh with high fat content. During the brining process, it is important to keep the brine agitated to maintain a constant concentration. The strength of the brine, the location of fish in the brine, the brine-to-fish ratio, and the temperature of the brining process all influence the rate of salt absorption. Sugar and other flavoring ingredients, coloring agents, and sodium nitrite may be added during the brining process.

Drying and Smoking

After brining, fish are hung or placed on racks for drying and smoking. Fish may be threaded through the head, gills, or mouth. They are then hung tail down on a rod. Other ways of hanging fish utilize hooks or nails, and racks. Fish are usually hooked through the head. Racks are used when there are no skeletal muscles in the fish pieces sufficient to support hanging. These pieces may include chunks, steaks, and fillets. Mesh bottom trays may be used in place of racks.

The exterior has a shiny appearance formed on the fish surface when the proteins dissolved in the brine solution dries during hanging. When properly cured and dried, the fish surface will have a smooth, dry, and shiny appearance. This characteristic helps smoke to be readily adhered to the exterior. If a glossy surface is not produced, the outer surface will emit coagulated body fluids. This imparts an unattractive appearance to the fish.

The drying process firms the fish flesh. Air circulation in the chamber, temperature, and the relative humidity of the air work to control the drying time. Fish should be dried in a cool area that has circulating air. Some processors use a smoking room and increase the temperature. Good air circulation and precise temperature control will prevent fish from being cooked. Protein must be allowed to set or to become denatured at low temperature. If air is too hot and circulating at a rapid rate, the surface of the fish will be damaged and improperly dried. Drying fish too quickly will denature proteins, causing the skin to harden. As a result, water will not evaporate. This process is referred to as case hardening.

Overdrying the fish surface also results in other problems. An overdried surface will crumble and inhibit sufficient smoke penetration along with limiting desirable smoke color formation. Improperly dried fish will result in an uneven smoke color, due to poor smoke absorption. This defect is described as being "streaky." Therefore, it is important to control the drying time. The longer the product is dried, the greater is the protein denaturation.

Although the traditional method of drying fish in the sun is replaced by modern dehydration processes, sun drying is still practiced in some cultures. The Japanese salt mackerel, sardines, cod, and other fish and dry them in the sun to produce "salt-dried" fish. Chinese and Indians still preserve fish by drying in a very simple, unscientific fashion, as practiced by their ancestors before the Christian era.

Different fish species require a specific processing time and temperature for proper smoke application. Generally, hot-smoking requires a short processing time compared to cold-smoking. The internal temperature of the fish is monitored during processing. A properly hot-smoked product has a moist consistency and is subject to spoilage. The finished product requires refrigeration.

The Association of Food and Drug Officials has recommended processing and packaging guidelines for hot- and cold-smoked fish to ensure safe products. Air-packaging hot-process smoked fish requires heating the fish to a continuous temperature of at least 62.8°C (145°F) throughout each fish for a minimum of 30 min. Fish must be brined to contain not less than 2.5% water phase salt in the loin muscle of the finished product. For hot-processed smoked fish to be vacuum-packaged or modified atmosphere-packaged, the same temperature and times are required. However, the loin muscle of the smoked fish must be brined to contain not less than 3.5% water phase salt in the finished product or the combination of 3.0% water phase salt and not less than 100 nor more than 200 ppm of sodium nitrite. Cold-processed smoked fish do not undergo heat processing but have similar packaging guidelines as detailed for hot-processed smoked fish.

Heat is generated in traditional gravity ovens or in smoke ovens. Gravity ovens use charcoal brisquets and gas burners. Smoke ovens use electricity, gas, or oil to produce heat. Smoke ovens can be automated to control drying, smoking, heating, cooling, humidity, and air flow.

Cooling

Recommendations published by the Association of Food and Drug Officials state specific guidelines for proper cooling of smoked fish. Packaging smoked fish when it is hot causes condensation to form inside the wrap. Therefore, cold- and hot-smoked products are cooled to 10°C (50°F) or below within 3 h postprocessing. The temperature of the products is further reduced to 3°C (38°F) or lower within 12 h. The latter temperature is maintained during the storage and the distribution of the smoked fish.

Fermented Fish

The principle of fermentation of fish relies on the addition of an adequate concentration of salt to degrade proteins and inhibit the growth of putrefactive microorganisms while promoting the growth of various desirable microorganisms. Fish fermentation does not necessarily utilize lactic acid bacteria as in fermented red meat sausages. Popular products in Southeast Asia are fish sauces and pastes made through natural fermentation. Fish are heavily salted and stored at ambient conditions for several months. Cathepsins, naturally occurring tissue enzymes, hydrolyze the fish protein. Low-salt products, such as "i-sushi," are fermented by various strains of *Lactobacillus*.

Softened Bone Products

Certain underutilized species are not used primarily because they are too small. These species are uneconomical to debone or fillet. A technique to soften bones enabled the Japanese to market small fish species. This process may be possible for commercial production of bone-softened fish products in the United States. Fish are cooked at a minimum temperature and pressure of 116°C (241°F) and 10 psi to soften the bones. Times and temperatures of processing are dependent on the bone type and the hardness. Collagen, which is the principle bone component, is denatured during thermal processing, thereby, loosening the supporting texture.

In Japan, sardine is the most used species for bone softening to manufacture ready-to-fry breaded products. A popular Japanese sardine product contains bone-softened sardine, breading, wheat flour, sesame, and salt. Fish are marketed with the addition of various flavors, such as curry, cream and parsley, and tomato. Other items include uncoated bone-softened fish. These items are sold directly for institutional use.

Dried Squid Products

The demand by American consumers for squid products is low, although popular in Europe and the Far East. A popular Japanese dried product is "surume." The squid is eviscerated to remove the ink sac, cartilage, and skin. The mantle, arm, and fins are dried. The product may be seasoned to produce a mild to sweet to spicy product. The texture may be crisp to chewy. The seasoning includes salt, natural sweeteners, spices, ascorbic acid, sugar, monosodium glutamate, and other flavoring agents.

The seasoning is placed with the squid for 4–6 h. During this time, fluids are released from the squid mix to form a solution with the seasoning. The squids are dried at 40°C (104°F) for 8–10 h, with the moisture content reduced to about 37–38%. This product called "daruma" can be marketed or further processed to form the finished product "saki-ika."

Daruma is pressed and roasted for 10–15 min between two heated plates. The fins and top of the mantle are manually removed and sold as "mimi." The mantle is fed into a roller press and flattened and shredded into strips about 3 mm (0.12 in.) in width. Alcohol diluted with water is sprayed over the squid and the shredded mantles are seasoned in polyethylene bag for 4 h. The seasoning is baked to form a coating on the mantles with an infrared drying unit. The moisture content is about 28% and the product is "saki-ika." Additional additives and seasoning in an ethanol base are added during packaging.

Convenience Products

Battered and Breaded

The major ingredient in a dry batter mix is flour, making up around 80–90% of the total weight. In a breading mixture, flour contributes 70–80% of the total ingredients. The major sources of flour are corn, rice, soy, and barley. Wheat flour is different from the other types because it forms a cohesive mass when hydrated and mixed.

Several major characteristics or functional properties distinguish the different breading. The range of the breading particle size is from 0.5-in. cubes to fine particles, passing through an 80-mesh standard sieve. Typical breading pass through a No. 5 to a No. 80 U.S. sieve. Particle size is the major factor that affects the appearance and texture of the coated food. A fine mesh will increase the ability of the batter to absorb liquid. A course coating will give a loosely adhering product that will fall off during handling and transport. A good breading size is in the range of U.S. No. 20 to 60 screens.

The browning rate is another important characteristic of breading properties. The amount of sugar in the coating determines the browning rate. The greater the concentration of sugar, the faster the browning rate will be, with a subsequent high processing rate. This allows the use of a lower frying time and/or temperature. There is less shrinkage in the product when the frying time is shortened and the temperature is reduced. In some products the sugar content is reduced to retard the browning rate. This situation applies to processing foods that are large or thick and require a long frying time. Balancing the color, texture, and cooking time can be controlled by varying the browning rate.

Another important characteristic is moisture and oil absorption. The particle size and porosity of the breading material determine how quickly moisture and oil are absorbed. A coating with a porous material will absorb and release moisture and oil faster than a densely coated product.

A variety of prepared breadings is used on battered foods to enhance the appearance and sensory qualities. Major types are cracker meal, bread crumbs, and oriental or Japanese crumbs. Cracker meal consists of a dough from flour, sugars, and salt, dried to approximately 35% moisture. The product is crumbled through a mill or grinder. This product is extensively used on fish and seafood products. Bread crumbs are also a dough mixture but followed by a fermentation step. It is more porous than the cracker meal, absorbing more oil and moisture. The crumbs cannot tolerate long frying times and tend to darken more quickly. The texture of bread crumbs is more tender than cracker meal. The splinterlike-shaped oriental or Japanese crumbs are low in density and very porous. The dough is baked in a special electrical oven. The resulting product has low density and low moisture content.

Production of Battered and Breaded Seafood Products

The world's annual production of frozen blocks may account for over 5.6 billion pounds (500,000 metric tons). Blocks are produced from minces or from mixtures of mince and fillet. Although mince blocks are a major commodity in international trade, they are only intermediates in the manufacture of the retail products, such as battered fingersticks, steaks, and cakes.

Approximately 50% of all frozen battered and breaded products in the United States is seafood. Precooked and raw fish portions are the most frequently coated products. Shrimp, fish sticks, and scallops follow closely behind. The application of batters achieved prominence in the 1960s, although breading had been used in the seafood industry before that time. In the mid-1960s "batter frying" became prevalent. Food is predusted with flour or dry batter mix, coated with a batter, and prefried to set the batter. The latter step imparts the desired frying oil content for texture and quality enhancements. The batters are either tempura (leavened) or nonleavened. The production rates for batter-fried products require half the time than for a breaded product. The expansion of battered products is limited by the inability to produce a microwaveable product that retains the crispiness.

There are several problems associated with battered products. There may be an uneven distribution of coating on the product. The batter may not coat the entire surface of the product, leaving large void areas. If the coating is applied too thinly, it will lose flakes or chip off. The other extreme is when coating is applied too thickly.

The product that is stored in a dry, cool environment will result in a higher-quality batter coating. The batter is mixed in cold water to increase the batter adhesion. Most processors maintain water temperature at 10°C (50°F) or lower. The batter is mixed to remove all unwanted lumps. Salt has been used in the predusting step to increase adhesion of the seafood product to the batter. Salt melts the ice that may be on the product. The formed liquid can hydrate the product, thereby improving the adhesion of the batter.

The oyster meat is usually breaded, sharp frozen at around −40°C (−40°F), packaged, and stored near −23°C (−10°F). Oysters will have a shelf stability of many months.

Selected References

AMI. 1982. *Good Manufacturing Practices; Number two; Fermented Dry and Semi-dry Sausage*. American Meat Institute, Arlington, VA.

Bechtel, Peter (ed.). 1986. *Muscle as Food*. Academic Press, Orlando, FL.

Bender, Arnold, *Food Processing and Nutrition*. Academic Press, New York.

deHoll, John and Jan deHoll. 1983. *Encyclopedia of Labeling Meat and Poultry Products,* 6th ed. Meat Plant Magazine, St. Louis, MO.

Giese, J. 1992. Advances in microwave food processing, *Food Technol*. 46(9):118–123.

Gillies, Martha. 1975. *Fish and Shellfish Processing*. Noyes Data Corporation, Park Ridge, NJ.

Judge, Max, Elton Aberle, John Forrest, Harold Hedrick, and Robert Merkel. 1989. *Principles of Meat Science,* 2nd ed. Dubuque Iowa, Kendall/Hunt Publishing Co.

Kulp, Karel and Robert Loewe (ed.). 1990. *Batters and Breadings in Food Processing*. St. Paul, American Association of Cereal Chemists, Inc.

Lawrie, Ralston, 1985. *Meat Science,* 4th ed. Pergamon Press, New York.

Long, Lucy, Stephan Komarik, and Donald Tressler, 1982. *Food Products Formulary, Volume 1 Meats, Poultry, Fish, Shellfish,* 2nd ed. AVI Publishing Co., Westport, CT.

Lopez, Anthony. 1981. *A Complete Course in Canning*. The Canning Trade, Inc.

Martin, Roy and George Flick. 1990. *The Seafood Industry*. Van Nostrand Reinhold, New York.

Martin, Roy and Robert Collette. 1990. *Engineered Seafood Including Surimi*. Noyes Data Corporation, Park Ridge, NJ.

Ockerman, Herbert. 1989. *Sausage and Processed Meat Formulations*. Van Nostrand Reinhold, New York.

Olson, Dennis. 1990. A new method for product formulation, Part I. *Meat & Poultry* 36(3):36–39.

Pearson, Albert and F. Warren Tauber. 1984. *Processed Meats,* 2nd ed. Van Nostrand Reinhold, New York.

Price, James and Bernard Schweigert (ed.). 1987. *The Science of Meat and Meat Products,* 3rd ed. Food & Nutrition Press, Inc., Westport, CT.

Romans, John, William Costello, Kevin Jones, C. Wendall Carlson, and P. Thomas Ziegler. 1985. *The Meat We Eat*. Danville, IL. The Interstate Printers and Publishers, Inc.

Rust, Robert. 1977. *Sausage and Processed Meats Manufacturing*. American Meat Institute, Arlington, VA.

Sikorski, Zdzislaw. 1990. *Seafood: Resources, Nutritional Composition, and Preservation*. CRC Press, Boca Raton, FL.

USDA. 1984. Control of added substances and labeling requirements for cured pork products; Updating provisions. *Fed. Reg.* 49:14856–14887.

USDA. 1988. Standards for frankfurters and similar cooked sausages. *Fed. Reg.* 53:8425.

6

Inspection

Daniel Scott Hale

Introduction

Food Safety Responsibility

The safety of the foods produced in the United States is the responsibility of everyone who comes into contact with it, from the farm to the dining room table. No matter how effective one segment of the food industry is in ensuring a safe food product, that safety can be compromised by the next segment in the food chain. It is the job of municipal, county, state, and national government agencies to oversee food production, distribution, procurement, and preparation to ensure food safety. Of the foods produced in the United States, it is well recognized that the meat industry is one of the most highly regulated of all food industries. A total of nine government agencies [Food Safety and Inspection Service (FSIS), Packers and Stockyard Administration (P&SA), Occupational Safety and Health (OSHA), Food and Drug Administration (FDA), Agricultural Marketing Service (AMS), Environmental Protection Agency (EPA), Consumer Product Safety Commission, Food and Nutrition Service, and the Animal, Plant and Health Inspection Service (APHIS)] serve as "watchdogs" to ensure meat and poultry presented to consumers is wholesome and safe. It must be recognized that the last line of defense is the consumer who has a distinct responsibility to handle foods in a safe manner. The government agency that plays the biggest part in overseeing the safe production of meat and poultry is the Food Safety and Inspection Service of the United States Department of Agriculture (FSIS-USDA), which administers a comprehensive system of inspection laws to ensure that meat and poultry products moving in interstate and foreign commerce for use as human food are safe, wholesome, and accurately labeled. The rules and regulations administered by the Food and Drug Administration (FDA) have a direct impact on the meat and poultry industry because it approves and polices the use of nonmeat ingredients (food additives) that are added to processed meat and poultry products as well as those used as processing aids. Additionally, the FDA tests, approves, regulates, and reviews the use of animal health products and chemicals used directly or indirectly in the processing of foods.

During the history of the United States, Congress has enacted laws designed to protect the public from food-borne illness. These early statutes focused on health hazards due to poor sanitation and economic fraud. Later, Congress enacted laws to control chemicals added to the food supply. Future laws will likely be designed to more closely regulate the nutritional quality and the microbiological safety of foods. Food safety regulation by the U.S. government, since 1977, has been intense with over 200 governmental legislative bills introduced to modify food safety laws. Tables 6.1 and 6.2 show some examples of food safety regulations passed by the United States Congress and placed under the authority of either FSIS or FDA.

Inspection History

Early in the history of the United States, it was recognized that there was a need to establish a minimum set of standards that the meat product must meet in terms of quality, packaging, and sanitation to be eligible for sale or barter. Probably the most concerted early efforts concerning classification and standardization in meats were those stimulated by and pertaining to the export trade, provisioning ships, and supplying foreign and domestic army contracts. An inspector was appointed in New York in 1857 to inspect and certify quantity and weight of all meats for exportation. By 1885 several packing companies had become quite large and had developed a nationwide distribution. The growth of the large packers led to pressure for national legislation to regulate sanitation aspects of meat products moving in interstate commerce. A series of decrees by European countries, against the importation of American meat products began in 1878 when England placed controls on imports of chilled beef and live animals. It was alleged that American products were packed under unsanitary conditions and that animals often were diseased. These actions were partly stimulated by fears of European slaughterers regarding competition from less expensive American livestock and meats that had begun to move into Europe in volume after 1875. These restrictions led to aggressive U.S. government and livestock industry efforts to control disease. Action taken at a cattlemen's convention in 1890, attended by 400 delegates appointed by the governors of 11 major cattle producing states, shows the wholesomeness mindset in place at the turn of the century. At that meeting, a vote was passed in favor of inspection of American meats by the general government, in hopes that this action would give "American meats such a guarantee of purity and healthiness that there would no longer be any excuse for restrictions against their importation into foreign countries."

Limiting trade of meat products from the United States to Europe by citing

Table 6.1. Major Inspection Laws and Events That Impacted the Way Meat and Poultry Are Inspected.

Year	Event of Inspection	Result
1906	*The Jungle* published (Sinclair)	Attention focused on safety of food from packers
1906/07	Federal Meat Inspection Act	Inspection required for all meat sold for interstate and foreign commerce
1926	Voluntary poultry inspection	Inspection required for all poultry sold for interstate and foreign commerce
1938	On-the-farm slaughter sold publicly	Commercial slaughter operations restricted to packing plants
1942	War-time intrastate meat inspection	Meat and poultry inspection extended to intrastate commerce until 1945
1957	Poultry Product Inspection Act	Antemortem, postmortem, and processing inspection required of poultry in interstate commerce
1958	Humane Slaughter Act	Humane slaughter required of animals providing products sold to federal agencies
1967	Wholesome Meat Act	Meat inspection extended to products in intrastate commerce by a state–federal cooperative program
1968	Wholesome Poultry Act	Inspection of poultry and poultry products required by USDA; poultry extended to intrastate commerce.
1978	Humane Methods of Slaughter Act	Inspection legislation amended to require that meat inspected and approved be only from livestock slaughtered by humane methods
1980	Total Quality Control Concept (TCC)	Control initiated in some plants on voluntary basis. FSIS begins shift from strictly inspection to setting process controls which prevent food safety problems.
1984	Streamlined Inspection System (SIS)	Field test which allowed greater plant involvement in the inspection process—discontinued in 1992.
1989	Performance Based Inspection (PBIS)	System of structured, computerized schedules now guides processing inspectors on what to inspect and when to inspect—revised in 1990
1990	Hazard Analysis and Critical Control Point (HACCP)	A concept introduced which is a highly specialized system for identifying points in the food production system where a hazardous or critical situation could occur and takes steps to keep those critical control points under control
1990	Nutrition Labeling and Education Act	Requires nutritional labeling of all foods under FDA jurisdiction
1994	Safe Food Handling Statement	Food-handling instruction labels are required by FSIS to be placed on packages of raw meat and poultry products.
1994	Mandatory Food Labeling	FSIS requires that all meat destined for retail meat counters have nutritional labels.

Table 6.2. Major Food Safety Regulation Laws and Events

Year	Food Safety Regulation	Result
1938	Food, Drug and Cosmetic Act	Prohibits interstate commerce of adulterated and misbranded food, drugs, devices, and cosmetics
1954	Pesticide Chemicals Amendments to the Food Drug and Cosmetic Act	Gave FDA jurisdiction over the use of pesticides on raw agricultural products
1958	Food Additive Amendment to Food Drug and Cosmetic Act	Gave FDA jurisdiction over the use of food ingredients and additives
1958	Delaney Clause in the Food Additive	Clause in the amendment prohibiting carcinogens in foods, establishing a zero tolerance
1960	Color Additives Amendment to the Food, Drug, and Cosmetic Act	Gave FDA jurisdiction to restrict the use of color additives to those listed by FDA as safe
1966	Fair Packaging and Labeling Act	Requires accurate information on package labels
1976	Vitamins and Minerals Amendment to Food, Drug and Cosmetic Act	Gave FDA authority to establish maximum levels of potency of vitamins and minerals and declared foods above those levels as drugs
1983	Spice irradiation	FDA cleared the use of irradiated spices in foods
1986	Irradiated foods	FDA cleared the irradiation of fruits, vegetables, and pork
1991	Irradiated poultry	FDA cleared the irradiation of poultry
1992	FDA Food and Ingredient Labeling	FDA requires mandatory labeling for foods and food ingredients under their jurisdiction

Source: Adapted from *Food Processing Magazine* (1990) Food Regulatory Chronology 1940–1990.

meat inspection discrepancies and livestock production practices has also been a recent occurrence. For example, in 1989, the European Economic Community (EEC) banned the importation of beef from the United States if that beef had been harvested from a steer or heifer given a growth promotant. These Food and Drug Administration approved growth promotants are used to enhance animal leanness and to improve the feed efficiency and daily weight gain of the cattle. The proper use of these growth promotants has been proven safe by the scientific community in the United States and Europe because properly used growth promotants result in very negligible changes in the chemistry of the muscle. In fact, it is impossible to examine meat products and determine if it came from treated or untreated animals. This scientific information led some U.S. officials to conclude that one reason the EEC took this action was to enact an artificial trade barrier.

Congress passed a general meat inspection act in March 1891, providing for the inspection of cattle. The legislation was expanded to include pork in 1894. The act was established primarily with foreign trade in mind. A decade later, the safety and wholesomeness of meat came under fire when Upton Sinclair's 1905 book *The Jungle,* with its description of poor sanitary conditions in the meat-packing industry, aroused a storm of public apprehension and protest. In part, as a result of this book, the Federal Meat Inspection Act was passed in 1906 requiring inspection of livestock and their carcasses destined for interstate and foreign commerce. Similar federal inspection was extended to poultry products in 1957 when the Poultry Product Inspection Act was passed by the U.S. Congress. In 1967 the Wholesome Meat Act and in 1968 the Wholesome Poultry Act were passed, updating the previous Acts to include mandatory inspection of meat and poultry to be processed and sold within the same state (intrastate inspection). Thus, all meat and poultry destined for sale to consumers must be inspected for safety and wholesomeness at the processing facility. These two Acts required state inspection systems equal to or better than the federal inspection system. In 1991, there were 27 states that conducted meat intrastate inspection and 23 states that performed intrastate poultry inspection. For states that have their own inspection system, the federal government assists those states by assuming 50% of the cost of operating the state system. Meat processing facilities that utilize state inspection systems are not permitted to sell and transport meat products across state lines.

The Poultry and Meat Acts passed by Congress gave FSIS the responsibility of implementing the inspection effort and of setting up procedures that fulfill the objectives of the Acts. These inspection procedures and rules can be found in the Code of Federal Regulations for Animal and Animal Products commonly referred to as the "Federal Regs."

Meat and Poultry Safety Inspection Regulation

The Food Safety and Inspection Service is the primary agency with the responsibility for meat and poultry inspection. Because meat and poultry inspection is mandatory, FSIS inspection programs are paid for by the U.S. taxpayer at a cost of less than $2.00 per person per year. If a processing facility is in production for more than 8 hours per day or during holidays, then the company is required to compensate the government for the cost of overtime inspection. The agency is divided into five major programs: Inspection Operations, Science and Technology, International Programs, Regulatory Programs, and Administrative Management. All except Administrative Management are directly involved with inspection.

The following is a description of the responsibilities of the divisions within FSIS that are directly involved with inspection activities:

- *Inspection Operations* oversees the inspection of all meat and poultry plants in the United States that move product across state lines, administers the federal–state cooperative inspection program, performs residue monitoring operations in plants, and coordinates FSIS action for handling emergency contamination problems.

- *Science and Technology* provides guidance to existing inspection services, develops and enhances the scientific basis for the agency's inspection programs, and refines and modernizes meat and poultry inspection systems, standards, and procedures.
- *International Programs* carries out requirements of the federal meat and poultry inspection laws to review the wholesomeness of imported meat and poultry products. This program includes the review of foreign inspection systems to ensure they are equivalent to U.S. standards, reinspects imported meats entering U.S. commerce, represents U.S. interests throughout the world to minimize regulatory impediments to meat trade, and coordinates the inspection and certification of meat and poultry products for export.
- *Regulatory Programs* provides FSIS management officials with an overview of the effectiveness of food safety and inspection programs by conducting systematic on-site and special reviews of FSIS programs. This group within the agency also directs agency compliance activities, reviews and approves labels destined to be used on federally inspected meats, and evaluates and sets standards for food ingredients, additives, and compounds used to prepare and package meat and poultry products.

Inspection Procedures for Meat and Poultry

Objective of Inspection

There are three general types of hazards that can occur and, thus, compromise the safety of the food supply (biological, chemical, and physical).

Diseases caused by microorganisms and parasites constitute potential public health problems. Biological hazards include brucellosis, tuberculosis, cysticercosis, and trichinosis. Meat products are randomly screened by FSIS for bacterial pathogens after production. Samples are sent to laboratories for microbial testing, particularly looking for *E. coli* 0157:H7, *Salmonella, Campylobacter, Yershina, Staphylococcus, Listeria, Clostridium perfringens,* and *Clostridium botulinum*. Certain molds, yeasts, and viruses can also compromise the safety of foods. Meat and poultry products are screened for molds and yeasts organoleptically; however, viruses are not screened.

Chemical hazards include natural and man-made environmental contaminants and potential animal health product residues found in livestock not withdrawn for the required time before marketing, as defined by FDA.

Physical hazards, such as bone chips, glass, plastic, and metal can also be a potential food safety concern, and their elimination from entering the food chain is a responsibility of the processing company under the supervision of meat and poultry inspectors.

Inspection Organization and In-Plant Activities

FSIS Inspectors

During 1991, in the United States there were 6400 meat and poultry processing plants under federal inspection. More than 7800 FSIS employees were responsible for conducting inspection in these plants. United States inspection activities are managed by a tiered network consisting of the FSIS headquarters in Washington DC, five regional offices, 26 area offices, and 188 inspection circuits. There are 7400 federal inspectors with regional, area, circuit, and in-plant responsibilities who are involved with the day-to-day inspection procedures. There are two types of inspectors: lay food inspectors (6200) and veterinary inspectors (1200). The latter are graduate veterinarians. There is an inspector-in-charge who is responsible for one or more plants within a circuit. Collectively, they inspected 6.9 billion livestock and poultry in 1992 (Table 6.3).

Table 6.3. The Number of Meat Animals and Poultry Inspected at Slaughter in Federally Inspected Plants, 1982–1992

Species	1982	1987	1991	1992
Cattle	33,260,932	34,811,000	29,619,712	30,759,499
Calves	2,647,362	2,779,200	1,463,005	1,352,864
Swine	80,593,850	76,387,900	81,297,724	89,210,132
Goats	79,291	159,000	190,955	224,704
Sheep and lamb	5,971,542	5,095,600	4,448,621	5,129,39
Equine	192,207	246,000	236,467	243,585
Other			2,180	3,688
Total	**122,745,184**	**119,478,700**	**117,258,664**	**126,923,811**

Class	1982	1987	1991	1992
Young chickens	4,079,196,000	4,927,454,000	6,145,776,555	6,368,648,885
Mature chickens	196,111,000	193,055,000	171,016,415	180,839,923
Fryer-roaster turkeys	6,309,000	5,164,000	2,607,173	1,403,436
Turkeys	153,602,000	216,489,000	273,540,739	275,801,223
Young turkeys	1,245,000	1,482,000	2,261,426	2,394,944
Mature turkeys	19,404,000	23,093,000	21,065,519	18,027,590
Ducks	984,000	1,555,000	5,484,840	5,036,099
Other				
Total	**4,456,851,000**	**5,368,292,000**	**6,621,752,667**	**6,852,152,100**

Source: USDA-Food Safety and Inspection Service.

FSIS Areas of Responsibility

The direct involvement of FSIS in the manufacture of meat and poultry products is extensive. The following is a list of in-plant inspection and in-plant related activities.

- Facilities construction, equipment approval, and operational sanitation
- Antemortem inspection
- Postmortem inspection
- Product reinspection and manufacturing
- Laboratory determination and assays
- Developing hazard analysis and critical control point plans
- Control and restriction of condemned products
- Marking (stamping) and labeling

Facilities Construction and Operational Sanitation

In order for a meat processing plant to be eligible for inspection it must meet FSIS regulations for construction, including water supply, plumbing, drainage, equipment, lighting, walls, floors, ceilings, partitions, doors, windows, rails, floor drains, and hand-washing basins. If all these standards are not achieved and maintained, then inspection will be withdrawn, which means the company can no longer function. For prospective new processing facilities the building plans and specifications must be approved by FSIS before construction begins. In addition to the criteria mentioned above, the product flow through the plant is also checked. For example, the design and construction of a meat processing facility must be done in a fashion to prevent fresh meat from cross-contaminating cooked meat. Additionally, all equipment that contacts product must have been approved by FSIS; and the source of water must be from a municipal system or otherwise shown to be potable. In 1990, the Facility, Equipment and Sanitation Division of FSIS reviewed blueprints of 3678 plants and 2893 drawings of equipment.

Prior to the beginning of the day's production at a processing facility, an examination of the establishment and premises is made by the inspector. The inspector examines the sanitary conditions, determines if facilities continue to meet building and equipment regulations, checks that adequate lighting is available for observing production and inspecting animals, carcasses, and product (minimum of 50 footcandles in all inspection areas), tests to determine if clean, hot water not less than 77°C (170° F) is available in sanitizing compartments, and looks for any other criteria that will impact the plant operations and safe and wholesome food-handling practices.

Antemortem Inspection

All livestock offered for slaughter in a federally inspected processing facility must be examined on the premises of the establishment by an inspector on the day and before slaughter. The animal is observed both in motion and at rest, in order to observe any conditions that may raise questions as to their general health. Animals suspected of disease or showing other conditions that might cause condemnation are retained, termed U.S. Suspect, slaughtered in a group, and more closely scrutinized during postmortem inspection. If during live animal inspection, animals have an abnormal temperature or show obvious symptoms of disease, then the animal is termed U.S. Condemned, and identified for elimination from the human food chain.

The following are some common antemortem conditions that would require livestock to be classified as U.S. Suspect or U.S. Condemned:

U.S. Suspect

- Crippled animals and nonambulatory animals or downers
- Livestock that have reacted to a tuberculin test
- Cattle with minor cancer eye (minor epithelioma)
- Cattle with minor brisket edema (anasarca)
- Hogs suspected of being affected with erysipelas
- Livestock affected by vesicular exanthema (blister on lips, tongue, and feet)
- Immature livestock

U.S. Condemned

- Dead or dying livestock
- Swine with temperatures greater than 41.1°C (106°F) and cattle, sheep, goats, and equine with body temperatures greater than 40.5°C (105°F) [The normal body temperatures of poultry are relatively high, 40.5°C (105°F) for chicken and 41.7°C (107°F) for turkeys, and therefore this criteria is not considered.]
- Livestock showing symptoms of certain metabolic, toxic, nervous, or circulatory disturbances, nutritional imbalances, or infectious or parasitic diseases including (but not limited to) anaplasmosis, ketosis, rabies, listeriosis, forage poisoning, generalized osteoporosis, acute influenza, generalized edema, advanced cancer eye (epithelioma)
- Swine with cholera (Other hogs from a lot that has had a hog condemned for cholera must also be held and slaughtered separately from the other hogs.)
- Goats that have reacted to a test for brucellosis

Animals that have been suspected of having been treated with or exposed to any substance that may leave a biological residue above approved FDA tolerance levels in edible tissue are identified as U.S. Condemned. These animals may be held under the custody of the inspector until the animal's metabolic processes have reduced the residue sufficiently to make the meat and edible by-products fit for human food, upon which the livestock would be released for slaughter.

In addition to assuring that all animals entering the slaughter facilities are healthy and fit for food, inspectors are responsible for enforcing the Humane Methods of Slaughter Act of 1978. Under the provisions of this act, all animals (except those slaughtered under the Kosher exemption) must be stunned prior to exsanguination and must be handled humanely by plant employees.

Postmortem Inspection

The most intense phase of meat and poultry inspection occurs during postmortem examination. The process of this inspection occurs under the supervision of a highly trained "vet inspector," who has expertise in the areas of anatomy, physiology, microbiology, and pathology of animals. Lay inspectors are trained to carefully observe the slaughter procedures and identify and retain carcasses (called U.S. Retained) and carcass parts that appear abnormal. It is the inspector-in-charge's responsibility to make the final decision as to whether a carcass or part is condemned. At the final inspection, the FSIS inspector will determine whether only the affected portion of the carcass or the entire carcass and parts should be U.S. Condemned. Table 6.4 shows the percentage incidence of livestock and poultry carcasses condemned in 1992.

Processing facilities that slaughter large numbers of livestock or poultry daily have multiple inspectors on site, usually located at three locations on the slaughter floor, (head, viscera, and final carcass inspection stations). Identification of the carcass and its parts (examples liver, lung, heart etc.) is maintained throughout the slaughter process in order to allow U.S. Retained and U.S. Inspected and Condemned carcasses to be traced to its carcass parts that have been removed during the process and vice versa.

Postmortem Condemnation

The following are leading causes for condemnation of livestock and poultry carcasses and carcass parts:

- *Contamination* is one of the leading causes for condemnation of livestock and poultry carcasses. It is the most common reason for carcass trimming.
- *Pneumonia* is a leading cause of condemnation of beef, lamb, and pork carcasses. An inflammation and deterioration of the lung tissue, pneumonia is caused by bacteria, viruses, fungi, parasites, and/or plant toxins.

Table 6.4. Number of Carcasses Inspected and Number of Carcasses Condemned by FSIS Inspectors in 1992

Species or Class	Amount Inspected	Amount Condemned	Condemned as a Percentage of Those Inspected
Cattle	30,759,499	150,417	0.49
Calves	1,352,864	23,258	1.72
Swine	89,210,132	205,236	0.23
Goats	224,704	1,373	0.61
Sheep	5,129,339	19,922	0.39
Equine	243,585	954	0.39
Other	3,688	11	0.30
Total livestock	**126,923,811**	**401,171**	**0.32**
Young chickens	6,368,648,885	61,548,308	0.97
Mature chickens	180,839,923	7,981,840	4.41
Fryer-roaster turkeys	1,403,436	6,991	0.50
Young turkeys	275,801,223	2,214,421	0.80
Mature turkeys	2,394,944	84,188	3.52
Ducks	18,027,590	278,638	1.55
Other	5,036,099	39,951	0.79
Total poultry	**6,852,152,100**	**72,154,337**	**1.05**

Source: USDA-Food Safety and Inspection Service.

- *Abscess/pyemia* leads to significant condemnation among beef, lamb, and pork carcasses. Abscesses are generally found in the form of pus pockets caused by an infection in the tissue. This is the leading cause for condemnation among pork carcasses. Liver abscesses are a leading cause of liver condemnation. Pyemia is a result of pathogenic bacteria in the circulatory system leading to the development of abscesses in carcass tissues and organs.
- *Septicemia* is the leading cause for condemnation of poultry carcasses. Septicemia is a form of blood poisoning and often results in pyemia.
- *Arthritis* which results in the inflammation of joints is the second leading cause of condemnation in hogs.
- *Caseous lymphadenitis* or the formation of a cheesy/pussy inflammation of the lymph nodes is the leading cause of condemnation of lamb carcasses.
- *Airsacculitis* is the leading cause of condemnation of poultry carcasses seen as an inflammation of the bird's air sacs.

Product Reinspection and Manufacturing

The jurisdiction of FSIS meat and poultry inspectors extends to the fabrication and further processing departments of a a packing plant. Each meat and poultry plant performs different processing functions. Some plants only slaughter livestock, whereas others conduct a variety of value-added processes, including fabrication of carcasses, producing ground meats and sausages, or smoking and curing meats. Therefore, it is important for each processing FSIS inspector to be fully acquainted with the details of good manufacturing processes. It is the inspector's job to make sure that these processes are carried out under sanitary conditions and to protect the consumer against the use of harmful substances in the formulations of products. All formulas and ingredients used in processing products are approved and filed with the inspector-in-charge. Only minor deviations are permitted. Each step in the cutting, trimming, boning, slicing, grinding, curing, smoking, cooking, canning, packaging, and rendering is under surveillance of meat processing inspectors. The inspector checks raw materials entering the processing facility which must bear the inspection label. Only federally inspected meat and poultry may enter another federally inspected plant. In addition, the inspector will randomly collect samples of final products for formulation, nutritional claim, residue, and microbiological laboratory testing.

Hazard Analysis and Critical Control Point

The Hazard Analysis and Critical Control Point (HACCP) concept has been used successfully in the canning industry (thermally processed, low-acid foods packaged in hermetically sealed containers) for over 20 years and now FSIS is moving toward reliance on HACCP in other areas of processing, as a form of safety regulation. The HACCP concept is a systematic approach to hazard identification, assessment, and control. A HACCP plan is implemented by plant personnel and monitored for effectiveness by FSIS inspectors. Traditional inspection is primarily focused on diseased animals and on observing problems with the final product, which is especially ineffective in identifying microbiological hazards in the final product before the food item is released to the public. HACCP on the other hand is a preventative system that attempts to keep microbiological, chemical, and physical hazards from entering the meat or poultry product at critical places along the production line at the processing facility.

In order for a HACCP plan to be truly effective in eliminating hazards from foods it should encompass every phase of production, from the time the animal is born, through the processing plant, to the consumer's table. Therefore, although a HACCP plan for the processing facility alone would minimize hazards, it would not totally eliminate food hazards.

Disposal of U.S. Condemned Meat and Poultry

Under the supervision of the FSIS inspector, the animals, carcasses, and carcass parts that have been condemned usually undergo severe heat treatment (tanked) which destroys any hazardous organisms. If the tanking process is done at another location, the condemned product is clearly marked with a denaturant which is poured over the product. All tanks and equipment used for rendering, preparing, or storing inedible product must be in rooms separate from those used to store and prepare edible product, and the FSIS guidelines require that there be no connection between inedible and edible product rooms. Failure to meet these criteria would result in FSIS withdrawing inspection, putting the meat or poultry processor out of business.

Laboratory Analysis

Thus far, we have discussed inspection procedures that are for the most part dependent on the inspector's ability to observe hazards using his senses. Additional inspection procedures involve the use of laboratory assays to detect biological and chemical hazards and to assure that product formulation is in compliance with FSIS regulations (Table 6.5). These laboratory analyses are performed on a random basis using a statistical sampling plan that permits a high degree of assurance that the products in the food chain are in compliance with safety and formulation regulations. FSIS tests products for certain types of bacteria (e.g., *Listeria* and *Salmonella*) and chemical residues. FDA approves the use of all livestock and poultry pharmacological agents, sets the time required before the animal may be marketed for slaughter, and sets the tolerance residue levels permitted to be found in the tissue after slaughter. These tolerance levels are set by FDA to provide a 100-fold margin of safety to the public. A residue level found above the approved tolerance level is referred to as a volative residue.

Standards of Identity and Composition

There are numerous meat and poultry products that can be manufactured using a multitude of meat and nonmeat ingredients. FSIS has set maximum, and in some

Table 6.5. Laboratory Samples Analyzed by FSIS in 1992

Category of Samples	Total
Food chemistry	49,185
Food microbiology and species	34,554
Chemical residues	157,422
Antibiotic residues	221,175
Pathology	10,612
Serology	466
Total	**473,414**

[a] Includes 106,133 SOS (Sulfa-On-Site) tests.

[b] Includes 117,858 STOP (Swab Test on Premises) and 79,666 CAST (Calf Antibiotic Sulfa Test) analyses.

Source: USDA, Food Safety and Inspection Service.

cases, minimum levels of these ingredients that can be used in the formulation of meat and poultry products. The use of nonmeat ingredients (e.g., nitrite) in meat and poultry products has been approved by FDA and are tested for compliance in the final product by FSIS.

In some products, FSIS also regulates the amount of fat, moisture, and protein in a product as well as the manner in which the product should be prepared. These regulations are referred to as product standards of identity or standards of composition. FSIS and FSIS approved laboratories randomly test products to determine if the company is in compliance. The following are three standards of identity or standards of composition found in the Code of Federal Regulations:

- "Fresh Pork Sausage" (9 CFR Part 319.141) is to be prepared with fresh and/or frozen pork and no pork byproducts, and the final raw product shall not contain more than 50% fat.
- "Ground Beef" (9 CFR Part 319.15) is manufactured using fresh or frozen beef; shall not contain added water, phosphates, binders, or extenders; if cheek meat is used it can not be greater than 25% of the product; and the final raw product shall not contain more than 30% fat.
- "Chicken Patties" (9 CFR Part 381.160) must consist of 100% chicken; with fat and skin not in excess of natural proportions; and the product may contain fillers and binders.

Standards for Preparation

In addition to setting standards for product formulation, FSIS also regulates the preparation of certain products. For example, FSIS specifies cooking temperatures for roast beef and ground beef patties, bacon and ham, and precooked chicken rolls. An example of such standards are those for trichina which is found in the Code of Federal Regulations (9 CFR Part 318.10).

Trichina or *Trichinella spiralis* is a parasite that can occur in fresh pork tissue. It is also found in rats, cats, dogs, bear (and other game), horses, and man. The incidence of *Trichina* is rare among domestic hogs, because of improved sanitary conditions, statutes requiring all garbage to be cooked to at least 100°C (212°F) for 30 min before feeding to swine, and a greater frequency of total hog confinement operations. Animals are infested with the parasite by eating an infected rat or when consuming the uncooked or improperly cooked infested tissue or viscera of other animals. The larvae pass to the small intestine of the host and remain there until they reach sexual maturity. After mating, the female penetrates the lining of the small intestine where she gives birth to the new larvae. These new larvae are carried by the circulatory system to striated muscle where they remain until they die, unless the cycle described above is repeated.

FSIS has set regulation for curing, heating, and/or freezing, which must be performed on all pork product that are not necessarily cooked at home or in a foodservice establishment before consumption. Bologna, frankfurters, hams,

Table 6.6. Example of Cooking and Freezing Schedules Published in the Code of Federal Regulations for the Treatment of Pork in Order to Ensure the Elimination of the Trichina Parasite

Cooking			Freezing		
Minimum internal temperature		Minimum	Maximum internal temperature		Minimum
°F	°C	time	°F	°C	Time
120	49.0	21 h	0	−17.8	106 h
122	50.0	9.5 h	− 5	−20.6	82 h
124	51.1	4.5 h	−10	−23.3	63 h
126	52.2	2 h	−15	−26.1	48 h
128	53.4	1 h	−20	−28.9	35 h
130	54.5	30 min	−25	−31.7	22 h
132	55.6	15 min	−30	−34.5	8 h
134	56.7	6 min	−35	−37.2	30 min
136	57.8	3 min			
138	58.9	2 min			
140	60.0	1 min			
142	61.1	1 min			
144	62.2	Instant			

Source: USDA-Food Safety and Inspection Service.

Canadian bacon, bacon used for wrapping around patties, and ground meat mixtures containing pork and other species are examples of products that must be processed in a manner which assures the death of the trichina parasite. Table 6.6 shows two time/temperature schedules approved by FSIS for the treatment of all pork products to ensure the elimination of the trichina parasite.

Marking and Labeling

FDA handles the labeling of all nonmeat/nonpoultry foods which make up 70% of the labels found in the grocery store. The other 30% of the labels which are placed on meat and poultry containers come under the jurisdiction of FSIS. Any word, statement, or other information required by FSIS to be on the label must be placed on the container in a prominent place and be written in a language that would allow it to be readily seen and understood. FSIS also specifies the size of the label on the container (for example, on a rectangular package the label must be the size of one entire side) and the size of the lettering on the label (for instance if the display panel is between 5 and 25 square inches, then the lettering must be 1/8 inch or greater). The label must be in English unless destined for Puerto Rico, in which case it can be printed in Spanish. Meat products destined for foreign countries may have labels printed in a foreign language. The following are the traditional FSIS label requirements (Fig. 6.1):

Figure 6.1. Traditional package labels including all the information required by USDA-FSIS (a); nutritional label required on all further processed meat and poultry products (b); and the safe food handling label required on all raw meat and poultry products (c).

Bevo's

ALL BEEF HOT DOGS

KEEP REFRIGERATED

INGREDIENTS: BEEF, WATER,
DEXTROSE, SALT, CORN SYRUP,
FLAVORINGS, SODIUM PHOSPHATES,
SODIUM ERYTHORBATE, SODIUM NITRITE

BEVO'S, INC.
REVEILLE, TX 77843

(a)

Nutrition Facts

Serving Size 1 cup (253 g)
Servings Per Container 4

Amount Per Serving	
Calories 260	Calories from Fat 70
	% Daily Value*
Total Fat 8g	**13%**
Saturated Fat 3g	**17%**
Cholesterol 130mg	**44%**
Sodium 1010mg	**42%**
Total Carbohydrate 22g	**7%**
Dietary Fiber 9g	**36%**
Sugars 4g	
Protein 25g	
Vitamin A 35% •	Vitamin C 2%
Calcium 6% •	Iron 30%

* Percent Daily Values are based on a 2,000 calorie diet. Your daily values may be higher or lower depending on your calorie needs:

	Calories:	2,000	2,500
Total Fat	Less than	65g	80g
Sat Fat	Less than	20g	25g
Cholesterol	Less than	300mg	300mg
Sodium	Less than	2,400mg	2,400mg
Total Carbohydrate		300g	375g
Dietary Fiber		25g	30g

Calories per gram:
Fat 9 • Carbohydrate 4 • Protein 4

(b)

Safe Handling Instructions

This product was inspected for your safety. Some food products may contain bacteria that could cause illness if the product is mishandled or cooked improperly. For your protection, follow these safe handling instructions.

Keep refrigerated or frozen.
Thaw in refrigerator or microwave.

Keep raw [meats or poultry] separate from other foods. Wash working surfaces (including cutting boards), utensils, and hands after touching raw [meat or poultry].

Cook thoroughly.

Refrigerate leftovers within 2 hours.

(c)

- The name of the product.
- List of ingredients, if there is more than one ingredient in the product formulation
- The name and place of business of the manufacturer.
- An accurate statement of the net weight.
- Products which have been prepared by salting, smoking, drying, cooking, chopping, and other common processes must be so described on the label, along with appropriate handling instructions (such as keep refrigerated).
- An official inspection legend with the official FSIS establishment number.

Each FSIS federally inspected plant is granted an establishment number which is for most purposes placed in the center of the official inspection mark. Each inspection legend is not only placed on the meat and poultry package label (Fig. 6.2) but also stamped on carcasses and carcass parts, including edible by-products. If a carcass or meat product is inspected and condemned, it is clearly marked as such and then disposed of in the required manner previously described (Fig. 6.3).

(a) (b) (c)

Source USDA-FSIS

Figure 6.2. The official inspection legend required to be shown on all labels for inspected and passed products from cattle, sheep, swine, and goats (size of the stamp may vary) (a); the inspection mark required to be shown on all labels for inspected and passed fresh or frozen poultry or processed poultry products (b); and the inspection mark for application to horse carcasses, carcass parts and horse meat food products (c).

U.S. INSP'D AND
CONDEMNED

Source USDA-FSIS

Figure 6.3. The "U.S. Inspected and Condemned" mark.

(a)

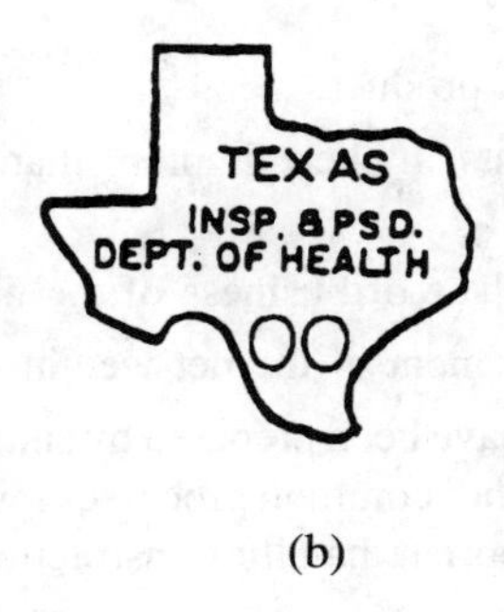

(b)

Figure 6.4. The inspection marks for meat and meat products inspected under Oklahoma (a) and Texas (b) state inspection.

Those meat and poultry products that are inspected under a state inspection system are also marked; however, the design of the seal used is distinctly different from the federal legend (Fig. 6.4). Similar to the federal inspection mark, the state mark also has the designated number of the processing facility.

Nutritional Labeling

Many foods at the grocery store are required by both FSIS and FDA to have nutritional information on the label. All packages of further processed meat and poultry products will be required to include labels that have, per cooked serving, information on calories, calories from fat, total fat, saturated fats, cholesterol, sodium, total carbohydrates, dietary fiber, sugars, protein, vitamin A, vitamin C, calcium, and iron. The long nutritional label format which applies to most further processed products also includes levels of certain nutrients required for persons on 2000 kcal and 2500 kcal daily intake diets (Fig. 6.1). Raw meat and poultry products are under a voluntary labeling system that does not require each container to have a nutritional label but recommends nutritional information be posted in a designated area of the grocery store or at the meat case. Raw and further processed products from seafood and game animals comes under FDA jurisdiction and, therefore, all containers must display nutritional information. Meat and poultry products destined for foodservice are not required to have nutritional labeling. Any meat or poultry product that makes a health claim, such as low fat or low sodium, is required to have a nutritional label on the container. A meat or poultry product can be labeled *low fat* if it has 3 g or less fat per 100 g of product and less than 30% of its total calories are from fat. A *low-sodium* meat item must have 140 mg or less sodium per 100 g of product. In addition to nutrient claims for fat, cholesterol, saturated fatty acids, calories, and sodium, claims regarding high protein, iron, Vitamin A, Vitamin C, and others can also be made if the meat or poultry product meets the requirements stipulated by FSIS.

Safe Food Handling Instruction Labels

FSIS's jurisdiction only covers the processing phase of meat and poultry. Although they do have the legal ability to extend inspection to wholesale clubs and retail grocery stores, because of budgetary and personnel limitations they have not exercised that option. In order to help extend proper food-handling procedures beyond the meat and poultry packer and processor, FSIS requires that all meat packages have safe-food-handling instructions. Food borne illness outbreaks on the west coast of the United States in 1993 prompted this label. All meat and poultry products not considered as "ready to eat" will require one of two labels on the product container. There is one safe-food-handling label for use on consumer packages (Fig. 6.1) and a different safe-food-handling label for meat and poultry products destined for foodservice. Both labels provide storage, sanitation and hygiene, and cooking and postcooking information. The only difference between the two safe-food-handling labels is the postcooking recommendations. The consumer product label says, "Refrigerate leftovers within 2 h," whereas the foodservice label says "Keep foods hot at 60°C (140°F) or higher. Immediately after service, refrigerate leftovers."

Imported Products Inspection

The safety of meat and poultry products from other countries is also under the jurisdiction of the Food Safety and Inspection Service. Before any product can be shipped to the United States, FSIS first determines if the meat inspection system in that foreign country is equivalent to the U.S. inspection system. In order to be certified to import meat to the United States, the country's inspection must be organized and administered by the national government, have the organizational structure and staffing so as to ensure uniform enforcement, have competent and qualified inspectors, possess the authority to enforce inspection laws, and uphold the inspection, sanitation, quality, species verification, and residue standards applied to products produced in the United States. After foreign packing plants have been certified, then USDA conducts spot inspections in those processing facilities to assure that the established inspection standards are being achieved.

In addition to a country and its processing facilities being certified to export to the United States, an import shipment of fresh and frozen meat products must also be accompanied by an inspection certificate. Except in Canada, the importer makes application for inspection of the product arriving at a point of entry into the United States. At the point of entry, each shipment is reinspected visually by FSIS personnel to determine if an import certificate is on hand and for label compliance. FSIS maintains a large computerized database on products from each country. Based on the history of compliance with FSIS rules and regulations, the computerized system advises inspectors when to select shipments for examination and when to obtain samples for microbiological, residue, and standard of identity testing. When meat and poultry products are shipped from Canada to the United States, under a "free-trade" agreement, the Canadian Inspection System contacts FSIS which randomly determines which lots of imported product will be held for inspection. The remainder of the Canadian shipments pass into the United States without being reinspected.

On January 1, 1991 there were 1370 foreign processing facilities from 29 countries, authorized to export products to the United States. During 1992, over 2.6 billion pounds of meat and poultry products were approved for entry (Table 6.7). Approximately 10.7 million pounds of meat and poultry products were denied entry into the United States (Table 6.7).

Table 6.7. Pounds of Meat and Poultry Product Imports Passed for Entry and Refused Entry into the United States in 1992.

Country of Origin	Pounds Refused Entry	Pounds Passed for Entry
Argentina	462,523	114,394,172
Australia	3,041,644	852,943,892
Belgium	90,196	9,459,764
Brazil	167,259	43,996,523
Canada	4,775,301	763,066,108
Croatia	1,332	2,523,909
Costa Rica	112	33,936,816
Czechoslovakia	0	34,296
Denmark	486,294	145,318,223
Dominican Republic	62,710	13,460,691
Finland	1,924	2,166,290
France	382	426,193
Germany	0	139,135
Guatemala	115,496	14,890,944
Honduras	100,155	35,245,531
Hong Kong	610	1,086,340
Hungary	266	14,699,412
Ireland	0	1,212,085
Israel	225	723,226
Italy	0	921,865
Japan	0	12,227
Mexico	47,552	986,374
Netherlands	74,613	16,218,585
New Zealand	1,122,522	502,724,221
Nicaragua	0	13,582,244
Poland	0	9,134,098
Romania	5,192	846,523
Slovenia	0	72,576
Sweden	21,005	6,220,853
Switzerland	0	70,927
United Kingdom	11	104
Uruguay	178,041	12,577,751
Yugoslavia	3,150	4,631,380
Total pounds	**10,758,515**	**2,617,723,278**

Source: USDA-Food Safety Inspection Service.

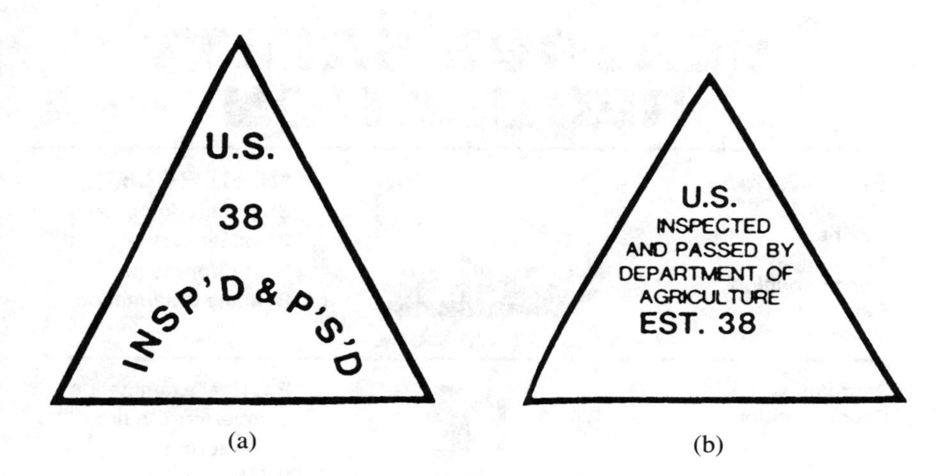

Source USDA-FSIS

Figure 6.5. Inspection mark for application to (a) exotic animals carcasses, primal parts, and cuts and (b) to all labels for meat and meat product packages.

Voluntary Inspection of Exotic Game Animals

In addition to the mandatory inspection programs for livestock and poultry, FSIS also has a voluntary inspection program for exotic game such as antelope, buffalo, deer, elk, and reindeer. Some state inspection services have implemented voluntary inspection systems for other species. For example, Texas has a voluntary inspection program for ratites (e.g., ostrich). This type of inspection is paid for by the processing facility requesting inspection. Generally, the facility and sanitation requirements are the same as those required for mandatory inspection of livestock, and each species has a set of antemortem and postmortem inspection procedures. Some county and municipal health departments and many restaurants and retailers serving exotic game meat require this type of inspection. Figure 6.5 shows the federal inspection mark placed on inspected and passed exotic game meat and meat products.

Seafood Inspection

At present, seafood inspection is under the jurisdiction of the United States Food and Drug Administration. However, the Department of Commerce conducts a voluntary inspection program, much as is done by FSIS on exotic game animal inspection. Primarily, the U.S. Department of Commerce, under this "fee for service" program, is moving toward a HACCP approach and randomly tests domestic processing facilities for cases of fraud and misleading packaging; how-

SEAFOOD SAFETY REGULATION

HAZARDS*		CONTROLS*
Chemicals Toxins Pathogens Decomposition Parasites Physical Objects	 **AT THE DOCK**	Shipment documentation Physical examination Laboratory tests Hygiene & sanitation
Pathogens Decomposition	 **HANDLING & STORAGE**	Physical examination Temperature & time monitoring Lab tests Hygiene & sanitation
Pathogens Decomposition Physical Objects Chemicals Parasites	 **PROCESSING & COOKING**	Temperature & time monitoring Processing controls Hygiene & sanitation
Pathogens Decomposition	 **PACKAGING & STORAGE**	Temperature & time monitoring Hygiene & sanitation Physical examination
Pathogens Decomposition	 **RETAIL (Model food code)**	Temperature & time monitoring Employee health Date-marking Consumer advisories

*Identification of hazards and the proper way to control those hazards will be specific to each seafood product and processing operation.

Companies will have to keep records at each critical step in the process. As part of its inspection process FDA will review the records.

Figure 6.6. FDA's proposed seafood safety regulation is modeled after the HACCP safe food production system, a prevention oriented approach embraced by the food industry.

ever, the agency also examines fish and seafood products for decomposition. The Food and Drug Administration has jurisdiction over the inspection of imported fish and seafood; as well as performs some microbiological testing of domestic products.

Concerned consumer groups are pressuring the U.S. Congress to require mandatory inspection of fish and seafood processing facilities. Given the volume and size of the fish and seafood harvested, it would be difficult for an inspector to identify food safety hazards using organoleptic properties (sight, smell, and feel) as is traditionally done with meat and poultry. The proposed FDA inspection procedures for fish and seafood would be built around a Hazard Analysis and Critical Control Point system (Figure 6.6). This HACCP inspection system would be administered by the Food and Drug Administration and would be separate from the voluntary program conducted by the Department of Commerce.

Selected References

Food Safety and Inspection Service, United States Department of Agriculture. 1990. *HACCP: HACCP Principles for Food Production, Hazard Analysis and Critical Control Point System National Advisory Committee on Microbiological Criteria for Foods*. U.S. Government Printing Office, Washington, DC.

Food Safety and Inspection Service, United States Department of Agriculture. 1990. *Nutrition Labeling of Meat and Poultry Products: FSIS Backgrounder*. U.S. Government Printing Office, Washington, DC.

Food Safety and Inspection Service, U.S. Department of Agriculture. 1992. *Meat and Poultry Inspection: 1992 Report of the Secretary of Agriculture to the U.S. Congress*. U.S. Government Printing Office, Washington, DC.

Food Safety and Inspection Service, United States Department of Agriculture. 1993. *Two-Track Approach to Risk-Based Inspection System: FSIS Backgrounder*. U.S. Government Printing Office, Washington, DC.

Kiekl, Elmer R. and V. James Rhodes. 1960. *Historical Development of Beef Quality and Grading Standards,* Research Bulletin 728, University of Missouri, College of Agriculture, Agricultural Experiment Station.

Office of the Federal Register National Archives and Records Administration. 1993. *Code of Federal Regulations: Animal and Animal Products,* 9, Part 200 to End. U.S. Government Printing Office, Washington, DC.

Stenholm, Charles W. and Daniel B. Waggoner, 1989. A congressional perspective on food safety, *J. Amer. Vet. Med. Assoc.* **195**(7):P 916–921.

The International Commission on Microbiological Specifications for Foods (ICMSF) of the International Union of Microbiological Societies. 1988. *Micro-Organisms in Foods 4: Application of the hazard analysis critical control point (HACCP) system to ensure microbiological safety and quality*. Blackwell Scientific Publications, Palo Alto, CA.

7

Grading

Daniel Scott Hale

Introduction

Meat Inspection Versus Meat Grading

It is important to differentiate between USDA Meat and Poultry Grading and USDA Meat and Poultry Inspection. Meat and Poultry Grading are voluntary programs performed by the Agricultural Marketing Service of the USDA. These grading programs segment carcasses and/or meat products into smaller, more homogeneous groups based on factors that estimate taste appeal of meat cuts, consumer acceptance, processing characteristics, and lean meat yield. Contrarily, Meat and Poultry Inspection are mandatory systems overseeing the production of safe and wholesome meat and poultry products and are conducted by the Food Safety and Inspection Service of the USDA. Grading is paid for by the meat processor; on the other hand, Federal Meat and Poultry Inspection programs are paid for by the U.S. taxpayer.

Objectives of Grading

Voluntary grading systems have been developed and implemented for the purpose of segmenting market cattle, sheep, hogs, and poultry into smaller, more homogenous groups. These grading systems are generally based on measurable traits, at the packing plant level, which predict important marketing and consumer acceptance characteristics of meat and poultry products. Grading was first developed primarily for purposes of market reporting. Livestock and poultry producers wanted to be informed about the price of "superior," "average," and "inferior" animals in the market place; to be assisted in making informed decisions regarding when and at what price they should offer their animals for sale, and to be assisted in making management decisions as to what to feed, how long to feed, and at what age to market their animals. From that beginning, the grades of livestock and poultry have served additional purposes, including packer-to-retailer communications, packer-to-food-service communications, and consumer marketing. Federal meat grading is administered by the Livestock Division and federal poultry grading by the Poultry Division of the Agricultural Marketing Service of the United States Department of Agriculture. These divisions serve two main functions: (1) to administer grades and (2) to serve as an unbiased third party between buyer and seller for the examination and certification of carcasses and cuts for grade and other factors.

History of Grading

Throughout the course of the history of the United States, there is evidence of attempts to classify livestock according to perceived quality differences. In the eighteenth century, in some instances, sales were based on a specified measurement of a fixed number of feet around the belly of the animal. Then, $1.00 was added or subtracted for every inch over or under this measurement.

Common terms that were used to describe cattle quality in the 1800s included westerns, native, fed Mexican, fed Texas, and pulp fed cattle carcasses. Some further quality distinctions were made in most markets such as extras, prime, choice, good, fair, and common. As grain feeding became more common in the early 1900s, special U.S. region classifications were rendered less useful. Greater uniformity and less seasonality, because of greater grain feeding and improved cattle selection, paved the way for uniform and nationally recognized livestock and poultry grading systems. Since 1930, there have been essentially two grade systems for carcasses in this country—a federal grading system (USDA) and a packer grading system (commonly referred to as house grades). Traditionally, the federal grades were applied largely to the better finished carcasses. The other carcasses were generally merchandised ungraded. Today, this phenomenon still occurs, and the term often used for the ungraded carcasses is "No Roll," because the carcass was not rolled or stamped with a USDA Grade.

History of Beef Grading

During the period from 1895 to 1920, livestock producers and the public generally held that the large meat packers were exploiting both producers and consumers. In 1914, Congress appropriated funds to establish the Office of Markets and Rural Organization for the purpose of developing a market reporting service. The government relied heavily on studies performed at the University of Illinois, under the leadership of Professor H.W. Mumford and Professor L.D. Hall. Using cattle at the Union Stockyard and carcasses at Chicago area meat packers, they undertook the task of formulating and testing a uniform set of standards for market cattle. In their studies, they used the terms Prime, Choice,

Good, Medium, and Common as grade designations, indicating gradation in the degree of finish and conformation. The government continued work on broadening the foundation of the original investigation and in 1923 issued tentative National standards for carcass beef (USDA Bulletin No. 1246, "Market Classes and Grades of Dressed Beef"). In 1924, Congress passed the United States Agricultural Product Inspection and Grading Act which authorized the federal grading of livestock and meat.

Although grading standards were developed for beef, they were not implemented until 1927. The following quote was indicative of the packer sentiment on the grading issue in the late 1920s, and interestingly the same quote may apply to some packer attitudes on grading today: "V.H. Munnecke, Armour vice-president, stated that all carcasses have to be sold, that the packer attempts to place every carcass where its particular degree of quality will bring the most money, therefore beef is in effect graded already."

An insignificant proportion of the total beef supply was graded the first few years, but the proportion of the top two grades that were graded was sizable. The percent of beef carcasses graded increased with the trend toward larger retail grocery chains and retail distribution centers. The USDA Yield Grading system was implemented in 1965. Today, 90% of the young fed beef supply (approximately 9–42 months of age) is either USDA Quality Graded, USDA Yield Graded, or both.

Official grades for veal were first established in 1928 and seven revisions have occurred, with the last amendment of the standards occurring in 1980. Until that time some veal carcasses were sold with the hide on, particularly in the northeastern United States. That change specified that grading of veal and calf could only occur after the hide is removed and only in an establishment where hide removal occurs.

Sheep and Hog Grading History

The grade standards for market hogs and slaughter sheep and lambs developed similarly to beef, but usually 1–5 years behind those of cattle and beef carcasses. Official standards for quality grading market sheep and ovine carcasses were developed and carcass stamping began in 1931. The sheep grading system has been revised five times. Yield grades were implemented in 1969 for use in conjunction with the quality grades on a voluntary basis, by users of the Federal grading service. However, yield grades of lamb remained virtually unused by the packing industry. The primary reason given for nonuse was that yield grades would be difficult to apply without significantly impairing packing plant production efficiency. In addition, a large percentage of lambs would qualify for the inferior yield grades. To make yield grades more workable, they were revised in 1992. The grading system was simplified by utilizing adjusted 12th rib fat thickness as the sole grading factor and requiring kidney and pelvic fat to be removed during the dressing process. The revision also required that all carcasses that are Quality Graded must also be Yield Graded. Today, most lamb carcasses (92.5%) are graded.

Tentative standards for grades of pork carcasses and fresh pork cuts were issued by the USDA in 1931. Similarly to lamb and beef, pork grades have been revised several times, with the last grade change occurring in 1985, at which time there was a major overhaul of the factors used in the cutability pork grades. These changes were made to attempt to create a usable system in the pork industry, given packing house production speeds of over 5000 hogs per day. Despite the change, most pork carcasses are not graded using the USDA system. Generally, each packer has an independent house grading system usually based on carcass weight and a single backfat measure taken at the last rib.

History of Poultry Grading

The Agricultural Marketing Act of 1946 authorized the development of grade standards and the identification and certification of class, quality, quantity, and condition of agricultural products. The standards for ready-to-cook poultry were added to the regulations in 1950, roasts were added in 1965, parts and certain products were added in 1969, and all provisions for grading live and dressed poultry were deleted from the standards in 1976. The additions and deletions mentioned are a result of the evolution of poultry production, processing, and marketing to a more integrated management system. Poultry companies are vertically integrated, owning the broiler from the egg to the grocery store and foodservice establishment. There is no need for live bird grades as no live poultry is merchandised. Grades are needed for the carcasses, carcass parts, and further processed poultry products offered for sale to retail and foodservice.

Veal and Calf Grading

Veal and Calf Versus Beef

Differentiation among veal, calf, and beef carcasses is made primarily on the basis of the color of the lean and secondarily using lean texture, color, shape, size, and ossification of bones and cartilage, characteristics of the fat, and the general contour of the carcass. The approximate corresponding chronological age of veal, calf, and beef is less than 3 months, 3–8 months, and 9 months or more, respectively.

Veal carcasses typically have a grayish-pink lean color, a velvety textured lean, soft pliable fat, and very red rib bones. Typical calf carcasses have a grayish red lean color, a flakier type of fat, and somewhat wider rib bones with less pronounced evidences of red color.

Conformation

Determining the final grade of veal or calf is dependent on conformation and quality. Conformation is supposed to be related to the proportions of lean, fat, and bone in the carcass. Conformation refers to the formation and contour of the carcass. Superior conformation is reflected in carcasses that are thickly fleshed and full and thick in relation to the carcass length and that have a plump, well-rounded appearance. In contrast, inferior conformation is reflected in carcasses that are very thinly fleshed and very narrow in relation to their length and that have a very angular, thin sunken appearance. Often a packer will sell whole veal and calf carcasses; therefore, the carcass appearance and shape is an important merchandising trait and is considered in grading.

Quality of Lean

Quality of lean in veal and in unribbed calf carcasses is estimated by using the following factors: overall maturity, the amount of feathering (thin fat streaks in the lean between the ribs), and the fat streaking in the flank (flank streaking). If a calf carcass is ribbed at the 12th to 13th rib interface, then the marbling in the ribeye should be used to evaluate lean quality instead of flank streaking and feathering. As the color becomes darker and bone maturity advances, increasingly higher flank streaking and feathering or marbling is required to qualify for each grade (Fig. 7.1).

Final USDA Veal and Calf Grades

The final USDA Veal or Calf Grade (Prime, Choice, Good, Standard, and Utility) is determined using a composite of the conformation and the quality of lean score. Superior quality of lean compensates without limit for deficient conformation on an equal basis. For example, a carcass with low Prime quality of lean and high Good conformation will have an average Choice final USDA Grade. Conversely, superior conformation is not allowed to compensate for inferior quality of lean score in the USDA Prime and USDA Choice grades. Therefore, a carcass with high Choice conformation and high Good quality of lean would have a final USDA Grade of high Good. In all other grades, conformation is allowed to compensate for quality of lean but only to the extent of one-third of a grade.

Beef Grades

Introduction

There are two types of beef grades which a carcass may receive: USDA Beef Quality Grades and USDA Beef Yield Grades. Under rules stated in the official

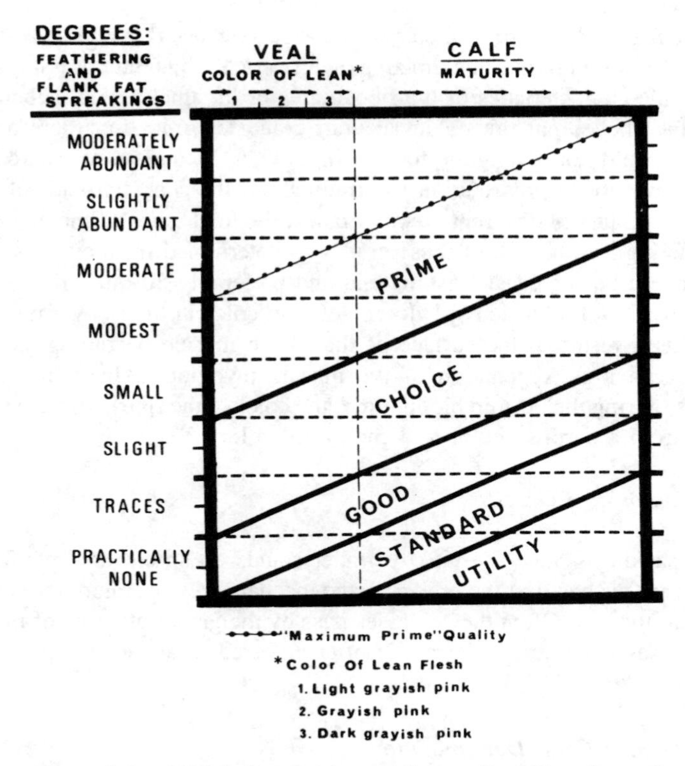

Figure 7.1. USDA chart for determining quality of lean for veal and calf carcasses.

standards, a carcass may be identified with a quality grade, a yield grade, or both.

Preparing the Carcass for Grading

To officially determine the grade of a carcass, it must be split down the back into two sides and one or both sides must be partially separated into a hindquarter and forequarter by cutting it with a saw and knife as follows: a saw cut perpendicular to both the long axis and split surface of the vertebral column is made across the 12th thoracic vertebra at a point which leaves not more than one-half of this vertebra on the hindquarter. After sawing, a knife cut exposes the ribeye cross section between the 12th and 13th rib. This procedure is commonly referred to as "ribbing." Beef carcass ribbing is normally done 21 h postmortem. Many large beef packing facilities are ribbing beef carcasses 48 h postmortem, because it is believed that the percent of carcasses qualifying for USDA Choice increases with prolonged chilling time. Carcasses are ribbed after a chilling period to maximize the bright cherry-red color of the ribeye lean and to allow the marbling in the ribeye to set up, thus maximizing quality grade. Slaughter dressing defects, that alter the characteristics of the ribeye area or the thickness of subcutaneous fat over the ribeye, may prevent an accurate grade determination; therefore, these carcasses would not be eligible for grading. When both sides of a carcass have been ribbed prior to presentation for grading and the characteristics of the two ribeyes would justify different quality grades, the final grade of the carcass shall reflect the highest of each of these grades as determined from either side. Carcasses should be ribbed at least 10 min and preferably 30 min prior to quality grading to allow for sufficient "bloom" of lean color in the ribeye muscle. The "bloom" is a result of the surface of the ribeye muscle becoming exposed to oxygen, allowing oxygenation of the muscle myoglobin. In young maturity carcasses, during the 30-min bloom time, the color of the ribeye muscle typically changes from a purplish color to a cherry-red color.

Application of Beef Grades

In large packing plants, the USDA grader stands on an elevated platform and carcasses are pushed past the grader by the mechanically operated, moving chain adjacent to the rail. Often the carcasses pass by the grader at a rate of more than 400 carcasses per hour. This rate is often referred to as "chain speed" in the industry.

Carcass Gender Class Determination

The first step in determining the quality grade of beef carcasses is to determine the gender classification of the carcass. There are five gender classes considered for the quality grading process based on evidence of maturity and apparent gender (Table 7.1).

Steer carcass characteristics include a rough and irregular shaped fat depot in the cod region, the presence of a relatively small pizzle eye (a white disk caudal to the aitch bone, which is the severed proximal portion of the penis), a relatively small pelvic cavity, and a diamond-shaped-appearing lean surface posterior of

Table 7.1. The Gender Classes of Beef Carcasses and the Eligible USDA Quality Grades

Gender Classes	USDA Quality Grades
Bull	Not eligible for Quality Grades
Bullock	Prime, Choice, Select, Standard, Utility
Steer	Prime, Choice, Select, Standard, Commercial, Utility, Cutter, Canner
Heifer	Prime, Choice, Select, Standard, Commercial, Utility, Cutter, Canner
Cow	Choice, Select, Standard, Commercial, Utility, Cutter, Canner

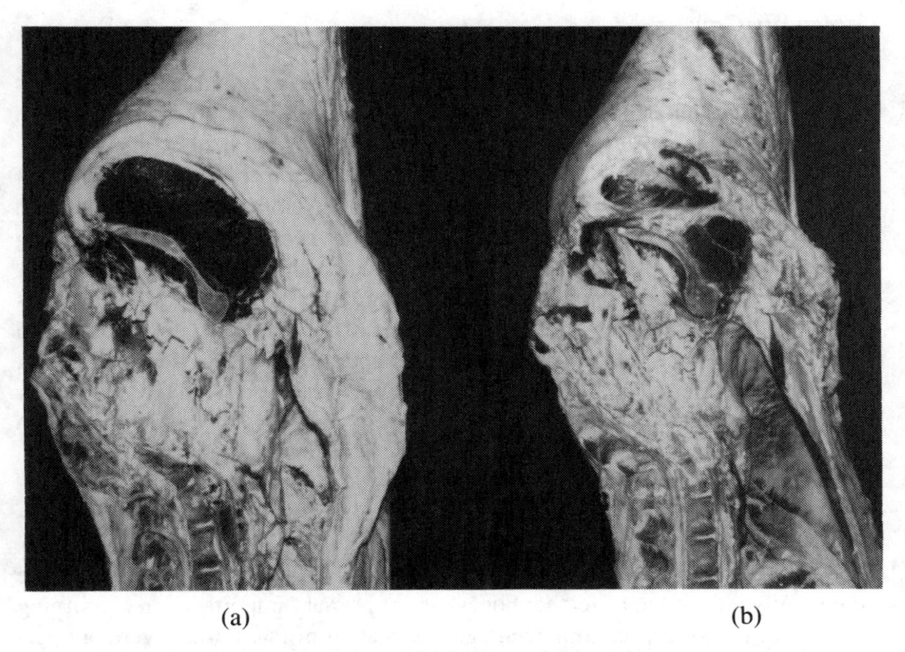

Figure 7.2. Gender Characteristics of (a) heifer and (b) steer carcasses.

the aitch bone (Fig. 7.2). In contrast to steers, *bullocks* and *bulls* are identified by a disproportionately developed round, a more prominent crest in the neck area, and a larger pizzle eye. The scrotal fat depot (compared to a steer's cod fat) of a bullock or bull is often smaller in size. The distinction between bullocks and bulls is solely on the basis of maturity. Bullock carcasses have an overall carcass maturity of "A" maturity, whereas bulls are older. The lean of bull carcasses is dark colored and coarse textured.

Heifer carcasses have a smooth-shaped, uniform fat depot in the udder region, the absence of the pizzle eye, and a slightly larger pelvic cavity and a straighter aitch bone than steers. In addition, heifer carcasses have a kidney-shaped-appearing lean surface above the aitch bone. A *cow* carcass has a large pelvic cavity, large prominent hip bones and a nearly straight aitch bone. Often the large udders on cow carcasses are removed, particularly those described by the industry as "wet bags."

Beef USDA Quality Grade

Bullock, steer, heifer, and cow carcasses are assigned quality grades based on subjective evaluations of three carcass traits—maturity (physiological), marbling (intramuscular fat), and muscle firmness.

Maturity Group

Animal maturity has been related to beef tenderness. As an animal becomes older, the amount of connective tissue in the muscle increases, and collagen within the muscle becomes less soluble. Therefore, an estimate of the maturity of the animal at the time of slaughter is a major factor considered in the Beef Quality Grading System. Maturity is described in physiological terms rather than based on chronological age, which is not known when cattle reach the packing house; however, Table 7.2 shows the approximate chronological age that corresponds to the five recognized beef maturity groups. Physiological maturity is determined by evaluating the size, shape, and ossification of the bones and cartilage of the carcass as well as the color and texture of the lean surface of the ribeye muscle at the 12th rib cross section. For skeletal maturity, particular emphasis is given to cartilage ossification differences along the vertebral column (sacral, lumbar, and thoracic vertebrae). As an animal matures, ossification (conversion of cartilage into bone) generally occurs at an early stage of development in the posterior region (sacral) and at progressively later stages of development in the anterior regions (lumbar and thoracic) of the vertebral column. Ossification amounts in the cartilaginous buttons of the split thoracic vertebrae (commonly referred to as feather bones) are of primary importance in classifying beef maturity. Reference to percentage ossification of the thoracic vertebra pertains to the top three thoracic buttons at the posterior end of the forequarter. The shape of the ribs and the appearance of the split chine bones may also be used to determine skeletal maturity.

To facilitate descriptions of differences in maturity of beef carcasses, the USDA recognizes five maturity groups—designated A, B, C, D, and E. For normal grading operations, any carcass within the "A" maturity group is designated only as "A-maturity," without regard to advanced maturity within that maturity group. However, for carcasses in the B, C, D, and E maturity groups, subdivisions of the maturity groups are made by percentage—unit positions

Table 7.2. The Bobby Beef, Veal, Calf, and the Five Beef USDA Maturity Levels, Corresponding Approximate Chronological Age, and Eligible USDA Quality Grades

Maturity Classification— Approximate Chronological Age	U.S.D.A. Quality Grades
Bobby Beef—Less than 3 weeks	
Veal—3 weeks to 3 months	Prime, Choice, Good, Standard, Utility, Cull
Calf—3–9 months	Prime, Choice, Good, Standard, Utility,
Beef: A—9–30 months	Prime, Choice, Select, Standard, Utility, Cutter, Canner
B—30–42 months	Prime, Choice, Select, Standard, Utility, Cutter, Canner
C—42–72 months	Commercial, Utility, Cutter, Canner
D—72–96 months	Commercial, Utility, Cutter, Canner
E—More than 96 months	Commercial, Utility, Cutter, Canner

within the group. For example, the very youngest carcasses eligible for the B-maturity group are designated as B^{00}, whereas the very oldest carcasses eligible for the B-maturity group are designated as B^{100}.

In the very youngest carcasses, considered as beef (A), the skeletal characteristics will appear as follows:

1. The cartilage buttons on the ends of the chine bones show no ossification, cartilage is evident on all vertebrae of the spinal column, and the sacral vertebrae show distinct separation between each vertebrae.
2. The split chine vertebrae usually are soft and porous and very red in color.
3. The rib bones have only a slight tendency toward flatness and are red in color.

Typically, as skeletal maturity advances, the following changes occur: Ossification changes become evident first in the bones and cartilage of the sacral vertebrae, then in the lumbar vertebrae, and still later in the thoracic vertebrae. In beef that is very advanced in skeletal maturity (E^{100})

1. All the split vertebrae will be devoid of red color, very hard and flinty, and the cartilage on the ends of all the vertebrae will be entirely ossified.
2. The rib bones will be very flinty, wide, and flat.

In beef carcasses, the color and texture of the lean also undergo progressive changes with advancing maturity. In the very youngest carcasses considered as beef (A), the lean will be very fine in texture and light grayish-red in color. In progressively more mature carcasses, the texture of the lean will become progressively coarser and the color of the lean will become darker red. The lean maturity cannot shift the overall maturity of the carcass more than one full maturity level from that indicated by its bones and cartilage. Moreover, carcasses designated greater than "C^{15}" skeletal maturity may not be moved into "B" overall maturity based on a more youthful lean color and texture in the ribeye.

Marbling Scores

Marbling (visible deposits of intramuscular fat) is the primary determinant of USDA quality grade after a maturity level has been assessed. There are 10 degrees of marbling: Very Abundant, Abundant, Moderately Abundant, Slightly Abundant, Moderate, Modest, Small, Slight, Traces, and Practically Devoid. (Fig. 7.3 shows 3 of the 10 marbling scores.) Marbling in beef has been shown to be related to tenderness and palatability.

Marbling and maturity are considered together using an official USDA Beef

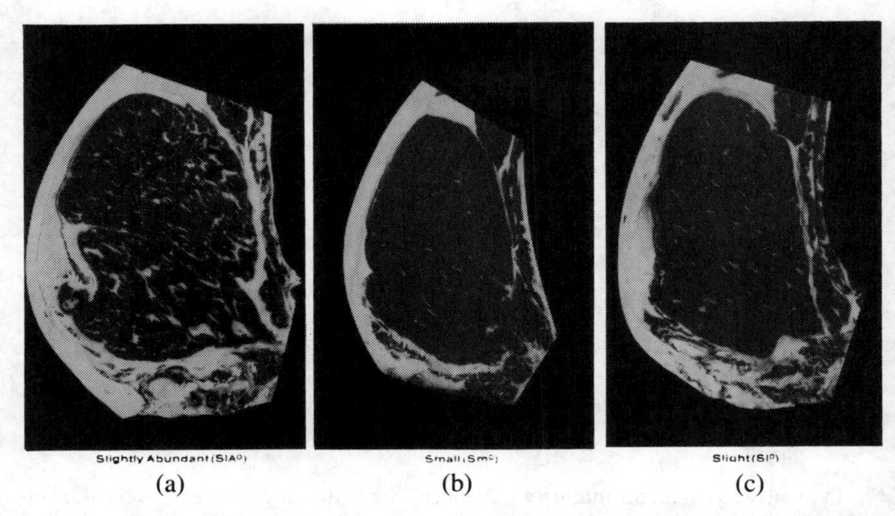

Figure 7.3. Minimum levels of marbling for "A" maturity group carcasses to qualify for USDA Prime (a), Choice (b), and Select (c). (From National Live Stock and Meat Board 1988.)

Quality Grading Chart (Fig. 7.4). Except for carcasses in the "A" maturity group, the marbling required increases with increasing maturity within a quality grade.

Lean Firmness

In addition to maturity and marbling, the official standards specify that the ribeye cross section display a minimum level of lean firmness. Superior firmness has no advantage in the grading system. The minimum level of lean firmness is moderately firm, slightly soft, moderately soft, and soft for the USDA Prime, Choice, Select, and Standard grades, respectively.

Dark Cutting Beef

Carcasses with dark-cutting beef are discounted in the USDA Quality Grading System. Dark-cutting beef is a result of a reduced glycogen content in the muscle prior to slaughter and is often associated with stress (possibly caused by mixing cattle, dramatic weather changes, and shipping fatigue) prior to slaughter. The reduced level of glycogen during rigor mortis prevents the muscle from reaching its normal postmortem pH of 5.4. Rather, the muscle pH of a dark cutter is generally above 6.0, which results in a higher water-holding capacity and more light absorbency than normal, thus, causing a dark lean color. The dark color of lean associated with "dark-cutting beef" can be present in varying degrees from that which is barely evident to so-called "black-cutters" in which the lean is actually black. Also, the lean from dark cutters has a "gummy or tacky"

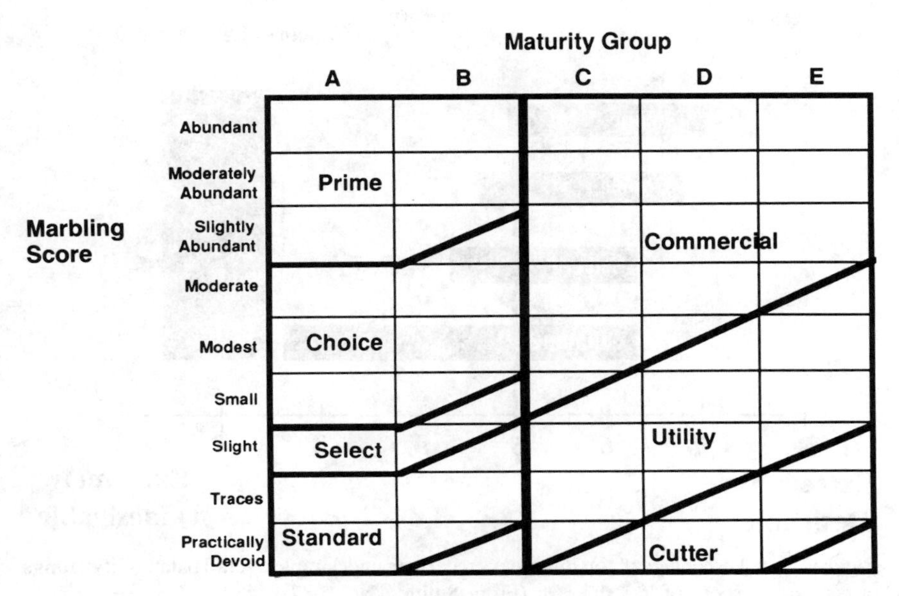

Figure 7.4. Official USDA Beef Quality Grading Chart with maturity groups and marbling score.

texture. Although there is little or no evidence which indicates that the "dark-cutting" condition has any adverse effect on palatability, it is considered in grading because of its impact on consumer acceptability. Depending on the degree to which this characteristic is developed, the final quality grade of carcasses which otherwise would qualify for USDA Prime, Choice, or Select grades may be reduced as much as one full grade. Among beef carcasses with the maturity and marbling characteristics of USDA Standard or Commercial grades, the final grade may be reduced as much as one-half of a grade. This condition is not considered in the USDA Utility, Cutter and Canner grades.

Industry Application of Quality Grades

Marbling or intramuscular fat is an economically important trait in the beef industry. The minimum amount of marbling required for "A" maturity carcasses to receive the USDA Choice Grade is "Small" (Fig. 7.4). A premium is placed on beef graded USDA Choice, because a majority of foodservice and retail establishments prefer USDA Choice beef. Therefore, at the packing plant a carcass that fails to reach that level of intramuscular fat is discounted accordingly, at times over $9.00 per 100 pounds of (45.4 kg) carcass weight. The actual price spread between USDA Choice carcasses and USDA Select carcasses is dependent on the supply and demand for both grades and fluctuates considerably during a given year.

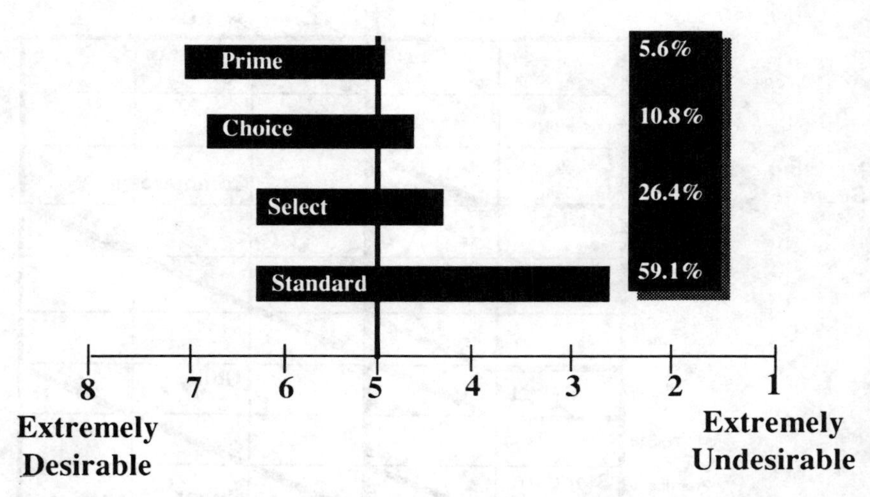

Figure 7.5. Percentage of top loin steaks receiving undesirable overall palatability ratings (a taste panel score of less than 5). (From Smith et al. 1987.)

A study examining trained taste panelist ratings of the palatability of 1005 top loin steaks of several different quality grades of beef showed that as quality grade decreased from USDA Prime to USDA Standard, the probability of a taste panelist receiving a steak with undesirable eating quality increased. Figure 7.5 shows the range in overall palatability ratings and the percent of top loin steaks receiving undesirable overall palatability ratings (an overall palatability rating of less than 5) reported in the study. There were over two times (59.1% versus 26.4%) as many undesirable palatability ratings for USDA Standard top loin steaks as for USDA Select top loin steaks. Even though there is overlap in the palatability rating given to steaks of the USDA Prime, Choice, Select, and Standard grades, the percent of undesirable steaks dramatically increases as USDA Quality Grade decreases. Retailers and foodservice operators often choose a particular USDA Quality Grade of beef for their establishment based on the likelihood of serving a desirable tasting steak to their customer.

Frequency Distribution of Quality Grades in the United States

The 1992 National Beef Quality Audit, surveyed 7375 carcasses for marbling score and maturity level in 28 major meat-packing facilities (Fig. 7.6). The last major beef quality audit, previous to 1992, was conducted in 1974. Results from the 1992 and 1974 quality audits showed that the percentage of carcasses that met the qualifications for USDA Choice and Prime decreased by 20% during that period (54% in 1992 versus 74% in 1974). Increased use of Continental and

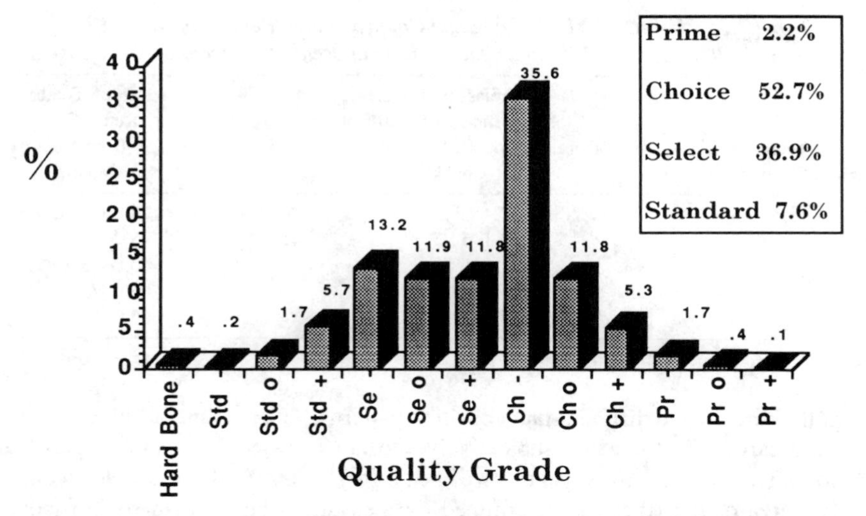

Figure 7.6. Quality grade distribution of carcasses surveyed in the National Beef Quality Audit (1992). Hard Bone: Commercial; Utility; Std: Standard; Se: Select; Ch: Choice; Pr: Prime.

Bos indicus breeding, cross-breeding, and fewer days fed high-concentrate diets in the feedlot may account for the decline in marbling score.

Beef Yield Grades

Introduction

Beef yield grades are a numerical representation of the percent boneless, closely trimmed retail cuts from the high-value parts of the carcass—the round, loin, rib, and chuck. However, they represent differences in the total yield of retail cuts (cutability). The USDA Yield Grades are 1, 2, 3, 4, and 5. Yield Grade 1 denotes the highest-yielding carcass and Yield Grade 5, the lowest (Table 7.3).

Unfortunately, percent retail yield cannot be determined directly on each carcass; therefore, the industry utilizes USDA Yield Grade to denote an estimate of the cutability of a carcass. The factors used to determine yield grade and subsequently derive an estimate of cutability are (1) the amount of adjusted 12th rib external fat thickness, (2) the percent of kidney, heart, and pelvic fat, (3) the area of the ribeye muscle at the 12th rib cross section, and (4) the hot carcass weight.

The amount of external fat is determined by measuring the thickness of fat over the outside of the cross-sectioned ribeye muscle at the 12th rib. Then, the grader adjusts this measurement to reflect unusual amounts of fat in other areas

Table 7.3. U.S.D.A. Yield Grade and Corresponding Percent of Retail Cuts from the Round, Loin, Rib, and Chuck and Total Retail Cuts from Beef Carcasses

Yield Grade	% of Carcass as Boneless, Closely Trimmed Retail Cuts from the Round, Rib, Loin, and Chuck	% of Total Retail Cuts from the Carcass
1	52.4 or greater	79.8 or greater
2	50.1–52.3	75.2–79.7
3	47.8–50.0	70.6–75.1
4	45.5–47.7	66.0–70.5
5	45.4 or less	65.9 or less

of the carcass and different fat deposition patterns. The amount of kidney, pelvic, and heart fat is evaluated subjectively and is expressed as a percentage of the hot carcass weight (this usually will be from 1% to 4% of the carcass weight). The area of the ribeye is determined by measuring the area (in square inches, using a dot-grid or planimeter) of the longissimus muscle at the 12th rib cross section. Carcass weight is the "hot" or unchilled weight (in pounds), usually obtained on the slaughter floor prior to entering the cooler (Fig. 7.7).

Final Yield Grade

The yield grade of a beef carcass is determined on the basis of the following equation:

$$\text{Yield Grade} = 2.5 + (2.5 \times \text{adjusted fat thickness, in.}) + (0.20 \times \%\ \text{kidney, pelvic, and heart fat}) + (0.0038 \times \text{hot carcass weight, lbs}) - (0.32 \times \text{ribeye area, in.}^2).$$

Note from the equation that fat thickness is the most important factor in determining yield grade. For every 0.1-in. increase in fat thickness the yield grade increases one-quarter (0.25) of a grade. It would require an increase of 1.25% kidney, pelvic, and heart fat, an increase of 66 lbs in hot carcass weight, or 0.8-in.^{2} decrease in ribeye area to create a similar one-quarter increase in yield grade. The corresponding "cutability" equation, representing the percent closely trimmed, boneless retail cuts from the round, loin, rib, and chuck is as follows:

$$\text{Cutability} = 51.34 - (5.784 \times \text{Adjusted Fat Thickness, in.}) - (0.462 \times \%\ \text{kidney, pelvic, and heart fat}) - (0.0093 \times \text{hot carcass weight, lbs}) + (0.74 \times \text{ribeye area, in.}^2).$$

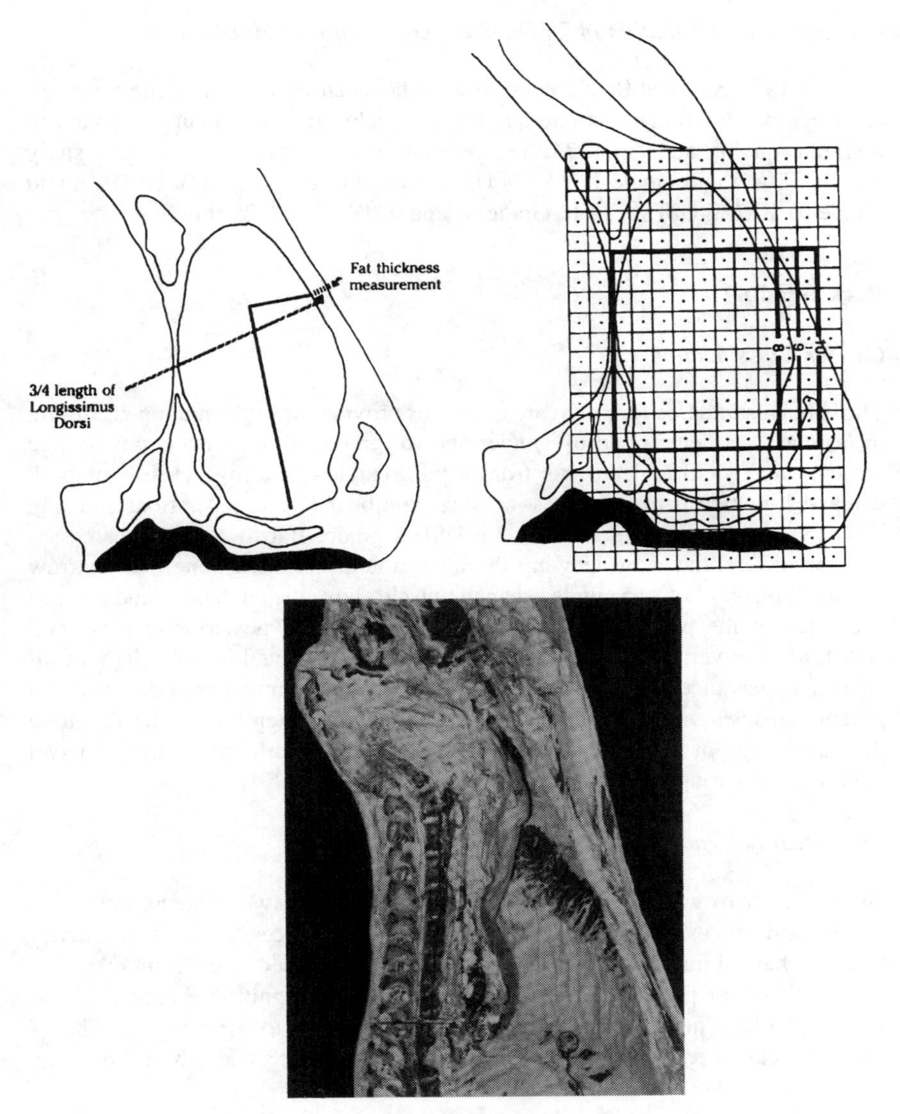

Figure 7.7. Twelfth rib fat thickness, ribeye area, and percentage kidney, pelvic and heart fat; three of the four factors used to determine USDA Beef Yield Grade. (From National Live Stock and Meat Board, 1988)

Frequency Distribution of USDA Yield Grade in the United States

In the 1992 National Beef Quality Audit, the mean yield grade of the carcasses reviewed was 3.16. That is one-quarter of a yield grade less than was found in a similar quality audit in 1974. Ten percent of the carcasses in the 1992 study were USDA Yield Grade 1, 33.9% USDA Yield Grade 2, 39.6% USDA Yield Grade 3, 13.6% USDA Yield Grade 4, and 2.9% USDA Yield Grade 5.

Pork Grading

Gender Classification

The official standards for pork carcass grades provide for segmentation according to (a) class, as determined by gender and (b) grade, which reflects quality and the expected yield of lean cuts from a pork carcass. The five classes of pork carcasses include barrow, gilt, sow, stag, and boar (Table 7.4). Boar and stag carcasses are not eligible to receive a USDA grade. Barrow and gilt carcasses may be differentiated by viewing the area around the aitch bone. The barrow carcass typically has a triangular-shaped muscle above the aitch bone and a pizzle eye (crus of the penis) adjacent to the aitch bone and posterior to the sacral vertebrae. Conversely, a gilt carcass has no pizzle eye and has a kidney-bean-shaped appearance to the exposed surface of the semimembranosus and the gracilis muscles on the inside of the ham, above the aitch bone. The fat along the navel edge of a gilt carcass is smooth, whereas the barrow carcass' navel edge fat has a rough appearance (Fig. 7.8)

Pork Quality Grades

Grades for barrow and gilt carcasses are evaluated for two characteristics: the quality and processing attributes of the lean and the percent of bone-in, closely trimmed ham, loin, Boston shoulder, and picnic shoulder (four lean cuts).

Two levels of pork quality for U.S. grades are recognized: Acceptable and Unacceptable. The Unacceptable designation is given to a carcass that has a condition called pale, soft, and exudative (PSE) lean, soft and oily fat, or if the

Table 7.4. Pork Gender Classes and Eligible U.S.D.A. Grades

Gender Class	U.S.D.A. Grades
Barrows	U.S. 1, 2, 3, 4, and Utility
Gilt	U.S. 1, 2, 3, 4, and Utility
Sow	U.S. 1, 2, 3, 4, Utility, and U.S. Cull
Stag	None
Boar	None

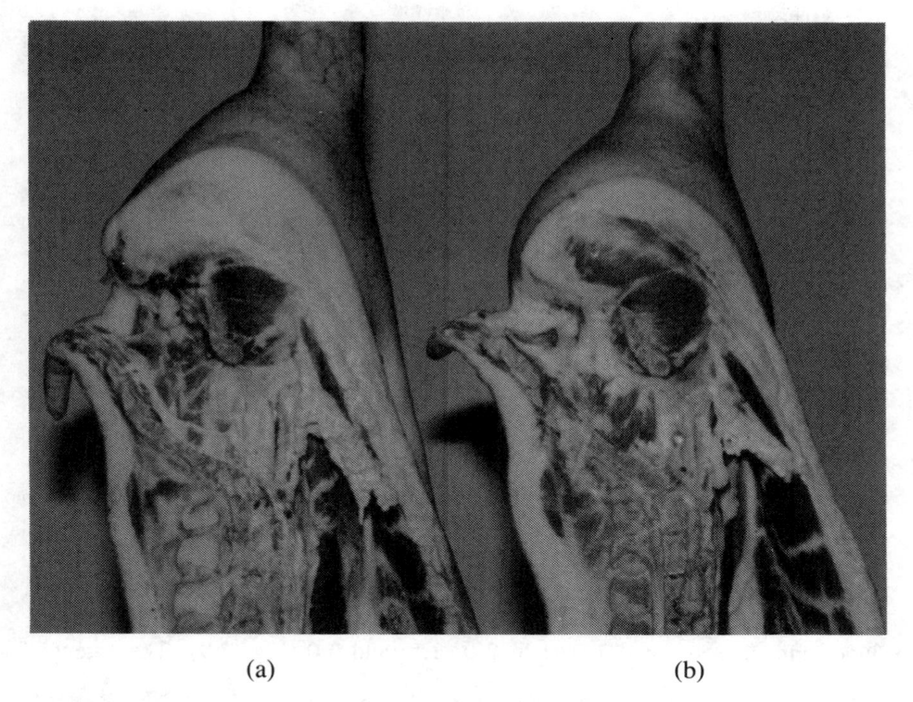

Figure 7.8. Gender characteristics of barrow (a), and gilt (b) pork carcasses.

belly is too thin to be utilized for bacon production [less than 1.52 cm (0.6 in.) thick].

PSE occurs as a result of an abnormal pH decline in the first few hours postmortem, a lower than normal final pH (less than pH 5.4), and, subsequently, a lower water-holding capacity, allowing more water loss and more light reflectance.

The lean surface of the longissimus muscle at the 10th rib cross section is best used to determine if the lean is pale, soft, and watery (exudative). However, in the commercial industry, pork carcasses are rarely ribbed, as a 10th rib cross section would segment the wholesale loin into two pieces. Therefore, when this surface is not available, the quality of lean is evaluated using the firmness of the fat and lean in the belly, the amount of feathering, and the color of lean between the ribs. The minimum levels for a carcass to qualify for Acceptable are a "Slight" degree of feathering, a fat that is "slightly firm," and a grayish-pink to moderately dark red colored lean between the ribs. Some pork processors are using hand-held pH probes to assist in identifying and screening for PSE pork.

Barrow and gilt carcasses that fall below the minimum quality standards (Unacceptable) are eligible for the U.S. Utility grade; those carcasses that meet or exceed the minimum quality standards (Acceptable) are eligible for one of four cutability grades.

Pork Cutability Grades

The cutability grade is a numerical representation (U.S. 1, 2, 3, and 4) of the percentage yield of bone-in, closely trimmed four lean cuts (ham, loin, Boston butt, and picnic shoulder). The expected yields, based on USDA cutting procedures, are as follows for the four cutability grades: U.S. 1—60.4% or greater; U.S. 2—57.4–60.3%; U.S. 3—54.4–57.3%; U.S. 4—less than 54.4%. The USDA cutability grade is determined using two factors: (1) last rib backfat thickness and (2) muscling score (Fig. 7.9).

The amount of external fat on a barrow or gilt carcass is the most important factor affecting the yield of lean cuts. As the amount of external fat increases, the yield of lean cuts decreases. External fat is measured at a point along the backfat opposite the last rib (measure includes skin) perpendicular to the skin surface. For skinless carcasses, 0.1 in. is added to the measure. The second factor considered in the cutability grading system is muscle score. There are three muscle scores described in terms and numerically in the standards—thick (3), average (2), and thin (1). These three muscle scores are often further subdivided into five groups—1.0, 1.5, 2.0, 2.5, and 3.0 (Fig. 7.9). The extent of

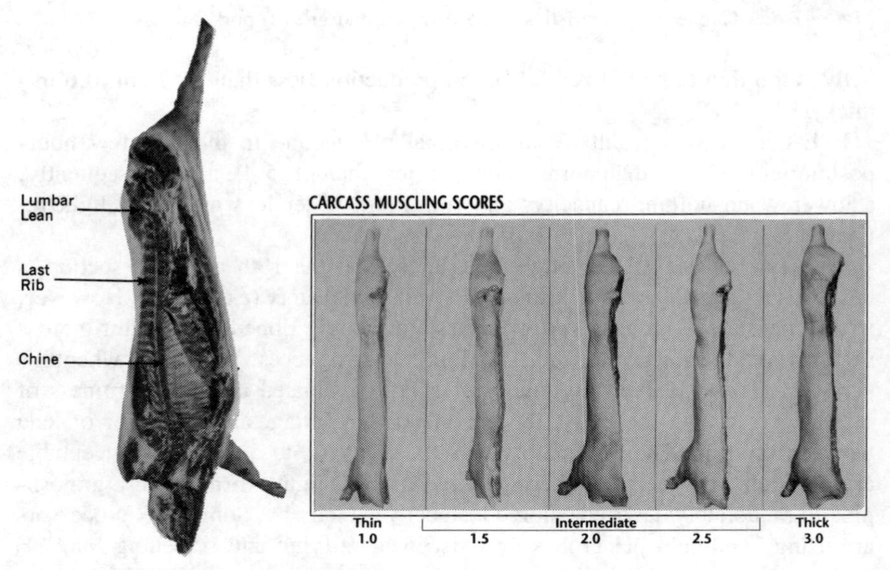

Figure 7.9. Last rib backfat thickness and muscle scores used to determine U.S. Cutability Grade for pork carcasses. (From National Live Stock and Meat Board 1988; National Pork Producers Council 1991.)

muscularity is primarily evaluated at the ham, ham–loin juncture, and loin, because these are the higher value lean cuts of pork.

Final Cutability Grade

The final pork cutability grade is determined using the following equation:

Cutability grade = (4.0 × last rib backfat thickness) − (1.0 × muscle score).

From the number derived by the equation, all fractions are dropped and the grade is not rounded to the nearest whole grade (e.g., 2.6 calculated grade, U.S. 2 reported grade). There is one exception to the use of this formula, which is that U.S. 1 grade may only be awarded to a carcass that has at least an "Average" muscling score.

Application of Grades

Few pork carcasses are sold intact. Instead, most pork carcasses are cut into wholesale cuts and sold as either fresh or cured and smoked meat items; therefore, U.S. grades are not used past the packer level. Hogs are most often slaughtered at a young age (5–6 months) and are fed a high concentrate diet for most of their life; therefore, there is more uniformity with regard to taste appeal among pork than in other livestock species and less need to segment pork cuts according to taste. Carcass fatness is of the utmost importance to the pork packer. The packer will trim the excess fat from a pork carcass, which in many places on the carcass is over 3.81 cm (1.5 in.), to less than 0.635 cm (0.25 in.) on the wholesale cuts before they ship the meat cuts to foodservice and retail. Although fatness is very important to the packer, the USDA pork grading system is rarely used. At the same time, more hogs than any other livestock species are sold to the packer on a grade basis rather than on a live basis. The grade that is used by packers is a house grading system, which is usually a premium/discount grid with fat thickness across one axis and weight down the other axis (Table 7.5). The price per 45.4 Kg (100 lbs) of carcass weight is often determined by identifying where the carcass weight and fat measures meet on the grid; and then this percentage discount or premium is multiplied by the base carcass price. Each packer has a different grid and, subsequently, puts different emphasis on carcass weight and last rib backfat thickness. The base carcass price and the percentage premiums and discounts posted on the grid may change with changes in market hog numbers and marketing conditions and with increases in packer competition.

Sheep and Lamb Grading

Introduction

Sheep and lamb grades, similar to beef carcasses, are based on separate evaluation of two general considerations: Quality Grade and Yield Grade. Grades for ovine

Table 7.5. An Example Price Grid Used by Pork Processors to Arrive at the Value of a Pork Carcass. Premiums and Discounts as a Percentage of the Base Carcass Price

Carcass Weight (lbs)	Last Rib Fat Thickness (in.)							
	0.7	0.8	0.9	1.0	1.1	1.2	1.3	1.4
147–153	104	104	103	102	101	100	100	99
154–161	103	103	102	101	100	99	99	98
162–168	103	103	102	101	100	99	99	98
169–175	103	103	102	101	100	99	99	98
176–182	102	102	101	100	98	98	98	97
183–190	101	101	100	99	98	97	97	96
191–197	101	101	100	99	98	97	97	96
198–204	100	100	99	98	97	96	96	95
205–212	99	99	98	97	96	95	95	94
213–219	98	98	97	96	95	94	94	93

Source: Hayenga et al. (1985).

carcasses are coupled, which means that if a packer chooses to have a USDA grader stamp the lamb carcasses with a quality grade, the appropriate yield grade must also be stamped on the carcass and vise versa. To be eligible for grading, ovine carcasses may not have more than 1.0% kidney and pelvic fat left in the carcass. The remainder must have been removed from the carcass prior to grading. Additionally, as with other species, the carcass has to be federally inspected and in a condition and dressed in a way where an accurate assessment of Quality and Yield Grade can be made by the USDA grader.

Gender Classification

Determination of ewe, wether, and ram carcasses is more complicated than in other livestock species, because the carcasses are not split into two sides. However, a wether and ewe carcass can be identified by the appearance of the fat deposit in the inguinal or mammary regions (cod/udder region). Wethers have a rough appearing deposit, which looks like 5–10 white marbles bunched together, whereas the same deposit in ewe carcasses is smooth and even in appearance. Ram carcasses are more muscular in the neck and shoulder regions, resulting in a heavy appearing foresaddle commonly referred to as a "bucky" appearance. The intact male has small amounts of scrotal fat and often an off-white, dingy colored appearance to the subcutaneous fat (Fig. 7.10). The quality standards are intended to apply to all ovine carcasses without regard to gender. However, carcasses from rams, with a bucky appearance, are discounted in quality grade. Depending on the extent of buckiness, discounts may vary from less than one half grade in carcasses from young lambs in which such characteristics are barely noticeable to as much as two full grades in carcasses from mature rams in which characteristics are very pronounced.

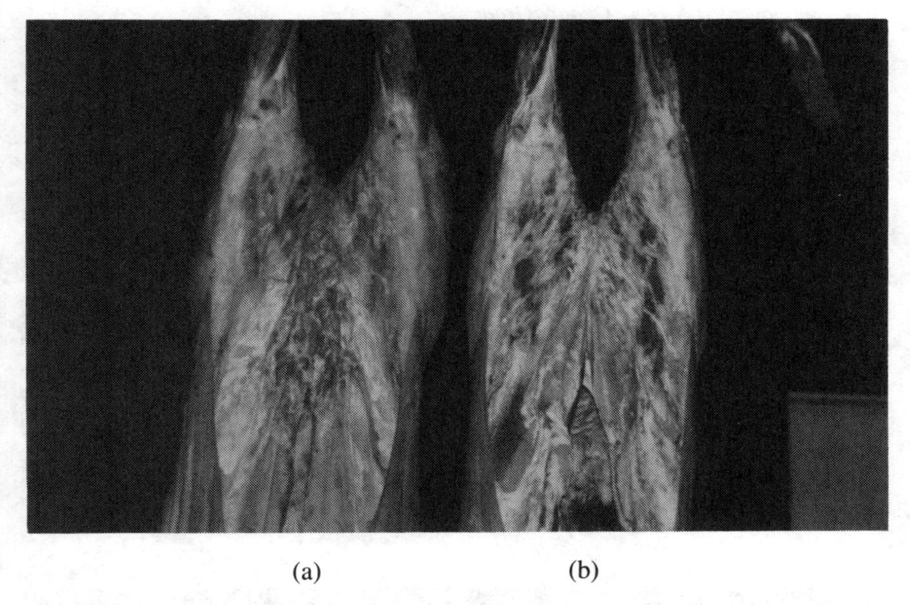

(a) (b)

Figure 7.10. Gender characteristics of wether (a) and ewe (b) lamb carcasses.

Maturity Classifications

The maturity of a lamb at the time of slaughter is a major factor in the quality grading system. Meat from older sheep tends to be less tender and have more off flavors. Ovine maturity is segmented into three divisions according to overall carcass maturity: lamb, yearling-mutton, and mutton (Table 7.6). Hothouse lambs are a subgroup of the lamb classification that is not recognized by the USDA Quality Grading system but is a common industry designation in some regions of the United States. Hothouse lambs are traditionally lambs born between September and January and then marketed at relatively light weights from 25 to 60 lbs. Some cities have hothouse lamb grading standards and subsequent grading terminology (Table 7.6). Two other subgroups of lambs are young lambs and old lambs also referred to as "A" maturity lambs and "B" maturity lambs. The packing industry sometimes refers to these lambs as spring born lambs and old-crop lambs.

To estimate the overall maturity, the parts of the carcass that are examined include break joints versus spool joints (Fig. 7.11), rib bones, color of the inside flank muscles, and lean texture. During the dressing process, the foot and pastern are removed from the carcass at a point between the proximal sesamoids and the metacarpal bone (foreshank). Immediately posterior to that point, on the end

Table 7.6. The Market Classes of Sheep and Lamb Carcasses and Subsequent U.S.D.A. Quality Grade

Classification of Sheep or Lamb	U.S.D.A. Quality Grade Eligible
Hothouse lamb—less than 3 months	Extra Fancy,[a] Fancy,[a] Good, Fair,[a] Plain[a]
Lamb A—maturity—3–8 months	Prime, Choice, Good, Utility, Cull
Lamb B—maturity—8–14 months	Prime, Choice, Good, Utility, Cull
Yearling-mutton—14–24 months	Prime, Choice, Good, Utility, Cull
Mutton—over 24 months	Choice, Good, Utility, Cull

[a] Not official federal grade names.

of the metacarpal, is a joint that resembles the appearance of a spool of thread ("spool joint") and immediately posterior to that point is the rough appearing tip of the metacarpal. In lambs, there is a thin layer of soft cartilage between the tip of the metacarpal and the spool joint, making it easy to break the spool joint from the metacarpal; hence, the rough edge of the metacarpal is referred to as the "break joint." As sheep advance in maturity, the thin layer of soft cartilage undergoes ossification and becomes bone. The ossification process fuses the

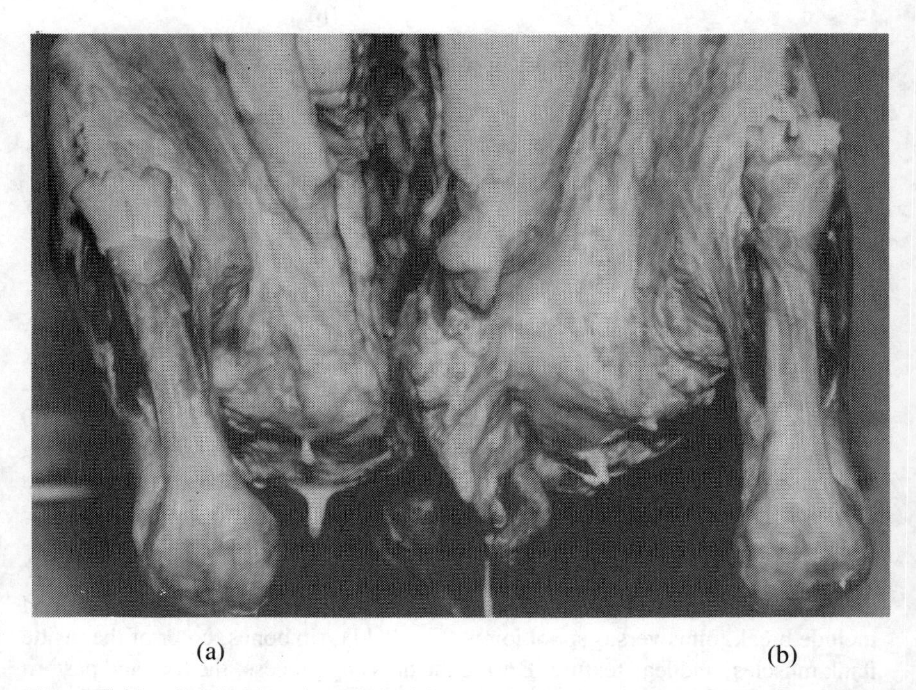

Figure 7.11. Break (a) and spool (b) joints on the metacarpal bones of the fore shanks used to determine the skeletal maturity of a lamb, yearling-mutton, or mutton carcass.

spool joint and the end of the metacarpal together, making it impossible to break the spool joint from the foreshank of the carcass. After the grader checks for the presence or absence of break joints, the ribs and flank are examined in order to further pinpoint the maturity level within the lamb, yearling-mutton, and mutton classifications.

Typical lamb carcasses tend to have slightly wide and moderately flat rib bones, both break joints, a light red colored lean in the flank, and a fine textured lean. By contrast, typical yearling-mutton carcasses have moderately wide and flat rib bones, a slightly dark colored lean in the flank, and slightly coarse textured lean. Yearling-mutton may have one, both, or no break joints. Mutton carcasses have wide, flat rib bones, two spool joints, and a dark colored, coarse textured lean.

Lamb Quality Grade

The Quality Grade of ovine carcasses is based on a composite evaluation of two major factors: conformation and quality score.

Conformation

Conformation is related to the thickness of muscle and the fullness of the carcass. It is often described in terms of carcass width and thickness in relationship to carcass length. The distribution and quantity of subcutaneous fat influences carcass conformation; however, fat levels in excess of that normally left on retail cuts should not be considered when evaluating conformation. It is reasonable to question why this factor is a standard for USDA Quality Grade, particularly as conformation and palatability of lean have not been linked in research. However, it is important to note that, unlike beef and pork, many lamb carcasses are not further processed into wholesale cuts at the packing house but rather are sold as carcasses. Therefore, the appearance and shapeliness of the carcass is an important marketing trait, particularly in the more traditional lamb markets on the east coast of the United States. Conformation levels are assigned to the nearest one-third of Prime, Choice, Good, Utility, and Cull (e.g., High Prime, Average Prime, Low Prime, etc.).

Quality of Lean

The overall carcass maturity, lean firmness and flank streaking (also referred to as flank lacing and fat streaking in the flank) are combined, using the chart in Fig. 7.12 to arrive at a quality score. Actually, the quality of lean is best evaluated by examining marbling and the color and texture of the lean surface of the ribeye at the 12th rib cross section. However, lamb carcasses are rarely ribbed in the packing plant carcass coolers, making this evaluation impossible. Therefore, the quality of lean is determined indirectly using flank streaking.

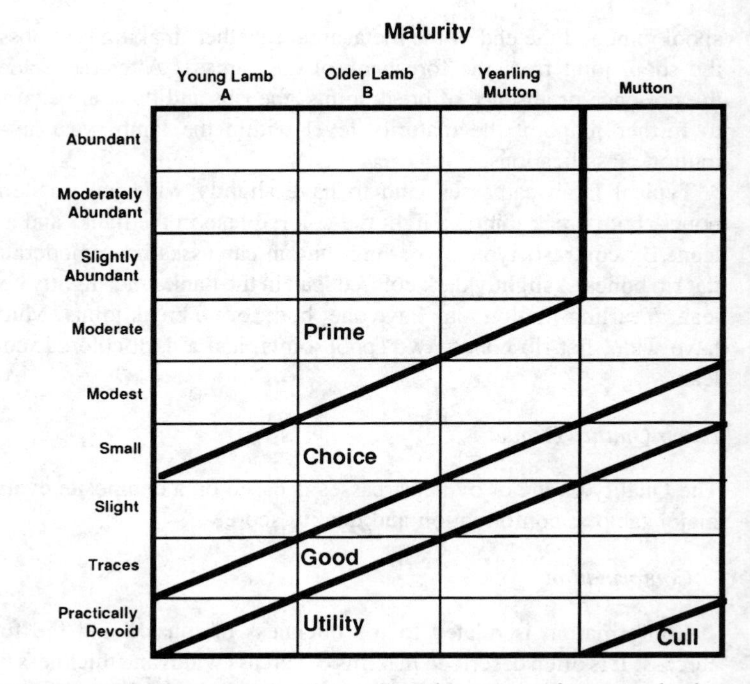

Figure 7.12. Grading chart used to derive the quality of lean score for young lambs, old lambs, yearling-mutton, and mutton.

Flank streaking is the streaky-appearing fat deposits, often shoe lace shaped, in the primary flank muscle (*Rectus abdominis*) and the secondary flank muscle (*Transverse abdominus*) (Fig. 7.13).

Final Quality Grade

To arrive at the final Quality Grade, the overall quality of lean score and conformation are examined together. The following rules are for USDA Prime and Choice carcasses:

- USDA Prime—Superior conformation will not compensate for deficient quality: therefore, a lamb must have a minimum quality score of low Prime to qualify for the USDA Prime Quality Grade. A superior quality score will compensate for inferior conformation equally. In order to qualify for USDA Prime, a carcass must have at least a low Choice conformation score.
- USDA Choice—Superior conformation may only compensate for inferior quality of lean given the following circumstances: Conformation is average Choice or higher and the quality score is at least high Good. Superior

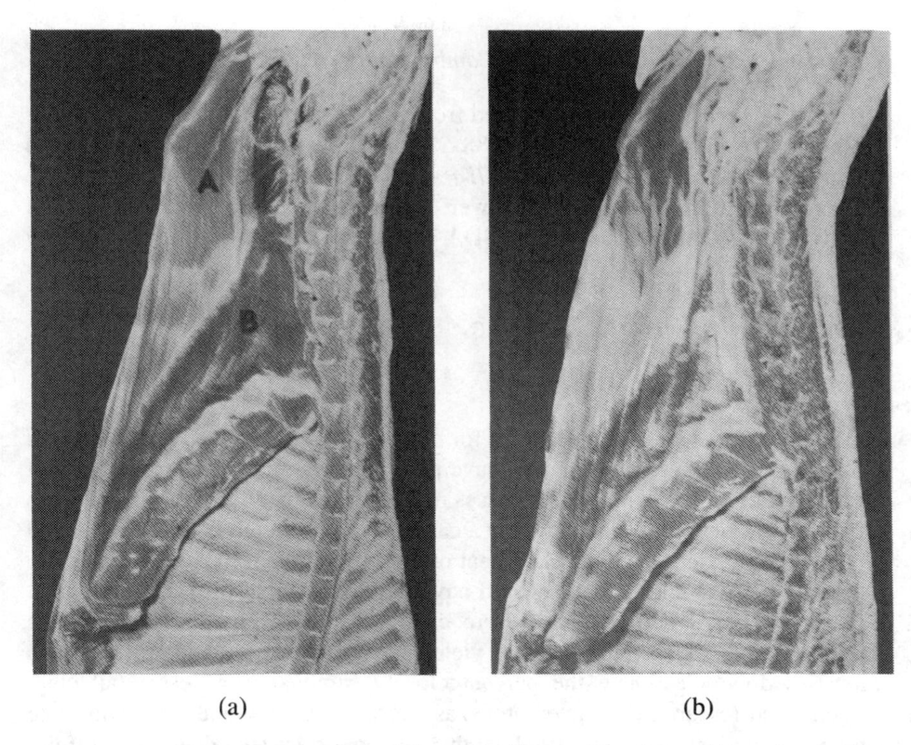

(a) (b)

Figure 7.13. Levels and location of flank streaking in the primary (A) and secondary (B) flank muscles used to determine the quality of lean or of a lamb carcass. (From National Live Stock and Meat Board 1988.)

quality may compensate on an equal basis for inferior conformation, to the extent of a combination of low Good conformation and high Choice quality score.

Lean Firmness and Fat Covering

In addition to quality score and conformation, the USDA standards also specify a minimum level of lean firmness for each USDA Quality Grade and a minimum external fat level for the USDA Prime and USDA Choice grades. The vast majority of lambs easily meet these standards. A thin covering of external fat is required because often carcasses are shipped long distances from the packer to the retailer. This thin fat covering helps to prevent dehydration and deterioration of the lean on the surface of the carcass. The minimum level of lean firmness is slightly firm and moderately firm for USDA Choice and USDA Prime carcasses, respectively.

Frequency Distribution of USDA Lamb Quality Grades

In 1987, carcass data were collected from 6224 lambs slaughtered in 11 major lamb packing plants. The majority (88.5%) of the carcasses in the study had a quality grade of USDA Choice, 9.7% were USDA Prime, and 1.8% of the carcasses graded USDA Good or lower. The primary reason for carcasses not qualifying for USDA Choice or USDA Prime was due to advanced maturity (i.e., yearling-mutton).

Sheep and Lamb Carcass Yield Grade

Introduction

There are five USDA yield grades for ovine carcasses (USDA 1, 2, 3, 4, and 5). Historically, these yield grades have provided a numerical representation of the percent of closely trimmed, boneless retail cuts that would be fabricated from the leg, loin, rack, and shoulder from a carcass (carcass cutability). USDA Yield Grade 1 would yield the highest percent of closely trimmed, boneless retail cuts, whereas USDA Yield Grade 5 would have the lowest yield of these retail cuts. The lamb yield grade standards were changed significantly in 1992. Unlike previous standards for lamb carcass yield grades, the 1992 standards for lamb carcasses do not estimate the percent closely trimmed, boneless retail cuts. Unpublished research conducted at Texas A&M University and Colorado State University indicate that the 1992 lamb yield grade system does successfully segment carcasses according to subcutaneous fat levels; however, the system is a poor predictor of carcass cutability. Although this is negative from the packer and meat processor standpoint, the 1992 system allows sheep producers to concentrate on one of the major problems of their products, which is excess fat. A positive point of the new system for meat retailers and lamb carcass wholesalers is that the kidney and pelvic fat must be removed prior to grading (up to 1% kidney and pelvic fat is permitted). Thus, the 0.454–2.268 kg (1–5 lbs) of fat often found in the kidney and pelvic region of a carcass is not shipped to the purchaser. In addition, this system is simple enough that it may be used at packing plants that often slaughter over 3000 lambs per day.

Yield Grade Factors

The yield grade of an eligible ovine carcass is based on the adjusted fat thickness over the ribeye at the 12th rib and 13th rib interface (Table 7.7). The subcutaneous fat measurement is taken at the midpoint of the ribeye (Fig. 7.14). Then, the grader adjusts that measure by examining the fat deposition over the remainder of the carcass.

The following equation could be used to convert adjusted fat thickness to USDA Yield Grade: Yield grade $= 0.4 + (10 \times$ adjusted fat thickness, in.).

Table 7.7. Ovine Carcass Yield Grades and Corresponding Adjusted Fat Thickness at the 12th Rib

USDA Yield Grade	12th Rib Adjusted Fat Thickness (in.)
1	0.00–0.15
2	0.16–0.25
3	0.26–0.35
4	0.36–0.45
5	0.46 or greater

Unlike beef, there is no corresponding cutability equation (percent closely trimmed, boneless retail cuts from the leg, loin, rack, and shoulder) that corresponds to the numerical yield grade reported in the Code of Federal Regulations.

Frequency Distribution of Lamb Carcass Yield Grade

The average lamb carcass USDA Yield Grade was 3.8 using the pre-1992 grading system and 3.3 using the 1992 grading system (determined using data collected during the 1987 National Survey of Lamb Carcass Cutability Traits). It is estimated that 20% more carcasses are being stamped USDA Yield grade 1 or USDA Yield Grade 2 under the 1992 yield grade system than under the previous yield grade system.

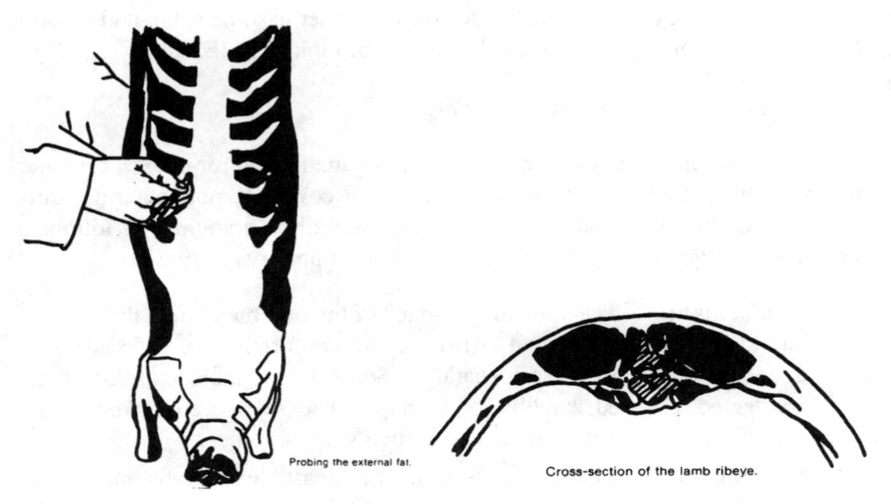

Figure 7.14. Fat thickness measurement over the ribeye muscle at the 12th Rib Crosssection used to determine the yield grade of a lamb carcass. (From *Laboratory Manual for Meat Service*.)

Application of Grades

Almost all lambs are quality and yield graded in lamb packing plants in the United States. Because of the number of lambs that a grader must examine each day, evaluation of yield grade is often done based on visual appraisal of overall carcass fatness and not utilizing mechanical probes to measure 12th–13th rib fat thickness.

Poultry Grading

Introduction

Poultry grades consist of classifying and sorting poultry based on conditions such as type, age, and gender and then on quality characteristics such as conformation, fleshing, processing defects, fat covering, and deformities.

The kinds of poultry that are eligible to be graded include, but are not limited to, chickens, turkeys, ducks, geese, pigeons, and guineas.

Poultry grades are a result of industry application and research that have identified certain characteristics in poultry that are desired by processors and consumers. Some of these characteristics include a good proportion of muscle to bone, adequate skin covering, absence of pinfeathers, and freedom from discoloration. For poultry value-added products, factors besides those already mentioned include the presence of tendons and blood clots as well as product specific factors. Less than 25% of all chickens and 20% of all turkeys are marketed as whole birds. The rest reach the consumer in some value-added form. The U.S. consumer grades for poultry are U.S. Grades A, B, and C.

Grading Factors for Carcasses and Parts

There are several factors examined to arrive at a final grade for poultry carcasses and parts, including conformation, fleshing, fat covering, pinfeathering, cuts, tears, broken bones, bruises, blemishes, and skin discoloration. The following are short descriptions of the factors considered in poultry grading.

- **Conformation**—The structure or shape of the bird may affect the distribution and amount of meat. Also, certain conformational defects may detract from the product's appearance. Some defects include breasts that are dented, crooked, knobby, or V-shaped; backs that are crooked; legs and wings that are deformed; and wedge-shaped bodies.
- **Fleshing**—The amount of muscle in the breast, leg, thigh, and back regions.
- **Fat Covering**—Fat in poultry is evaluated entirely by accumulation under the skin.

- **Pinfeathers**—Two types of pinfeathers are considered in grading—protruding and nonprotruding. Protruding pinfeathers have broken through the skin and may or may not have formed a brush. Nonprotruding pinfeathers are evident but have not pushed through the outer layer of skin. Poultry must be free of protruding pinfeathers that are visible to a grader during examination of the carcass at normal operating speeds.
- **Cuts, Tears, and Broken Bones**—The number and extent of these types of defects permitted depends on their location on the bird. These defects detract from the appearance of the carcasses and parts.
- **Skin Discoloration**—Skin that dries out may become discolored, which detracts from consumer appeal. Companies marketing poultry in areas that prefer yellow skin will scald birds at lower temperatures between 51.1 and 52.2°C (124 and 126°F), thus leaving the cuticle (bloom) on the chicken and enhancing its yellow appearance. In order to produce a whiter appearance, the birds are scalded at higher temperatures from 55.5 to 57.8°C (132–136°F) to remove the cuticle. Removing the cuticle enhances the bonding of the batter to the chicken skin when processing.
- **Bruises and Blemishes**—Bruises in the flesh or skin are permitted only to the extent that there is no coagulation or clotting (discernible clumps of red cells). Blue or green bruises must be removed before grading. Some breeds of turkeys may have a condition that occurs called "blue back," which causes a discoloration over the back and/or wing. This condition is considered in assessing the total area of discoloration.

U.S. Classes for Poultry

The first step in examining poultry carcasses and cuts is to classify the bird with regard to poultry type, age, and gender. Below are descriptions of chicken, turkey, and duck classes. There are also classifications for geese, guineas, and pigeons.

Chicken

- Rock Cornish Game Hen or Cornish Game Hen is a young immature chicken (5–6 weeks of age) weighing not more than 2 lbs, prepared from a Cornish chicken or Cornish cross chicken.
- Rock Cornish Fryer, Roaster, or Hen is a progeny of a cross between a purebred Cornish and purebred Rock chicken. No weight requirement is specified.
- Broiler or Fryer is a young chicken (under 13 weeks of age) of either sex.

- Roaster or Roasting Chicken is a young chicken under 5 months of age, of either sex. The breastbone cartilage is somewhat less flexible than that of a broiler or fryer.
- Capon is a surgically unsexed male chicken (usually under 8 months of age).
- Hen, Fowl, or Baking or Stewing Chicken is a bird that is a mature female chicken (over 10 months of age).
- Cock or Rooster is a mature male.

Turkeys

- Fryer Roaster Turkey is a young immature turkey (usually under 16 weeks of age) of either sex.
- Young Turkey is usually under 8 months of age.
- Yearling Turkey is a fully mature turkey under 15 months of age.
- Mature Turkey or Old Turkey (Hen or Tom) is a mature turkey of either sex in excess of 15 months of age.

Duck

- Broiler or Fryer Duckling is a duck under 8 weeks of age of either sex.
- Roaster Duckling is a young duck under 16 weeks of age of either sex.
- Mature Duck or old Duck is a mature duck over 6 months of age of either sex.

Official Standards for Poultry Grades

Below is a description of the A and B Grade standards for poultry. Carcasses and parts that do not meet the requirements for A or B Grades may be of C quality if the flesh is substantially intact.

Conformation

A— Free from deformities that detract from its appearance.

B— The carcass or part may have moderate deformities, such as dents, curved or crooked breasts or misshapen legs and wings, which do not materially affect the distribution of flesh or the appearance of the carcass or part.

Fleshing

A— The carcass has a well-developed covering of flesh considering the kind of poultry, the class, and the part. The breast should be long and deep, the leg well fleshed and moderately thick and wide, and the drumstick and thigh are well fleshed and moderately thick.

B— The carcass has a moderate covering of flesh considering kind of poultry, the class, and the part. The breast has a substantial covering of flesh, the leg is fairly thick and wide, and the drumstick and thigh have a sufficient amount of flesh to prevent a thin appearance.

Fat Covering

A— Depending on the class, the carcass or part has a well-developed layer of fat in the skin.

B— There is sufficient fat in the skin to prevent a distinct appearance of flesh through the skin, especially on the breast and leg.

Defeathering

A— The carcass or part has a clean appearance especially on the breast. It is free of visible pinfeathers and hairs.

B— The carcass or part may have a few nonprotruding pinfeathers or vestigial feathers that are scattered widely enough not to appear numerous.

Disjointed and Broken Bones

A— Parts or carcasses must be free of broken bones and a carcass must not have more than one disjointed bone.

B— Parts may be disjointed, but are free of broken bones. The carcass may have two disjointed bones or one disjointed bone and one broken bone.

Applications of Poultry Grades

Poultry carcasses and parts are initially graded by plant employees specifically authorized by USDA to perform this task. Subsequently, the product is check-graded by a USDA grader to assure compliance with applicable grade standards. Each carcass and part, including those used in preparing a poultry food product bearing a grade-mark, must be graded and identified in an unfrozen state on an individual basis. The usual procedure is to examine the fleshing and fat covering factors first. Then the intensity, aggregate area, and location and number of defects must also be evaluated. The final quality rating is based on the factors with the lowest rating. In other words, if the requirements for "A" quality are met in all factors except one, and this factor is "B" quality, the final grade designation would be "B."

Objective Measurements of Cutability and Meat Quality

Introduction

This section will describe potential nondestructive and noninvasive technologies that may result in a commercially applicable grading instrument that will objectively, and with enhanced accuracy, predict aspects related to carcass value. It would be preferable for the instrument to be applied to the animal prior to hide, skin, or pelt removal, thus eliminating the impact that dressing defects could cause in the prediction of carcass value. Additionally, an instrument must be able to operate within the current production speeds and under the environmental conditions experienced in a processing facility (e.g., cold temperatures and high humidity).

Ultrasound

There are three ultrasonic technologies that may be useful in the development of an instrument grading machine; real-time ultrasound (B-mode), A-Mode, and velocity of sound. There are numerous research projects, in progress, examining the potential of ultrasound in measuring marbling, maturity, tenderness, and carcass cutability. A potential drawback to all ultrasound technologies is that they cannot differentiate lean color and would not be able to screen for dark cutters.

B-Mode/Real Time

Real-time ultrasound provides images of the cross section of an animal or carcass almost instantaneously. Ultrasonic images are a record of sound waves interacting with the physical properties of tissues as high-frequency sound, emitted through a transducer, travels through the animal or carcass cross section. When the sound signal encounters an interface between two tissues (such as fat and muscle), some sound reflects back to the transducer at varying rates depending on the tissue density. The reflected signals picked up through the receiver in the transducer are amplified and can be displayed on a television monitor. Digital processing allows for these acoustical signals to be "mapped" as image brightness or gray-scale pictures. Computer image enhancement research is being conducted to obtain a clearer picture and to more accurately compute the ribeye area and fat thickness automatically for carcass cutability assessment. This same technology may be used to predict marbling in the longissimus muscle on a carcass before the hide is removed on the slaughter floor. The white specks often seen in an ultrasound image may be due to marbling (sometimes referred to as speckle) and/or are a result of electrical "noise." Filtering systems are being developed to eliminate the "noise" so that intramuscular fat may be accurately quantified.

A-Mode

Similarly to real-time ultrasound, this technology sends high-frequency sound signals through a cross section of the animal; however, A-Mode, rather than displaying a two-dimensional image, is a one-dimensional representation of the reflected signal in which the horizontal axis represents time or distance and the vertical axis represents amplitude. The distance between amplitude peaks represents the thickness of the tissue being measured. Therefore, this technology could measure the subcutaneous fat layer at numerous anatomical locations of a live animal or a carcass, including the 12th–13th rib interface. It could also be used to estimate the depth of the longissimus and other muscles.

Velocity of Sound

Ultrasonic waves travel more quickly through muscle than fat tissues. Moisture content of the tissue influences the rate at which sound travels through a tissue. The sound signal passes through muscles more quickly, in part, because it has a higher moisture content. Conversely, sound travels slower through a tissue like fat which has a low moisture content. Measuring the velocity of sound as it passes from a sound transmitter on one side, through an animal or carcass cross section, to a sound receiver on the other side may be useful in segmenting carcasses according to leanness.

Elastography

Elastography also utilizes ultrasonic technology. However, instead of analyzing how sound signals change as they pass through tissues, as with A-mode and B-mode ultrasound, elastography conversely uses ultrasound to measure changes in tissues resulting in stress placed on the surface of the cross section. Elastography uses ultrasonic pulses to track the internal displacement of small tissue elements in response to an externally applied stress, which is then interpreted as tissue density. Elastography works on the principle that when an elastic medium, such as biological tissues, is compressed by a constant stress applied from one direction, all points in the tissue experience a resulting level of strain along the axis of compression. A harder tissue which is less elastic will demonstrate less strain than a softer tissue which is more elastic. It is believed that this technology may be able to differentiate muscle elastic properties which are related to tenderness.

Video Image Analysis

Video Image Analysis utilizes light reflectance and intensity from a surface to differentiate colors and to quantify shapes. The technology is based on the use of a video camera to obtain a video image, which is fed through an analog/

digital converter, and then is processed by a computer. This technology could be used to differentiate animal or carcass contours (conformation) or measure ribeye area and fat thickness at the 12th rib cross section of a chilled carcass. It also could recognize color variations in the lean and marbling score in the ribeye muscle.

Wholebody K^{40} Counting

The K^{40} technique utilizes two principles to measure the muscle content of a living animal and carcasses. First, approximately 67% of the potassium found in an animal is in its muscle and second, potassium contains a fixed proportion of radioactive atoms that constantly release small amounts of energy which can be quantified. If the potassium is found predominantly in fat-free tissue, then lean body mass can be calculated by determining the body potassium levels. Approximately 0.012% of the naturally occurring potassium in muscle is the radioactive isotope K^{40}, which emits gamma radiation at 1.46 MeV. The potassium level in the body can be estimated by measuring the intensity of 1.46 MeV gamma emissions from the body. The "K^{40} Counter" has a large tunnel-shaped chamber in which the animal is held stationary for counting. Animals must be held off feed for 24 h (because of potassium levels in different feedstuffs) and thoroughly washed before being placed in the chamber. It is important that the chamber be in a facility that shields the "K^{40} Counter" from background radiation. These constraints and the cost of the machine and installation may prohibit the use of this machine for commercial practice. This concept could be used to evaluate the composition of boxed beef and other meat products.

Optical Probes

Optical lean and fat probes are currently being used extensively in European and United States pork-packing plants. These probes measure the reflectance of muscle and fat when inserted into a region of the carcass. Obviously, this is an invasive evaluation technique, but it is a nondestructive method of evaluation. In some European countries, measurements at multiple probe sites are used in their grading systems. Research is currently underway to develop optical probes that use fluorescence of connective tissue to estimate connective tissue content within the muscle.

Bioimpedance Analyzer

Bioimpedance operates on the principle that there are conductivity differences between muscle and fat. As carcass fat increases, the impedance of the flow of electricity also increases. Composition can be estimated by measuring resistance and reactance of constant levels of AC (alternating current) electrical impulses as they pass through a tissue. In actuality, fat acts like an electrical insulator and lean serves to conduct the electrical impulses.

TOBEC/Electromagnetic Scanning

Similar to Bioimpedance, TOBEC (Total Body Electrical Conductivity) is also based on conductivity difference between lean and fat tissue. TOBEC consists of a coil of copper wire wrapped around a large Plexiglas tube. This forms a scanning chamber. Current is applied to the coils creating a 2.5-MHz electromagnetic field. This type of field is similar (but on a much larger scale) to when a copper wire is coiled around a nail and then connected to a 6-V battery creating a magnet.

Carcasses or meat cuts enter the chamber and pass through this electromagnetic field and absorb energy from the electromagnetic field emitted by the coils. The amount of energy absorption that occurs while the meat product is in the chamber serves as an index of the conductive mass of the carcass or meat cut. Fat-free mass is 20 times more conductive than fat; therefore, the conductivity index is highly related to lean tissue mass. Another term used for machinery that utilizes this technology is Electronic Meat-Measuring Equipment (EMME).

Subjective Versus Objective Grading of Carcasses

Subjectivity of Grading Systems

This chapter has described the current grading systems for beef, pork, lamb, and poultry. All grading systems utilize, at least in part, subjective evaluations of carcass traits to arrive at an estimate of carcass cutability or meat quality. Even some objective measures, such as the ribeye area of a beef carcass, are in reality determined by visual appraisal because of the rapidness in which the USDA grader must collect and then assimilate the grading factors to arrive at a final grade. Production speeds in beef, pork, and lamb processing plants may exceed 5000 animals per day and over 150000 per day in poultry-processing plants. Given the time constraints on the USDA grader and the fact that all traits (with the exception of carcass weight) are visually obtained, the accuracy level of graders to obtain the correct grade is remarkable. In 1980, the USDA-AMS evaluated the accuracy and uniformity of the USDA Beef Quality and Yield Grading systems to see if grades could be accurately determined at rapid chain speeds. They reported the error rate was 7.3% for quality grades and 11.6% for yield grades, on a national basis.

Objectively Grading Carcasses

The meat and poultry industries want to move all grading to a more objective system. This would potentially improve the accuracy of assessing the final grade of a carcass. It would also allow for different carcass measurements to be obtained, resulting in a better estimate of meat taste appeal and carcass cutability.

For example, marbling score and maturity are the best subjective measures that can be evaluated under the current processing plant production system to evaluate beef tenderness, juiciness, and flavor. However, an objective instrument that could measure percent intramuscular fat in the lean, along with, connective tissue amount, collagen cross-linking, and muscle fiber tenderness might be a much more accurate tool for segmenting meat according to palatability differences.

Another example is beef yield grading. Although yield grading was implemented in 1965, it is still the best available system to segment carcasses according to carcass cutability. However, ribeye area, a fat measure at the 12th rib, kidney, pelvic, and heart fat, and carcass weight do not account entirely for the lean, fat, and bone differences that are found in the general cattle population. This equation has a particular problem quantifying bone differences, and fat partitioning differences among carcasses (particularly seam fat). An objective instrument might be able to evaluate locations other than the traditional 12th–13th rib cross section for bone and fat partitioning differences. Possibly seam fat in the round or chuck, brisket fat, rump fat, and bone circumference could be measured, or maybe an instrument will scan the entire carcass to measure total carcass composition. The use of an instrument may totally change the existing grading system.

Within the last couple of years, there has been an even greater emphasis on the development of carcass evaluation tools. Even with this emphasis, instrument grading in the livestock and poultry industries will most likely not be routinely used until after the year 2000.

Selected References

Cross, H.R. and K.E. Belk. 1992. Objective measurements of carcass and meat quality. In: *38th International Congress of Meat Science and Technology Proceedings,* Clermont-Ferrand, France, Vol. 1, Pages 127–134.

Forrest, J.C. 1991. Update on current carcass evaluation technique. In: *44th Reciprocal Meat Conference Proceedings,* Page 121.

Hayenga, M.L., B.S. Grisdale, R.G. Kauffman, H.R. Cross, and L.L. Christian. 1985. A carcass merit pricing system for the pork industry, *Am. J. Agric. Econ.* V 67(2):315–319.

Kiehl, E.R. and V.J. Rhodes. 1960. *Historical Development of Beef Quality and Grading Standards,* Research Bulletin 728, University of Missouri, College of Agriculture, Agricultural Experiment Station.

Kinsman, D.M. 1989. *Veal: Meat for Modern Menus.* National Live Stock and Meat Board, Chicago.

National Cattlemen's Association, 1992. *Improving the Consistency and Competitiveness of Beef*. National Cattlemen's Association, Denver.

National Live Stock and Meat Board. 1988. *Meat Evaluation Handbook*. National Live Stock and Meat Board, Chicago, 1922.

National Pork Producers Council, 1991. *Procedures to Evaluate Market Hogs,* 3rd ed. National Pork Producers Council, Des Moines, IA.

Ophir, J., R.K. Miller, H. Ponnekanti, I. Cespedes, and A.D. Whittaker. 1992. Elastography of beef muscle. In: *38th International Congress of Meat Science and Technology Proceedings,* Clermont-Ferrand, France, Vol. 1, Pages 153–160.

Smith, G.C., J.W. Savell, H.R. Cross, Z.L. Carpenter, C.E. Murphey, G.W. Davis, H.C. Abraham, F.C. Parrish, Jr., and B.W. Berry. 1987. Relationship of USDA quality grades to palatability of cooked beef. *J. Food Qual.* **10** 269–286.

Tatum, J.D., J.W. Savell, H.R. Cross, and J.G. Butler. 1989. A national survey of lamb carcass cutability traits, *Sheep Ind. Develop. J.* **5**(1):23–31.

Topel, D.G. and R. Kauffman. 1988. *Live Animal and Carcass Composition Measurement*. Pages 258–269. Designing Foods, National Research Council, National Academy Press, Washington DC.

United States Department of Agriculture—Agricultural Marketing Service, 1980. *Official United States Standards for Grades of Veal and Calf Carcasses*. Code of Federal Regulations, Title 7, Ch. I, Pt. 53, Section 107-53.111.

United States Department of Agriculture—Agricultural Marketing Service, 1984. Standards for grades of barrow and gilt carcasses and for slaughter barrows and gilts. *Fed. Reg.* 49(242):48669–48676.

United States Department of Agriculture—Agricultural Marketing Service, 1989. *Official United States Standards for Grades Carcass Beef*. Code of Federal Regulations, Title 7, Ch. I, Pt. 54, Section 102-54.107.

United States Department of Agriculture—Agricultural Marketing Service, 1989. *Poultry-Grading Manual*. Agricultural Handbook Number 31, U.S. Government Printing Office, Washington, DC.

United States Department of Agriculture—Agricultural Marketing Service, 1990. *Meat, Prepared Meats and Meat Products: Grading, Certification, and Standards*. Title 7, Subtitle B, Ch. I, Subchapter C, Part 54.

United States Department of Agriculture—Agricultural Marketing Service, 1992. Standards for grades of lamb, yearling mutton, and mutton carcasses and standards for grades of slaughter lambs, yearlings, and sheep, *Fed. Reg.* 57(97):7CFR, Part 53 and 54.

United States Department of Agriculture—Science and Education Administration Agricultural Research, 1980. *An Evaluation of the Accuracy and Uniformity of the USDA Beef Quality and Yield Grading System*. Report to the Office of the Inspector General.

8

Meat-Animal Composition and Its Measurement

Robert G. Kauffman and Burdette C. Breidenstein

Composition: Description, Definition, Expression, and Variation

Composition is the aggregate of ingredients, their arrangement, and the integrated interrelationship which form a unified, harmonious whole. Figure 8.1 is an example that represents the composite average of cattle, hogs, and sheep and includes the major parts. Because animals are intentionally raised to produce meat for humans, the greatest emphasis is the musculature and its relationship to everything else. The proportion of the animal's musculature is related to several criteria, but the three most important are visceral proportions (primarily affected by contents of the alimentary canal and pregnancy), fatness, and muscling (as expressed by muscle/bone). Variations in fatness and muscling are illustrated by the exhibits shown in Fig. 8.2., and practical averages of carcass composition (including poultry, fish, and venison) are included in Table 8.1. Since Fig. 8.1 represents a composite of beef, pork and lamb, Figs. 8.3–8.6 are included to illustrate how each of these three species' parts are divided into muscle, fat, and bone. These are averages that are highly dependent on method of cutting, stage of growth, and heredity.

There are at least five different arithmetic approaches to expressing quantitative composition:

1. Weight of part, organ, or tissue expressed as a percentage of live body weight, preferably on an empty body weight (free of ingesta and excreta) basis.
2. Weight of the part expressed as a fraction of empty body weight and that fraction adjusted to weights of different animals when discussing growth potential. This comparison can be made at different empty body weights as well as at different ages. The work of Ellenberger et al. demonstrates this.

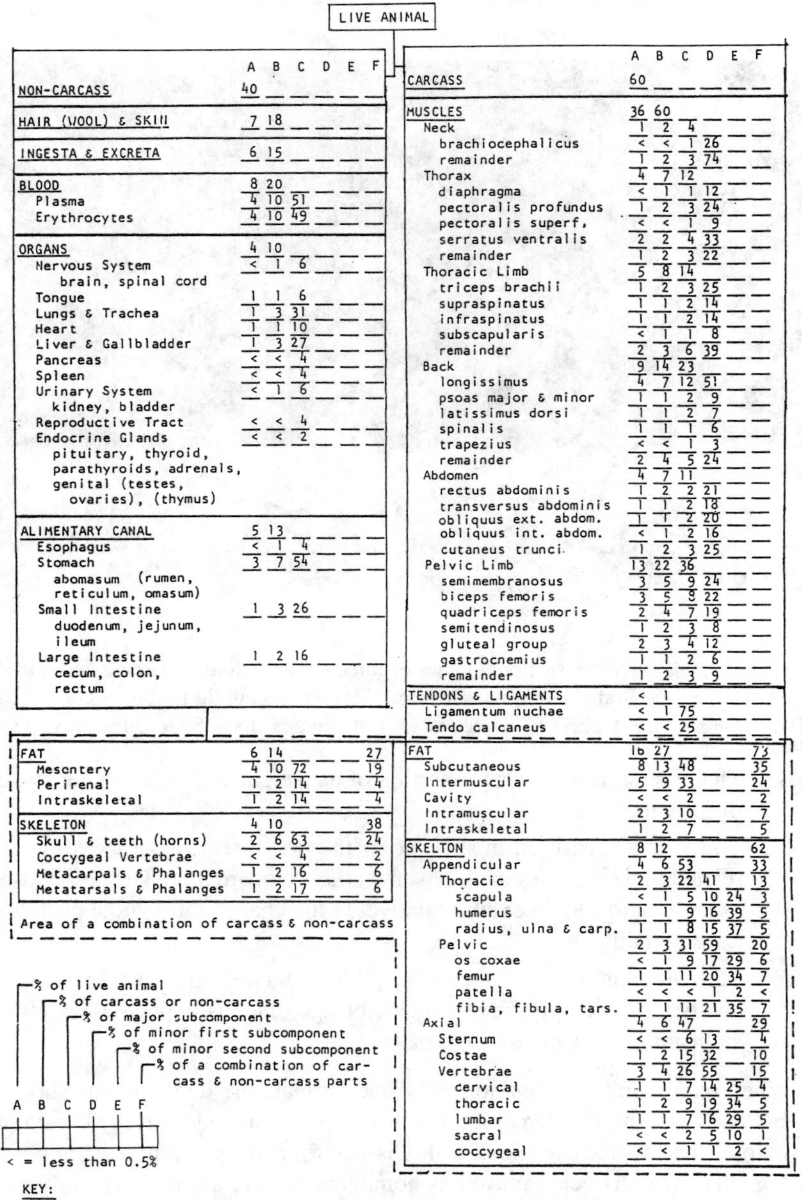

KEY:
The data represents all meat animals. Prototype was the bovine animal. Each number following a part is its percent of larger parts by weight. Example: scapula is <0.5% of live animal, 1% of carcass, 5% of carcass skeleton, 10% of appendicular skeleton, 24% of thoracic limb skeleton and 3% of total skeleton. Sum of percents within a group = 100%. Parts in parentheses may be absent due to species, sex or stage of growth.

Figure 8.1. Composition of a meat animal: Relating proportions of each component to larger components.

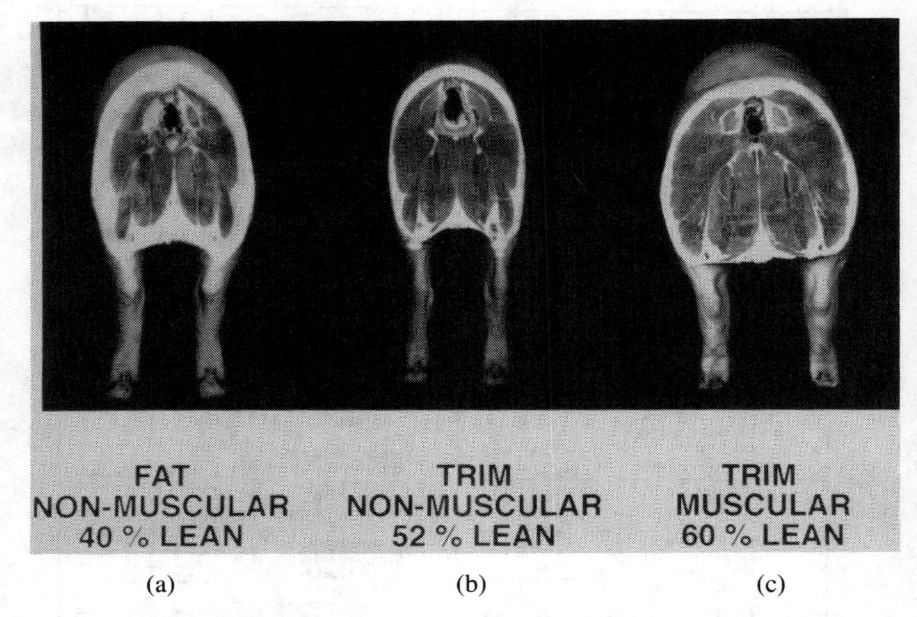

Figure 8.2. Transverse sectional views of three combinations of fatness and muscling of Porcine pelvic limbs and approximate percentage of muscle in the carcass of each. (Photograph courtesy of E. A. Kline and L. L. Christian, Iowa State University, Ames.)

3. The part expressed as a proportion of its weight at earlier or later stages of growth.
4. The part is measured in any one of the first three ways (above) but is related to the measurement of some other standard part. The part chosen as a standard should exhibit relatively little change throughout postnatal development.
5. Measurement of a part can be plotted against the measurement of another part, empty body weight, or age. This is expressed on either an arithmetic or logarithmic basis.

When expressing composition of a live animal, the denominator should be empty body weight. For carcasses, the warm weight should be used. The numerator should be fat-free muscle and this is determined by the five steps outlined in Fig. 8.7. The dissected muscle is homogenized and a sample is analyzed for lipid by ether extraction (or by chloroform–methanol to include membrane lipids). Because lipid extracted from muscle does not represent the connective tissue and fluids also associated with adipose tissue (fat), the lipid percentage should be adjusted as shown in Fig. 8.7. Once the quantity or proportion of fat-free muscle is determined, it can be readjusted to a fat-standardized basis by arbitrarily deciding what proportion of fat the muscle should contain, and then dividing fat-free muscle by 1.00 minus the fraction of desired fat level. [Example: If fat-

Table 8.1. Average Compositional Variations among Species of Meat Animals.

	Beef	Veal	Pork	Venison	Lamb	Tom Turkey	Broiler Chicken	Farm-raised Catfish
Live weight, kg	550	160	110	70	50	15	2	7
Proportion of live weight								
Non-carcass, %	38	46	27	42	48	18	23	37
Carcass Skin, %	—[1]	—	5	—	—	9	9	—
Carcass Fat, %	17	7	23	10	17	6	7	—
Carcass Bone, %	10	15	9	8	10	17	22	12
Carcass Muscle, %	35	32	36	40	25	50	39	51
Total	100	100	100	100	100	100	100	100
Dressing %	62	54	73	58	52	82	77	63
Carcass Muscle/bone	3.5	2.1	4.0	5.0	2.5	2.9	1.8	43

[1]included with non-carcass component

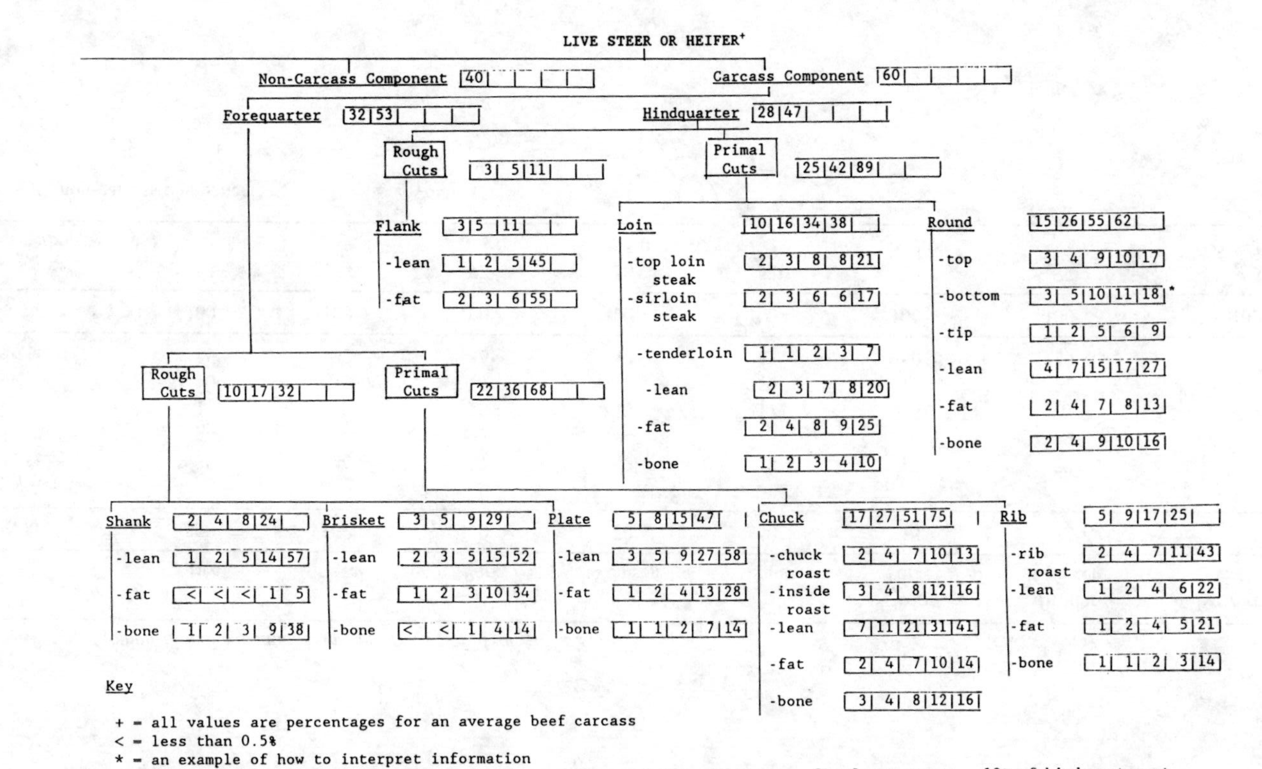

Key

\+ = all values are percentages for an average beef carcass
< = less than 0.5%
* = an example of how to interpret information
The bottom round cut of a wholesale beef round represents 3% of live wt., 5% of carcass wt., 10% of hindquarter wt. 11% of hindquarter primal cut wt. and 18% of round wt. Sum of percentages within a group in last column = 100%.

Figure 8.3. Average proportions of lean, fat, and bone in commercial cuts of young beef.

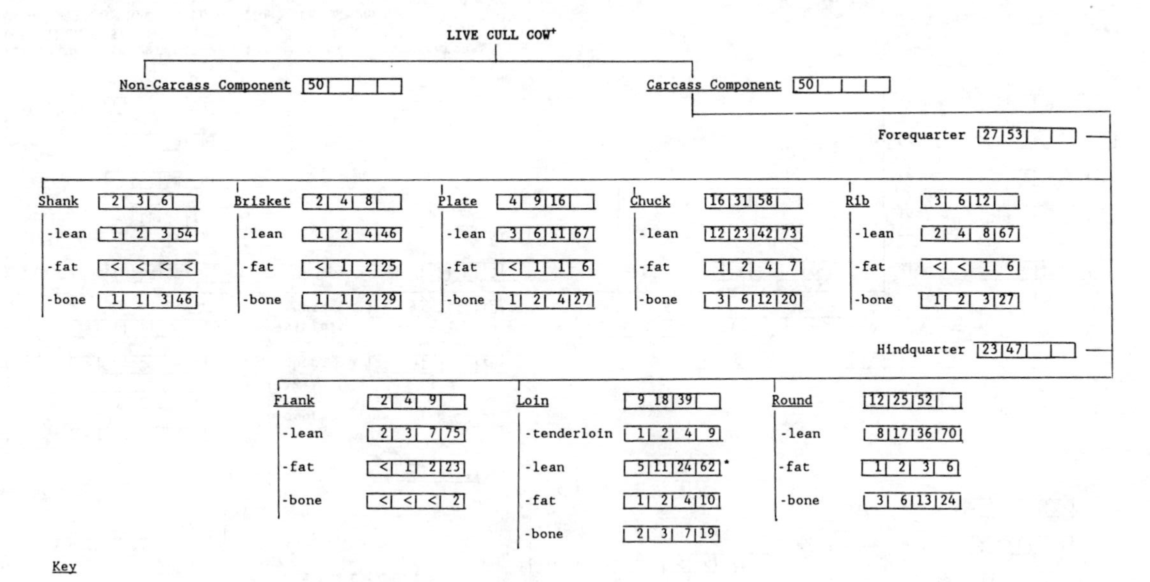

Key

+ - all values are percentages for an average cutter cow carcass
< - less than 0.5%
* - an example of how to interpret information
The lean component (excluding tenderloin) of a wholesale cow loin represents 5% of live wt., 11% of carcass wt., 24% of hindqu. and 62% of loin wt. Sum of percentages within a group in last column = 100%.

Figure 8.4. Average proportions of lean, fat, and bone in commercial cuts of cow beef.

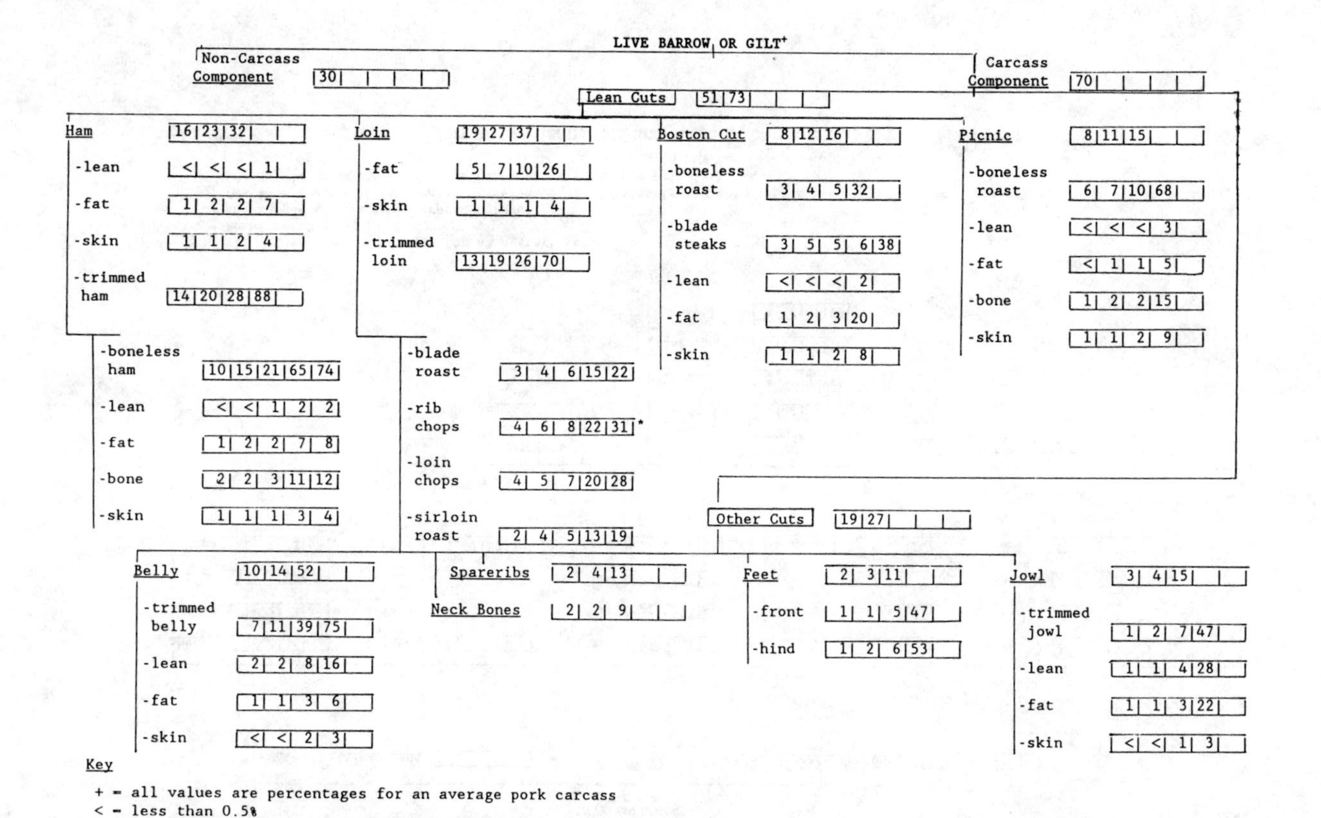

Figure 8.5. Average proportions of lean, fat, and bone in commercial cuts of pork.

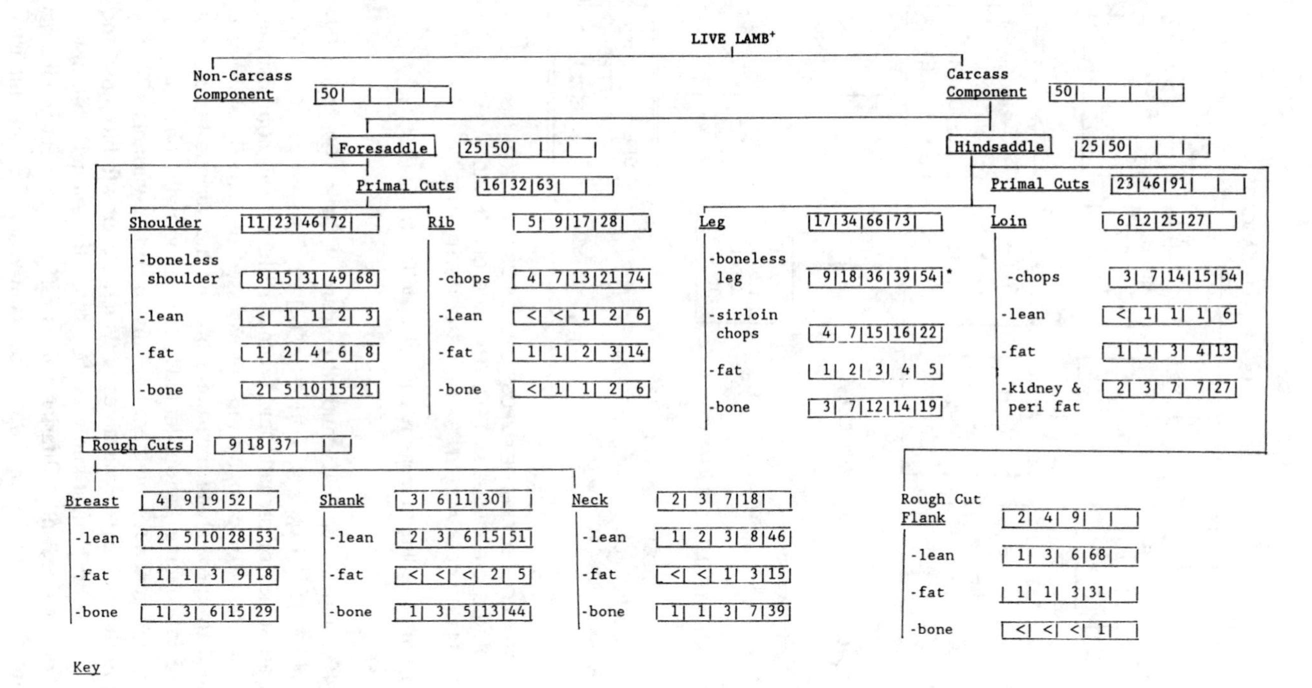

Key

+ = all values are percentages for an average lamb carcass
< = less than 0.5%
* = an example of how to interpret information
The boneless leg represents 9% of live wt., 18% of carcass wt., 36% of hindsaddle wt., 39% of primal cut wt. and 54% of wholesale leg wt. Sum of percentages within a group in last column = 100%.

Figure 8.6. Average proportion of lean, fat, and bone in commercial cuts of lamb.

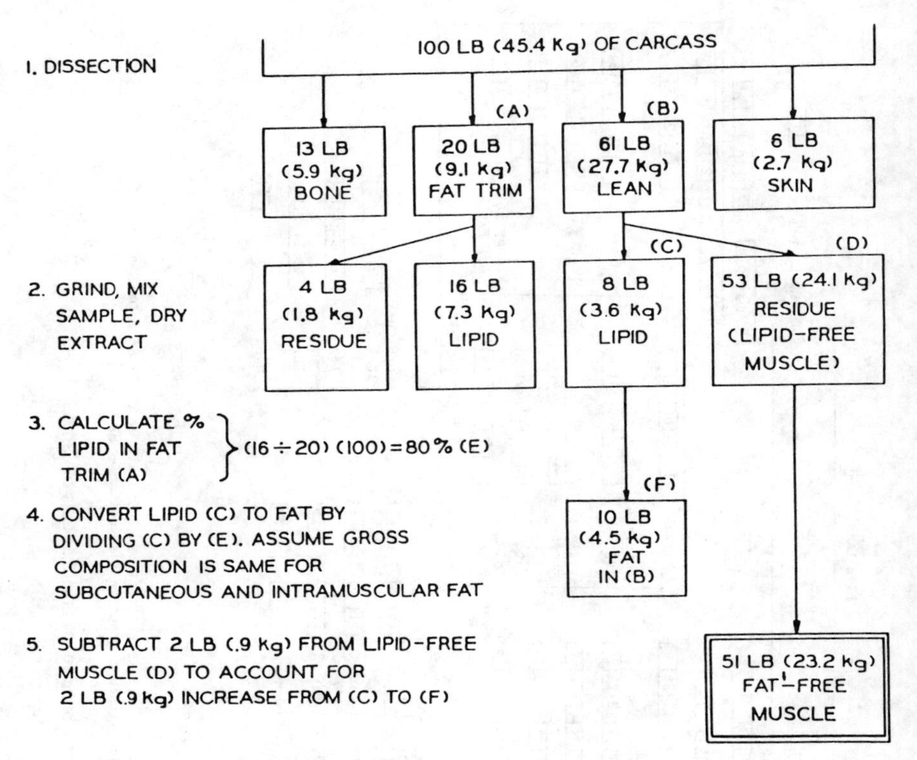

Figure 8.7. Outline of procedure to measure and calculate fat-free muscle.

free muscle = 54%, and the desired standard content of fat is 10%, then fat-standardized muscle = [54/(1.00 − 0.10)] = 60%.]

All animals at all stages of prenatal and postnatal life vary in composition. Therefore, when variations of composition are assessed objectively and accurately, there is need to know about measurements and the various factors that affect them. These influencing factors include stage of growth, physiological maturity, nutrition, and genetics which are not only important variants affecting composition, but in themselves are variables in which the growth biologist, the nutritionist, and the geneticist must know as they study the animal's biology.

The economic, qualitative, and utilitarian values assigned to meat-producing animals depend greatly on their composition. In recent years, the meat industry has made dramatic progress in reducing fat content of domestic animals as an overt response to consumer demand for lean meat. Consequently, economic pressures have been placed on the livestock–meat industry to breed, grow, feed, process, and distribute animals and their products that contain less fat and more lean. Nevertheless, many carcasses still contain more than one-quarter of their weight as fat.

The proportion of muscle in a live animal varies from less than 35% to nearly 50%. In addition to stage of growth, nutrition, and genetics, several other factors contribute to variations in body composition. They include contents of the alimentary canal, pregnancy, presence of abnormalities, and condition of the skin or fleece. All of these collectively complicate the accurate measurement of body composition when it is assessed on a live animal weight basis. Nevertheless, it is important to seek methods that reliably estimate body composition because of its contribution to the ultimate value of meat animals.

Compositional Interrelationships

Figure 8.8 is included to emphasize that muscle and bone are related and that this is primarily influenced by animal shape (conformation) and stage of growth. These two components may increase or decrease together or in opposite directions. Both are affected by fat content and by proportion of the noncarcass (visceral) portion (sometimes called the "fifth quarter"). However, there is evi-

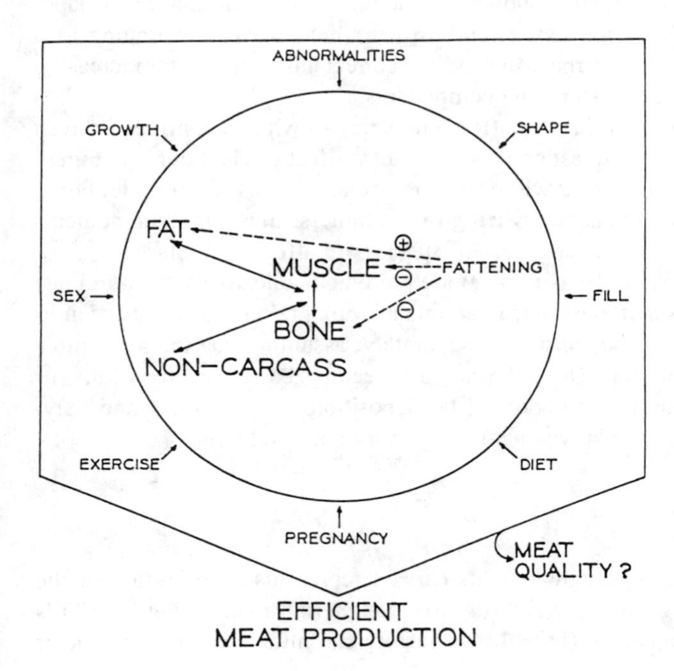

Figure 8.8. Compositional interrelationships of meat animals.

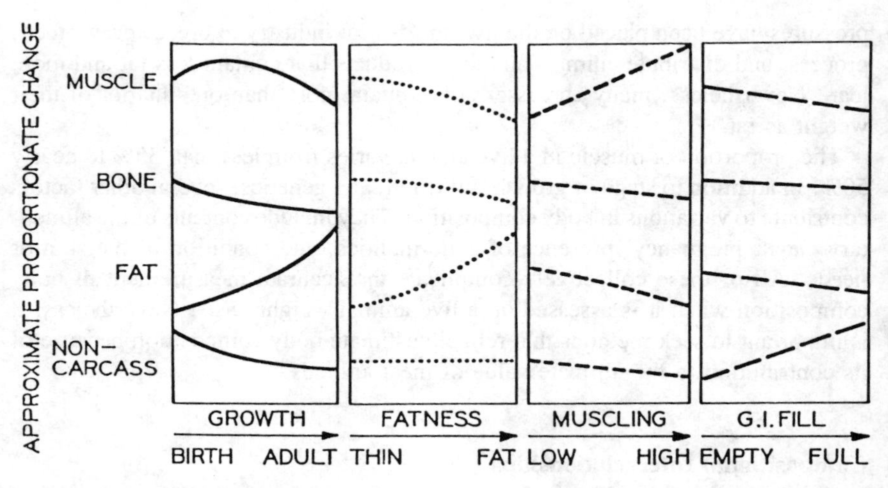

Figure 8.9. Influence of growth, fatness, muscling, and fill on relative composition of live meat animals.

dence that degree of fatness per se does not greatly influence the proportions of the noncarcass component.

Carcass compositional interrelationships are constantly under the influence of one or more of the eight factors shown outside the circle (abnormalities, shape, fill, diet, pregnancy, exercise, sex, and growth), but extent of fattening as a result of diet and growth is perhaps the one most direct influential factor increasing fat and decreasing muscle and bone components.

These interrelationships directly affect efficiency in which quantities of meat are produced, but there is question as to how they affect product quality. Figure 8.9 summarizes the effects that will occur, on a relative basis, for muscle, bone, fat, and noncarcass components, when growth, fatness, muscling, and contents of the alimentary canal (fill) change. All species are affected similarly.

Even though four variables (Fig. 8.9) are presented collectively to coordinate the interrelationships that exist, each variable affecting relative amounts of individual components is to be considered separately, assuming that the other three variables are held constant. Nevertheless, it is recognized, for example, that as growth proceeds, muscle vs: bone and fat deposition increase simultaneously. Each curve represents a relative change . . . not an absolute one.

Dressing Percentage

When a meat animal is slaughtered, its carcass represents a proportion of the live weight. Dividing carcass weight by live weight and multiplying by 100 is dressing yield or percentage (DP). Many factors can influence DP. Any factor that increases the denominator (live weight) and/or decreases the numerator (carcass weight) lowers DP, whereas the reverse raises DP. As the carcass represents the major value of the live animal and the noncarcass components have only minimal value, DP is of ultimate importance to farmers and meat packers when livestock are marketed on a live weight basis. When livestock are purchased on a carcass merit basis, then DP is not as important. However, more than half of all market livestock are purchased on a live weight basis and so DP continues to be a dominating force in affecting their economic worth.

At least 13 factors affecting DP can be categorized into 4 broad groups; species (ruminants versus monogastrics, slaughter method, fleece, hide, horns), sex (males versus females, pregnancy), environment (carcass temperature, fill, condemnations), and composition (stage of growth, muscling, fatness). Twelve of these factors and their relative importance to DP are included in Fig. 8.10.

Species influences DP in about five ways. Ruminants have proportionately larger digestive tracts than monogastrics, thus yielding a lower DP. Because of traditional approaches of transforming each species into food, different methods of slaughtering have evolved that affect carcass weight. Skins, hides or fleeces, and feet of cattle and sheep are removed, whereas only the hair on hogs (except when infrequently skinned), feet and feathers on poultry, and scales and heads on fish are removed. Perirenal (kidney) fat and kidneys may or may not be removed, which is yet another procedural effect.

Sheep having long (often soiled) fleeces have lower DPs. Some species (cattle) are more subject to accumulating excessive mud and manure on their hides, thus lowering DP. Finally, the presence or absence of horns and tails affect DP. Some breeds of cattle are either naturally polled (Angus) or have been dehorned, and it is common for market hogs and lambs to have had their tails docked to prevent tail biting and fly strikes. Because of these various species effects, the following DPs can be expected: fish—63%, poultry—80%, swine—73%, fed cattle—62%, veal—56%, and fed lambs—52%.

Sex influences DP in at least two ways: lactating females having large wet udders, and mature intact males (especially rams) have their udders or testicles removed during slaughtering, thus lowering DP. Females in the latter stages of pregnancy have considerably lower DPs (Ellenberger et al. 1950).

Environmental factors are important. One important factor is content of the gastrointestinal (GI) tract, which includes undigested feed (ingesta) and feces (excreta), commonly referred to as "fill." Injury and disease of the live animal and sanitation of the slaughter process reduces DP. If bruises, grubs, or filth are removed from the carcass, or if a portion of the carcass is diseased, condemned, and destroyed (these conditions not shown in Fig. 8.10), then final carcass weight is less. The ultimate loss of marketable value occurs when the entire carcass is condemned, resulting in DP being reduced to zero. Finally, weighing errors (both live and carcass) and condition of carcass (hot and wet or chilled and dry) affect DP.

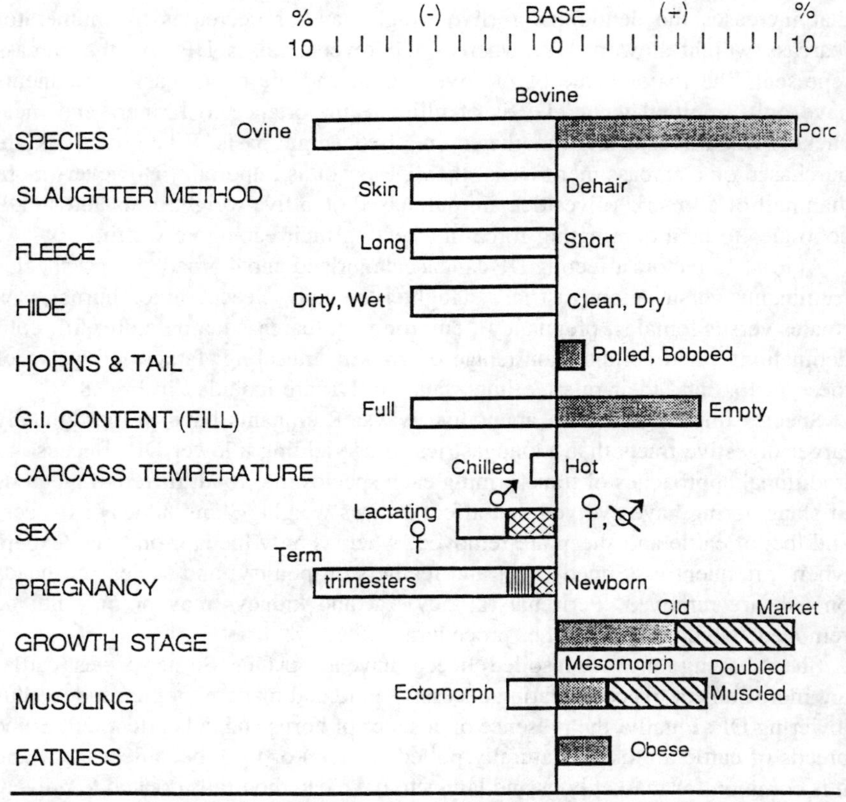

[1] Some traits are interrelated such as species, slaughter method and fleece.

[2] Condemnations (due to bruising, disease, contamination, etc.) and weighing errors are not included but could change yields slightly to excessively.

Figure 8.10. Traits influencing variation in dressing percent of meat animals.

Compositional factors are important. Stage of physiological maturity influences DP. Newly born animals have proportionately larger skeletons (especially skulls) and digestive tracts compared to the carcass, and as they grow, muscle development increases at a faster rate as compared to other components. Because most of the musculature remains with the carcass, DP increases as the animal matures. Heavily muscled animals (mesomorphs as compared to ectomorphs) and especially double-muscled ones (cattle) have higher DPs because their thoracic and abdominal cavity contents are relatively smaller. Finally, degree of fatness slightly affects DP. As the animal fattens, the proportion of carcass fat to noncarcass fat remains relatively constant. Regardless of how fat a carcass is, the proportion of total animal fat in the carcass portion is about 85%, whereas the noncarcass fat content is the remainder, or 15%. If DP was also 85%, then DP would not change. However, because a hog's DP is about 73%, increases in fatness increase the DP slightly, and for lamb, slightly more. For market weight hogs, every increase of 1% total body fat increases DP by 0.1%. In market weight lambs, the DP increases by 0.35% because the ratio of carcass fat to total fat is higher than the carcass to live weight ratio.

Specific Methods to Measure Composition

Table 8.2 provides a subjective and relative comparison of 18 methods to illustrate their versatility and appropriateness. Subjective projections indicate accuracy, cost, speed and ease of application, extent of use, major advantage(s) and disadvantages, and specific comments unique to each method. This abbreviated approach provides information to assess measures of composition. The following discussions elaborate on and explain the principles of the methods included in Table 8.2.

A. Visual Appraisal

1. Scores and grades. Visually assessing carcasses is a commonly used technique to estimate composition. A major problem with visual evaluation is distinguishing between muscling (shape or contour that is directly related to muscle/bone ratio) and fatness. Therefore, visual assessments of muscling are likely to be more effective as indicators of muscle deposition within a narrow range of fatness and particularly when fat levels are low.

Visual predictions of composition are difficult to describe and standardize so they can be consistently applied over time. However, the European Community (EC) has developed photographic standards to subjectively assess degrees of muscling and fatness for beef, pork, lamb, and veal carcasses, and they appear to be reasonably effective in classifying composition.

B. Physical Methods

2. Weight. Carcass or cut weight is only useful when combining it with other independent measurements (i.e., fat depth) to ensure that it is considered when large variations occur among exhibits. The weight of an untrimmed, bone-in carcass or cut only accounts for variation in percent lean when fatness and muscling are constant. Therefore, the only time when weight alone becomes a reliable predictor of composition is when homogeneous products are evaluated. This occurs in poultry and fish but rarely in the larger meat-animal species.

3. Linear measures. A simple, inexpensive ruler to measure fat thickness (FT)

Table 8.2. Relative Comparison of Methods to Measure Composition in Meat Animals

Method	Accuracy	Initial Cost of Equipment	Speed	Application	Extent of Current Use in United States	Advantage	Disadvantage	Other Considerations
A. VISUAL APPRAISAL								
1. Scores and grades	Moderate	Low	Rapid	Simple	Extensive	Rapid and simple	Difficult to Standardize	Should use photographs or other standards
B. PHYSICAL METHODS								
2. Weight	Low	Moderate	Rapid	Simple	Extensive	Rapid and simple	Other factors influential	Best when weight range is large
3. Linear measures ruler	Moderate	Low	Rapid	Simple	Extensive	Economical	Accuracy	Best used with other traits
4. Density	Moderate	Moderate	Slow	Simple	Research	Accurate when M/B is constant	Speed	Specific gravity in water or gas
5. Areas (grid, planimeter, ultrasonics)	Low	Low	Rapid	Simple	Modest	Rapid	Accuracy	Best used with other traits
6. Coring	Moderate	Low	Moderate	Complex	Research	None	Anatomical orientation	Experimental
7. Cutting tests	Moderate	Low	Very slow	Simple	Tests only	Accuracy	Speed	Degree of trimming, presence of bones affect accuracy
8. Dissection	Very high	Low	Extremely slow	Complex	Research	Accuracy	Speed	Experimental

C. ELECTRICAL METHODS								
9. Light reflectance	Moderate	Moderate	Rapid	Simple	Extensive	Rapid	Cost, anatomical orientation	Also used to detect muscle color
10. Video imaging	Moderate	High	Rapid	Simple	None	Standardized	Cost	Experimental
11. Ultrasonics	Moderate	High	Rapid	Simple	None	Measures fat depth	Anatomical orientation	Several types available
12. Electromagnetic scanning (TOBEC)	Very high	High	Rapid	Simple	Slight	Accuracy	Implementation	Could be used as a standard
13. Bioelectrical impedance	Moderate	Low	Moderate	Simple	None	Inexpensive	Slow	Experimental
14. Anyl-ray	High	High	Rapid	Simple	None	Accuracy	Sampling	Used in processing
15. Computerized tomography	High	Extensive	Slow	Complex	Research	Accuracy	Cost, speed	Experimental
16. NMR	High	Extensive	Slow	Complex	Research	Accuracy	Cost	Experimental
D. CHEMICAL METHODS								
17. Potassium-40	High	Extensive	Slow	Complex	None	Accuracy	Cost, speed	Experimental
18. Proximate analysis	Very high	Moderate	Slow	Complex	Tests	Accuracy	Speed, cost	Experimental

Note: This information is based on research, review of literature, experience, and opinions of the authors, and advice obtained from colleagues.

and length and width of the longissimus thoracis (LT) is used. These measurements are acceptable but not perfect predictors of body composition.

Reflectance probes to measure fat and muscle depths are applied to pork carcasses. The Fat-O-Meater and Hennessy-Chong probes are examples, and both are used commercially. These instruments measure reflectance of the muscle and fat components when a probe is inserted into the loin section of the carcass. The cost of equipment is relatively high but not excessive if used for grading large numbers of carcasses. It is simple to use, and data are electronically recorded.

4. Density. This Archimedean principle suggests that a body displaces a volume equal to its own. For carcasses, density (or specific gravity) is expressed in relation to the density of a reference standard, usually water at 20°C.

The rationale for estimating fatness or muscling from density is based on the assumption that the carcass is a two-component system, each possessing different but constant densities. If the densities of the components are known, the proportions of the two components can be estimated from the density of the carcass. The two components are fatty tissue and the fat-free carcass. The fat-free carcass is assumed to be constant in composition for mature animals. However, water content in the fat-free carcass is not constant in young, growing animals; therefore, the density measurements are not likely to be as accurate for predicting their composition. The commonly reported density values for fat, lean, and bone are 0.9, 1.1, and 1.8, respectively.

A major problem in determining density is the measurement of volume. Water displacement, which is the simplest method, has disadvantages. Air-displacement procedures, such as helium dilution, are even more complicated.

Carcass specific gravity accounts for about 80% of the variance in proportions of carcass fat and muscle. However, it is assumed that muscle/bone ratio is constant. A carcass having a low ratio will reflect a higher density (because of the greater proportionate quantity of high-density bone) and thus an inflated predictable proportion of lean. However, if the muscling (an indicator of muscle/bone) of a group of carcasses is similar and the degree of fatness varies considerably, then specific gravity is a useful predictor of composition.

5. Areas. For any constant weight or size of animal or carcass, the size of individual or groups of muscles is related to lean content. For a given frame size of an animal, muscles having a larger area are likely to be associated with more muscling and, consequently, a higher muscle/bone ratio. Fatness is clearly the one most important factor affecting composition, but when muscling is represented by cross-sectional muscle areas, equations used to predict composition are usually improved (Fahey et al.). A good example where muscle area and fat depth are applied is in determining beef carcass yield grades. The areas of intermuscular fat deposits would be another use of areas in estimating composition.

6. Coring. This technique has been used exclusively for semifrozen carcasses in which a sharp metal coring devise is used to obtain soft tissue originating from various anatomical locations of the carcass. Each "plug" of soft tissue is thoroughly mixed and analyzed chemically for lipid, protein, water, and ash. This procedure lacks usefulness because (1) bone is not considered, (2) the difficulty in obtaining representative samples, (3) difficulty in standardizing the location for sampling, and (4) time required to complete the analysis.

7. Cutting tests. Perhaps the most practical and economically meaningful measure of composition of meat animals and their carcasses is to cut carcasses into either wholesale or retail cuts as they are to be merchandized. To standardize the amount of trim, the amount of bone to be removed and the anatomical locations from which the cuts must be made is difficult. Standard procedures have been established; however, a major limitation is the time-effort required to satisfactorily obtain reliable information. Many meat companies routinely conduct cutting tests to establish value and yield of products purchased as well as to monitor production processes within the company.

If cutting tests are needed, all separable fat and bone should be removed and economic values be applied so that a value index is established. A more productive exercise is to muscle dissect a specific cut or one side. If there is a uniform medial separation, right and left sides are bilaterally symmetrical and are constant in their proportional distribution of parts regardless of fatness and muscling, especially if they are at a similar stage of growth.

8. Dissection. Of all procedures employed to assess the "true" composition of animals, this method is the most universally accepted. It is also one of the most time-consuming and expensive. Simply stated, the principle involved is in knowing gross anatomy and having the patience and care to separate each component, prevent weight loss through evaporation and drip, and weighing and recording accurately. If time and expense are limited and one is willing to sacrifice accuracy, dissection of a representative portion of the side is an alternative (9–11 beef rib section requires 2 h versus 30 h for the entire side).

C. Electrical Methods

9. Light reflectance. (See 3. Linear Measures.)

10. Video imaging. This method has been studied as a replacement for or supplement to the subjective visual assessment of grading carcasses. It can determine areas, contours, and even composition of the surface of a sectioned muscle (marbling content). The concept is based on the use of a video camera to obtain a video image (a numerical array of gray values) through an analog/digital converter. The values are manipulated by a computer. Some difficulties include (1) time required to develop image analysis procedures associated with carcass anatomy, (2) technical problems associated with daily use of electronic equipment in the harsh slaughterhouse environment, (3) standardization of the video inspection process (relative position of the object, background contrast, and lighting), and (4) development of software with the capability to handle carcasses with large variations in size.

11. Ultrasonics. This is based on the principle of high-frequency sound signals passing through tissues, and when an interface between two tissues is encountered, sound waves are reflected. A pulse generator conveys electrical pulses that are converted into sound signals in the transmitter. These signals are passed through the tissues until they are reflected at each interface. The reflected signals are detected by the receiver and can be amplified and shown in a visual form on an oscilloscope. Variations in the time taken for the reflected signals to return to the transmitter–receiver are used to measure variations in the distances of the boundaries between tissues. The "A" mode ultrasonic instruments display echo amplitude against time which is shown on the screen as peaks superimposed on a time baseline. The distance between the peaks represents the thickness of the tissues being measured. For "B" mode instruments, the signals are shown on a cathode ray tube as a series of bright spots. The thickness of the tissues is represented by the distance between successive bright spots.

These instruments have single transducers that move across the carcass on a track. As the transducer moves, a picture is reconstructed on a cathode ray screen. Real-time instruments produce a practically instantaneous picture by rapid electronic switching from element to element. The interpretation of results for "B" mode instruments and real-time scanning usually requires the tracing of depths and areas from pictures. This is usually accomplished by using planimeters linked to microprocessors or computers. Ultrasonic scanning predicts composition with a degree of accuracy similar to that of the corresponding cut surface measurements on the carcass.

12. Electromagnetic scanning. This principle relies on eddy currents induced in the carcass by an alternating magnetic field produced by a current passing through a coil surrounding the object. The eddy currents generate a magnetic field that can be detected by a change of impedance in the coil. The method depends on conductivity of muscle being higher than that of fat, on a constant tissue temperature.

The total body electrical conductivity (TOBEC) instrument currently is being used commercially to estimate the composition of meat-animal carcasses and boxed meat (see Fig. 8.11). According to Forrest et al. (1991), the procedure is accurate in estimating composition of warm pork carcasses (Coefficient of Determination = 95% with low residual deviations), and it can be used to identify composition of carcass components (i.e., hams) as well. A serious limitation is that carcasses or cuts must be introduced to TOBEC in a horizontal position, which is not compatible with vertically suspended carcasses in conventional processing slaughter lines.

13. Bioelectrical impedance. The resistance and reactance is measured when a constant current is passed through a biological mass. Terminal plethysmographs introduce a constant current that provides a deep homogenous electrical field in

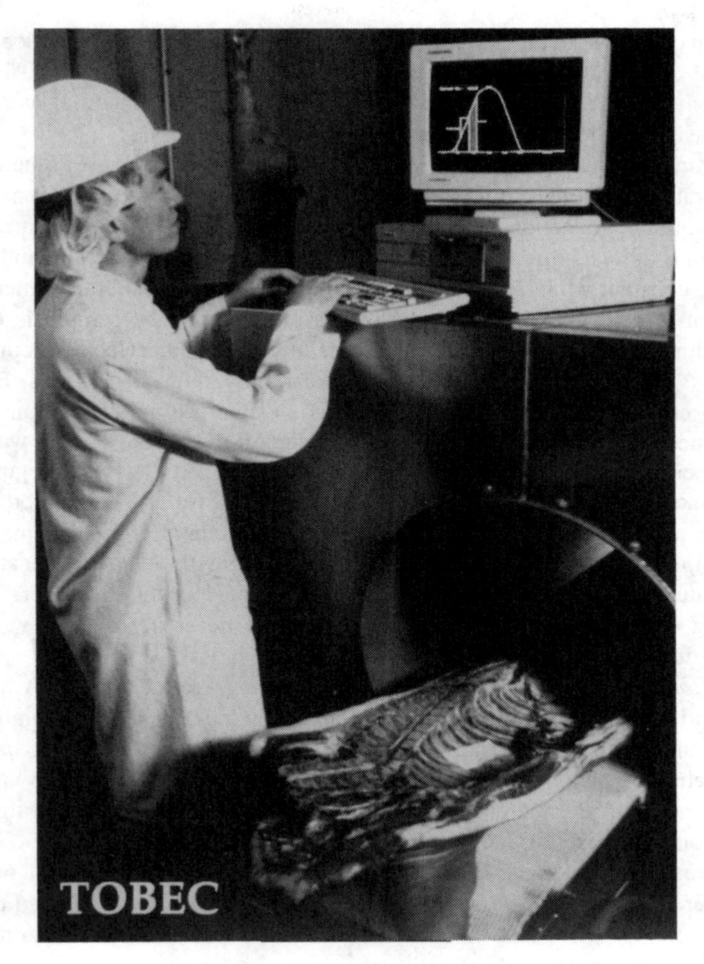

Figure 8.11 Measuring lean content in a pork carcass with electromagnetic scanning (TOBEC). (Photograph courtesy of J. C. Forrest, Purdue University, West Lafayette, IN.)

the carcass. An alternating current of 800 μA at 50 kHz is introduced into the carcass via transmitter terminals and received by detector terminals. The basis of impedance is the determination of voltage drop as an electrical current passes through a conductor. The conductance or resistance is dependent on the geometric configuration, volume of the biological conductor, signal strength, and frequency of the applied current. Assuming a relatively constant geometry for carcasses studied of one species and using a constant electrical signal impedance, resistance and reactance should be dependent on body composition. Muscle is less resistant than fat.

When compared to fat-free carcass composition determined by chemical analysis, and adjusting for weight, resistance values account for over 80% of the variation in fat-free mass. Therefore, this technique has potential as a rapid, noninvasive method for estimating composition of carcasses.

14. Anyl-ray. This is based on X-ray attenuation as an index of tissue fatness. It generates electromagnetic waves of a character which are very sensitive to absorption and reflection or backscatter by the elements in ground meat. The radiation is directed through a sample where size, shape, compaction, and weight must be constant while sample composition varies. When calibrated intensity of radiation is held constant, the energy passing through the sample is directly proportional to sample composition. A carefully mixed and selected ground meat sample is subjected to a minute amount of carefully controlled X-rays. Because lean absorbs more X-rays than fat, there is a difference in energy transmitted. The penetrating rays are collected by a radiation-measuring device which, in turn, energizes a calibrated digital percentage fat meter. It is used regularly to determine fat content in ground meat used for processing. The method is fast, requires a sample that can be reutilized for processing after measurement, and has a high degree of accuracy. The instrument correctly evaluates meat at any temperature, provided it can be properly compacted in the container. Further grinding or the addition of warm water may be necessary to achieve proper compaction for frozen samples.

15. Computerized tomography. The Nobel Prize was awarded to A.M. Cormack and G.N. Houndsfield for the development of the computerized tomography (CT) technique, sometimes referred to as CAT scan. The concept is based on presentation of anatomic areas of the carcass by computed synthesis of an image from X-ray transmission data obtained in different directions through the plane under consideration. An X-ray tube rotates around the carcass and the computer reconstructs, from a series of pictures, a slice through the carcass. CT numbers for different tissues are fat = −100 to 0, muscle = +30 to +100, and bone = +400 to +500. The main limitations to CT are expense and the time required to obtain an estimate.

16. Nuclear magnetic resonance (NMR). The NMR method for estimating composition is based on a strong static magnetic field and pulsed radio waves that induce resonance of protons in the carcass. The signals emitted are a reaction of the tissues to the high-frequency disturbance. Therefore, they are a product of the matter itself, with intensities depending on the proton spin densities and molecular structures. The NMR signal does not continue indefinitely. Environmental influences cause the individual flipped magnetic moments to go out of phase and return to the orientation they had before the radio-frequency pulse was applied. The time required to reestablish original conditions has been defined as spin-lattice relaxation time T_1 and spin-spin relaxation time T_2. Procedures to determine T_1 are known as inversion recovery, and for T_2 as spin-echo methods. Both systems produce a data matrix of the size 128-128 or 256-256 that contains in X-ray CT the normalized Houndsfield units ranging from −1000 (air) to more than +1400 (compact bone). There are several ways to produce images. On a device with seven colors, the total data space in a matrix is subdivided into seven regions, with each region representing a different color. Fat, muscle, bone, and connective tissue are always presented if the total data space is mapped onto seven colors.

NMR imaging has great potential, but very limited data are available on its usefulness for predicting composition of carcasses. The equipment is extremely expensive, and the method is very complex; its future will depend on the amount of resources available for its development.

D. Chemical Methods

17. Potassium-40. Estimating carcass composition by use of K-40 is feasible because of the direct relation of potassium to lean body mass and its indirect relation to fat. Potassium is the only single element found in body tissues in significant amounts that has any predictive value of composition. It is found primarily in muscle fibers, and, thus, carcass potassium is indicative of total lean. Potassium is not found in adipose tissues in any significant quantities, and, therefore, if potassium is present in the fat-free tissues as a constant percentage, then a value for carcass potassium can be converted to a weight of lean body mass.

Approximately 0.012% of all naturally occurring potassium is made up of the radioactive isotope ^{40}K. The rest is composed of the stable isotopes ^{39}K and ^{41}K. Because ^{40}K emits gamma radiation at 1.46 MeV, the intensity of 1.46 MeV from the carcass can be captured electronically and used to estimate total potassium content. From these values, total protein and lean mass can be estimated, assuming that the mass of protein or muscle versus potassium is constant.

18. Proximate analysis. For animal tissues, the primary chemical components used as a follow-up to physical dissection are moisture, protein, lipid, and ash. The procedures for these are described in the A.O.A.C. A major concern in the determinations is adequate mixing and sampling of the tissues to be analyzed. Another limitation is in chemically analyzing bone because of the difficulty in grinding and sampling. Also, mixing ground components of soft tissues creates problems of fat collecting on the sides of the mixer. Finally, when moisture is being assessed, large errors will occur when muscles are soft and watery such as the pale, soft, and exudative (PSE) condition in pork.

There is a significant relationship between chemical components of beef carcass sides and the meat sawdust from sawing through the frozen round, loin, rib, and chuck at repeated intervals. The reliability of the sawdust procedure for predicting carcass composition is greatest from frozen carcasses sawed every 2 cm, followed by the meat sawdust from cutting frozen carcasses into retail cuts. Chemical composition of meat sawdust can provide a good estimate of the chemical composition of carcasses when frozen samples are collected. The method is simple but time-consuming, and slight devaluation of the carcass occurs.

Composition and Quality Interrelationships

When evaluating meat-animal carcasses and their parts, both composition and quality are important. Leanness (as contrasted to fatness) is virtuous, but by itself, it fails to meet ultimate expectations of consumers. The nutrient density of muscle is more acceptable in leaner meat because nutritive value is a part of quality. Therefore, composition from this perspective affects quality. However, quality is more than nutrient value. Wholesomeness, appearance, water-holding capacity, and palatability are other quality traits covered in greater detail in other chapters.

Marbling contributes to juiciness and flavor. However, more marbling reduces nutrient density. Furthermore, carcass fatness is related to marbling. The association is not strong, but fatter carcasses usually exhibit muscles containing more marbling. The paradox is that meat animals are fed to heavier weights, for longer times, and to ultimately less favorable "feed-to-meat" ratios so that muscles contain more marbling to satisfy consumer demands. This negative relationship between marbling and leanness is difficult to compromise and is one of the reasons why beef and lamb carcasses may be too fat.

In pork, trim and heavily muscled carcasses appear to be more susceptible to the PSE (pale, soft, exudative) and PSS (porcine stress syndrome) conditions. Such carcasses yield bulging hams and shoulders and large loin muscles with minimum quantities of fat. Nevertheless, their muscles often shrink excessively during processing. Fresh pork cuts that are exceptionally lean may be pale, soft in texture, and watery, all of which detract from appearance and, ultimately, their acceptance by the consumer.

Therefore, quality of all meat products must be considered along with composition when overall value is assessed. The conflict between composition and quality continues to challenge scientists to discover new genetic combinations, different feeding and management programs, and more satisfactory postmortem processing technologies to ensure an ideal meat product that meets consumer demands.

Acknowledgments

The authors appreciated being invited to prepare this chapter. Also, they thank the many scientists that contributed resources and advice, and are especially indebted to the Department of Meat & Animal Science and the College of Agricultural and Life Sciences, University of Wisconsin for providing financial assistance and support.

Selected References

A.O.A.C. 1990. In: *Official Methods of Analysis,* 15th ed., K. Helrich; ed. Association of Official Analytical Chemists, Arlington, VA.

Berg, R.T. and R.M. Butterfield. 1976. *New Concepts of Cattle Growth.* John Wiley & Sons, New York.

Ellenberger, H.B., J.A. Newlander, and C.H. Jones. 1950. *Compositions of the Bodies of Dairy Cattle,* Bulletin 558, Agricultural Experiment Station, University of Vermont, Burlington.

Elsley, F.W.H., I. McDonald, and V.R. Fowler. 1964. The effect of plane of nutrition on the carcasses of pigs and lambs when variations in fat content are excluded, *Anim. Prod.* 6:141.

Fahey, T.J., D.M. Schaefer, R.G. Kauffman, R.J. Epley, P.F. Gould, J.R. Romans, G.C. Smith, and D.G. Topel. 1977. A comparison of practical methods to estimate pork carcass composition. *J. Anim. Sci.* 44:8.

Forrest, J.C., C.H. Kuei, W. Chen, R.S. Lin, A.P. Schinckel, P. Walstra, H. Kooper, and M.D. Judge. 1991. Electromagnetic scanning: carcass evaluation. In: *Electronic Evaluation of Meat in Support of Value-Based Marketing.* Pages 85–111. Purdue Meat Science Industrial Advisory Consortium and Purdue University, West Lafayette, IN.

Kauffman, R.G. 1971. Variation in gross composition of meat animals. *Proc. Recip. Meat Conf.* 24:292.

McMeekan, C.P. 1941. Growth and development in the pig with special reference to carcass quality characters. *J. Agric. Sci.* 31:1.

9

Pathogenic Microorganisms and Microbial Toxins Associated with Muscle Foods

Jennifer L. Johnson

Introduction

Foodborne diseases are estimated to cost the United States approximately 8.4 billion dollars a year (Todd 1989), an astonishing fact when one considers that most cases of foodborne disease are preventable. Indeed, most food scientists agree that microbiological contamination of foods poses a far greater threat to public health than do pesticides and additives—two items at the top of any list of consumer concerns. An estimated one in three Americans suffers at least one episode of foodborne enteric disease annually, and figures are almost certainly underestimated. Contaminated muscle foods have been implicated in 44–61% of foodborne disease outbreaks reported since 1973 (Bryan 1988; Bean and Griffin 1990). Many of these illnesses are due to growth of pathogens and/or toxin formation during periods of temperature abuse at the point of food preparation and consumption.

Factors Influencing Our Perception of "Important" Microorganisms

Historical Factors

At any given time and place, the designation of certain microorganisms as "important" pathogens is determined by a particular combination of factors including history, geography, technology, culture, economy, demography, and what may be called "sensational" factors. Up until the mid-1970s, the list of recognized pathogens important to muscle foods was a short one: *Clostridium botulinum, Staphylococcus aureus, Salmonella, and Clostridium perfringens*. Salmonellae, staphylococci, and *C. perfringens* had been identified in food poisoning cases linked to animal products for decades, whereas the recorded association of *C. botulinum* with meats goes back more than 1000 years.

The 1980s saw the "emergence" of a number of foodborne pathogens which previously had not been recognized as being transmitted by muscle foods. In 1982, for instance, *Escherichia coli* 0157:H7 was implicated in cases of foodborne disease linked to fast-food hamburgers. Later, the transmission of *Vibrio* spp., *Campylobacter jejuni, Yersinia enterocolitica,* and *Listeria monocytogenes* by muscle foods was documented. Awareness of dinoflagellate-derived toxins in shellfish also increased in the 1980s.

Cultural and Ethnic Factors

An organism regarded as an important foodborne pathogen in one part of the world may not share that distinction in another region due to cultural or ethnic differences in food preparation and consumption habits. For instance, the Japanese custom of eating raw seafood means that *Vibrio* spp. is a much more important pathogen in that country than it is in countries where the inhabitants typically cook seafoods before consumption. The rising popularity of uncooked seafood (especially shellfish) in the United States has likely contributed to the increasing number of foodborne illnesses attributed to seafoods.

Geographical Factors

Although cultural food preferences are easily understood in relation to the incidence of various foodborne diseases, geographical differences in the distribution of pathogenic microorganisms are much harder to understand. The prevalence of pathogenic *Y. enterocolitica* on pork, for example, is much higher in northern Belgium than it is in the center of the Netherlands—two regions having similar animal husbandry practices and geographically separated by a distance of approximately 100 miles.

Geographical distribution of foodborne pathogens becomes an important consideration when muscle foods are imported. The types of pathogens found on imported product may not be the same as those found on domestic product. In particular, pathogens such as *E. coli* and various nematodes may be more common on muscle foods imported from some countries than from others. Seafood and other muscle food products imported from less-developed countries may contain high levels of pathogens due to poor hygiene, inadequate treatment of human sewage, and insufficient refrigeration.

Economic Factors

Certainly, any organism regulated by federal or state agencies must be considered important. The most obvious examples of this are *Salmonella* in ready-to-eat muscle foods and *C. botulinum* in low-acid canned products. In these products, *Salmonella* and *C. botulinum* are regarded as avoidable contaminants or adulterants, and contaminated product may be recalled from distribution and retail outlets. Recent regulations prohibiting the presence of *L. monocytogenes* in ready-to-eat muscle food products have done much to bring that organism into the public eye.

In general, raw muscle food products are subject to much less regulation than ready-to-eat products with regard to pathogens. This is due, in part, to the prevailing notion that most pathogens present on raw muscle foods will be destroyed during proper cooking. Sometime in the 1960s, according to Foster (1989), a court ruled "that *Salmonella* in raw poultry (and, by inference, in other raw animal products) was an unavoidable defect. Thus neither government nor industry has been impelled to contribute the necessary effort into developing ways to avoid contamination of raw animal products with *Salmonella*." Only recently have government and industry programs begun to address the very complex issue of pathogen contamination of raw animal products. Our current inability to produce raw muscle foods that are absolutely pathogen-free is partly due to economic limitations on the production of pathogen-free live animals.

The potential for economic loss or adverse publicity to industry determines, in part, the relative importance of a particular pathogen. Cases of botulism due to commercially processed foods, although rare, generate a tremendous amount of adverse publicity. This publicity generally spreads beyond the implicated manufacturer and causes economic hardship across the whole spectrum of companies processing that particular product. When these costs are added to medical costs, the average cost per case of foodborne disease is highest for *C. botulinum*, followed by *Salmonella typhi* and *L. monocytogenes* (Todd 1989). All told, bacterial and viral diseases account for approximately 84% of all foodborne disease costs in the United States.

Technological Factors

Advances in epidemiological and microbiological methodology, both in clinical and food applications, have greatly contributed to our understanding of foodborne disease. Our ability to link a given organism to both disease and an implicated food product has improved tremendously over the last 10 years. This new methodology offers unprecedented opportunities to investigate the contributions that animals, humans, and the plant environment make to pathogen contamination of the final product.

It seems likely that the "emergence" of foodborne pathogens such as *E. coli* 0157:H7 is partially due to improvements in methodology. Techniques such as enzyme-linked immunoassays, nucleic acid hybridization assays, and the polymerase chain reaction have been developed and popularized by clinical microbiologists, then adapted for use in food microbiology. Such methodology will continue to revolutionize the way we detect foodborne pathogens.

Sensational Factors and Changing Demographics

Various "sensational" factors also contribute to the perceived importance of a given microorganism. The morbidity (disease rate) and mortality (death rate) associated with a pathogen, together with the severity of the disease state, contribute to the perception of that organism's importance. Pathogens that cause a particularly dramatic disease state (e.g., the bloody diarrhea accompanying *E. coli* 0157:H7 infections) or affect a vulnerable segment of the population (i.e., the elderly or infants or immunocompromised people) will be regarded as being more important than organisms that cause unremarkable illnesses in ordinary healthy people.

Changing demographics also influence the importance of a particular foodborne pathogen. The number of elderly and immunocompromised people in our population continues to rise, a fact that definitely reverberates within the food industry. Loss of immune function, whether caused by age, drugs, or underlying disease (especially cancer and AIDS), renders an individual more susceptible to pathogens such as *Salmonella, L. monocytogenes, Aeromonas hydrophila,* and *Toxoplasma gondii*. In general, most deaths associated with foodborne pathogens occur in individuals having predisposing conditions. The case-fatality rate for foodborne disease outbreaks in nursing homes, for instance, is approximately 10 times that of outbreaks occurring in other settings.

Changes in our eating habits are sure to have an impact on the incidence of foodborne disease. An increasing number of meals eaten outside the home shifts the burden of safe food preparation to a transient, often poorly educated, population of food handlers. This fact, together with the large numbers of people served in restaurants and institutional settings, adds up to an increased potential for large-scale foodborne disease outbreaks.

Foodborne Disease and Muscle Foods

Cases and Outbreaks of Foodborne Disease

Despite unprecedented technological advances in the field of food processing, the incidence of foodborne disease worldwide appears to be increasing. Foodborne disease statistics tabulate "cases" (instances in which epidemiological evidence links an individual's case of foodborne disease to a particular food) and "outbreaks" (occurrence of two or more cases of a disease, associated in time and place so as to implicate a common food). A majority of foodborne outbreaks occurs between late spring and early fall, but sporadic cases of foodborne illness do not generally have a seasonal distribution and occur throughout the year. Sporadic cases of foodborne disease are much more common, and more likely to be underreported, than are cases that are part of an outbreak.

Foodborne disease statistics include only those cases of illness that are reported to public health authorities, often through a passive reporting system. The true number of cases (both reported and unreported) may be as much as 10 to 25 times higher than the number of reported cases. This underreporting is due to a number of factors including the existence of mild self-limiting disease conditions that do not require medical attention and are, therefore, not reported, the common belief that one merely has "a case of that flu that's going around," and the levels of bureaucracy inherent in our municipal–state–federal reporting system. With the exception of botulism, sporadic cases of foodborne disease are usually not reported to the surveillance system.

Recent years have seen a decrease in foodborne surveillance activities, meaning that the actual incidence of unreported foodborne disease is probably even higher than in previous years. This decreased surveillance is largely due to the increasing demands that the AIDS epidemic has placed on the limited resources available to state and federal health agencies (Bean and Griffin 1990).

Even so, estimates of the number of cases of foodborne illness in the United States range from more than 12 million to 81 million annually. A sizable portion (30–60%) of both cases and outbreaks of foodborne disease are considered to be "of unknown etiology," meaning that epidemiological evidence links a food source to disease, but laboratory confirmation cannot be obtained. An inability to identify the causative organism may reflect the unavailability of suspect foods, poor sample collection techniques, insensitive laboratory methods, incorrect medical diagnoses, or inappropriate recovery procedures. In some instances, an investigation may not even involve laboratory analyses, further contributing to the number of foodborne illnesses of unknown etiology.

Based on the onset time of the illness (the time interval between consumption and symptoms of foodborne disease), cases or outbreaks of unknown etiology can be broken down into four subgroups: onset times of < 1 h generally indicate chemical poisoning (including histamine poisoning and paralytic shellfish poisoning); 1–7 h indicate preformed toxin (probably *S. aureus* poisoning); 8–14 h, probable *C. perfringens* food poisoning; or > 14 h, indicative of other infectious agents or toxic agents.

Table 9.1 presents a listing of important foodborne pathogens together with estimates of the number of cases of foodborne disease and approximate costs incurred in the United States per year. Data compiled by the Centers for Disease Control and Prevention (CDC) indicate that bacterial pathogens are responsible for the largest number of cases and outbreaks of foodborne disease of known etiology (66% and 92%, respectively), followed by chemical agents (including the various seafood toxins), which account for 26% of the outbreaks and 2% of cases. Parasites cause approximately 4% of outbreaks and $< 1\%$ of cases, and viruses cause 5% of both outbreaks and cases (Bean et al. 1990).

Information presented in Table 9.1, like any statistics on foodborne disease, is at best an estimate. As such, there may be differences of opinion over the

Table 9.1. Estimated Number of Cases and Total Costs of Selected Types of Foodborne Disease in the United States per Year

Disease Agent	Number of Cases	Estimated Annual Cost[a] (US $)
Salmonella[b]	2,960,000	3,991,000,000
Staphylococcus aureus	1,555,000	1,516,000,000
Clostridium perfringens	652,000	123,000,000
Campylobacter spp.	170,000	156,000,000
All *E. coli*	69,000	223,000,000
Vibrio vulnificus	29,000	37,000,000
Listeria monocytogenes	25,000	313,000,000
Vibrio cholerae/Parahemolyticus	13,000	13,000,000
Clostridium botulinum	270	87,000,000
Toxoplasma gondii	1,437,500	445,000,000
Trichinella spiralis	40,000	144,000,000
Viruses	216,000	337,000,000
Seafood intoxications	58,260	125,000,000
Unknown etiology	5,217,000	529,000,000

[a]In 1985 dollars.

[b]Species other than *S. typhi*.

Source: Adapted from Todd (1989).

reliability of a particular figure. Reporting of foodborne disease statistics typically occurs 3 or more years after the data were collected, meaning that current information is not now available. *Campylobacter* spp., for instance, are actually much more frequent causes of foodborne disease than published reports indicate because most advances in detection methodology have been made in the last 2–3 years.

Typically, foodborne disease statistics reflect only gastroenteritis, meaning symptoms of diarrhea, vomiting, nausea, and abdominal pain. Recent evidence indicates that foodborne pathogens may cause a number of additional sequellae including arthritis, atheroschlerosis, autoimmune thyroid disease, and neural disorders (Archer and Young 1988). The chronic nature of these disorders and a multitude of other contributing factors make it nearly impossible to determine the contribution which foodborne pathogens make to such chronic disease conditions.

Foodborne Intoxications and Infections

Foodborne diseases can be divided into two categories: foodborne intoxications and foodborne infections. Foodborne intoxications are true cases of "food poisoning" in that a susceptible human consumes food containing a toxin or poison. For intoxications of bacterial origin, the following sequence of events must occur: The microorganism must be present in the food, conditions must be suitable for its growth, growth must result in large numbers of the organism, toxin must be produced, and the food must be consumed by a susceptible host. The presence of viable microbes is not necessary for disease. Due to the presence of preformed toxin in the food sample, the onset times of foodborne intoxications can be very short, with symptoms occurring within a few hours after consumption. Pathogenic bacteria involved in foodborne intoxications include *S. aureus, Bacillus cereus,* and *C. botulinum*. Foodborne intoxications can also be caused by ingestion of ciguatoxin, scombrotoxin (histamine), and a variety of dinoflagellate-derived toxins.

Foodborne infections occur when a susceptible host consumes food that is contaminated by viable organisms capable of colonizing the host's gastrointestinal tract or being transported across the intestinal lining. Disease may result when only a few microorganisms are ingested, as in the case of *Shigella, Campylobacter,* and some strains of *E. coli*. Foods containing high levels of such organisms, however, are more likely to cause disease than foods having only a low level of contamination. Because microbial growth in the intestinal tract usually precedes colonization, transport, and tissue damage, foodborne infections are characterized by relatively long onset times, ranging from >12 h to several weeks. Microorganisms responsible for foodborne infections include human enteric viruses, *Salmonella, Campylobacter* spp., *C. perfringens, L. monocytogenes, Y. enterocolitica,* and various parasites.

Typically, most cases of a particular foodborne disease are caused by a relatively small number of clones that are either especially virulent (having a high degree of pathogenicity) or very common in the environment. These clones are well suited to causing disease and, as the disease state generally results in release of large numbers of these organisms into the environment (i.e., in vomitus and diarrhea), are likely to become predominant. Such organisms are so virulent that ingestion of only low numbers may be sufficient to cause disease. The infectious dose (ID; number of organisms necessary to cause disease) of a particular agent varies with the organism, the health of the host (especially immune status and age), and the nature of the contaminated food. The ID of *Shigella* may be as low as 10 organisms, whereas the ID of *Campylobacter* may be >100 organisms and that of *C. perfringens,* > 100,000 organisms.

Pathogenic microorganisms can be divided into three groups on the basis of host characteristics. The first group includes organisms such as *Salmonella, Campylobacter,* and the Norwalk virus that are known to cause disease in healthy adults. Organisms belonging to a second group are not common causes of disease in healthy adults but most often affect especially susceptible subgroups of the general population. This group includes *L. monocytogenes* and *Toxoplasma gondii* which tend to infect the very young, the very old, pregnant women, and immunocompromised individuals. The third group includes organisms that are of uncertain pathogenicity for humans; these organisms will be discussed later in the chapter.

Table 9.2 presents a brief summary of common symptoms of the various foodborne illnesses that are mentioned in this chapter. Due to space limitations, detailed descriptions of foodborne bacterial pathogens, together with extensive information on disease conditions, virulence, and recovery procedures, will not be presented here. For more information, the reader is directed to the list of selected references at the end of this chapter.

Association of Muscle Foods with Foodborne Disease

Unfortunately, muscle food products are often implicated in both foodborne infections and intoxications. For the period 1977 to 1984, Bryan (1988) calculated that seafoods were responsible for 24.8% of foodborne outbreaks, red meats 23.2%, and poultry 9.8%. Data compiled by CDC for the years 1973 to 1987 indicate that beef was responsible for 10% of cases of foodborne illness, turkey 6%, pork 5%, chicken 3%, shellfish 3%, and finfish 2% (Bean and Griffin 1990). Table 9.3 presents information on some important foodborne pathogens, muscle food products implicated in disease, and probable causes of contamination.

Foodborne disease statistics indicate that seafoods are responsible for approximately half of all outbreaks and roughly a quarter of all cases of known etiology involving muscle foods. Individual cases of disease generally implicate shellfish and contamination with viruses and *Vibrio* spp., whereas outbreaks are usually due to seafood toxins (especially ciguatoxin and scombrotoxin). Examination of these data reveals that a considerable number of cases and a smaller number of foodborne outbreaks could be prevented if raw seafoods were not consumed. In fact, the incidence of foodborne disease attributable to cooked seafoods is relatively low.

Unrecognized Foodborne Pathogens and Zoonotic Agents

Muscle food products are host to a wide variety of bacterial species, some of which may represent unrecognized human pathogens. *A. hydrophila,* a potential pathogen increasingly recovered from the stools of patients with gastroenteritis, is common on raw muscle foods of all types. Another potential pathogen, *Plesiomonas shigelloides,* is often recovered from oysters and minimally processed seafood dishes like salt mackerel and cuttlefish salad. Other microorganisms such as *Streptococcus* and *Yersinia pseudotuberculosis* have been regarded as pathogens for years, but definitive evidence of their involvement in foodborne disease is still lacking. Strains of *B. cereus* capable of producing enterotoxin at 7°C (45°F) have been recovered from muscle foods, but their involvement in foodborne illness remains to be proven.

Opportunistic pathogens which may be present on muscle foods include *Citrobacter, Enterobacter* spp., *Klebsiella pneumoniae,* and *Pseudomonas aeruginosa.* Pathogenic strains of *Proteus, Providencia, Moellerella, Hafnia alvei,*

Table 9-2. Characteristics and Symptoms of Some Foodborne Illnesses that May Be Transmitted by Muscle Foods

Organism/Illness	Onset Time	Duration	Major Symptoms[a]
Campylobacter spp.	2–5 d	2–10 d	Diarrhea, fever, cramps
Clostridium botulinum	12 h–5 d	Varies	Nausea, vomiting, impaired swallowing, speech and respiration, double vision
Clostridium perfringens	8–24 h	24–48 h	Diarrhea, cramps
E. coli 0157:H7	3–9 d	2–9 d	Severe cramps, bloody diarrhea
Listeria monocytogenes	?	?	? (may include mild fever, malaise, diarrhea)
Salmonella	6–72 h	1–7 d	Nausea, diarrhea, fever, cramps, vomiting
Shigella	1–7 d	2 d–2 wks.	Diarrhea, fever, abdominal pain
Staphylococcus aureus	0.5–8 h	6–24 h	Nausea, vomiting, cramps, diarrhea
Vibrio spp.	6 h–5 d	2–12 d	Diarrhea, nausea, severe abdominal pain, vomiting
Yersinia enterocolitica	1–2 d	1–3 d	Fever, abdominal pain, diarrhea
Hepatitis A virus	15–50 d	Weeks–months	Fever, Malaise, anorexia, nausea, abdominal pain
Norwalk-like virus	12–48 h	24–48 h	Vomiting, diarrhea
Trichinella spiralis	2–28 d	Weeks	Nausea, vomiting, diarrhea, fever, muscle soreness, profuse sweating
Histamine poisoning	Few minutes–hours	< 12 h	Tingling, itching, flushing, nausea, vomiting, rash, headache
Ciguatera	Few hours	Hours–months	Abdominal pain, nausea, vomiting, diarrhea, muscular aches, tingling and numbness, metallic taste, chills, paralysis
Paralytic Shellfish Poisoning	< 1 h	Few days	Tingling, burning, numbness and other neurological symptoms, rash, fever, paralysis

[a] List of symptoms is not all-inclusive, and symptoms of a particular condition may vary among individuals.

Table 9.3. Association of Bacterial Pathogens with Muscle Foods Frequently Implicated in Disease with Probable Causes

Pathogen	Muscle Foods Frequently Implicated in Disease	Probable Cause
Clostridium botulinum	Fermented seafoods,[a] fish	Improper fermentation, cold smoking, temperature abuse
Staphylococcus aureus	Ham	Contamination of cooked product by food handler, possible temperature abuse
Clostridium perfringens	Roast beef, turkey	Temperature abuse (improper cooling)
Salmonella	Roast beef, turkey, chicken	Temperature abuse, cross-contamination, insufficient heating
Campylobacter spp.	Poultry, pork	Insufficient heating, cross-contamination
Yersinia enterocolitica	Pork	Consumption of raw meat, insufficient heating, cross-contamination
E. coli 0157:H7	Beef	Insufficient heating
Listeria monocytogenes	Ready-to-eat product	Contamination of finished product, possibly by environmental source
Vibrio spp.	Seafood	Consumption of raw seafood
Shigella	Ready-to-eat product	Infected food handler
Hepatitis A Virus	Seafood, ready-to-eat product	Consumption of raw seafood, infected food handler
Histamine	Mahi-mahi	Temperature abuse
Ciguatera	Amberjack, jack	Consumption of toxic diatoms by fish
Paralytic shellfish poisoning	Scallops, mussels, clams, oysters	Consumption of toxic diatoms by shellfish
Trichinella spiralis	Pork, wild game meat	Consumption of raw product, insufficient heating
Toxoplasma gondii	Pork, other meats	Consumption of raw product, insufficient heating

[a] Including fermented fish and marine mammal parts produced by Native Americans in Alaska.

Kluyvera, Morganella, Serratia, and *Bacillus* spp. (other than *B. cereus*) have been found on muscle foods, but their involvement in foodborne disease is uncertain.

Zoonoses are diseases that are transmitted between animals and humans. Foodborne zoonoses important in developing countries include those caused by *Bacillus anthracis* (anthrax), *Brucella* spp. (brucellosis), *Mycobacterium* spp. (tuberculosis), *Coxiella burnetti* (Q fever), *Leptospira* (leptospirosis), *Erysipelothrix rhusiopathiae* (erysipelas), and *Francisella tularensis* (tularemia). These zoonoses are generally not transmitted by food in developed countries, but rather through occupational exposure.

Humans employed by farm animal industries are exposed to a variety of nonfoodborne pathogens during the course of their work. Veterinarians and individuals working in slaughterhouses, tanneries, and rendering plants are particularly at risk from pathogenic microorganisms associated with animal hides, intestines, and various tissues. Organisms that can be classified as "occupational hazards" include *Brucella, Mycobacterium tuberculosis, Bacillus anthracis,* and *Francisella tularensis*. Additionally, certain rickettsia and chlamydia can be transmitted via the respiratory tract. Individuals engaged in harvesting and processing seafood may contract *Vibrio vulnificus* through puncture wounds, cuts, and scrapes incurred during handling finfish and shellfish.

Microbiology

Classes of Microorganisms

It is not the intention of this chapter to present detailed information on basic microbial physiology and classification. Instead, the main classes of microorganisms will be introduced, and further emphasis will be placed on those that are important causes of foodborne disease associated with muscle food products. For background information on the characteristics of various classes of microorganisms, the reader is directed to any general microbiology textbook.

Most of the microorganisms discussed in this chapter are widespread in the natural environment and are, therefore, commonly found on animals, fish, and shellfish. Because the gastrointestinal (GI) tract of any animal is an open-ended system, microorganisms residing in the GI tract may be regarded as still being in the "outside" environment. Such organisms may be either harmless commensals or potentially harmful pathogens.

Bacteria

Bacteria are single-celled organisms that reproduce by cell division. These organisms may be either unintentional contaminants (spoilage or pathogenic organisms) of muscle foods or intentional additives such as the lactic acid starter cultures used in the manufacture of fermented products. Bacteria are probably the single most important class of pathogens occurring on muscle foods, and much of this chapter will be devoted to a discussion of their occurrence and control.

Bacteria are typically classified on the basis of their cell-wall composition as indicated by reaction to various stains (especially the Gram stain). Although this may seem an irrelevant classification criterion to the muscle food scientist, bacterial cell-wall composition is related to pathogenicity, as well as resistance to sanitizing agents, heat, freezing, and a host of other things. Generally, gram-positive bacteria are more resistant to chemical agents and physical processes than gram negatives because a thick layer of peptidoglycan surrounds their cell wall, thereby protecting the sensitive cell membranes.

The group of gram-negative bacteria includes pathogens such as *E. coli,* the darling of geneticists and probably the best-characterized bacterial species, *Salmonella* and *Campylobacter,* which are leading causes of human diarrhea, *Shigella,* an organism specifically adapted to humans and other primates, *Y. enterocolitica,* a pathogen associated with consumption of undercooked pork, and *Vibrio,* a marine bacterium common on seafoods. Gram-negative organisms involved in spoilage of muscle foods include *Pseudomonas, Alcaligenes, Enterobacter* spp., *Flavobacterium,* and the *Moraxella-Acinetobacter* group.

Gram-positive pathogens of importance to muscle foods include *C. botulinum* which produces the most lethal proteinaceous toxin known to man, *S. aureus,* an organism common on human skin which is capable of producing a heat-stable toxin on some processed meats, *L. monocytogenes,* a cold-tolerant pathogen which is widespread in nature, and *C. perfringens,* which grows rapidly in meats and gravies subjected to temperature abuse. Important gram-positive spoilage organisms include lactic acid bacteria, the predominant spoilage organisms on vacuum-packaged meats, and *Brochothrix thermosphacta,* a common spoilage organism on fresh red meats.

Some bacteria are capable of forming spores, small highly resistant structures that can remain dormant for long periods of time. Under the proper conditions, spores may germinate and develop into viable, actively growing vegetative cells. Spore-formers important to muscle foods include *C. perfringens, C. botulinum,* and *B. cereus.*

Yeasts

Like bacteria, yeasts are single-celled organisms, but yeasts reproduce by budding rather than by symmetrical cell division. Although yeasts are commonly found on muscle foods, they generally play only a very small role in the ecology of these products. Under normal circumstances, bacterial growth on a muscle product proceeds at such a pace that the slower-growing yeast cells are outcompeted. That is, the bacteria metabolize the available nutrients and multiply to high levels before significant growth of yeast cells can occur. Spoilage by yeasts will only occur on products that are preserved in such a way as to make them unsuitable for bacterial growth.

Yeasts that are pathogenic to humans are rarely found on muscle foods. For this reason, further discussion of yeasts will be omitted.

Molds

Molds are multicellular organisms characterized by a tangled, fuzzy mass of growth (mycelium). Large numbers of spores may be released by molds during asexual reproduction and carried by winds for considerable distances. As with yeasts, growth of molds on fresh muscle foods generally does not proceed at an appreciable rate due to a lack of competitiveness with bacteria. Once a product is dried or preserved in some other manner that lowers the amount of available moisture, mold growth can occur. Growth of mold on muscle foods may be desirable, with the molds contributing to flavor development and tenderness. This is true of some *Aspergillus* and *Penicillium* spp. on fermented dry sausages and country-cured hams, and of the masses of *Thamnidium elegans* mycelia ("whiskers") sometimes visible on the surface of extensively aged beef carcasses.

Certain molds are capable of producing toxic metabolites known as mycotoxins (including aflatoxin, ochratoxins, and sterigmatocystins). It is possible for animals fed a diet of mycotoxin-contaminated grain to accumulate these toxins in organ (especially liver and kidney) and muscle tissues. Generally, an absence of aflatoxin in liver and kidney means that no aflatoxin residues will be present in muscle. There are few data available on this phenomenon, but it appears that pigs (the animal considered most prone to mycotoxin accumulation) accumulate only very low levels of aflatoxin in edible tissues even when fed corn contaminated with high levels of aflatoxin (Stubblefield, Honstead, and Shotwell 1991). Under experimental conditions, then, animal tissues may accumulate sufficient aflatoxin to pose a threat to human health, but it seems unlikely that this would occur under practical conditions.

A variety of mold species, including *Aspergillus* spp., *Penicillium* spp., *Scopulariopsis, Rhizopus, Cladosporium,* and *Mucor,* is commonly found on muscle foods. Although mold species capable of mycotoxin production may be isolated from muscle foods, the production of mycotoxins on those products is not a common occurrence. Research has shown that mycotoxins are most likely to form on cured meat products (fermented sausages and dry-cured hams) during aging at temperatures above 15°C (59°F) and at relative humidities above 75%. As mycotoxin contamination of muscle foods does not appear to be of great concern, molds will not be discussed further.

Viruses

Unlike the microorganisms mentioned previously, viruses require a live host for multiplication, and often require a specific host. For instance, viruses causing diseases in farm animals are usually incapable of infecting humans. Also, viruses for which there are no receptors in the human digestive tract are unlikely to cause illness when ingested with food. For this reason, viruses causing hepatitis B, genital herpes, and AIDS are not believed to be transmissible by foods.

Although viruses cannot multiply on a food product, they may still be present, usually due to fecal contamination by an infected food handler. When ingested by a suitable host, viable foodborne viruses may multiply in the host's tissues and cause illness. In general, only durable viruses capable of maintaining viability both inside and outside the GI tract are transmitted by foods. Viruses do not "die" in the strict sense because they are never actually "alive." A virus is a simple construct of nucleic acid and protein, and multiplication is possible only in conjunction with a suitable living host.

Inactivation ("attrition") of viruses can occur upon exposure to heat, sunlight, certain bacterial enzymes, and various chemicals, but the conditions necessary for complete inactivation vary depending on the type and level of virus. Characteristics of the surrounding medium (food or water) also influence virus survival. Fats and proteins stabilize viruses, and viruses present in perishable foods often outlast the food.

Due to environmental differences, in particular, growth in waters contaminated with human sewage, fish and shellfish are more commonly contaminated with human enteric viruses than are red meats and poultry. Consumption of raw and improperly cooked shellfish is frequently implicated in cases of viral enteritis and is probably the main cause of foodborne viral disease. Enteric viruses believed to be transmitted primarily by the fecal–oral route include poliovirus, echovirus, coxsakieviruses, various enteroviruses, "small round viruses," hepatitis A and non-A/non-B type, Norwalk virus, calicivirus, astrovirus, reovirus, rotavirus, adenovirus, and Snow Mountain Agent. It is widely believed that most cases of foodborne viral disease could be prevented if all fecal contamination of food and water were eliminated.

Our knowledge of foodborne viruses is still incomplete, a situation complicated by severe difficulties in isolating and culturing these agents. Heptatitis A and Norwalk-like virus, the two most important types of foodborne viruses, are seldom detectable because they do not cause perceptible infections in laboratory host systems. We do know that active virus particles can be shed in human feces for a number of weeks after clinical disease symptoms have disappeared.

Parasites

Parasites of primary importance in muscle foods include nematodes like *Trichinella spiralis* and the anisakids, cestodes (tapeworms) such as *Taenia* and *Diphyllobothrium,* and protozoa like *Toxoplasma gondii*. Certain parasites, especially anisakids and cestodes, pose a special threat to ethnic groups who are accustomed to consuming raw muscle foods. In many cases, the risk of contracting a parasitic foodborne disease is directly dependent on the host species and the area; this is especially true of fish.

The American pork industry has been battling *Trichinella spiralis* for a number of years, and the prevalence of that parasite in hogs has been reduced to around 1 per 1000. This figure is somewhat deceptive as the prevalence of trichina-infected hogs varies in different parts of the country and epidemiological evidence indicates that a large proportion of infected hogs is processed by small slaughter plants. Regardless, the organism remains a public health concern. Regulations mandate the cooking of garbage and food scraps fed to swine and the use of certified trichina-free pork in uncooked pork sausages. Trichinosis may also be transmitted by wild animal meat containing encysted *Trichinella*. Evidence indicates that some *Trichinella* strains isolated from mammals in the Artic may be more resistant to freezing than strains recovered from hogs. This means that freezing schedules sufficient to inactivate *T. spiralis* in pork may be inadequate for destruction of the parasite in meat from Artic animals.

Other nematodes important in muscle foods are the anisakids, a group which includes *Anisakis simplex* (also known as the herring worm) and *Phoconema decipiens* (a.k.a. the cod worm). Anisakids have a three-stage life cycle requiring marine mammals, crustaceans, and fish as hosts. As with *Trichinella,* humans serve only as incidental hosts, and the anisakids can complete their life cycle without infecting a human. Anisakids are contracted by humans through the consumption of raw or insufficiently cooked fish.

Cestodes include *Taenia solium* (found in hogs and humans), *Taenia saginata* (the beef tapeworm), *Taenia cysticercus* (the human tapeworm), *Diphyllobothrium latum* (found in freshwater fish) and *Diphyllobothrium uris* (found in salmon and other fish whose life cycles include both salt water and freshwater). Tapeworms are contracted by eating raw or insufficiently cooked infected muscle. Although foodborne disease caused by cestodes is rare in developed countries, it is a significant problem in some developing countries.

Toxoplasma gondii is arguably the most economically important parasite in the United States. A protozoan similar to the malaria parasite, *T. gondii* requires carriage by cats in order to complete its life cycle. Serological evidence indicates that up to 50% of the American population is infected with *Toxoplasma,* probably due to ingestion of raw or undercooked meat from infected animals. Cysts of *T. gondii* may occur in muscle of any domestic or wild warm-blooded animal, but pork is believed to be the major cause of toxoplasmosis in humans.

Other parasites such as *Giardia lamblia* and *Entamoeba histolytica* may be transmitted by muscle foods following contamination by water or infected workers. These organisms do not infect the food animal, however, and muscle foods are infrequently implicated in these types of foodborne disease.

Transmissible Encephalopathy Agents

Several types of transmissible encephalopathies are known to occur in humans and animals. These include Creutzfeldt–Jakob disease in humans, scrapie in sheep, and bovine spongiform encephalopathy (BSE; also known as "mad cow disease") in cattle. The BSE agent is one of the most publicized microbial pathogens of recent memory. The infective agent responsible for causing BSE appears to be very similar, if not identical, to that which causes scrapie in sheep.

The outbreak of BSE that resulted in the death or destruction of more than 25,000 cattle in Great Britain during the late 1980s is attributed to changes in rendering practices of meat and bone meal. The current hypothesis is that discontinuation of a solvent extraction process in 1981/1982 produced meat scraps (tankage) and bone meal with excessive levels of fat which protected the infective agent from inactivation during the heating process (Wilesmith, Ryan, and Atkinson 1991). This may have allowed survival of sufficient levels of the agent to cause clinical disease during the lifetime of cattle because the incubation period of BSE is related to the dose of the agent. It is unlikely that the BSE agent is a recent variation of the scrapie agent, but rather it represents a cattle "adapted" strain of a scrapielike agent which has already been present in the cattle population for some time.

The exact nature of the agents which cause BSE and scrapie in animals and Creutzfeldt–Jakob disease in humans is not known. Presently, there is no conclusive evidence to suggest that BSE is a zoonotic disease that can be transmitted from animals to humans. As a precaution, however, Great Britain has banned the use of bovine brains, spinal cord, thymus, spleen, and tonsils for human food and for animal feed. There is little published information on the occurrence of the BSE or scrapie agent in muscle, but the agent seems to concentrate in lymphatic and nervous tissue and bone marrow. Muscle which has been separated from these tissues is thought to pose little risk.

Bacterial Toxins

Endotoxins and Exotoxins

Toxic materials associated with bacteria have various origins. Endotoxins are bacterial cell components that cause a toxic reaction when introduced into a host by ingestion or by another route. The classical bacterial endotoxin is the lipopolysaccharide components of gram-negative bacteria. Cell walls of some gram-positive bacteria (including *L. monocytogenes*) may contain similar lipopolysaccharides, but the role of those compounds in disease is unknown.

An exotoxin is a substance produced during bacterial growth and excreted into its environment. Well-known exotoxins include the heat-stable toxin produced by *S. aureus* and the heat-labile exotoxin produced by some types of *C. botulinum*. Toxins such as those produced by *S. aureus* are classified as enterotoxins because they act in the GI tract of the human host. The majority of pathogenic bacteria involved in foodborne disease produces enterotoxins which figure prominently in the disease process. A notable exception is *C. botulinum* toxin which is a neurotoxin.

Scombrotoxins and Biogenic Amines

Certain by-products of bacterial growth, such as the histamines produced on histidine-rich fish muscle by certain types of spoilage bacteria (including *Morganella morganii, Klebsiella pneumoniae,* and some lactobacilli), may also be involved in foodborne disease. Rapid growth of histamine-producing organisms may occur when fish are held without proper refrigeration. Even after refrigeration of temperature-abused fish, residual enzymatic activity will continue, resulting in histamine formation in the absence of microbial growth. Ingestion of fish containing high levels of histamines may cause the so-called "scombroid fish poisoning" which is associated with both scombroid fishes (tuna, mackerel, bonito, and saury) and nonscombroid fishes (including herrings, sardines, anchovies, mahi-mahi, and bluefish).

Biogenic amines (including histamine) may be found on a variety of foods including raw and fermented meat products. While useful as indicators of microbial spoilage of some products, these compounds can build up to levels that are toxic to humans. Members of the Enterobacteriaceae, in particular, are capable of decarboxylating a variety of amino acids to form biogenic amines. Tyramine, for instance, can be formed during the fermentation and ripening of salami and accumulate to levels that could be troublesome to tyramine-susceptible individuals.

Dinoflagellate- or Algal-Derived Toxins

Tetrodotoxin (Pufferfish) Poisoning

Tetrodotoxin poisoning results when toxic components of pufferfish are consumed. Tetrodotoxin is one of the most potent natural toxins known, and high concentrations may be present in the internal organs, skin, and spawn of numerous species of pufferfish. The origin of tetrodotoxin is in dispute, and the toxin may not be of dinoflagellate or algal origin. Recent Japanese studies suggest that it may be a bacterial toxin produced through an interaction between symbiotic bacteria (possibly of the family Vibrionaceae) and the host. Despite extensive regulations governing the preparation of pufferfish by only specially trained chefs, a number of deaths attributable to tetrodotoxin still occurs each year in Japan and China.

Ciguatera

Ciguatera poisoning is caused by the ingestion of fish that have fed on toxic dinoflagellates, a type of microscopic marine planktonic algae. Finfish implicated in ciguatera poisoning include a wide variety of reef fish and bottom-feeding fish including grouper, barracuda, red snapper, and sea basses. Rapid population explosions (or "blooms") of toxic dinoflagellates may occur in tropical and subtropical waters; the toxins are then assimilated through the food chain. It appears that larger fish may be more ciguatoxic than smaller fish of the same species, either as a result of greater accumulation or slower clearance of the toxins. Currently, no control measures exist for ciguatera poisoning, in part because of the occurrence of ciguatoxic hot spots even within a particular reef or area. Ciguatera poisoning is generally not lethal, but symptoms may recur over a period of years.

Paralytic Shellfish Poisoning

The accumulation of toxins involved in paralytic shellfish poisoning (PSP) begins with the ingestion of toxic dinoflagellates (especially *Gonyaulax* spp.) by shellfish such as scallops, mussels, clams, and cockles. Although various species of dinoflagellates may be involved in these cold-water blooms, the toxins are known collectively as saxitoxins. Saxitoxins, like ciguatoxin, are present at higher levels in internal organs than in muscle tissue, and dark shellfish meat contains more saxitoxin than lighter-colored meat. Monitoring programs for PSP include sampling shellfish from various beds and analysis for saxitoxin by mouse bioassay. Shellfish beds found to have hazardous levels of saxitoxin are quarantined and closed to all harvesting. Existing surveillance programs have been quite successful at decreasing the number of cases of PSP due to commercially harvested shellfish. Consumption of toxic shellfish collected from closed areas by recreational fishermen, however, remains a serious problem. PSP is a potentially life-threatening condition.

Neurotoxic Shellfish Poisoning

Neurotoxic shellfish poisoning, sometimes called brevetoxic shellfish poisoning, originates with the dinoflagellate *Ptychodiscus brevis,* the organism responsible for red tides along the Gulf of Mexico and Florida coasts. As with PSP, filter-feeding molluscs ingest the toxic dinoflagellates and concentrate the toxin in their tissues. Toxic blooms of this dinoflagellate are associated with massive fish kills which serve to alert shellfish harvesters and monitoring agencies so that harvesting areas may be closed. It is important to note that clearance of the toxins from the shellfish will require additional time beyond the end point of the red tide bloom.

Miscellaneous Shellfish Poisonings

Diarrhetic shellfish poisoning is also caused by dinoflagellate toxins and has been associated with short-necked clams, mussels, and scallops. The internal organs of the shellfish seem to contain the majority of toxin, and avoidance of these tissues should reduce the risk of diarrhetic shellfish poisoning. At the present time, diarrhetic shellfish poisoning does not appear to be a problem with domestic shellfish, but imported shellfish may contain the toxin.

Amnesic shellfish poisoning is another name for shellfish poisoning caused by domoic acid. As with the other types of shellfish poisoning, this is due to an accumulation of a diatom-derived toxin in the edible tissues of mussels, clams, crabs, and anchovies. Outbreaks of this severe disease have occurred in Canada and the northwestern United States, and shellfish is routinely tested for domoic acid.

A syndrome known as "crab poisoning" has been reported in Japan. In this disease, crabs which have fed upon remnants of toxic animals become toxic themselves.

A short-lived hallucinogenic condition associated with the consumption of mullet and a variety of reef fish has been reported in Hawaii. The cause of this syndrome is unknown.

Microflora of Muscle Foods

Characteristics of Muscle Foods That Influence Microflora

Strictly speaking, bacterial "growth" refers to an increase in cell numbers (multiplication) rather than to an increase in cell mass or size. Bacterial growth follows a well-established pattern once bacteria are introduced onto meat or into a culture medium. An initial period of adjustment (called the "lag" phase) occurs during which the microorganisms adapt to their new surroundings. The duration of the lag phase varies with the substrate and the temperature. Next comes a period of rapid, exponential growth termed the "log" phase. Once nutrients become depleted and toxic metabolites accumulate, the organisms enter into a "stationary" phase in which growth ceases and some death occurs; additional death occurs during the "reduction" phase. These various growth phases are illustrated in Fig. 9.1.

Product characteristics and storage parameters not only influence the rate of microbial growth on muscle foods but also determine which type of microorganism will predominate. Following is a general discussion of some of the factors that influence the microbial communities present on muscle foods.

Temperature

Bacteria may be classified on the basis of temperature tolerance or optima. Psychrotrophic bacteria can grow at temperatures below approximately 10°C (50°F), but their optimum growth temperature is higher. This group includes *Y. enterocolitica*, *L. monocytogenes*, and *A. hydrophila* as well as nonproteolytic types of *C. botulinum*. Another group of microorganisms, the psychrophiles, grow only at temperatures below 10°C (<50°F). A wide variety of organisms

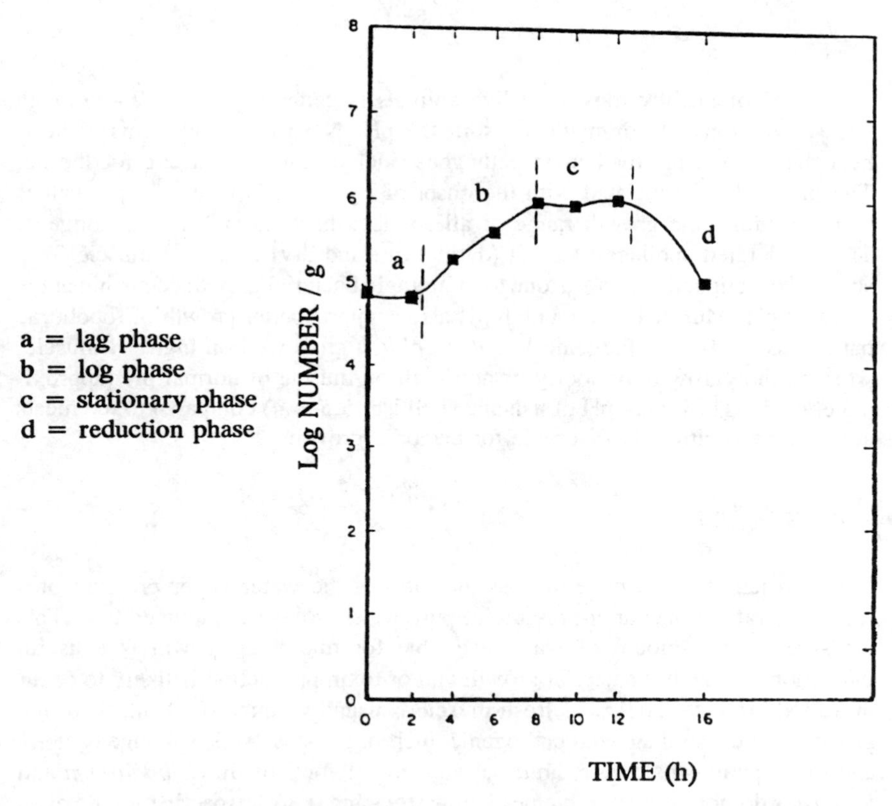

Figure 9.1. Growth curve of *Staphylococcus aureus* on barbecued pork after inoculation and storage at 35°C (95°F). (Adapted from N. G. Marriott and D. P. Fedorko, *J. Foodserv. Sys.* **5:** 243 (1990), with permission of Food & Nutrition Press, Inc.)

are termed "mesophiles," meaning that they grow best at temperatures around 37°C (98.6°F)—normal human body temperature. Mesophilic bacteria include *Salmonella, E. coli, Campylobacter,* and other pathogens that originate in the GI tract of man and animals. Due to a requirement for elevated growth temperatures, thermophilic organisms play only a limited role in the microbiology of refrigerated products such as muscle foods.

Temperature is probably the most important parameter affecting microbial growth. Proper meat storage temperatures of less than 5°C (41°F) will largely restrict growth on muscle foods to psychrotrophic and psychrophilic bacterial species. This does not mean that other organisms are destroyed at this temperature, merely that their growth is inhibited due to a prolonged lag phase. A good rule of thumb is that meat should be stored at the lowest possible temperature below 5°C (41°F) without freezing.

pH

The pH of healthy muscle in live animals is generally around 7.4, though stress conditions can dramatically shift the pH. Not surprisingly, this value is near the optimal pH for human pathogens, including those that are foodborne. The pH decline associated with the onset of rigor mortis results in pH values that are within the growth range of all spoilage bacteria and most pathogens. The accelerated spoilage of DFD (dark, firm, and dry) meat and muscle from stressed or fatigued animals is due to more rapid bacterial growth occurring at the elevated pH. Muscle having a high pH also supports better growth of foodborne pathogens. In fact, pathogenic *Y. enterocolitica* grow well on high pH muscle, whereas they grow only poorly or not at all on muscle of normal pH (approximately 5.5). The higher pH of fish and shellfish (5.5–6.4) compared to red meats and poultry is similarly favorable for bacterial growth.

Water Activity

Water activity (a_w) is defined as the ratio of the water vapor pressure of a food substrate to the vapor pressure of pure water at the same temperature. This measure of the amount of water available for microbial growth is a useful indication of whether bacterial growth and/or toxin production is likely to occur in a food product. The a_w of fresh muscle is usually around 0.99, ideal for the growth of most spoilage and pathogenic microorganisms. Most spoilage bacteria and many pathogens require an a_w of above 0.91 for growth. *C. botulinum* and *S. aureus* do not grow at a_w values below 0.95 and 0.86, respectively, and even higher a_w values are needed for toxin production.

Oxidation-Reduction Potential and Atmosphere

Bacterial growth may occur under a variety of atmospheres. Bacteria that require oxygen are termed "aerobes," those that cannot grow in the presence of oxygen are "obligate anaerobes," and organisms that can grow under either atmosphere are described as "facultative."

The surface of muscle packaged in oxygen-permeable wrap is highly aerobic, but diffusion of oxygen into the muscle is limited to only a few millimeters. Multiplication of strict anaerobic bacteria on muscle cannot occur until the highly oxidative conditions have been reduced following rigor mortis. Once this drop in oxidation–reduction potential occurs, the large numbers of sulfhydryl groups found in muscle proteins maintain a reduced atmosphere. Raw muscle foods packaged under vacuum will effectively have an atmosphere of carbon dioxide due to muscle respiration and exhaustion of oxygen.

Nutrient Availability

Microbial metabolism is largely geared toward the utilization of simple, low-molecular-weight compounds such as carbohydrates, nucleotides, amino acids, and peptides. Glucose is the preferred substrate of most of the spoilage and pathogenic bacteria important to muscle foods. Microbial affinities for a particular substrate will largely determine which organisms will grow on a product. In this respect, substrate affinity is a more important determinant of survival than is growth rate.

Muscle having a pH >6.0 [e.g., dark, firm, and dry (DFD) beef] generally contains very limited quantities of glucose, and this may influence the growth of spoilage and pathogenic bacteria. Little information is available on the growth of bacteria on low pH muscle [e.g., pale, soft, and exudative (PSE) pork], but spoilage appears to be similar to that of muscle having a normal postmortem pH.

Meat is a good substrate for microbial growth, being rich in both simple compounds and B vitamins. Utilization of more complex substrates such as proteins will occur only when the supply of simple compounds has been exhausted. Generally, glucose depletion at the muscle surface occurs due to growth of spoilage bacteria. Some diffusion of simple compounds from deep muscle to the surface does occur, but this is usually not significant.

Bacteria growing aerobically on muscle have sufficient substrate for oxidative metabolism; aerobic growth ceases due to oxygen limitation. Under anaerobic conditions, growth of organisms like lactobacilli, *Brochothrix,* and clostridia ceases due to substrate depletion. Growth of facultative microorganisms (including pathogens like *Salmonella, L. monocytogenes,* and *E. coli*) typically stops when nutrients become depleted. Microbial metabolism related to cell maintenance continues however, even in the absence of growth.

Far from acting independently, muscle temperature, pH, water activity, oxidation–reduction potential, and nutrient availability all have interactive effects on the growth of a particular microorganism. For instance, at a suboptimal temperature, bacterial growth ceases at a higher water activity than would normally be the case. Chapter 14 should be consulted for more information on this subject.

Conditions required for toxin production by an organism like *S. aureus* are generally more restrictive than those that permit growth. Temperature, pH, and atmospheric conditions that are just barely adequate to allow growth will typically be too stringent to allow the organism to produce toxin.

Concept of Microbial Community on Product

At any given time, several types of microorganisms are present on a nonsterile muscle food. This conglomeration of spoilage organisms and pathogens is best thought of as a microbial community. Figure 9.2 shows a community consisting

Figure 9.2. Scanning electron micrograph of beef inoculated with *Staphylococcus aureus, Salmonella,* and *Campylobacter jejuni*. (Photograph courtesy of Peter Cooke.)

of *S. aureus, Salmonella,* and *C. jejuni* which has been inoculated onto beef. Conditions typically favor one segment of the community over the others, but a variety of complex interactions continue to exist between the various microbial groups. These interactions affect not only growth and survival but may also impact on the virulence of the resident microorganisms.

Interaction Between Members of Community

Foodborne pathogens are generally considered to be poor competitors compared to spoilage organisms. *S. aureus,* for instance, is capable of competitive growth only on preserved products such as ham which possess an altered microbial flora. On raw muscle foods, staphylococcal numbers are unlikely to reach the threshold level of 10^6–10^7 staphylococci/gram of product necessary for production of hazardous levels of toxin.

Apart from outright competition, one organism may inhibit another due to production of peroxides, acids, and various volatile compounds (Mattila-Sandholm and Skyttä, 1991). Additional inhibitors produced during microbial growth include antibiotics and bacteriocins—substances that inhibit growth of competitors while allowing growth of the producers. Whereas antibiotics are typically directed against unrelated microorganisms, bacteriocins are proteinaceous compounds which inhibit organisms closely related to the producer. The acids (principally lactic acid) and bacteriocins produced by lactic acid bacteria are thought to contribute heavily to the inhibition of *L. monocytogenes* observed on vacuum-packaged muscle foods.

Maintenance and Transfer of Virulence Factors

The classification of an organism as "pathogenic" reflects an all-or-none capability for causing disease. Virulence, however, reflects the degree of pathogenicity associated with a particular strain and is related to infectivity and morbidity. Among members of a particular pathogenic bacterial species, virulence varies depending on the presence or absence of virulence factors like plasmids, outer membrane proteins, infection by bacteriophages, and presence of specific chromosomal genes. Of these various virulence factors, plasmids are the best understood. The other virulence factors have only recently been investigated and information is still scarce. *Y. enterocolitica* is an interesting example of a bacterial species that contains both pathogenic organisms (a majority of which belong to serotypes 0:3, 0:9, and 0:8) and nonpathogenic organisms (often referred to as "biotype 1" or "environmental" *Y. enterocolitica*). In a further complication, a pathogenic *Y. enterocolitica* isolate can lose its virulence plasmid during isolation procedures, thereby becoming an avirulent *Y. enterocolitica!*

Bacterial virulence factors are maintained during bacterial residence in the intestinal tract of a host animal. Indeed, passage through a host is a widely used means of selecting for highly virulent organisms. Although bacterial plasmids are easily lost during routine lab culturing, they appear to be much more stable in a host, particularly if some sort of selective pressure (e.g., antibiotic administration) favors the survival of plasmid-containing organisms. A *Salmonella newport* outbreak linked to consumption of ground beef provides retrospective evidence that plasmids carrying characteristic antibiotic resistance patterns can be selected for by administration of antibiotics to livestock. Microorganisms carrying those genetic determinants persisted on the meat and caused human disease following consumption.

Little is known about the influence that different muscle substrates and competitive flora have on the maintenance and possible transmission of virulence factors. There are reports that *L. monocytogenes* grown at refrigeration temperatures is more virulent than listeriae grown at higher temperatures. It is not known whether this actually occurs when *L. monocytogenes* grows on muscle foods. Mattila,

O'Boyle, and Frost (1988) revealed that, under experimental conditions, transfer of plasmids between bacteria does not occur on meat held at 4°C (39°F). The effect of food storage conditions on the maintenance of virulence plasmids is not known.

Role in Determining Shelf-life and Safety

Destruction of the spoilage flora of a particular muscle food, often regarded as beneficial, can have undesirable consequences for food safety. For instance, growth and toxin formation by *C. botulinum* E is more common on irradiated vacuum-packaged fish than on nonirradiated vacuum-packaged product, largely due to a lack of competitive flora (Cann et al. 1966). Products such as surimi are nearly sterile after processing and must be carefully protected from contamination with pathogens. The near absence of competitors on surimi means that growth of any contaminating pathogens will be unrestricted. Further, processing steps used in the production of surimi remove soluble "fishy" components which would normally signal spoilage. In the absence of detectable organoleptic spoilage, a consumer is much more likely to eat product containing unacceptably high levels of bacteria.

The normal postmortem pH of muscle is typically on the lower end of the preferred pH range of many pathogens. Growth of some spoilage organisms, especially those like *Pseudomonas* which deaminate amino acids, may result in an increase in the pH of the muscle food. Removal of the pH barrier could permit growth of a pathogen that is capable of utilizing the remaining substrates.

Origin of Microflora During Processing Steps and Control Measures

Microorganisms present on a muscle food product represent an accumulation of organisms picked up at different stages in the production process. A proportion of these organisms was associated with the live animal, coming from either the environment or the feed. Additional organisms may have been introduced during transport of the live animal, perhaps as a result of comingling with other animals. Such organisms present on the animal at the time of harvesting or slaughtering represent what may be called the "primary microflora" of the muscle food product. During harvesting or slaughtering procedures, primary microflora may be transferred from the exterior surface of the animal or from the intestinal tract onto the muscle surface.

After harvesting or slaughtering, additional microorganisms that contaminate the muscle food represent a sort of "secondary microflora." In many cases, secondary flora originate in the processing facility, either from the environment and equipment, or contamination by human operatives, or improper product storage. Even with optimal processing conditions, some microbial contamination will occur. Subsequent to leaving the production facility, product may be further contaminated during distribution and while in the consumer's possession.

Because it is not possible to wholly eliminate contamination and subsequent deterioration of muscle foods, great care must be taken during the storage and preparation of muscle foods. All harvesting, slaughtering, handling, storage, and food preparation practices should be designed to minimize contamination and prevent or retard microbial growth. Generally, if muscle foods are handled so as to minimize adverse microbial changes, then nonmicrobial deterioration will also be minimized.

Source and Control of Microorganisms at Production Level

Animals

Animal carriage of pathogens has been well studied, but application of this knowledge in the production of food animals has been sporadic at best. Although the concept of "specific pathogen-free animals" is well known, it has generally been applied only to animal pathogens that cause significant (and costly) disease in livestock, and not to human pathogens that may be carried by animals without causing overt disease. Examples include the localization of pathogenic *Y. enterocolitica* in the tonsils of swine and the presence of *Campylobacter jejuni* in the intestines of chickens. Neither condition appears to be overly detrimental to the animal host, but contamination of the resultant muscle foods has serious implications to human health. It is unlikely that a concerted effort toward production of "human pathogen-free animals" will be made unless financial incentives (whether industry or government sponsored) are available to defray the cost of maintaining such systems.

An additional consideration is the vertical transmission of pathogens such as *Salmonella* from breeder chicken flocks to multiplier flocks to broilers. The production of *Salmonella*-free live fowl is an enormous challenge, and there are questions about whether such a goal could be achieved at any cost. It is known that young animals are more susceptible to intestinal colonization by pathogens than are older animals, and it would seem prudent to direct attention in that direction.

The existence of asymptomatic animal carriers of *Salmonella* and *Campylobacter* is particularly problematical. In some cases, the pathogen is not excreted on a daily basis, but only under conditions of extreme stress. This makes detection of carriers difficult and treatment next to impossible. Animal density is thought to be a major factor contributing to the prevalence of salmonellae and other enteropathogens.

Feed

The feed supply of captive-reared animals (including those grown under aquaculture) can be a significant source of pathogens. In particular, insufficiently heated milled feeds containing contaminated animal proteins and feeds that have been contaminated subsequent to heating are major sources of pathogens. This route is well proven in the case of *Salmonella* and is strongly suspected in the transmission of the BSE agent to cattle. Pelleting feeds results in lower microbial loads compared to nonpelleted feeds, but proper time/temperature processing conditions must still be followed to produce acceptable feed. It is possible, for instance, to initiate *Salmonella* infection in chickens when the feed contains as few as one organism/gram.

New approaches toward feed decontamination are being developed; the use of formic, propionic, and other organic acids, for instance, appears promising. Acid addition to animal feed can alter the animal's intestinal flora, reducing pathogen carriage and subsequent carcass contamination. Addition of certain carbohydrates to animal feeds can achieve a similar effect by reducing adherence of some enteropathogenic bacteria to intestinal epithelial cells.

Surprisingly, there are few published reports that grazing cattle on manure-fertilized pasture results in an increased prevalence of *Salmonella*. This should not be interpreted as meaning that the problem does not occur. Transmission of the nematode, *T. spiralis,* to hogs through the feeding of raw or insufficiently cooked garbage is thought to be a major source of that organism. Proper ensilage conditions in the production of fermented grain/grass feed must be maintained so as to prevent the transmission of large numbers of *L. monocytogenes* to ruminants. Consumption of toxic dinoflagellates by fish and shellfish results in the flesh of those animals becoming toxic to humans.

Antibiotics, probiotics, and competitive exclusion

Various adjuncts added to animal feed or administered directly to the animal may have an impact on the carriage of pathogens. The administration of antibiotics to animals at subtherapeutic levels may contribute to the development of multiply-resistant human pathogens. Due to this concern, some countries differentiate between antibiotics allowed for human and animal use. Seafood produced under aquaculture often receives antibiotic treatment, as illustrated by tetracycline-resistant strains of *A. hydrophila* recovered from farmed catfish and their environment. Improper administration of antibiotics may adversely affect an animal's normal intestinal flora, providing new opportunities for pathogen colonization.

Another regimen that has been proposed for the reduction of pathogen carriage is the administration of probiotics. Probiotics are microbial cultures that stabilize the intestinal flora and contribute to improved animal health and/or growth. The majority of commercially available probiotics contain lactobacilli, streptococci, and/or bifidobacteria. Probiotics are not widely used, in part because the growth-promoting effects are highly variable and data collected during field trials are not convincing. There is currently very little evidence to suggest that probiotics are effective at reducing pathogen carriage.

Competitive exclusion, the administration of harmless microbes capable of colonizing the intestinal tract and excluding pathogens, shows terrific promise for the reduction of pathogens in food animals. This process is being used in certain countries for poultry where it is often referred to as the "Nurmi concept." Administration of intestinal flora taken from older pathogen-free birds to eggs or newly pipped chicks at the hatchery will result in early establishment of a protective, competitive gut flora. This reduces carriage and excretion of pathogens such as *Salmonella, Campylobacter*, pathogenic *E. coli, Y. enterocolitica, C. botulinum* type C, and *C. perfringens*. Competitive exclusion is unlikely to confer complete protection against pathogens but it can greatly reduce the incidence of pathogen-carrying animals entering slaughter and processing plants. There is some preliminary evidence to indicate that competitive exclusion may work with domestic animals other than fowl.

Water

Although domestic animals may acquire pathogens from poor-quality drinking water or from water contaminated with animal feces, that source is considered to be less important than feeds.

In the production of fish and shellfish, however, water quality is the most important factor in determining the microbial quality of the final product. In many cases, the original source of waterborne pathogens is pathogen-laden raw human sewage and improperly treated wastewaters. Run-off from pastures and feedlots may also contaminate surface waters and serve as a source of pathogenic bacteria. With respect to human enteric viruses, primary contamination of shellfish due to polluted water is responsible for an overwhelming majority of cases of foodborne viral disease. Contamination of seafoods with *Vibrio* spp. and *A. hydrophila* may not be related to fecal contamination, as those bacteria are natural inhabitants of marine environments. With regard to pathogen contamination of seafood, the risk of human infection by viruses is higher than that associated with bacteria.

Shellfish such as oysters, clams, mussels, and scallops use a process known as filter-feeding to filter out nutrients from the surrounding water. In so doing, the shellfish also retain waterborne pathogens. A process called "depuration" allows filter-feeding animals an opportunity to gradually purge themselves of pathogens. In short, shellfish are gathered from open waters, then transferred to enclosed pools or ponds where water quality can be closely monitored and controlled. The concentration of pathogens in the flesh and GI tract of those animals gradually decreases as a result of the new environment. Due to the expenses involved in maintaining depuration facilities, a practice called "relaying" is more popular in the United States. Relaying involves a transfer of shellfish to an unpolluted area where they are allowed to filter-feed in clean water. Both depuration and relaying will decrease the level of bacterial pathogens of fecal origin, but such procedures are of questionable efficacy for the removal of viruses and naturally occurring marine organisms like *Vibrio* spp. Shellfish that have been subjected to depuration or relaying must not be regarded as pathogen-free, as depuration/relaying may be inadequate or subsequent recontamination can occur.

Environment

The environment in which an animal lives, whether natural or man-made, is a major source of pathogenic microorganisms. Many studies have documented the transmission of *Salmonella* from one group of animals to another through improperly cleaned pens and cages. For this reason, an all-in/all-out system of animal management is recommended with thorough cleaning and sanitation of facilities between batches of animals. Reuse of *Salmonella*-free litter in the production of broiler chickens, however, can positively influence the birds' intestinal flora by providing a source of nonpathogenic competitive flora.

An additional factor influencing pathogen carriage and excretion is stress. Stressful conditions may result in a suppression of the harmless gut flora and an increase in numbers of enterobacteriaceae. Animals stressed by loading, transportation, starvation, or introduction into a pen of unfamiliar animals will have a higher excretion rate of *Salmonella* (and possibly other pathogens) than animals protected from such stresses.

Pests

Pests such as rodents, flies, mites, beetles, and birds may carry pathogens and transmit them to animals. Much effort has been spent on enclosing animal housing so as to deny access to pests, but this is a challenge of enormous magnitude. Rats are thought to be a major factor contributing to the perpetuation of *T. spiralis* in American hogs. Exclusion of cats from animal feeds and bedding materials will prevent deposition of cat feces and possible contamination due to *T. gondii* oocysts present in the feces.

Source and Control of Microorganisms During Harvesting and Slaughtering

The harvesting/slaughtering process offers numerous opportunities for a muscle food to become contaminated with microorganisms. There exist tremendous differences in the slaughtering and dressing processes used for the various animal species, necessitating separate mention of some treatments used for cattle/sheep, hogs, poultry, fish, and shellfish.

Farm Animal Species

Slaughtering and dressing processes used for farm animal species involve a large number of manipulations by workers and equipment. Each manipulation carries the potential for spreading pathogens from one carcass to the next. The modification of equipment on the slaughter line is thought to be less important than worker hygiene and linespeed, at least as far as the slaughter of cattle, sheep, and hogs is concerned. In the opinion of Roberts and Hudson (1987), there is "increasing evidence that the economic pressures to slaughter large numbers of animals at high rates are detrimental to the hygienic status of the carcasses."

It is commonly believed that deep, interior muscle of healthy live animals is essentially sterile. The sterility of muscle after death, however, is a much more complex issue, and there are numerous conflicting reports on whether agonal (occurring during dying) and/or postmortem bacterial invasion of muscle tissue occurs.

The microbiological quality of muscle and organ tissues from diseased animals (whether symptomatic or asymptomatic) is questionable and depends on the disease condition. Abscess formation, particularly in animals that have been vaccinated under poor hygienic conditions, is not uncommon and deep-muscle abscesses may escape detection by slaughterhouse inspectors. Abscesses must be carefully trimmed away so as to avoid contamination with the millions of microorganisms (including enterotoxigenic *S. aureus*) present in abscessed tissue.

Introduction of bacteria into the animal's bloodstream may occur during the exsanguination process, particularly when an already-contaminated knife is used to pierce a dirty skin or hide. It is nearly impossible not to introduce some organisms during this process, but most of these organisms will be confined to internal organs (particularly spleen and liver) and muscle contamination is thought to be rare. Following the death of an animal, the immune system remains active for a short period of time (approximately 1 h after death) and is capable of destroying small numbers of introduced microorganisms.

Hide removal on cattle, veal, and lamb carcasses contributes heavily to the bacterial load on superficial carcass tissues. Hides and hooves, together with the slaughterhouse environment, are major sources of both psychrotrophic spoilage organisms and pathogens such as *Salmonella*. Fecal contamination may spread pathogens onto carcass surfaces unless extreme care is taken to avoid contact with the contaminated outer surface of the hide.

Another major source of contamination is the hands and knives of workers, especially as few workers take the time to wash and sanitize as frequently as would be desirable. It is important to remember that significant carcass contamination may be present even though the carcass surface is not visibly dirty. This is the rationale behind the preevisceration carcass washing procedures now under consideration for use in plants slaughtering cattle and sheep.

In the United States, hog hair is typically removed from carcasses by scalding in hot water followed by a singeing operation. An additional step may be removal of residual hair by a scraping or polishing machine, or by humans. Although scalding may reduce the microbial load on the carcass, it also provides an entry for contaminated scald water into the respiratory system due to voluntary or involuntary respiration. Aspiration of scald water can be avoided by leaving sufficient time between exsanguination and scalding to allow for the cessation of movement.

Singeing will greatly reduce the level of microbial contamination, but only on those portions of the carcass directly exposed to the flames. The area around the feet and in the skin folds around the neck may require special attention. Scraping or polishing operations add contaminants to the carcass unless machine paddles, knives, and hands are cleaned and sanitized regularly. Unfortunately, most scraping and polishing equipment is very difficult to clean, and microorganisms multiply rapidly on the hair and skin remnants left on the paddles. A carcass-washing procedure prior to scalding operations has been proposed as one means of reducing the level of contamination in the scald tank and on subsequent equipment.

Skinning hog carcasses may be one way around the contamination associated with current scalding and singeing practices, and semiautomated machinery can be used for this purpose. With today's processing practices, removal of the skin from hams and bellies poses much less of an acceptance problem than in earlier years.

Surface contamination of feathers and feet of fowl is a major source of microorganisms (including *Salmonella*) in poultry processing plants. These microbes, together with organisms from fecal matter, are introduced into the scalding tank together with the fowl. When operated at "hard scald" temperatures (60°C; 140°F), scalding tanks can actually effect a reduction in the number of pathogens and other microorganisms on poultry skin. Alkalinizing scald tank water will also reduce contamination as will the use of counterflow scalding tanks where the carcass first encounters dirty water then clean water. Due to water uptake by poultry skin, spray washing operations down the processing line are recommended in order to replace dirty water retained by the carcass with clean water. Spray rinse operations will remove *Salmonella* and other bacteria before they can attach to skin or connective tissue.

The defeathering step which follows scalding often reintroduces considerable contamination. The defeathering machine (or plucker) removes feathers using flexible rubber fingers which are often colonized by large numbers of microorganisms. Defeathering machinery is notoriously difficult to clean and sanitize properly, and the biomass which builds up during the work day may not be effectively removed during clean up operations. Prompt and continuous removal of feathers from both scalding and defeathering equipment will help to reduce contamination.

Evisceration of the various farm animal species provides ample opportunity for the contamination of carcass surfaces by fecal matter from a ruptured intestine or stomach. Observation of proper feed withholding times is necessary so that excessive intestinal fill levels do not result in an increased chance of rupturing the GI tract. The GI tract of cattle, sheep, and hogs is generally closed off at both ends before removal from the body cavity. Closure can be done by tieing, rodding, or by a frozen anal plug such as that used in Denmark. In all cases, the worker doing the eviscerating must exercise extreme care to prevent contaminated hands and knives from contacting the carcass surfaces.

Historically, fecal matter has been regarded as the main source of pathogens contaminating carcass surfaces. However, considerable evidence shows that slaughterhouse environments are an equally important source of contamination, especially if worker hygiene and cleaning and sanitation practices are substandard.

In this respect, any processing steps that can be conducted without extensive contact with human operatives or equipment would be expected to decrease contamination. High-pressure jets that use air or water are being evaluated for use in skinning, evisceration, and splitting operations. This would minimize contact with contaminated equipment and workers as would use of automated, self-sanitizing equipment. Continuous spray-washing of evisceration equipment with chlorinated water (⩾20 ppm chlorine) is quite effective at reducing contamination on surfaces in contact with carcasses. When hot water baths are used to sanitize knives and other equipment, water temperatures must be ⩾82°C (180°F) and immersion time must be ⩾10 s.

Rinsing carcasses with chlorinated water and organic acids (principally lactic, acetic, and citric acids) will reduce surface microbial contamination, but only when used in conjunction with good manufacturing practices. A significant reduction in the numbers of *Salmonella* on poultry carcasses, for instance, may be accomplished by using chlorine (10–50 ppm) in the chill system, and spraying carcasses with tri-sodium phosphate solution or 1–2% lactic acid solutions before air chilling. Careful optimization of acid or chlorine concentration in rinse water is necessary to prevent discoloration of the carcass surface, and the corrosiveness of decontamination solutions and possible formation of mutagenic chlorine reaction products are additional considerations. As mentioned earlier, mechanized spraying is preferred to hand-operated spraying operations.

Poultry chill tanks can be a major venue for cross-contamination of poultry carcasses if the chilling system is not properly designed and maintained, and water quality is not closely monitored. Air chilling of poultry in a manner similar to that used for red meat carcasses has been recommended as a means of reducing cross-contamination, and the relative merits of water chilling versus air chilling continue to be debated. When air chilling is used, air quality and the cleanliness of the chillers become of primary importance. A chiller temperature as close as possible to −1.5°C (29°F) should be used, and it is important that temperature is set relative to product temperature and not incoming air temperature. The tendency toward using "wet chilling" (relative humidities near 100% in the chiller) should be resisted as reduced dessication losses come at the expense of greater microbial growth. For the same reason, carcasses should be spaced so that surface-to-surface contact is avoided.

Fish and Seafood

The majority of the risks associated with microbial contamination of seafood relate to the ocean environment and are largely beyond the control of the seafood harvester. The first step at which control can be exercised is in the selection of the area to be harvested. Areas known to be experiencing blooms of dinoflagellates are to be avoided at all costs, and state public health authorities should be consulted for the most current information on closures. Harvesting certain species at given times of the year may also have to be avoided in a particular locale. The main problem is that known "hot spots" may persist for extended periods of time or may change altogether. Commercial fishermen are more likely to avoid quarantined areas than are recreational or subsistence watermen and fishermen. Ideally, seafoods should be tested for toxins while still on the boat, but existing methodology is inadequate for that purpose.

The second control step in the harvesting of seafoods is rapid chilling to temperatures at or below 1°C (34°F). This may be accomplished by the addition of ice, super-chilled water, or brine so as to decrease the temperature of fish muscle to <15°C (59°F) and preferably to <10°C (50°F) within 4 h. Rapid chilling immediately following harvesting will control pathogen growth (especially *Vibrio* spp.) and prevent the formation of histamine which leads to scombroid fish poisoning. As with many of the seafood poisonings, histamine poisoning is particularly likely to occur among recreational or subsistence fishermen who lack proper refrigeration facilities. Under no circumstances should seafood or freshwater fish be harvested then left on the deck of the boat without refrigeration. Such practices are common on "day boats"—boats that return to the dock late in the afternoon carrying unrefrigerated fish. Improper refrigeration greatly increases the hazard posed by pathogenic bacteria and histamine poisoning; viruses, parasites, and dinoflagellate-derived toxins are not influenced.

Harvesting procedures for fish may or may not call for prompt exsanguination and evisceration depending on the species and the area being harvested. Herring, cod, and other fish that may be infected with anisakids should be eviscerated as soon as possible in order to prevent possible migration of worms from the intestine to muscle. Shellfish harvesting typically consists only of collecting the shellfish, with all further processing done at the point of food preparation or consumption.

Source and Control of Microorganisms During Fabrication and Processing

Contribution of the Environment

Microorganisms found in fabrication and processing facilities are mainly a reflection of those organisms that are found on incoming raw product. Due to widespread use of refrigerated rooms, growth of microorganisms within fabrication and processing facilities is mostly limited to psychrotrophic organisms. The temperature of product entering the cutting room should be ≤7°C (45°F), and the room itself should be as close to 10°C (50°F) as possible.

The existence of populations of "resident flora" in fabrication and processing plants has only recently been extensively investigated. It appears that a microbial population (including pathogens such as *S. aureus* and *L. monocytogenes*) can evolve which is more resistant to cleaning and sanitizing agents than are non-adapted organisms of the same species. The resident organisms are much more likely to contaminate product than are nonadapted strains of the same species. This may account for the fact that many of the strains of *L. monocytogenes* found on meat are distinct from the organisms that are common on animals.

Strict adherence to proper sanitation and good handling practices will decrease the microbial load on fresh muscle foods. Obtaining a significant reduction under practical day-to-day conditions takes careful attention to details and an ongoing commitment by all personnel involved in sanitation, cutting, fabrication, and processing operations.

Processes Contributing to Contamination

Most contamination of whole muscle foods will be limited to the surface. Changes in muscle structure incurred during rigor mortis may, however, facilitate penetration of muscle by microorganisms. Bacteria on a cut meat surface may penetrate into the muscle, sometimes moving between the various connective tissue layers and muscle fibers.

The majority of contamination incurred during cutting operations comes from contact with contaminated tables and workers. Indeed, fabrication and cutting operations can be a more significant source of contamination than carcass dressing procedures. The air jets currently being evaluated for carcass and meat cutting would greatly decrease contamination, but their efficacy is, as yet, unknown. Computer-guided deboning offers similar advantages but that technology is still largely untested.

Comminution processes such as grinding and flaking are critical fabrication operations. Comminution not only exposes muscle to contamination carried by workers, surfaces, and equipment, but also increases surface area, releases muscle fluids, and allows for greater oxygen availability and penetration. All of this adds up to increased contamination and accelerated growth of microbial contaminants.

Production of many ready-to-eat muscle food products involves a heating step to destroy microorganisms and retard enzymatic activity. Regardless of the type of heating process applied, the potential for subsequent contamination does exist. In the case of product such as roast beef heated in plastic bags, chilling may involve immersion of the packaged product in cold water. Defective bags can allow contamination by the cooling water, and additional microorganisms may be introduced when the bags are opened to drain off fluid, during resealing, or during the application of a new bag.

Postcooking manipulations such as slicing, casing removal, and packaging must be done in the most hygienic manner possible, and all opportunities for cross-contamination between cooked product and raw product must be avoided. For this reason, segregation of processing areas into separate spaces for handling of raw and cooked product is recommended. Partial physical separation, however, will do little good if the areas share a common air source and workers are allowed to move between areas.

Recent experiences with the control of *L. monocytogenes* in processing plants have taught us a great deal about environmental contamination and contamination occurring during postcooking manipulations (postprocessing contamination). Operations such as removing inedible casings from frankfurters and removing metal molds from loaves are critical because of the creation of a moist, warm microenvironment perfect for the growth of listeriae. Ideally, such operations should be conducted in an area removed from slicing and packaging machines.

Similarly, packaging operations should, whenever possible, be done with a minimum of handling. Sanitation of surfaces which encounter ready-to-eat product is critical, as is worker hygiene. Packaging should never be regarded as a substitute for refrigeration or other means of preservation.

Additives as a Source of Contamination

Additives combined with muscle foods during processing operations can serve as a source of microbial contamination. Due to the processing given additives such as sugar and spices, contamination is largely limited to spore-forming bacteria.

Control of Microorganisms

Detailed information on preservation practices applied to muscle foods will be covered in Chapter 14. Accordingly, only a very cursory discussion will be presented here.

Control measures used at the fabrication stage include curing, freezing, heating, irradiation, and other preservation processes. Freezing muscle foods has gained widespread acceptance as a means of inactivating *T. spiralis* and *T. gondii*, and the time/temperature combinations required have been carefully determined. Other parasites including anisakids and cestodes are also inactivated by freezing.

Irradiation is another effective means of inactivating parasites, and relatively low pasteurization-level doses (0.3–1 kGy) are effective. Irradiation at 2–7 kGy will destroy most gram-negative pathogenic bacteria (including *Salmonella*) and reduce numbers of spoilage organisms. Commercial use of irradiation worldwide has been thwarted by adverse consumer reactions, negative publicity, and concern over the development of unacceptable "off" odors and flavors. If implemented, commercial irradiation of poultry and seafoods would do much to reduce the incidence of pathogens on raw muscle foods. Available evidence indicates that this would ultimately decrease the incidence of foodborne disease. Roberts (1985) estimates that irradiation of 81% of the chickens processed in the United States could result in a net benefit of $186,000,000 to $498,000,000 per year due to decreased public health costs associated with poultry-borne salmonellosis and campylobacteriosis.

At-home preservation of muscle foods is often done by poorly educated people having little knowledge of proper practices and using inadequate equipment. Traditional preservation practices used by certain ethnic groups have been implicated in several cases of foodborne disease, including foodborne botulism.

Control of Microorganisms During Distribution, Storage, and Display

Refrigeration is the process most often relied upon to control the growth of microorganisms during distribution of muscle foods. Recent advances in knowledge have indicated, however, that complete reliance on refrigeration for the control of pathogens is not prudent. Whenever possible, product formulation should include one or more "hurdles," designed to act in concert with refrigeration during product storage to slow or prevent the growth of pathogens (see Chapter 14).

Several caveats on the use of refrigeration to control pathogens should be considered. First, the interior temperature of refrigerators set at 5°C (41°F) may fluctuate, and higher temperatures are not uncommon. Periods of mild temperature abuse (temperatures between 5 and 12°C (41 and 54°F) may allow the growth of pathogens such as *Salmonella, S. aureus, B. cereus,* some *E. coli,* and *V. parahemolyticus*. Second, a number of pathogenic bacteria are capable of competitive growth at 5°C (41°F). These organisms include *L. monocytogenes, Y. enterocolitica, A. hydrophila,* enterotoxigenic *E. coli,* and nonproteolytic *C. botulinum* types E, B, and F. Third, low temperatures may actually prolong the survival of pathogens such as *C. jejuni*.

Temperature abuse and time abuse of perishable muscle food products are the main problems encountered during distribution. Chilled product should be loaded at an internal temperature <4°C (39°F) and frozen product should have an internal temperature below −18°C (0°F). Refrigeration equipment should be capable of maintaining these product temperatures even when ambient temperature is above 30°C (86°F). Upon delivery, perishable product should immediately be moved into chillers or freezers and never left on a loading dock.

Display cases used in retail outlets should maintain fresh-product temperature at just above the freezing point. Furthermore, product should be stacked in such a manner that the temperature of the uppermost package is within specifications. Strict adherence to stock rotation and careful monitoring of product expiration dates is also important.

In general, voluntary action by commercial fish distributors has worked well to keep traditionally toxic fish species out of the marketplace. In Hawaii, for instance, amberjack is not sold commercially because it is known to have a high incidence of ciguatoxin.

Source and Control of Microorganisms at the Point of Consumption

Education of Food Handlers

American consumers have, by and large, grown complacent due to easy access to an abundant food supply and the widespread belief that eating should be a risk-free venture. Even if it were possible to produce food products having no element of risk, hazard could still be introduced at the consumer level. It is up to the consumer to take responsibility for ensuring the safety of product from the time of purchase until consumption.

Studies by the CDC indicate that the following are the most common causes of foodborne illnesses: improper storage or holding temperature, poor personal hygiene of food handler, inadequate heating, time lapse between cooking and serving, and cross-contamination between raw and cooked products. These trends hold true for red meats, poultry, and nonmolluscan seafoods. As mentioned above, in the case of the seafood toxins (ciguatera toxin, paralytic shellfish poison, etc.), the food itself is unsafe and illness cannot be prevented by handling or preparation procedures, including normal cooking and processing. Inadequate heating can be a problem with shellfish and crustaceans, as traditional cooking practices are often insufficient to inactivate all bacterial and viral pathogens. For instance, consumers are often told to cook shellfish until the shells open. Unfortunately, although the shells of mussels and clams may open after as little as 1 min, complete destruction of pathogens may require 5 or more minutes of additional cooking.

Temperature abuse, especially of cooked muscle food products, is a major contributor to foodborne disease. Such abuse may result when cooked foods are left at room temperature after serving or when food is inadequately or slowly cooled. The common practice of placing uneaten portions of cooked roasts directly into an already-crowded refrigerator results in only slow cooling. Core temperatures of such roasts may remain within a favorable bacterial growth range for hours before cooling occurs. Piling slices of cooked meat into a deep pan then ladling hot gravy on top will result not only in slow cooling but also provide the reduced atmosphere favored by pathogens such as *C. perfringens*. Pathogens commonly associated with temperature-abused muscle foods include *Salmonella, S. aureus,* and *C. perfringens*. The often repeated warning, "keep cold foods cold and hot foods at >60°C (140°F)" is still good advice. Temperature abuse has little significance when foods are contaminated with pathogens such as viruses and parasites which are unable to multiply outside a living host.

The hygiene of food preparers is an obvious factor in food safety. Infrequent hand washing, particularly after using toilet facilities, can lead to the contamination of muscle foods by a variety of microorganisms. Transmission of *Shigella* and human enteric viruses (especially hepatitis A virus) often occurs when ready-to-eat or cooked muscle products are contaminated in this manner.

One area of consumer education that deserves special emphasis is warning at-risk individuals about the consumption of raw seafoods. This at-risk group includes the immunocompromised, those with liver disorders or diabetes, and the elderly. Attempts to require a warning label on packages of raw seafood have met with heated argument from seafood producers and the issue remains unresolved. Certainly, a concerted educational effort could be made through health care professionals and community health programs. Advising consumers to remove the viscera of shellfish and crabs before consumption, for instance, will reduce exposure to toxins (e.g., toxins responsible for amnesic and paralytic shellfish poisonings) that concentrate in those organs.

Cross-contamination

Cross-contamination usually refers to the transfer of pathogens from raw muscle foods to cooked, ready-to-eat product. A typical example is placing cooked poultry on the same, unwashed plate that held the raw product. Surfaces and equipment used for the preparation of raw muscle foods should always be thoroughly washed with hot soapy water and well rinsed before being used for cooked or ready-to-eat foods. Similarly, human hands can be responsible for cross-contamination. A recent case of *Y. enterocolitica* infection in an infant in the United States was traced to a food handler who had been preparing raw hog chitterlings then gave a cookie to the child without first washing her hands.

Environment

The environment of home and institutional kitchens can serve as a source of microbial contamination for foods. Hygiene may be lax, pests (particularly flies) may be present, and food handlers may be ill yet still engaged in food preparation activities. Infected boils and cuts on the hands of food preparers may contaminate foods and food preparation surfaces with pathogenic *S. aureus*.

An often-neglected source of contamination can be the refrigerator which may harbor psychrotrophic pathogens such as *L. monocytogenes, Y. enterocolitica,* and *A. hydrophila*. Every home refrigerator should have an accurate thermometer, and the temperature should be kept below 4°C (39°F). Foods that are to be frozen should be thoroughly chilled first, then placed into the freezer. Consumers must remember that refrigeration does not stop microbial growth and freezing does not kill all microorganisms present on food products.

It is fitting that this chapter should end with a discussion of the consumer's role in preventing foodborne disease. Improper handling at the point of food preparation and consumption can single-handedly wipe out all the efforts exerted at earlier stages in the food chain!

Selected References

Ahmed, F.E. (ed.). 1991. *Seafood Safety/Committee on Evaluation of the Safety of Fishery Products*. National Academy Press, Washington DC.

Archer, D.L. and F.E. Young. 1988. Contemporary Issues: diseases with a food vector. *Clin. Microbiol. Rev.* **1**(4):377–398.

Bean, N.H. and P.M. Griffin. 1990. Foodborne disease outbreaks in the United States, 1973–1987: pathogens, vehicles, and trends. *J. Food Protect.* 53:804–817.

Bean, N.H., P.M. Griffin, J.S. Goulding, and C.B. Ivey. 1990. Foodborne disease outbreaks, 5-year summary, 1983–1987. *J. Food Protect.* 53:711–728.

Bryan, F.L. 1988. Risks associated with vehicles of foodborne pathogens and toxins. *J. Food Protect.* 51:498–508.

Cann, D.C., B.B. Wilson, J.M. Shewan, T.A. Roberts, and D.N. Rhodes. 1966. A comparison of toxin production by *Clostridium botulinum* type E in irradiated and unirradiated vacuum packed fish. *J. Appl. Bacteriol.* **29**:540–548.

Doyle, M.P. (ed.). 1989. *Foodborne Bacterial Pathogens*. Marcel Dekker, Inc. New York.

Foster, E.M. 1989. A half century of food microbiology. *Food Technol.* 43(9):208–216.

Institute of Food Technologists' Expert Panel on Food Safety and Nutrition. 1988b. Virus transmission via foods. *Food Technol.* **42**(10):241–248.

Johnson, J.L., M.H. Hinton, and J.G. van Logtestijn. 1991. Meat safety and the assurance of public health; considerations and concerns. In: *The European Meat Industry in the 1990's*. F.J.M. Smulders, ed. Pages 167–192. Audet Tijdschriften B. V., Nijmegen, The Netherlands.

Lederberg, J., R.E. Shope, and S.C. Oaks, Jr. (eds.). 1992. *Emerging infections—microbial threats to health in the United States*. National Academy Press, Washington, DC.

Mattila, T., D. O'Boyle, and A.J. Frost. 1988. Inability to transfer antibiotic resistance in *Escherichia coli* on meat surfaces. *J. Food Sci.* 53:1309–1311.

Mattila-Sandholm, T. and E. Skytta. 1991. The effect of spoilage flora on the growth of food pathogens in minced meat stored at chilled temperatures. *Lebensm.-Wiss. Technol.* 24:116–120.

Murrell, K.D. 1985. Strategies for the control of human Trichinosis transmitted by pork. *Food Technol.* 39(3):65–68, 110–111.

Pearson, A.M. and T.R. Dutson. 1986. *Meat and Poultry Microbiology. Advances in Meat Research, Vol.* 2., AVI Publishing Co., Westport, CT.

Roberts, T. 1985. Microbial pathogens in raw pork, chicken, and beef: benefit estimates for control using irradiation. *Am. J. Agric. Econ.* 67:957–965.

Roberts, T.A. and W.R. Hudson. 1987. Contamination prevention in the meat plant: the standpoint of the importing country. In: *Elimination of Pathogenic Organisms from Meat and Poultry,* F.J.M. Smulders, ed. Pages 235–250 Elsevier, Amsterdam

Smith, J.L. 1991. Foodborne toxoplasmosis, *J. Food Safety* 12:17–57.

Smulders, F.J.M. (ed.). 1987. *Elimination of Pathogenic Organisms from Meat and Poultry*. Elsevier, Amsterdam.

Stubblefield, R.D., J.P. Honstead, and O.L. Shotwell. 1991. An analytical survey of aflatoxins in tissues from swine grown in regions reporting 1988 aflatoxin-contaminated corn. *J. Assoc. Off. Anal. Chem.* 74:897–899.

Taylor, S.L. 1988. Marine toxins of microbial origin, *Food Technol.* 42(3):94–98.

Todd, E.C.D. 1989. Preliminary estimates of costs of foodborne disease in the United States. *J. Food Protect.* 52:595–601.

Wilesmith, J.W., J.B.M. Ryan, and M.J. Atkinson. 1991. Bovine spongiform encephalopathy: epidemiological studies on the origin. *Vet. Rec.* 128:199–203.

10

Chemical Residues in Muscle Foods

William A. Moats

Introduction

Reports from the 19th century and earlier indicate that adulteration of foods with fillers, coloring agents, and preservatives was widespread. Some of these adulterants such as copper sulfate and boric acid were quite toxic by present-day standards. After many years of effort, beginning as early as 1889, the U.S. Congress passed the first Pure Food, Drug, and Cosmetic Act in 1906 under the leadership of Dr. Harvey W. Wiley. Under Dr. Wiley's direction, additives to foods were tested for toxicity, and use of those found to be a health hazard were banned or restricted. Today, the use of intentional food additives is well controlled by the U.S. Food and Drug Administration. However, since Dr. Wiley's day, it has become recognized that a variety of unintentional additives may inadvertently contaminate foods including muscle foods.

Chemical residues in foods may be defined as any chemical substance not normally present or present at levels above what would be normally or naturally present in food. Thus, animal tissues normally contain trace amounts of heavy metals, certain naturally occurring hormones, and radioactive K^{40}. The principle chemical residues are man-made substances that are not natural constituents of animal tissues and elevated levels of heavy metals, also usually a result of human activity. Certain natural toxic substances produced by plants, molds, and fungi or other microorganisms may carry over into muscle foods. Naturally high levels of mercury found in certain wide-ranging deep-sea fish such as swordfish and tuna are also of concern.

In surveys of consumer attitudes toward various food-related issues, chemical residues such as hormones and pesticide residues rank just below nutritional value as consumer concerns. The European Economic Community (EEC) has restricted imports of U.S. beef because of concern about hormone residues from hormone implants (banned in the EEC) which are widely used in beef production in the United States. Production of so-called "natural" beef, without use of antibiotics and hormones, has been advocated as a means of allaying consumer concerns about possible residues of animal drugs. Consumer attitudes about residues, whether justified or not, can have a significant effect on consumption patterns of muscle foods.

The types of compounds that may occur as contaminants in muscle foods are summarized in Table 10.1. Since 1906, a great many new chemicals and drugs have been used in various aspects of agricultural production. Some of these have the potential to end up as residues in meat. These include insecticides and herbicides used in the production of feedstuffs as well as compounds used directly in animal production such as antibiotics and growth promotants.

In addition, it has been recognized that environmental contaminants may be ingested by animals and be deposited as residues in meat. Use of nitrates and nitrites in meat curing may result in formation of low levels of nitrosamines which are potentially carcinogenic.

Residues of Concern in Muscle Foods

Pesticides Used for Crop Production

Included in this broad category are insecticides, acaracides, herbicides, and fungicides used primarily in crop production, including production of feedstuffs for animals. Some insecticides are also used for control of flies on livestock. Prior to the discovery of the insecticidal properties of DDT in 1939, the principal insecticides used were arsenicals and other inorganic compounds. A few insecticides of botanical origin such as pyrethrin and rotenone as well as petroleum products were used. The principal fungicides were inorganics such as lime-sulfur and paris green. More recently, a great variety of synthetic herbicides, fungicides, acaracides, and insecticides have been introduced. These compounds may occur as residues in feedstuffs ingested by animals and, thus, have a potential to

Table 10.1. Residues of Concern in Muscle Foods

Chemicals used in crop production	Environmental contaminants
Insecticides and acaracides	Polychlorinated biphenyls
Herbicides	Dioxin
Fungicides	Heavy metals
	Radionuclides
Chemicals used in animal production	Wood preservatives
Antibiotics	Mycotoxins
Hormones and other growth promoters	
Repartitioning agents	Other
Insecticides	Nitrosamines
	Components of packaging materials

contaminate animal tissues. Most are rapidly metabolized by animals and either broken down or excreted. The occurrence of residues of most of these compounds in muscle foods is, therefore, unlikely. An exception is the chlorinated hydrocarbon group of insecticides which includes DDT, lindane, heptachlor, aldrin, dieldrin, and chlordane. Soon after these compounds were introduced, it was discovered that they were very persistent in the environment. Furthermore, when ingested by animals, they accumulated in fatty tissues rather than being broken down and excreted. Thus, low levels in feedstuffs could be concentrated to significant levels in animal tissues. In a biological food chain, concentrations could ultimately reach lethal levels. The compounds were not only stable in the environment but were readily dispersed in dust or in the vapor form. Thus, residues were found in tissues of penguins and sea life living in areas as remote as Antarctica. Because of concerns about the toxicity and environmental damage caused by chlorinated hydrocarbon insecticides, the use of these compounds has been severely restricted, if not totally banned in the United States since the 1970s. However, some uses are still permitted in some other countries. Background levels in the environment have gradually decreased following the restrictions on use. However, detectable levels of several chlorinated hydrocarbons are still found in some foods of animal origins, although at levels well below tolerance limits. Inadvertent feeding of forages treated with chlorinated insecticides can result in unacceptable levels in animal tissues. These residues are very persistent. Holding livestock long enough for residues to be depleted may not be economical.

Chemicals and Drugs Used in Animal Production

This category includes antibiotics and other drugs used to control diseases and parasites, natural and synthetic anabolic hormones, natural and synthetic growth hormones, and repartitioning agents such as β-agonists and tranquilizers that increase the ratio of lean to fat in animals. Insecticides (organophosphates and pyrethrins) used in fly control around livestock are generally of low mammalian toxicity. They also are readily metabolized so that they do not leave significant residues.

Antibiotics and Other Antibacterials

These compounds are widely used therapeutically and are also sometimes added to livestock feeds at low levels as growth promotants. The relative persistence of antibiotics in animal tissue varies widely depending on the compound and the method of administration. The sulfonamides have created the greatest residue problems, especially in swine. Even accidental contamination of feeds with sulfonamides has resulted in significant residues in tissues. Aminoglycoside antibiotics (e.g., neomycin, streptomycin) and tetracycline antibiotics are also rather persistent in animal tissues. Retention of others such as the β-lactam group may vary greatly depending on the formulation, dosage, and mode of administration. Use of low levels in feeds as growth promotants is less likely to result in significant residues than therapeutic use.

Hormones and Repartitioning Agents

Both natural and synthetic anabolic agents have been administered to livestock to improve feed efficiency. Other types of drugs have also been used to reduce the ratio of fat to lean. These include certain β-agonists, such as clenbuterol, and some tranquilizers such as diazepam. All of these compounds can produce residues in animal tissues. Animal tissues contain low levels of naturally occurring anabolic steroids. Administration of these naturally occurring steroids to animals does not result in levels above those naturally present. Some synthetic anabolic agents such as diethylstilbestrol have been banned in the United States because of carcinogenic potential. The European Economic Community (EEC) has banned use of anabolic agents and β-agonists. There is, however, evidence of widespread illegal use in the EEC. Use of growth hormones (somatotropins) has been very controversial but does not appear to result in tissue residues above those naturally present.

Environmental Contaminants

Certain industrial chemicals have been found to have a widespread potential for environmental contamination. One of the most serious has been polychlorinated biphenyls. These were once used as electrical transformer oils, in copy paper, and in many other uses. These compounds are chemically similar to the chlorinated hydrocarbon insecticides and behave in much the same manner in the environment. Like the chlorinated hydrocarbon insecticides, these compounds are concentrated biologically in fatty tissues of animals and may be concentrated to toxic levels. Although use of polychlorinated biphenyls has been phased out, residues remain in the environment, especially in stream and lake sediments. Some fish remain significantly contaminated in polluted areas. Because of widespread environmental contamination, small amounts may sometimes be found in animal tissues.

Dioxins are somewhat similar types of chlorinated compounds. They are byproducts in the production of certain herbicides and are also formed by combustion of organic matter. Low levels are widespread in the environment and they are also concentrated in fatty tissues. Dioxins have been found to be potent carcinogens in tests with rodents. However, epidemiological studies indicate that they are at most weakly carcinogenic to humans.

Heavy metals are potential contaminants of animal tissues. For terrestrial animals, heavy metals may contaminate forage in the form of dust from industrial operations such as smelters, manufacturing operations using heavy metals, or combustion of trash containing heavy metals. Crops grown on land treated with sewage sludge may contain elevated levels of heavy metals. In the past, crops grown near highways contained elevated levels of lead from road dust. This may no longer be a problem since leaded gasoline was phased out. Because animals preferentially excrete ingested heavy metals, levels in animal tissues are lower than in the diet and seldom, if ever, reach hazardous levels.

Industrial discharge of wastewater frequently contains elevated levels of heavy metals that accumulate in river and lake sediments. Mercury is converted to methyl mercury which is accumulated in the tissues of fish and other aquatic organisms to dangerous levels. A notable instance of this occurred in Minamata Bay in Japan which was contaminated with mercury from an industrial operation. Numerous serious illnesses and some deaths were reported in persons consuming seafood from the bay. Swordfish have natural levels of mercury in muscle approaching levels considered hazardous.

Animals ingesting radionuclides may contain significant levels in their tissues. Radionuclides were widely distributed in the environment from atmospheric testing of A- and H-bombs until a ban on atmospheric testing was instituted in 1963. Levels were particularly high immediately downwind from nuclear tests. Since the ban on atmospheric testing in 1963, the principal source of radionuclides in the environment has been leakage from nuclear reactors used for manufacturing weapons-grade materials and from power plants. Levels have been generally low and are significant only in a few cases near nuclear plants. An exception was the Chernobyl disaster in the Soviet Union (now Ukraine) in which the explosion of a nuclear power plant scattered radioactive debris over a wide area of central and eastern Europe. In some cases, levels in animal tissues reached levels considered hazardous.

Pentachlorophenol, a wood preservative, has been found in cattle tissues, apparently as a result of cattle chewing on treated wood or ingesting treated wood shavings used as litter. Livestock may ingest mycotoxins from moldy feeds. Although these are more or less toxic to animals, they are metabolized by animals and thus are not likely to present a significant residue problem.

Other Sources of Residues

Nitrates and nitrites are used in curing meats. These compounds produce the characteristic color and flavor of cured meats. They also serve as antioxidants and inhibit the growth of bacteria including the deadly *Clostridium botulinum*. However, use of nitrates and nitrites in meat curing has been severely criticized because of the finding that these compounds can react with amines to form nitrosamines which are considered to be carcinogens. Cooking of cured meats may enhance nitrosamine formation. Levels of added nitrate and nitrite have been reduced in recent years. The other principal source of nitrate in the diet is green vegetables. Nitrate can be reduced to nitrite in the gut and react endogenously with amines to form nitrosamines. The additional risk from cured meats must, therefore, be considered relative to other dietary sources of nitrate and nitrite.

Plasticizers and unreacted monomers in plastic packaging materials may migrate into packaged meats. Nitrosamines from rubber netting used to package hams can migrate into the meat.

Harmful Effects of Residues

Is public concern about residues justified? What are the potential harmful effects of residues? These include acute toxicity, chronic toxicity, mutagenicity, teratogenicity (birth defects), carcinogenicity, allergic responses, and hypersensitivity. Residues of drugs such as tranquilizers, β-agonists, and hormones may exert their pharmacological effects in human consumers. The antibiotic, chloramphenicol, is known to cause a rare but fatal disorder known as aplastic anemia.

Residues in animal tissues are rarely if ever at levels that would be acutely toxic, for the obvious reason that such high levels would probably kill the animal. However, chronic toxic effects of low levels of residues are of concern. These effects may be subtle and difficult to determine if such effects are significant or indeed if they occur at all. Neurotoxicity of heavy metals such as lead and mercury is an example of chronic toxicity.

Severe allergic responses may occur in individuals sensitized to compounds such as the penicillins. Although allergic responses to foods are not uncommon, identification of the specific food component responsible is a difficult and time-consuming procedure. Thus, it is difficult to establish that allergic responses are, in fact, caused by antibiotics. A few cases of allergic responses to penicillin residues have been documented. As little as 10 μg of penicillin G can cause an allergic response in sensitive individuals. This is equivalent to a residue level of 0.1 ppb in 100 g of meat or milk.

There is also concern about the possible carcinogenicity of a variety of compounds including nitrosamines, some environmental contaminants, and pesticides.

Regulatory Aspects of Residues

In the United States, the Environmental Protection Agency (EPA) is responsible for establishing tolerances for pesticide residues in muscle foods; the Food and Drug Administration (FDA) is responsible for establishing tolerances or acceptable levels for animal drugs and environmental contaminants. Criteria are also developed for use of animal drugs and pesticides in animal production in a manner that will not result in unacceptable levels in animal tissues. These include setting approved treatment protocols and withdrawal (holding) times after treatment.

Tolerances for residues are based on toxicity testing in animal (usually rodents) and are set with a 100- or 1000-fold safety margin. An exception is compounds that are carcinogenic in animal tests. In this case, a so-called "O" tolerance must be set according to the provisions of the "Delaney clause" of the Food, Drug and Cosmetic Act of 1959. The validity of extrapolating tests for carcinogenicity done with high dosage of compounds to residue levels has been questioned.

It is neither possible nor necessary to test for all types of potential residues. The United States Department of Agriculture's Food Safety and Inspection Service (FSIS), which is responsible for monitoring residues in red meat, has established criteria for ranking compounds on the basis of potential occurrence as residues and the risks to human health. Ranking is based on factors such as the probability of occurrence of residues of the compound and the toxicity of the compound or its metabolites. If the compound is selected for monitoring, a practical test must be available. About 400 compounds are considered as potential residues under this program.

Manufacturers of animal drugs and pesticides are required to supply a suitable analytical method for residues of the drug when the product is registered for use. These methods are not always practical for routine residue monitoring, especially when multiresidue approaches are needed. Some of the methods supplied may become obsolete with advances in methodology for residue analysis. For example, many antibiotics were originally approved for use with relatively nonspecific microbiological assays. Test procedures can be broadly classified into screening, determinative, and confirmatory procedures although the distinctions among the three are sometimes blurred.

Screening procedures are rapid simple methods which can be used to establish that a product is free of a particular type of residue or residues. They usually require little or no sample preparation. Generally, screening procedures are not specific or quantitative enough to be used as the sole criteria for rejection of a product but can be used to clear products which are free of residues. A great variety of approaches to screening tests has been used. Some screen for a number of compounds in a single test. An example would be tests for microbial inhibition that test for the presence of antibiotics and other antimicrobial compounds. However, no test organism responds equally to all antibacterial compounds. Further testing is required to distinguish specific compounds. Other approaches include enzyme inhibition for penicillin (Penzym) and some pesticides (cholinesterase inhibition), a variety of immunoassays, and a bacterial receptor assay, marketed commercially as the Charm test. Immunoassays can be relatively specific but may be suitable for only one or a few compounds in a single test. Times required and costs of screening procedures are quite variable. Tests using agar plates are inexpensive but usually require overnight incubation. Some procedures require less than 1 h but are more expensive and labor intensive. These include immunoassays and the Charm receptor assay. Screening tests are not available for many compounds.

Determinative procedures generally require more expensive equipment and a laboratory setting. Residues must be extracted from the food substrate. For analysis of low levels of organic compounds, some type of chromatographic separation is usually necessary. With chromatographic methods, it is frequently possible to determine a number of compounds in a single procedure. Determinative procedures provide quantitative information and usually give reasonably certain identification of residues. Chromatographic methods include gas liquid chromatography (GLC) for volatile compounds, high-performance liquid chromatography (HPLC), which is useful for all types of compounds, thin-layer chromatography (TLC), which can be an inexpensive alternative although quantitation is less accurate, and supercritical fluid chromatography (SFC), which is an emerging technique. Some chromatographic methods require little sample preparation and, thus, are comparable to many so-called screening procedures in speed and simplicity, while giving more information. However, further cleanup of sample extracts is frequently required, sometimes coupled with derivatization of analytes to improve detectability. Such procedures are sometimes lengthy.

Special detectors can be used to enhance selectivity of GLC methods. These include the electron capture detector for chlorinated insecticides, the nitrogen–phosphorus detector for carbamate and organophosphate insecticides, and the thermal energy detector for nitrosamines.

In certain cases, additional confirmation of the identity of residues is needed, especially when regulatory agencies prosecute a violator in court. Mass spectrometry is the standard of proof in such cases. However, interfacing HPLC with mass spectrometry is difficult and requires expensive equipment. Diode-array spectrophotometry is a simpler and less expensive alternative. Penicillins can be confirmed by rerunning a sample after treatment with penicillinase.

Selected References

Hecht, H., 1988. Residues in meat and associated problems. *Fleischwirtsch* 68:873–877.

Oehme, F.W. 1973. Significance of chemical residues in United States food-producing animals. *Toxicology* 1;205–215.

Moats, W.A., 1986. *Agricultural Uses of Antibiotics,* ACS Symposium Series: American Chemical Society, Washington, DC.

Pullen, Michael M. 1990. Residues in meat and health, in: *Advances in Meat Research.* A.M. Pearson and T.R. Dutson, eds. Vol. 6, Pages 135–156. Elsevier, New York.

United States Department of Agriculture. 1992. Food Safety and Inspection Service. *Compound Evaluation and Analytical Capability,* National Residue Program Plan Jeffrey Brown, ed. U.S. Government Printing Office Washington, DC, 1992.

11

Quality Characteristics

Rhonda K. Miller

Quality characteristics of muscle foods are influenced by muscle structure, chemical composition, chemical environment, interaction of chemical constituents, postmortem changes in muscle tissues, stress and preslaughter effects, product handling, processing and storage, microbiological numbers and populations, and meat cookery. Whereas muscle food quality is influenced by the physical characteristics of the muscle, quality is product-specific and is actually a measurement of acceptability by the consumer. This chapter will concentrate on the quality factors of muscle foods that are related to sensory characteristics, the physiological mechanisms related to quality characteristics, and how to measure these quality characteristics.

Factors Influencing Quality Characteristics of Muscle Foods

Muscle color, fat content, connective tissue, muscle fiber characteristics, and storage conditions and temperature have a significant influence on muscle food quality. It is obvious that discolored meat, meat containing either an unexpected excess or low amount of fat, tough meat, meat spoiled due to high initial contamination or to the effect of storage conditions, or meat that has been subjected to temperature abuse would be considered low-quality product.

Muscle Color

Perception of muscle color, either raw or cooked, influences the human perception of product acceptability. Colors of muscle foods vary depending on species, muscle function within the animal, age of the originating animal, and storage conditions.

The range in color for muscle foods from different species of animals and the corresponding myoglobin content within the muscle are presented (Table 11.1). As myoglobin content increases, the muscle food increases in color intensity from white or pink to very dark red; therefore, myoglobin content is directly related to final muscle color. Higher myoglobin content in beef muscle is the major factor that differentiates the bright, cherry red color of beef when compared to the lighter color of pork or poultry meat. Muscles within a species and even within a carcass also vary in color. The classic example of white or dark chicken meat illustrates the variation in muscle color. Muscles vary in myoglobin content based on the physiological role of the muscle. High-use muscles, such as the leg muscle in chicken and other species, have a higher myoglobin content due to the need for myoglobin to store and deliver oxygen in the muscle. As animals within a species increase in age, myoglobin content increases within the muscle tissue (Table 11.1) and thus, meat from older, mature animals is darker than meat from younger animals. Therefore, muscle color has been used as an indication of maturity and quality.

Myoglobin is the major pigment in meat, accounting for 50–80% of the total pigment. However, hemoglobin, the major color pigment in blood, can also contribute to muscle pigment. Hemoglobin content of meat is strongly influenced by conditions immediately prior to and during exsanguination at slaughter. Conditions that restrict blood removal during exsanguination, such as preslaughter stress, inadequate severing of the artery or vein used for exsanguination, extended time between stunning and exsanguination, and improper suspension of the carcass during exsanguination, influence the volume of blood removed from the animal. Improper bleeding during exsanguination results in a greater amount of hemoglobin retained within the muscle tissues. Additionally, suboptimal conditions during exsanguination result in reddish discoloration of the fat, referred to as blood spotting or splattering, which is undesirable. Other meat pigments, such as cytochromes, catalase, and flavins, exist within muscle foods and influence meat color, but only to a very minor extent.

Meat color, although strongly influenced by myoglobin concentration, also is affected by handling and storage prior to presentation to or consumption by the consumer. The anaerobic environment created during vacuum-packaged storage versus the use of modified atmosphere packaging where oxygen, nitrogen, and/or carbon dioxide can be used as the storage environment affects meat color. Additionally, the length of time meat is held in storage and temperature during storage influence muscle color. These factors affect the oxidation state of myoglobin as discussed in Chapter 3. The partial pressures of oxygen, nitrogen, and carbon dioxide influence the pigment state of myoglobin. As partial pressure of one compound increases, a higher amount of that compound is present to influence the oxidation state of myoglobin. The lack of oxygen in vacuum-packaged meat systems converts beef exposed to the atmosphere from a bright, cherry red color to a purple red color in the vacuum-package (the conversion of oxymyoglobin to deoxymyoglobin). The replacement of the traditional merchandising of beef

Table 11.1. Relationship Between Species of Origination of Muscle Foods, Raw Meat Color, Muscle Foods Myoglobin Content, and Major Factors Influencing Quality of Meat

Species of Origin of Muscle Food	Animal Age	Myoglobin Content (mg/g)	Visual Color	Major Factors Influencing Quality Listed in Decreasing Order of Importance Within a Species
Beef	12 days	0.70	Brownish pink	Tenderness, juiciness and flavor
	3 years	4.60	Bright, cherry red to dark red	
	> 10 years	16–20		
Lamb	Young	2.50	Light red to dark red	Flavor, juiciness and tenderness
Poultry dark meat	8 weeks	0.40	Dull red	Flavor, juiciness, tenderness
	26 weeks (females)	1.12		
	26 weeks (males)	1.50		
Fish dark meat species		5.3–24.4	Dull red to dark red	Flavor, juiciness, texture
Turkey dark meat	14 weeks (female)	.037	Dull red	Flavor, juiciness, tenderness
	14 weeks (male)	0.37		
	24 weeks (female)	1.00		
	24 weeks (male)	1.50		
Pork	5 months	.30	Grayish pink	Juiciness, flavor, tenderness
Poultry white meat	8 weeks	0.01	Grayish white	Flavor, juiciness, tenderness
	26 weeks (females)	0.08		
	26 weeks (males)	0.10		
Turkey white meat	14 weeks (female)	0.12	Dull red	Flavor, juiciness, tenderness
	14 weeks (male)	0.12		
	24 weeks (female)	0.25		
	24 weeks (male)	0.37		
Fish white meat species		0.3–1.0	Grayish white	Flavor, juiciness, texture

retail cuts, such as steaks and roasts placed in a Styrofoam tray and overwrapped with polyvinyl chloride, with vacuum-packaged beef steaks and roasts in the retail meat case has not been successfully implemented. Consumers perceive purple-colored beef cuts to be either from older animals that might be tough, to be from meat that is at the end of its shelf-life due to microbial spoilage, or to be from meat that has been temperature-abused which resulted in dark-colored lean from metmyoglobin formation. However, the vacuum-packaged cuts are purple based on the lack of oxygen and the conversion of the myoglobin pigment to deoxymyoglobin in vacuum-packaged beef, not because of inferior-quality meat. The successful merchandising of vacuum-packaged beef subprimals is one of the major challenges for the beef industry. Meat from other species such as pork, poultry, and fish do not face the same marketing challenge, as these meats are lower in myoglobin content, and change in meat color upon vacuum-packaging is not as dramatic.

Storage temperature affects muscle color due to its effect on the rate of chemical reactions and its influence on microbial growth. The appearance of sulfmetmyoglobin, a green color, is the result of sulfur present from bacterial growth coming in contact with the ligand of myoglobin. In muscle foods when meat, fat, or skin begin to discolor, it is an indication that the product is reaching the end of its storage life and, therefore, meat quality is deteriorating.

Fat Content of Muscle

Meat fat content has been highly related to quality as fat content has been shown to affect flavor, juiciness, and tenderness of meat. For example, the USDA Beef Carcass Grading Standards (USDA 1989) utilize the amount of intramuscular fat within the Longissimus dorsi muscle at the 12th–13th rib interface along with animal age to determine the quality grade of beef carcasses. Therefore, beef carcasses containing a higher level of evenly distributed intramuscular fat are eligible for a higher-quality grade. A carcass with a higher-quality grade would be expected to produce meat with more desirable palatability (juiciness, tenderness, and flavor) than the meat from carcasses of lower-quality grades. Additionally, increased levels of marbling have been associated with decreasing the variability in palatability of beef. However, depending on the species of origin, the factors that drive quality attributes of meat vary (Table 11.1). For example, juiciness is a major quality issue in pork as a lack of juiciness is usually due to overcooking. Therefore, pork muscle that lacks marbling is considered low quality, as marbling provides a protective barrier against overcooking. An abundant amount of marbling is considered low quality in pork due to the negative consumer appeal associated with muscle containing a high fat content. Pork muscle having marbling between practically devoid and moderately abundant, the upper and lower limits in the pork grading standards, is acceptable.

Documentation of the role of intramuscular fat in meat palatability has been extensively studied. Savell and Cross (1988) developed the concept identified as the Window of Acceptability for beef (Fig. 11.1). This concept illustrates the relationship between chemical lipid (intramuscular fat) in the lean of muscle tissue and the overall palatability of the cooked muscle. When fat content is less than 3%, palatability declines below an acceptable level. High fat content also can be associated with a negative perception of quality. As fat content exceeds 7.3%, issues related to increased consumption of fat and the relation of fat intake to coronary heart disease, obesity, or some forms of cancer affect consumers' perception of acceptability.

Intramuscular fat and juiciness are positively related. As fat content increases, the human perception of juiciness increases. Fat content influences juiciness through direct and indirect effects. The feel of juice released in the mouth upon mastication or within the first bite and the subsequent stimulation of the salivary gland by fat influences the perceived juiciness of a product. Additionally, meat products with a higher amount of intramuscular fat will sustain the feel of juiciness in the oral cavity. Fat content also has an indirect effect on juiciness by providing an insulatory effect for meat during cooking. High temperatures used in cooking degrade protein which results in the release of water from the meat. Fat conducts heat at a slower rate than lean tissue, and lean tissue is about 75% water; therefore, fat slows down the migration of heat and decreases the shock effect of high heat on protein degradation and moisture release. The result is that meat containing higher fat content does not cook as rapidly; therefore, the amount of water and fat lost during cooking is reduced. Savell and Cross (1988) stated that "fat may affect juiciness by enhancing the water-holding capacity of meat, by lubricating the muscle fibers during cooking, by increasing the tenderness of meat and

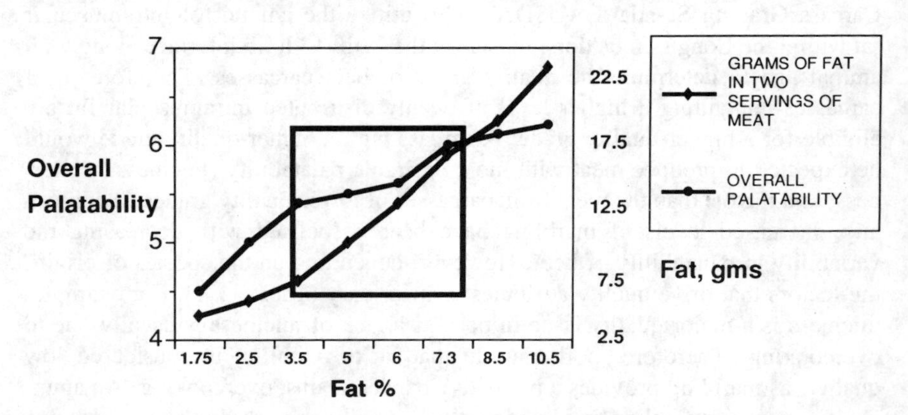

Adapted with permission from Designing Foods: Animal Product Options in the Marketplace. Copyright 1988 by the National Academy of Sciences. Courtesy of the National Academy Press, Washington, D.C.

Figure 11.1. The Window of Acceptability.

thus the apparent sensation of juiciness, or by stimulating salivary flow during mastication."

The relationship between fat content and tenderness has been extensively examined with sufficient evidence to dispute or to support that tenderness of meat is influenced by fat content. Four hypotheses for a tenderness/fatness relationship have been reported. The four hypotheses are the bulk density (bite effect), the lubrication effect, the insurance theory, and the strain theory.

The bulk density effect is based on the density differences between low-density fat and higher-density heat-denatured protein. As fat increases as a percentage or proportion of the unit volume of meat, the overall density of a given bite of meat decreases; thus, meat that has a higher fat content is more tender.

As fat content within the muscle increases, intramuscular fat is deposited between muscle fibers (intramuscular fat), and fat stores within the muscle fiber increase. Therefore, muscle fibers are surrounded or bathed in fat to a greater extent in meat with a higher fat content. Upon cooking and mastication, fat is released, which stimulates salivation and the perception of juiciness and tenderness so that meat with a higher amount of fat is perceived as juicy and tender because of the lubrication effect of fat.

The insurance theory is based on the premise that muscles with higher fat content have a greater amount of insurance against the negative effects of overcooking or high heat on protein denaturation. Fat, which conducts heat at a slower rate than lean, acts to protect and insulate against the effect of extremes in high heat. During cooking, proteins are denatured. As proteins denature, water is released and protein-to-protein interactions are affected. Lower heat results in less severe changes during cooking which is manifested by lower moisture loss and less structural change in the protein component.

The strain theory relates to the effect of increased fat on the strength of connective tissue within muscle tissues. As muscles increase in fatness, lipid is deposited in the perivascular cells interspaced within perimysium muscle connective tissue. As fat deposition increases, the strength of connective tissue decreases and the subsequent meat is more tender. The electron micrograph in Fig. 11.2 illustrates how adipose tissue is embedded within the perimysial connective tissue in a beef muscle.

Other theories propose that intramuscular fat is indirectly related to tenderness. The fact that intramuscular fat and meat tenderness have a low to moderate relationship to meat tenderness supports the validity of an indirect effect. Generally, marbling is the last fat depot to be deposited within the animal system. Animals fed high-concentrate diets to escalate growth, a common practice referred to as the feedlot feeding of young beef cattle prior to slaughter, have greater rates of protein and lipid accretion and therefore, have higher rates of lipid accumulation in the various fat depots. This increase in live weight results in increased amounts of internal (seam fat), external (subcutaneous fat), and mar-

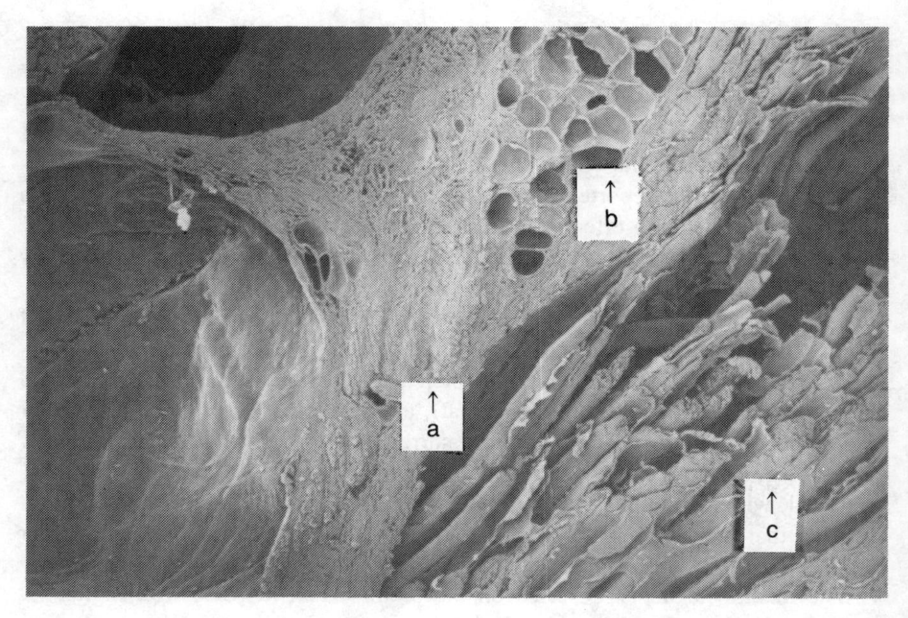

Figure 11.2. Electron micrograph which illustrates where perimysial (a) connective tissue and (b) adipose cells are located in relationship to (c) muscle fibers in cooked beef semimembranosus muscle.

bling in combination with increases in muscle mass. As external fat and muscle mass increase, carcass postmortem chilling rates slow. Therefore, carcasses with less fat chill rapidly and are more susceptible to cold-induced toughening that has been directly related to meat tenderness. Additionally, feeding high-energy density diets for at least 70 days prior to slaughter for young animals has been shown to increase heat-labile collagen solubility. As collagen solubility increases, a greater proportion of collagen can be degraded during cooking, therefore removing or decreasing the influence of connective tissue on tenderness. Increased collagen solubility also is an indication that protein synthesis and degradation, also defined as protein turnover, increase in animals fed high-energy diets. Increased protein turnover could directly or indirectly influence meat tenderness through the influence of increasing the activity of proteolytic enzyme systems involved in protein degradation. So, marbling has been hypothesized to affect meat tenderness as follows: (1) Marbling is an indication of length of feeding of a high-concentrate diet prior to slaughter, and feeding of high concentrate diets affects meat tenderness by increasing outside carcass fatness and muscle mass which reduces cold-induced toughening of meat immediately postmortem, and by increasing protein turnover which increases proteolytic enzyme activity and collagen solubility; and (2) marbling influences palatability through the bulk density, lubrication, insurance, and strain theories. Then, it is not surprising that the relationship between fat content and tenderness is difficult to define. It is this author's hypothesis that each of the aforementioned factors influences the relationship between meat tenderness and intramuscular fatness. However, the magnitude of the influence of each factor varies depending on the species, environment, and genetic influences on the subsequent muscle food. For example, pork is inherently more tender than beef, as pork degrades rapidly postmortem due to greater proteolytic enzyme activity. However, two beef steers that are slaughtered at the same age and after the same number of days on a high concentrate diet and who produce carcasses containing the same level of marbling could differ in tenderness. These tenderness differences could be attributed to slight changes in the collagen component, proteolytic enzyme activity, or pH decline postmortem due to the influence of chilling rate on the development of rigor mortis.

Flavor

Meat flavor is the result of compounds stimulating the olfactory and taste receptors in the oral and nasal cavity of humans. These chemical compounds can vary in concentrations due to the influence of heat on the chemical structure, the degree of oxidation, the initial level of compounds, and the interactions between compounds. The muscle system can be divided into the lean portion and the lipid or fat portion, with each component contributing to meat flavor. Meat flavor is composed of (1) meatlike flavor derived from water-soluble reducing sugars and amino acids, (2) species-specific flavors which are due to differences in fatty acid composition and aromatic, water-soluble compounds that are stored in lipid depots of the animal; and (3) off-flavor development as the result of oxidation of lipid double bonds, defined as lipid oxidation or autooxidation, and other degradation processes.

Hornstein and Wasserman (1987) reported that basic meat-flavor precursors are water-soluble compounds. To determine the origin of meatlike flavor, defined as the characteristic flavor of meat regardless of species, fluids were obtained from raw beef lean. These fluids had a typical meat aroma. The fluids were found to contain sugars, salts, amino acids, and low-molecular-weight polypeptides. However, when the individual compounds were heated, typical meat flavors were not expressed; however, when the compounds were heated in combination, typical meat flavors were identified. Therefore, meatlike flavor compounds are water soluble and the combination or interactions of these compounds during heating are responsible for meatlike flavor. Meat extracts containing higher amounts of amino-acid-like compounds and carbohydrates are more intense in meatlike flavor.

Characteristic muscle food flavor also is associated with individual species. When comparing meat from different species for acceptance of flavors but from animals of similar maturity and fatness, young goat meat differed in flavor and was less desirable in flavor than pork, was less tender than pork, beef, or lamb, and was less satisfactory overall than pork, beef, or lamb. The species effect on muscle food flavor is most likely expressed through the genetic control of lipid composition and metabolism, as fatty acids, triglycerides, and phospholipids contribute to the species flavor of muscle foods.

The effect of animal sex on muscle food flavor has been highly related to the effect of testosterone which is produced in intact males. Testosterone has been shown to increase muscle growth and decrease intramuscular lipid deposition. As lipid deposits increase within muscle tissue, the characteristic fat-associated flavor increases and meatlike flavor decreases. As the U.S. consumer becomes accustomed to consuming lamb and beef that has a minimum amount of intramuscular fat, the reduction in fat induced by the production of noncastrated males could very likely result in consumer-perceived differences in flavor. The effect of "boar-odor" in pork from intact males has been well documented. Boar-odor has been related to the presence of 5α androst-16ene-3one, which is a metabolite of testosterone. This odor has been described as a urinelike, sweaty aromatic that develops when boar fat is heated. It is not known if the presence of 5α androst-16ene-3one is a breakdown product of testosterone or a secondary product produced in the animal system related to testosterone metabolism. Boar-odor also has been detected in meat from gilts and castrated males. Boar-odor is most prevalent during cooking and varies in intensity. In lamb, castration (rams versus wethers) has little or no effect on lamb flavor, depending on the age at the time of slaughter. As length of time past puberty increases, rams tend to have more intense flavor than wethers or ewes of the same age. Meat from bulls and steers has been shown to not differ and also to differ in flavor. Bulls have been shown to be more susceptible to long-term preslaughter stress than steers and heifers. As meat from animals subjected to long-term preslaughter stress, defined as dark, firm, and dry (DFD) meat, is higher in pH, some of the effect of sex on flavor may be due to a pH effect. Meat with a high pH, greater than 5.6–5.8, has been described as musty/moldy or more intense in beefy flavor, cowy/grainy, or serumy/bloody flavor aromatics. Additionally, intact males deposit less fat throughout the body and within muscle. Meat from animals with lower amounts of fat tends to have more meaty flavors and less fat flavors. Therefore, differences in meat flavor between intact males and castrated males or females can be partially attributed to differences in fat content levels, a pH effect, and the effect of testosterone-related products contained in the muscle.

The relationship of increased animal age and changes in muscle foods flavor has been examined. The meat from older beef animals is more intense in beef flavor than the meat from younger animals. Cowy/grainy and serumy/bloody flavor aromatics, and metallic feeling factor tend to increase in intensity in beef derived from older animals. Serumy/bloody and metallic descriptors have been associated with increased levels of myoglobin in meat from older animals. In lamb leg roasts, flavor intensity decreased as animals increased in weight and age; however, these lambs were slaughtered at a maximum live weight of 61 kg (134.2 lbs). The traditional mutton-flavor associated with meat from mature sheep is a classic example of how some flavor attributes concentrate in meat from older animals. Diet also can contribute to flavor differences in meat from young versus older animals. As older animals traditionally have a higher likelihood of being fed grass- or forage-based diets prior to slaughter, the effect of these diets on meat flavor can contribute to differences in flavor of meat from older animals. For example, beef from older animals is more intense in flavor than meat from younger animals. However, beef from young versus mature beef steers fed a high-concentrate corn diet for 185 days prior to slaughter did not differ in flavor desirability. The similarity in meat flavor from young and mature animals could be attributed to a predominant corn diet. In another study, beef cows were fed a high-concentrate diet for 0, 28, 56, or 84 days prior to slaughter. The carcasses from these cows were E maturity, the oldest maturity classification in the USDA beef grading standards. Beef steaks from these cows were more intense in the beefy flavor associated with young beef as time on a high-concentrate corn diet increased.

The effect of diet on the flavor of meat from ruminants has been examined. In lambs, meat from animals fed roughage diets had lower flavor scores than lambs fed high-energy diets. The deposition of compounds from the diet can be deposited in the lipids of the animal. Interestingly, changes in meat flavor that are associated with feeding high-concentrate diets are highly correlated to increased levels of muscle fatness. Therefore, many of the changes in flavor can be associated with subsequent changes in fatness as well as flavor associated with compounds deposited in lipids. Melton (1983) summarized that the corn in beef high-concentrate finishing diets can be partially or totally replaced by corn silage, a combination corn silage and alfalfa, alfalfa hay, or a combination of alfalfa hay and timothy hay without significant changes in beef flavor. However, when corn was replaced in beef cattle, diets with dietary sources other than high-quality hay or grains, beef flavor was affected. Recent research from my laboratory has shown that beef steers fed either a corn, corn/barley, or barley based high-concentrate finishing diet 102–103 days prior to slaughter to a final live weight of approximately 495 kg did not differ in cooked beef flavor intensity. The greatest effect of diet on beef flavor can be seen when beef fed corn-based diets where compared to beef from cattle fed on pasture. Meat from pasture-finished cattle usually is less desirable in flavor than meat from grain-finished cattle. Melton (1983) pointed out that the differences in acceptance of cooked meat flavor between grass-fed and grain-fed cattle were approximately 1 point on an 8-point scale or the prevailing description of grain-fed cattle was slightly to moderately desirable, and beef from grass-fed cattle was rated slightly desirable to slightly undesirable. In the aforementioned studies, beef cattle produced on pasture-finished diets were lower in fat than cattle from grain-finished diets; therefore, the effect of beef flavor could have been confounded with fatness level. However, when cattle fed either pasture-finished or grain-finished diets prior to slaughter and the subsequent cattle are slaughtered at a similar level of fatness, beef from pasture-finished cattle was still less desirable than meat from cattle fed grain-based diets. Therefore, flavor differences are not only due to differences in fatness, but flavor is influenced by the deposition of compounds (derived from the feed source) in the fat component of the animal system. Bromegrass and bluestem; bluegrass and clover, fescue, orchardgrass, and clover, fescue alone; flint hills grass; native range grass; forage sorghum; orchard grass and clover; oats, rye, and ryegrass; millet or coastal Bermuda grass; and Bermuda grass-clover and sudan grass resulted in lower flavor ratings for beef from cattle fed these pasture-types prior to slaughter. To remove the undesirable flavor effect from meat of pasture-fed cattle, supplementation of grain while the cattle are on pasture is suggested. For young cattle, removing the cattle from pasture and feeding them grain-based diets for 90–100 days prior to slaughter will delete the negative effect of pasture-feeding on beef flavor.

In nonruminants, the effect of diet can affect meat flavor substantially. Feeding growing pigs a high-oleate diet increased the oleate content in muscle and adipose tissues. Modification of beef fatty acid composition by dietary means is more difficult in ruminants as microflora biohydrogenate unsaturated fatty acids. If nonruminant fatty acid profiles are modified, meat flavor and palatability could be altered. Cooked *Longissimus* chops derived from pork fed a high-oleic sunflower-oil-containing diet were juicier, had higher tenderness scores but similar flavor attributes as chops from pork fed a traditional swine diet prior to slaughter. However, chops from pork fed a high-oleic sunflower-oil-containing diet had higher unsaturated fatty acid composition than the chops from the pork fed a control diet. Slight alterations in the fatty acid composition of pork may not significantly alter flavor but did affect meat palatability. Chops from pigs fed canola oil had 46% more off-flavor, lower flavor-quality scores, and lower overall palatability than pork chops obtained from pigs fed either control, animal fat-, safflower oil-, or sunflower oil-based diets. In summary, the alteration of the fatty acid composition in nonruminants can be accomplished by altering the animal diet; however, changes in fatty acid profile to decrease the fatty acids associated with increasing human serum cholesterol is not substantial enough to impact the overall health aspects of consuming these altered products.

Tenderness

Connective Tissue

Tenderness of muscle foods strongly influences and drives consumer's perceptions of acceptability and quality of muscle foods. Muscle tenderness can be segmented into the influence of connective tissue and muscle fiber characteristics. Total amount of connective tissue from within different muscles of the same animal affects tenderness between muscles. For example, the muscles from the round that are used in locomotion of the animal are higher in connective tissue content than muscles from the loin that are mainly used for structural support. Muscles have been generally classified into tender, intermediate, and tough categories. Tender muscles are the *Psoas major, Infraspinatus, Gluteus medius, Longissimus dorsi,* and *Triceps brachii;* intermediate muscles are the *Biceps femoris, Rectus femoris, Adductor, Semitendinosus,* and *Semimembranosus;* and tough muscles are the *Deep pectoral, Latissimus dorsi, Cutaneous trunci, Trapezius,* and *Superficial pectoral.* Between animals, the age-associated increase in collagen cross-linking or decreased collagen solubility significantly affects muscle tenderness. As animals increase in age, their meat is tougher. Identification or measurement of animal age, either through ossification of the vertebral column of an animal, identification of dentition, or evaluation of animal chronological age has been used as a bench mark for the effect of connective tissue on meat quality between animals.

Muscle Fiber Characteristics

The relationship between the contractile state of muscle fiber proteins and meat quality, especially meat tenderness, has been extensively investigated. Biochemical and physical conditions present when a muscle proceeds through rigor mortis affects the final contractile state (sarcomere length) and tenderness of the muscle. The phenomenon of cold-shortening negatively affects meat tenderness due to the effect of shorter sarcomere lengths in muscle that are chilled too rapidly during rigor mortis and the interruption of biochemical tenderization by rapid temperature decline. Sarcomere length is measured as the distance between the two Z-lines within a sarcomere. Z-lines are rigid structures that have to withstand the forces applied during contraction, and Z-line structure and density have been related to muscle fiber tenderness. In cold-shortened meat, Z-line density increases within a given quantity of meat. As long as the Z-line has not been disrupted by either degradation or fragmentation from contractile proteins in supercontracted meat, increased density of Z-lines has been related to increased meat toughness. Additionally, enzyme activity is usually very dependent on temperature. As temperature declines postmortem, the activity of enzymes involved in myofibrillar degradation, the calpain and calpastatin protease system, is decreased. Therefore, myofibrillar degradation, which has been related to improvements in muscle fiber tenderness, is reduced.

Tatum (1980) discussed the effect of dietary nutrition prior to slaughter and the subsequent effect on cold-shortening and muscle tenderness. As carcasses increase in fatness and weight due to high-energy diets, the susceptibility of the carcass to rapid postmortem chilling is reduced. These effects are due to the insulatory effect of fat and slower heat removal from a greater muscle mass.

Other researchers have demonstrated that lean carcasses chilled more slowly had more tender muscles than those chilled rapidly.

During refrigerated postmortem storage, improvement in meat tenderness, commonly called meat aging, occurs. The major factor responsible for postmortem improvement in meat tenderness is degradation of muscle proteins. Proteolysis of muscle postmortem has been attributed to either Ca^{2+}-dependent proteases (calpains), the level of the inhibitor of calpains, called calpastatin, lysosomal enzymes, or the synergistic action of both proteases. The physiological changes in postmortem muscle that highly correlate with improvement of meat tenderness are degradation of Z-lines, troponin-T, titin, and desmin, and the appearance of a 95,000-Da and 30,000-Da components. The role of calpains, calpastatin, and lysosomal enzymes have a major impact on meat tenderness and they are discussed in Chapter 3.

The injection or infusion of calcium chloride into pre-rigor meat has been shown to improved meat tenderness and, therefore, final meat quality. By increasing the level of intracellular calcium by 0.3 M immediately, postmortem in sheep carcasses, calcium-mediated biological processes, or the activation of calcium-dependent proteases are stimulated. A significant improvement in tenderness was seen in cooked meat from lambs that were infused with calcium chloride at 24 h postmortem when compared to noninfused lambs at 24 h and 7 days postmortem. This research documented that postmortem proteolysis, measured by gel electrophoresis, is a calcium-mediated event. A 0.3 M solution of calcium chloride and water was injected into strip loins, top sirloins, and top rounds from E-maturity cows immediately postmortem and improved the tenderness, as measured by Warner–Bratzler shear force, by 41.1, 40.1, and 15.3%, respectively. These substantial improvements in meat tenderness indicate that the injection of calcium chloride into pre-rigor meat from mature animals can offset the negative effect of physiological age through the activation of calcium-mediated proteolysis using the calpain system inherent in muscle tissue. Additionally, the majority of the improvement in tenderness was due to the degradation of the myofibrillar component of the muscle. Therefore, the myofibrillar component plays a significant role in modulating tenderness of muscle foods.

Microbial Level and Types

The influence of microbiological flora in the stability and shelf-life of muscle foods is discussed in Chapter 9. In general, with increased microbial growth, meat quality attributes can be affected. The development of off-colors, green and brown, are indications of high microbial levels. With increased microbial numbers, a slimy or slippery exudate on the meat surface or a cloudy white exudate in the meat purge develop. Odors such as putrid, sour, or sulfur are indicators of microbial growth in muscle foods. Visual and odor evaluation of muscle foods, cooked or raw, are the most reliable, rapid indications of the quality attributes.

Preparation of Muscle Foods by Foodservice or Consumer

The method of cooking and preparation of muscle foods significantly influences the color, fat content, muscle fiber characteristics, collagen content and solubility, and juiciness of muscle foods. These factors, in turn, influence the sensory properties of the cooked product. The influence of heat or meat cookery on muscle foods is discussed in Chapter 16.

Methods of Identifying Quality Characteristics of Muscle Foods

The assessment of product quality at the point of production, prior to purchase or consumption by the consumer, is important in assuring the reduction or elimination of consumption of low-quality muscle foods and increasing the likelihood of repeat consumer purchases. As the volume of purchases directly affects the profitability of muscle foods, the assessment of muscle food quality is directly related to the profitability of meat companies. Quality assessments can be either evaluated by the human as an evaluation instrument or by using instrumentation to determine the quantitative value.

Quality Term

A two-step system for assessing food quality was discussed by Civille (1991). The system first assesses the sensory properties of the food and those properties that are critical to quality, and in the second step, the quality properties are measured as to conformance. In the first phase, the sensory attributes of the product are defined by a Quality Team. The Quality Team, made up of quality control/assurance, research and development, or sensory professionals, conducts a category survey to define product attributes. Samples, up to 100–300, of the defined product from different production times and locations are sampled to assure that all attributes of the product are identified. It is important that the Quality Team evaluates products that possess normal production variation so that the full scope of product attributes is defined. For muscle foods, the scientific community and industry personnel have previously defined some common product attributes as listed in Table 11.2. These attributes, although not all the sensory attributes of muscle foods, can be used as predetermined significant characteristics or as a bench mark for the development of a more complex list that is product dependent. The Quality Team evaluates the appearance, flavor, and texture of the survey samples and develops in a full list of descriptors for each category, the range of each attribute, and the average intensity for each attribute. After the Quality Team has defined the attributes, the range of variation for all attributes

Table 11.2. A Summary of Commonly Used Product Quality Attributes Used in the Evaluation of Muscle Foods

Quality Attribute	Description of the Quality Attribute
Appearance/visual	
Color	The perception by the eye of light waves that includes blue (400–500 nm), green and yellow (500–600 nm), and red (600–800 nm) and can be defined by hue, value, and chroma. Color can be defined by overall color or an estimation of positive and negative color attributes.
Texture	A visual evaluation of the texture attributes of a product such as surface smoothness, consistency of texture
Fat content	An estimation of the amount of internal or external fat content of a muscle food, such as marbling, subcutaneous fat trim level, or amount of seam fat
Aroma	The volatile compounds detected within the nasal passage by the olfactory system; perceived by smelling (drawing air into the nasal and oral cavities).
Flavor	
Aromatics	The volatile compounds detected within the nasal passage by the olfactory system; perceived by mastication of the product in the mouth with subsequent passage of the volatiles into the nasal passage.
Taste	The basic tastes of salty, sweet, sour and bitter that are perceived by gustation through stimulation of taste buds mainly located on the surface of the tongue by water soluble compounds in the mouth.
Mouthfeel/feeling factors	The perception of senses in the mucosa of the eyes, nose, and mouth through the stimulation of the trigemenal nerves such as perception of heat, burn, cold, metallic, and astringency.
Aftertastes	Residual effects from a food after swallowing that result in identification of flavor aromatics, tastes, or feeling factors.
Oral texture	The structure or inner makeup of a product determined through the kinesthetic senses in the muscles of the tongue, jaw, or lips, or the tactile feeling properties in the tactile nerves in the surface of the lips, tongue, or mouth and measured as geometrical particles or moisture properties.

is documented by a descriptive attribute panel. The attributes then are used in a large-scale consumer study to determine how attribute variations affect consumer acceptance. This method provides a sound method of relating descriptive analysis data with consumer information. The correlation between descriptive sensory attributes and consumer acceptance identifies the relationship and a range for a descriptive attribute that is acceptable to the consumer. Key consumer terms can be closely related or correlated to sensory attributes that provide a mechanism of identifying consumer attributes with strong recognition for the descriptive sensory team. For example, in muscle foods, the cooked beef/brothy aromatic is considered a positive flavor characteristic in cooked beef. Cooked beef/brothy has been positively and highly related to consumer acceptance, flavor desirability, preference for beef flavor and beef-flavor intensity (0.95, 0.97, 0.83, 0.96, respectively). In cooked beef roasts, a difference of 0.5–1.0 on a 15-point scale in cooked beef/brothy, as determined by a trained descriptive attribute panel, affected consumer preference by 1 point on a 9-point hedonic scale. Additionally, variation of one attribute can strongly influence consumer perception of other attributes. For example, putrid or sulfur odors are easily detected at very low concentrations by trained and consumer panelists. Slight changes in putrid or sulfur odors negatively affect consumer preference. Also, as putrid or sulfur odors increase, other odors, such as beefy/brothy, may be masked. Therefore, increased intensity of an undesirable attribute can significantly affect consumer like or acceptance.

The second phase of assessing food quality is to set up a system for conformance that includes a sampling plan for monitoring production, defining final specifications, and consumer validation. The system for assessing food quality as defined by Civille (1991) provided a statistically based, consistent mechanism for establishing quality parameters and linking quality parameters to consumer acceptance. Additionally, this system provides a method of unbiased monitoring of quality attributes by descriptive attribute sensory panels and validation with consumer sensory panels. By using a scientific-based method of assessing quality, the value or effect of new technology or product development on the quality and consumer satisfaction of the product can be assessed.

Human Determinations

Color can be evaluated by a trained panel using a predetermined scale and can be a viable determinate of meat quality. Listed in Table 11.3 are examples of commonly used scales. However, color scales are often criticized for inconsistency in defining color descriptors. In my laboratory, color chips or photographs that closely illustrate or anchor the panelists to the descriptors are used to define the reference points for color descriptor scales. Color chips purchased at a local paint store that closely simulate the variation in color of the samples are useful in anchoring a color scale. If the panel can agree upon a standard color and its descriptor, a photograph can be taken of the sample and the photograph can be used as an anchor in the color scale. However, accurate color in the photograph needs to be verified by the panel prior to use of the photograph as an anchor. Color scale development sessions are held with the panelists. Sensory panelists used for color evaluation should be selected and screened as described in Chapter 12, as not every individual can determine differences in color attributes. In these sessions, products are presented to a minimum of five panelists. Products should represent the range of products that will be used in the study or the range in quality evaluation if color is being used as a quality control tool. For example, if meat color is going to be evaluated in a shelf-life study of fresh beef steaks,

Table 11.3. Color Evaluation Scales Used to Determine Color Characteristics of Muscle Foods

Muscle Food	Visual Color Scale
USE FOR CHARACTERIZATION OF OXYGENATED PIGMENT LEAN COLOR	
Beef[a]	8=Extremely bright cherry-red
	7=Bright cherry-red
	6=Moderately bright cherry-red
	5=Slightly bright cherry-red
	4=Slightly dark cherry-red
	3=Moderately dark red
	2=Dark red
	1=Extremely dark red
Lamb[a]	8=Extremely bright red
	7=Bright red
	6=Moderately bright red
	5=Slightly bright red
	4=Slightly dark red
	3=Moderately dark red
	2=Dark red
	1=Extremely dark red
Pork[a]	8=Extremely bright grayish-pink
	7=Bright grayish-pink
	6=Moderately bright grayish-pink
	5=Slightly bright grayish-pink
	4=Slightly dark grayish-pink
	3=Moderately dark grayish-pink
	2=Dark grayish-pink
	1=Extremely dark grayish-pink
Turkey and/or poultry light meat	8=Extremely dark greyish pink
	7=Dark grayish pink
	6=Moderately dark grayish pink
	5=Slightly dark grayish pink
	4=Slightly light grayish pink
	3=Moderately light grayish pink
	2=Light grayish pink
	1=Extremely light grayish pink
Turkey and/or poultry dark meat	8=Extremely dark dull red
	7=Dark dull red
	6=Moderately dark dull red
	5=Slightly dark dull red
	4=Slightly light dull red
	3=Moderately light dull red
	2=Light dull red
	1=Extremely light dull red
USE FOR CHARACTERIZATION OF DEOXYGENATED PIGMENT LEAN COLOR	
Beef or lamb or pork[a]	8=Extremely bright purple-red or purplish-pink
	7=Bright purple-red or purplish-pink
	6=Moderately bright purple-red or purplish-pink
	5=Slightly purple-red or purplish pink
	4=Slightly dark purple or purplish-pink
	3=Moderately dark purple or purplish-pink
	2=Dark purple or purplish-pink
	1=Extremely dark purple or purplish-pink

[a] Adapted from Hunt et al. (1991).

beef steaks in the form used in the study should be presented to panelists. Therefore, steaks that represent the shortest to the longest length of shelf-life should be included for training. Additionally, steaks that have been subjected to a variety of treatments to simulate the testing conditions should be presented. The panelists then describe the color descriptors, such as beef, pork, or lamb color, intensity of color, degree of doneness, amount of discoloration, and surface discoloration, that represent the products; then a scale is defined. The panel leader will assist with descriptor definitions based on previous knowledge of color descriptive scales used in similar studies. After the color descriptors and the scale are initially defined, the panel leader obtains either photographs that represent the scale or paint chips that closely assimilate the color descriptors. The panelists work with the photographs or paint chips until each panelist and the panel leader agree with the descriptors and the anchors for each descriptor. During panel ballot development, the panel leader assures that the gradation or space between descriptors and the anchors are equal across the scale. The use of photographs or color chips also can be used in product specifications to standardize quality parameters across suppliers in industry applications. The greatest problem when working with photographs is that reproduction of exact colors may be difficult, but with new photographic processing technology, photographs can be altered to reflect exact colors by working with a photographic services laboratory that does its own color processing. Photographs do fade with time and must be replaced to maintain consistent anchoring for color descriptors. After the ballot development sessions are complete, sufficient training sessions must be conducted to assure proper use of the scale and to reduce variability among panelists. At the initiation of each evaluation day, conducting a warm-up exercise where panelists evaluate a sample similar to the test samples using the defined scale will help reduce panel drift and assist the panel leader in ascertaining if all panelists are using the scale properly. Intermittent retraining or correlation sessions must be conducted to prevent panel drift and improve consistency between panelists. The panel leader should monitor panelist scores on a daily basis. As variation between panelists and inconsistent evaluation of the warm-up sample increases, retraining sessions are warranted. Some controversy exists as to how many panelists are needed to make a valid evaluation of color. A minimum of five panelists should be used to evaluate color, and the mean of the panelists should be the value reported and used for determining color of the sample. When less than five panelists are used to evaluate color, any one panelist can significantly influence the overall mean. Optimally, five to eight panelists will provide reliable color evaluation.

Texture can be evaluated in muscle foods; however, the correlation of human texture measurements to quality in muscle foods has not been high. Coarseness of texture, actually a visual evaluation of the smoothness of the lean cut muscle surface, has been extensively used to evaluate the lean surface of beef, lamb, and pork muscles. The coarseness of the lean surface of the muscle, either on the carcass as used in the USDA beef and lamb grading standards, or on the primal cut surface as would be used to evaluate the quality of beef, lamb, or pork, has been related to quality. As the muscle increases in coarseness of texture, the muscle tends to be tougher. Another example of texture scales is the evaluation of meat from animals that have been subjected to preslaughter stress. When the live animal was subjected to either long-term or short-term preslaughter stress, the resultant meat can be dark, firm, and dry, or pale, soft, and exudative, respectively. Evaluation of the firmness or softness of meat can provide a method of segmenting meat from these animals. In Table 11.4 examples of commonly used texture scales are presented.

The human evaluation of fat, either the total amount of fat, or external fat, or a specific fat depot, or intramuscular fat, contained within a muscle food has been extensively used to determine muscle food quality. Examples of exterior fat amount as a quality characteristic in muscle foods are determining if steaks were trimmed to an average of 0.34 cm of exterior fat or if the fat remaining on a skin-on chicken breast contained the side muscles and the associated fat as defined by a specification. The amount of visible fatness in a meat product has been related to consumers' perception of the total amount of fat in the product and its acceptance from a diet/health standpoint. To effectively use an estimation of the fat content within a muscle food, identification of the fat parameters has to be determined. In a whole-muscle steak product, the amount of exterior fat that is acceptable has to be defined. The scale then can be defined and the evaluators can be trained to recognize varying levels of exterior fat. A common use of scales is to estimate intramuscular fat content, such as the use of marbling scores to determine the amount of intramuscular fat in beef carcasses. The amount of marbling, as defined by three trained evaluators, and the chemical amount of fat in the *Longissimus dorsi* muscle at the 12th rib of beef carcasses have been highly correlated ($r = 0.88$). The use of marbling scores to predict muscle fat content and the subsequent palatability of muscle have been extensively evaluated. Although low to moderate correlations between marbling score and meat palatability and fat content have been reported, more effective, rapid, and repeatable estimations of muscle fat content and palatability have not been found. As previously discussed, intramuscular fat contributes to palatability but does not explain all the variability in palatability.

The human evaluation of the visual appearance of muscle foods, such as surface dryness, sliminess of the meat surface, amount of purge in a package, and tightness of the package in a vacuum-packaged product, can be highly related to the quality and shelf-life of muscle food products. Defining scales and standards of reference for each visual appearance attribute of interest can be an effective method of monitoring product quality and shelf-life. Examples of visual appearance scales to evaluate muscle food quality are presented in Table 11.5.

The odor, or the identification of the aromatic characteristics of products by smelling, can indicate storage life, product stability, presence of product

Table 11.4. Texture Evaluation Attributes Used to Determine the Texture Characteristics of Muscle Foods

Muscle Food	Texture Attributes
Beef/lamb/pork/turkey/poultry	
Whole muscle	Juiciness
	Overall tenderness
	Muscle fiber tenderness
	Connective tissue amount
Processed meats	Springiness
	Cohesiveness
	Cohesiveness of mass
	Juiciness
	Hardness
	Denseness
	Fracturability
	Chewiness
	Overall tenderness
Fish	Flakiness
Silver hake or whiting, Atlantic mackerel, white hake, cusk, goosefish or monkfish, Atlantic pollack, tilefish, Atlantic wolffish, winter or blackback flounder, weakfish, gag or grouper, haddock, Atlantic halibut, swordfish, Atlantic cod, bluefish, striped bass, and dab[a]	Firmness
	Moistness
	Chewiness
	Mouth-drying
	Fibrousness (stringiness)
	Oily mouth-coating
	Adhesiveness (sticks to teeth)
	Cohesiveness of the mass (chewed sample holds together in mouth)
Shrimp[b]	Roughness (surface)
	Hardness
	Friability
	Cohesiveness
	Toughness
	Chewing resistance
	Wetness
	Juiciness
	Fibrousness
	Graininess
Multiple species[c]	Firmness
	Cohesiveness
	Gel persistence (at five chews)
	Rigidity

[a] Adapted from Sawyer et al. (1988).

[b] Adapted from Solberg, Tidemann, and Martens (1986).

[c] Adapted from Hamann and Lanier (1986).

Table 11.5. Visual Appearance Attributes Used to Monitor the Visual Appearance Characteristics of Muscle Foods

Muscle Food	Visual Appearance Attributes
Beef/lamb/pork/turkey/poultry	Lean color intensity based on attributes listed in Tables 9.1 and 9.3
	Deoxygenated pigment of lean color
	Percentage of lean or fat discoloration
	Amount of purge or purge color
	Fat color
Fish	Whiteness[a]
Silver hake or whiting, Atlantic mackerel, white hake, cusk, goosefish or monkfish, Atlantic pollack, tilefish, Atlantic wolffish, winter or blackback flounder, weakfish, gag or grouper, haddock, Atlantic halibut, swordfish, Atlantic cod, bluefish, striped bass, and dab	Darkness
Rainbow trout[b]	Whiteness/blackness
	Chromaticness (chromaticity)
	Hue—red and yellow
Shrimp[c]	Natural red color[c]
	Yellowish discolor
	Dehydration
	Shininess/translucence
	Smoothness, surface
	Natural shape

[a] Adapted from Sawyer et al. (1988).

[b] Color to be evaluated using 9-point scales where increasing intensity of each parameter was given increasing values; adapted from Skrede et al. (1989).

[c] Adapted from Solberg, Tidemann, and Martens (1986).

contaminates, or identify proper or improper spice formulations of a muscle food product. Off-odors used as an indication of microbiological growth are rotten egg (sulfur), lactic soured, ammonia in fish, or putrid. Odors, such as cardboard, painty, and livery aromatics, also can be used as an indication of fat oxidation of the muscle. Product contamination can be identified by odor detection. Examples of odors that indicate product contamination are ammonia (i.e., refrigeration leaks), chlorine (i.e., inadequate washing), or the amount of pepper aromatic in a product that should not contain that spice odor. Odor evaluation in a spiced product can be an indication that the product has been properly or improperly formulated. For example, in a muscle food product that contains garlic in the spice blend, identification or quantification of the garlic odor aromatic can be used as an indication that the spice blend is at the proper level in the product. To develop scales for odor or aromatic characteristics of muscle food products, the Quality Team can be used to identify the odors or aromatics of interest. The

Quality Team then can define the range of acceptable levels and determine threshold levels for each desirable and undesirable odor or aromatic. Sensory panelists should be selected and screened as defined in Chapter 12 to assure that panelists can detect and quantitate differences in attributes. Pure references of each odor need to be presented to the individuals used for odor determinations. After each panelist can readily identify each reference, then panelists should be trained on varying concentrations for each reference. This allows each panelist to establish a threshold for each odor and to quantitate concentration differences for each odor parameter. After panelists can identify each odor and can differentiate between different concentrations of each odor, the panelists need to be presented with the actual product with differing concentrations of the odor added. The protocol for identifying odors then needs to be defined. An example of a protocol is as follows. If the odor panel is identifying the aromatics associated with microbiological spoilage in vacuum-packaged, retail fish steaks, the panels may make two odor evaluations of the product. The first evaluation may be immediately upon opening the package which could be very important for many products as consumers may make their initial determination of quality immediately upon opening the package. The second odor evaluation may occur after a defined time after opening the packaged product to provide an estimation of the product odor without the vacuum-packaged odor of the product interfering with the odor evaluation. The second evaluation relates to the odor that consumers would perceive immediately prior to cooking the product. Odor evaluations of muscle food products also could be defined at a specified point during a processing procedure or at a critical point in the production or merchandising process. Examples of some odor evaluation scales used for muscle foods are presented in Table 11.6.

Instrumental Determinations

The relationship between instrumental measures of quality attributes can be determined following the Quality Team approach as discussed for human determinations; however, the instrumental measurement method is substituted for the descriptive attribute sensory panel. Identification of the instrumental method is the first step. Four different measurement methods used to evaluate quality characteristics of muscle foods are described.

For the evaluation of tenderness in whole-muscle meat products, the American Meat Science Association (AMSA 1978) recommends use of the Warner–Bratzler shear force machine. An Instron fitted with a Warner–Bratzler shear head also is used to determine texture differences in whole-muscle meat products. To conduct a Warner–Bratzler shear force test, the whole-muscle meat product is cooked using standardized cooking procedures. After cooking, the product is cooled to room temperature and multiple cores, either 2.54 cm or 1.27 cm, are removed from the cooled sample. The cores are removed so that the fiber direction

Table 11.6. Odor or Aromatic Evaluation Attributes Used to Determine the Odor/Aromatic Characteristics of Muscle Foods

Muscle Food	Odor/Aromatic Attribute	
Beef	Cooked beefy/brothy; cooked beef fat; serumy/bloody; grainy/cowy; carboard; painty; fishy; livery/organy; soured (grainy); medicinal; putrid; acidic/soured	
Lamb	Cooked lamb; cooked lamb fat; serumy/bloody; grainy; carboard; painty; fishy; livery/organy; soured (grainy); medicinal; mutton/musty; putrid; acidic/soured	
Pork	Cooked pork; cooked pork fat; serumy/bloody; grainy; carboard; painty; fishy; livery/organy; soured (grainy); medicinal; boar taint; putrid; acidic/soured	
Turkey	Cooked turkey; meaty; brothy; liver/organy; browned; burned; cardboard; painty; fishy; soured (grainy); putrid; acidic/soured;	
Poultry	Chickeny; meaty; brothy; liver/organy; Browned; burned; cardboard; painty; fishy; soured (grainy); putrid; acidic/soured;	
Fish		
Silver hake or whiting, Atlantic mackerel, white hake, cusk, goosefish or monkfish, Atlantic pollack, tilefish, Atlantic Wolffish, winter or blackback flounder, weakfish, gag or grouper, haddock, Atlantic halibut, swordfish, Atlantic cod, bluefish, striped bass, and dab[a]	Delicate or fresh fish; heavy or gamey fish; fish (old fish); sweet; briny, salty; sour; seaweed; bitter; fish oil; buttery; nutty; musty; ammonia metallic; shellfish	
Cod, haddock, and whiting[b]	Salty; sweet; milky; musty; cardboard; boiled cabbage; meaty (boiled); metallic; chickenlike	
Shrimp[c]	*Odor*	*Aromatic/flavor*
	Total odor intensity	Total flavor intensity
	Fresh, typical	Fresh, typical flavor
	"Factory"	Sweet flavor
	Stale	"Factory"
	Fishy	Stale
	Ammoniacal	Fishy
	Urinelike	Acrid/sharp
	Sour	Sour
	Old seaweedlike	Metallic
	Sulfide	Cardboardlike
	Cardboard-like	Stockfishlike
	Stackfishlike	Rancid
	Rancid	Mudlike
	Mudlike	Salt
		Aftertaste

[a] Adapted from Sawyer et al. (1988).

[b] Adapted from Love (1988).

of the muscle is parallel with the length of the core. Multiple cores are removed from each sample to account for the variation within the sample and to identify the area of highest shear force. By identifying the dorsal/ventral, medial/lateral location of individual cores, the locations of worst "cold-shortening" can be determined within a muscle food. The core is placed in the Warner-Bratzler shear head so that the length of the core is horizontal to the shear head (Fig. 11.3). The core then is sheared across the grain and the total pounds or kilograms of force are determined. The core is sheared in the approximate center to reduce the effect of the outer, dehydrated, hard surface that is present after cooking. Meat cores can be sheared twice and the average of the two values obtained from each core is recorded as the shear force value for that core. The Warner–Bratzler shear force machine provides a value in pounds or kilograms that is the total amount of force required to shear a cooked, standard size core of meat. A Warner–Bratzler shear force head can be attached to an Instron Universal Testing machine and meat cores can be evaluated as previously described. The Instron machine can chart the curve during the entire compression of the core (Fig. 11.4). Data can be obtained that describe the area of the curve for compression. These data are total area under the curve, peak area which is the area under the curve from the point of compression to the point of maximum or peak force, fail area or the area under the curve from point of peak force to the completion of compression, peak slope which is the slope of the line from the initiation of the compression to the point of maximum or peak force, peak elongation or the length of the line from the initiation of compression to the point of maximum or peak force, peak force which is the maximum amount of pressure required to segment a standard core of meat, fail slope or the slope of the line from the point of peak force to the completion of compression, and fail elongation or the

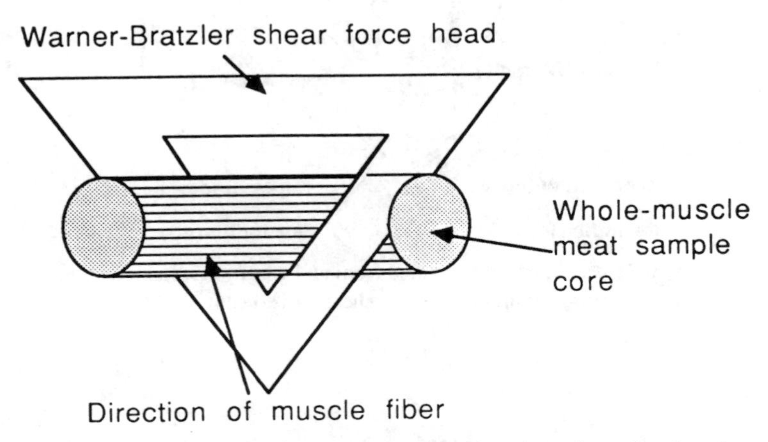

Figure 11.3. Diagram of the Warner–Bratzler shear force head and a whole-muscle meat core inserted into the Warner–Bratzler shear force head.

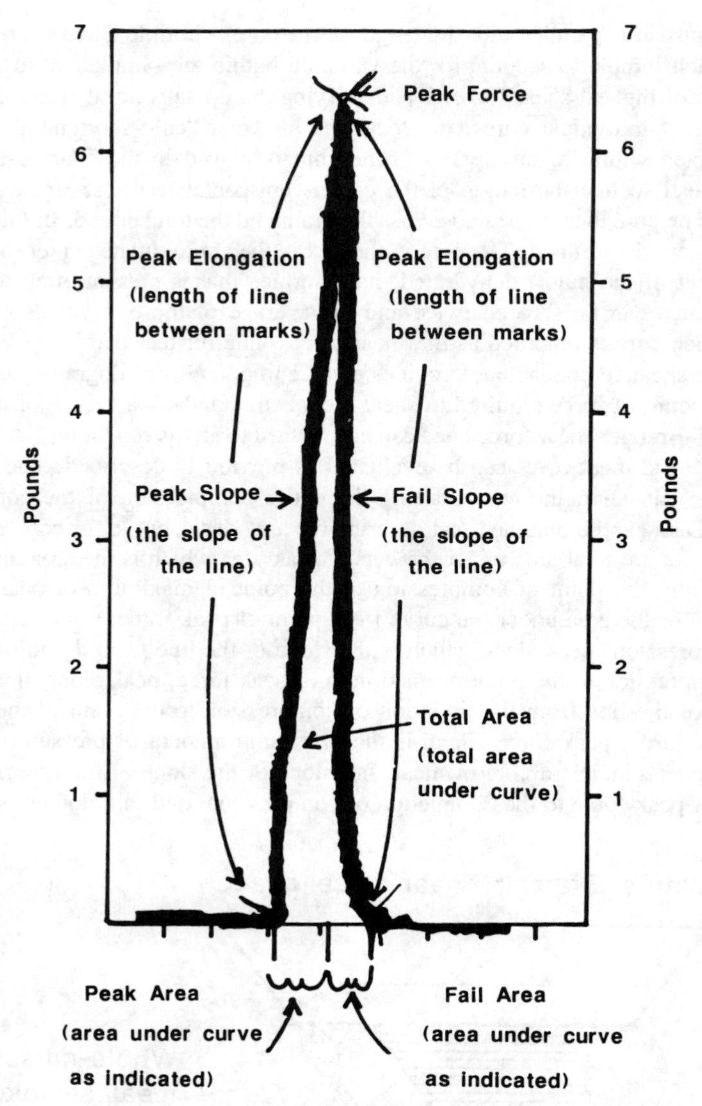

Figure 11.4. A curve generated from an Instron Universal Testing machine using a Warner–Bratzler shear force head.

length of the line from the point of peak force to the completion of compression. Correlations between components of the curve, except for peak force, and sensory characteristics for texture or tenderness have not been high; however, the parameters of the curve may describe variation among treatments.

Shear force values or peak force from the Instron can be used to identify if whole-muscle meat products contain a high amount of variability in total shear force. Differences in shear force values can be used to determine if differences exist in total force between meat samples. Warner–Bratzler shear force values have been criticized for not accounting for all the texture characteristics of muscle foods. Whereas Warner–Bratzler shear force values provide information on the total force required to segment a piece of meat, these values do not explain the fracturability, cohesiveness of mass, springiness, number of chews required to segment a meat sample, initial juiciness, sustained juiciness, connective tissue amount, or muscle fiber tenderness that would be determined using sensory panels. However, Warner–Bratzler shear force values have been highly correlated to overall tenderness of whole-muscle meat samples with correlation coefficients ranging from −0.01 to −0.94.

For processed meats and fresh muscles not suited for coring such as chicken breast, the Allo–Kramer shear force is a common method of objectively measuring texture attributes. The Allo–Kramer shear head can be installed on an Instron Universal Testing Machine or the head can be purchased with a Food Technology Corporation Texture Test System (referred to the Allo–Kramer Shear Press in the past). The Allo–Kramer shear device consists of a removable load cell that is approximately 6.67 × 7.30 × 6.35 cm (Fig. 11.5). At the bottom of the load cell are 0.32-cm-wide metal bars spaced 0.32 cm apart. A standard weight of meat is placed in the load cell and the lid is closed. The meat is cut into standard squares, usually 2.54-cm squares, and a standard number of squares are evaluated. The load cell is placed on a platform in the machine under the ram that has 25.72-cm-wide by 6.99-cm-long metal bars that match up with the lid and the bars in the load cell. The blades are forced down through the load cell at a constant speed (called ram speed) and at a constant distance. A standard load cell (i.e., 50 kg), the load range of the load cell (i.e., 0–50 kg), and the chart speed (i.e., 200 mm/min) need to be standardized within an experiment or usage. During compression, the meat sample is compressed and then extruded in the load cell. Compression can range from partial compression, such as 50%, to full compression. If a partial compression is used, one or two compressions per sample can be recorded to simulate one- or two-bite work–force compression curves. Compression curves are recorded on chart paper.

Texture Profile Analysis can be conducted on an Instron Universal Testing Machine or the Food Texture Analyzer (Tecture Technologies Corp., Scarsdale, NY) for muscle foods. The sample is placed on the Instron machine or Food Texture Analyzer and compressed twice using a single blade. Texture attributes can be calculated from the compression curves as follows: (a) Hardness is calcu-

Figure 11.5. An Allo–Kramer shear head attached to an Instron Universal Testing machine.

lated as the peak force during the first compression cycle; (b) cohesiveness is the ratio of force area during the second compression to that during the first compression; (c) springiness is the height that the meat recovers during the time that elapses between the end of the first bite and the start of the second bite; (d) gumminess is the product of hardness times cohesiveness; and (e) chewiness is the product of gumminess times springiness.

New technologies that provide nonintrusive measurements of rheological properties of meat are continually being examined. Bulk modulus of elasticity, which is defined as stress/strain, could be applied to meat samples to examine the elastic properties of muscle tissues. The bulk modulus of elasticity could provide textural information for whole-muscle and restructured muscle tissues. Bulk modulus of elasticity is determined by placing the item of interest under a compression plate attached to an Instron Universal Testing Machine. A known compression is applied to the tissue and the bulk modulus of elasticity, or hardness of the tissue, is obtained. This method provides an overall elastic value for the tissue. Ophir et al. (1991) introduced a new technology called elastography which uses ultrasound waves to evaluate the elastic properties of biological tissues nonintrusively and two dimensionally. Elastography could provide information related to the structure of whole-muscle or restructured meat samples. Although the development of this technology has recently been initiated, it is possible that information could be extracted on the structural characteristics of muscle foods.

Chemical analysis is an accurate determination of fat or lipid content in muscle foods. AOAC defines approved methods for determining lipid content. Petroleum ether or diethyl ether is used as a solvent to extract the lipid from homogenized meat samples to determine the total or percentage of lipid. The ether extraction method can utilize either the Soxhlet apparatus or Goldfisch apparatus as the extraction mechanism for refluxing ether through a meat sample to extract the lipid component. An alternate fat extraction method, the Folch method, uses chloroform methanol to extract the phospholipid component of the muscle tissue as well as the other lipid components from homogenized meat samples. As the Folch method extracts the lipid and the phospholipid component from muscle tissues, and the ether extraction method extracts only the lipid component, slightly higher percentage values for total lipid are obtained from the Folch method. Muscle contains about 0.5–1.0% phospholipids.

Rapid methods for determining fat content of muscle foods also have been developed. The CEM fat and moisture extraction system (CEM Corp., Indian Trail, NC) has been AOAC approved as a method for determining fat and moisture content of muscle foods. The CEM method utilizes microwaves to heat and dry homogenized meat and methylene chloride to extract fat from the dried meat sample. Approximately 5 g of homogenized sample is used for a CEM determination. The advantages for using the CEM method are that both moisture and fat can be determined within 20 min and that the results are comparable to AOAC methods for accuracy. Disadvantages of the CEM method are that the equipment is expensive, if the sample is not homogenized sufficiently, a 5-g subsample may not be representative of the whole sample, and it is not a productive method to evaluate large numbers of samples.

Other methods for determining fat content of muscle foods include numerous modifications of the Babcock method, the Hobart method, and the Analray (Analray Corp., The Kartridg Pak Co., Davenport, IA). These methods are not AOAC approved, but they are rapid and lend themselves for easy application in plant situations. The original Babcock method utilizes sulfuric acid to digest 9 g of a homogenized meat sample, and then the fat is separated from the digested sample by heat and centrifugation. Water is added until the fat layer that is floating on the top of the sample is contained in the graduated neck of a Paley–Babcock bottle. The amount of fat is determined by the difference from the bottom to the top of the fat layer as determined on the graduated neck of the bottle. Advantages of this method are that the procedure requires little capital investment and is a good indicator test. The disadvantages are that sulfuric acid is a very strong acid and the associated dangers of handling a strong acid, 9 g of sample may not be representative of the whole sample, and the technique is accurate only to the nearest 1–2% of lipid. The Babcock method is a good indicator method to assure compliance to a range in fat, such as monitoring that frankfurters are less than 30% fat. The Hobart machine utilizes a ground, donut-shaped sample of meat. The fat within the sample then is melted out using a heat source. The difference between the beginning weight and the fat extracted weight determines the percentage of fat within the sample. The Anal-ray is a machine that sends a X-ray through a standard quantity, 5.9 kg, of ground meat. Fat tissue defracts X-rays, so as the fat content of the standard quantity of ground meat increases, fewer X-rays are relayed through the meat sample. The difference between the amount of X-rays that are transmitted versus the amount that the sensor identifies at the opposite end of the sample is used to calculate the percentage of fat within the sample. Advantages of using the Anal-ray machine are that it is a very rapid method that takes less than 5 min to evaluate each sample, the machine is in one contained unit that is not very expensive, the larger sample size is more likely to be representative of the sample being tested, and the machine is fairly rugged so it will hold up in industrial environments. The disadvantages of using the Anal-ray for fat determinations are that there are different standards for pork versus beef, if the machine is not calibrated correctly the results will be unreliable; technician error can be an issue as the same person should prepare and pack the samples each time to assure even compaction, if the sample is packed too tight or loose, readings will vary; the Anal-ray readings should be routinely compared to CEM or analytical methods to assure accuracy.

Instrumental evaluation of color, which either estimates the color of the sample or the amount and/or state of pigments in muscle foods, can be determined either by pigment extraction and quantification or by using reflectance methods. Pigment extraction requires homogenization of the sample, extraction of pigments, usually the heme-containing pigments oxymyoglobin, deoxymyoglobin, or metmyoglobin, and quantification of the pigments through absorbance or transmittance of light at a defined wavelength. Pigment extraction methods effectively separate pigments, but this method tends to overestimate oxymyoglobin and metmyoglobin content while underestimating deoxymyoglobin. These phenomena are due to the effect of extraction, exposure to oxygen, and measurement on the state of the pigments. A common criticism of the pigment extraction method is that when the entire sample is homogenized, the values obtained do not reflect the pigments present on the exterior surface. To circumvent this, a thin layer or multiple layers of the surface, about 1–2 mm thick, should be removed from the sample and pigments extracted from these portions, as layers of myoglobin chemical forms vary in thickness. Pigment extraction also is a good method for determining total heme or total myoglobin pigment; however, this method does not provide information on the stability of the pigments.

The reflectance method uses a defined light wave that is directed toward the surface of a meat sample. The amount of reflectance at standard wavelengths provides information on the color and pigments of the surface. The advantages of using reflectance methods are that multiple measures can be obtained on the same sample and that changes over time can be monitored on the same sample. Color is composed of three quantitative components. Hue, which is the name of the color such as red, green, blue, and yellow, is based on differences in wavelength, chroma is the strength or intensity of the color, and value is the lightness of the color from white to black. With reflectance, quantitative values of color can be obtained. In muscle foods, CIE (Commission Internationale de l'Eclairage, or International Commission on Light) X, Y, and Z values, defined as tristimulus values, are related to color where X relates to red, Y is an estimation of green, and Z indicates blue color. The XYZ system does not account for lightness or value of color; however, the percentage of Y is a measurement of lightness. Color values also can be reported as tristimulus coordinates of *x, y,* and *z* that are the percentage of red, green, and blue estimated in a sample surface from X Y Z values. *x, y,* and *z* values are calculated as follows: $x = X/(X + Y + Z)$;$y = Y/(X + Y + Z)$; $z = Z/((X + Y + Z)$. Hunter L, *a* and *b* Color Solid values, developed by Hunter Laboratory, Inc., Reston, VA are commonly used to evaluate the color of muscle foods. L-values, which range in value from 0–100, indicate lightness with 100 equal to white and 0 equal to black. Positive *a*-values quantitate red color and negative a-values are an indication of green. Positive b-values are a determination of yellow and negative *b*-values indicate blue color. *L a* b-values can be calculated from X Y Z-values as follows: $L = 10(Y)^{1/2}$; $a = [17.5\ (1.02\ X - Y)] \div Y^{1/2}$; $b = [7.0\ (Y - 0.947\ Z)] \div Y^{1/2}$. Reflectance at defined wavelengths, ratios of reflectance, or K/S values (absorbence coefficient divided by scattering coefficient) can be used to estimate the amount of oxymyoglobin, deoxymyoglobin, or metmyoglobin in muscle samples.

Fresh meat color can be determined from reflectance measurements. K/S ratios can be obtained at different wavelengths. To calculate the amount of each pigment, the following equations can be used: Deoxymyoglobin, % = K/S 474 ÷ K/S 525; metmyoglobin, % = K/S 572 ÷ 525, and oxymyoglobin, % K/S 610 ÷ K/S 525. The reflectance ratio of 630/580 nm is an evaluation of changes in red color, or as the ratio increases, the muscle sample is redder.

Measurement of cured meat color usually centers around the evaluation of pink pigment and an estimation of the fading of the pink pigment to brown. The reflectance ratio of wavelength 650/570 nm is used to measure the intensity of cured meat color. It is important that the reflectance of a cured meat sample be completed immediately after exposure of the surface to air. Use of this reflectance ratio minimizes the effect of pigment concentration and has been highly correlated to visual evaluations of color. The *a*/*b* ratio is an indication of shifts from pink to tan or from red to maroon red to brown. Determination of the total heme pigment and the percentage conversion to nitrosohemochrome has been a viable method of evaluating cured meat color. A 90% pigment conversion to nitrosohemochorme would be good, 80% conversion would be intermediate, and values of

45–60% conversion would be indication of poor conversion, exposure to oxygen, or that the product is reaching the end of its shelf-life. However, the method is very sensitive to light and care must be taken to minimize light exposure.

Reflectance measurements at wavelengths of 545, 580, and 630 nm have been highly correlated to degrees of doneness in cooked meat. Wavelengths of 630 and 580 nm have been especially useful in determining changes from red to brown due to increased degree of doneness. Brown color is difficult to measure, so most methods of measuring cooked meat color concentrate on the loss of red color that results with increased cooking. Cooked color can also be monitored using CIE Lab-values, *a*/*b* ratio, hue angle, and saturation index.

The American Meat Science Association suggested the following measurements as methods to determine color in meat: (1) L^*, a^*, and b^* values, (2) *a*/*b* ratios, hue angle, saturation index, (3) reflectance ratios as follows: (a) fresh meat color and color change = 630 ÷ 580 nm, (b) cured meat color and fading = 650 ÷ 570 nm, (c) cooked meat color changing from red to brown = 630 ÷ 580 nm, (d) estimation of deoxymyoglobin = K/S 474 ÷ K/S 525, (e) estimation of metmyoglobin = K/S 572 ÷ K/S 525, (f) estimation for oxymyoglobin = K/S 610 ÷ K/S 525.

As with any instrumental measurement method, sample preparation is critical to obtaining repeatable, reliable results. The amount of time that the sample is exposed to air, the fiber orientation of the surface where reflectance measurements are being determined, the amount of surface measured, the percentage discoloration of the surface and how much of that surface is evaluated by the measurement tool, how representative the sample is of the whole, and the type of film used to wrap the meat, as this affects the way that light waves travel through the sample, should be considered prior to instrumental determinations of color.

In muscle foods, two major types of instruments are used to measure muscle color. The colorimeter is an instrument that uses three or four colored lights to match the color of a sample. Colorimeters are good for determining small color differences between samples. Spectrophometers use light from one source and evaluate the spectral reflectance curve of the sample. Spectrophometers provide absolute values for a sample and are a good method for determining differences in color between samples.

The Hunter Colorimeter (Hunter Associates Laboratory, Inc., Reston, VA) is a commonly used objective determination of muscle color. The Hunter Colorimeter determines L, *a,* and *b* or L^*, a^* and b^* values based on 1976 recalculations. Hunter color values can be obtained on raw and cooked meat and for whole-muscle or processed meats. Differences in these values provide an analytical method of determining color differences. The Hunter Colormeter is easy to standardize and to obtain measurements. A sample protocol must be defined that includes how to prepare the meat and where the Hunter values are to be obtained on the surface of the sample so reliable Hunter values can be obtained. Then, the machine is standardized and the meat is placed over an orifice on the Hunter

Colorimeter. A cover is placed over the meat to block external light sources. The light waves are reflected off the surface of the meat placed over the orifice. Different size orifices are available and the size of the orifice should coincide with the surface sample of the product being evaluated. Reflectance spectrophotometer is another method of instrumentally evaluating muscle color. A spectrophometer equipped with a reflectance attachment can be used to obtain CIE X, Y and Z-values.

Instrumental measurements of flavor, taste or odor involves quantitating volatile compounds using either (1) a gas chromatograph (GC) with or without a mass spectrophotometer, (2) infrared or nuclear magnetic resonance, or (3) liquid chromatography. The basic premise of the gas chromatograph is that volatile compounds can be quantitated by instrumentation. Volatile compounds are those released from foods that may come into contact with the olfactory senses by passing through the back of the mouth and into the nose by way of the retronasal passages. The olfactory senses are a great funnel. As compounds pass through the olfactory senses, these compounds are collected and amplified for sensing. The olfactory senses are capable of interpreting highly dilute chemical concentrations. Research has indicated that there is no apparent limit to the number of smells that can be discriminated and identified by humans, provided familiarization or training is provided. Some flavor compounds can be detected by humans in parts per billion, whereas instrumental detection is not as sensitive. Instrumental quantification of volatiles can provide information on the physical state of a muscle food, but changes in volatile compounds may or may not be related to human perceived changes in flavor and ultimate quality of those products.

The GC is instrumentation that is used to separate and quantify compounds from a given quantity of material. Early flavor chemists utilized exit ports (sniff ports) to sniff compounds as they were eluted from the GC. Use of this methodology allowed the identification or correlation between a chemical compound and how it was perceived by the human. By utilizing the human sense of smell with chemical quantification of compounds, the relationship between the sensitivity of the human olfactory senses and compound quantification could be established. The addition of the use of mass spectrometry to the GC/sniff port system allows for the identification of specific compounds that are separated and the chemical characterization of muscle foods can be identified. Marker compounds, compounds that when they change are highly correlated to quality changes in the muscle food, can be used to track acceptability of products.

Selected References

Aberle, E.D., E.S. Reeves, M.D. Judge, R.E. Hunsley, and T.W. Perry. 1981. Palatability and muscle characteristics of cattle with controlled weight gain. Time on a high energy diet. *J. Anim. Sci.* 52:757.

AMSA, 1978. *Guidelines of Cookery and Sensory Evaluation of Meat*. American Meat Science Association, National Live Stock and Meat Board, Chicago, IL.

ASTM. 1978. *Compilation of Odor and Taste Threshold Values Data*. American Society for Testing and Materials, Philadelphia.

Batzer, O.F., A. Santoro, and W.A. Landmann. 1962. Identification of some beef flavor precursors. *J. Agric. Food Chem.* 10:94.

Batzer, O.F., A. Santoro, M.C. Tan, W.A. Landmann and B.S. Schweigert. 1960. Precursors of beef flavor. *J. Agric. Food Chem.* 8:498.

Bowling, R.A., G.C. Smith, Z.L. Carpenter, T.R. Dutson, and W.M. Oliver. 1977. Comparison of forage-finished and grain-finished beef carcasses. *J. Anim. Sci.* 45:209.

Bowling, R.A., J.K. Riggs, G.C. Smith, Z.L. Carpenter, R.L. Reddish, and O.D. Butler. 1978. Production, carcass and palatability characteristics of steers produced by different management systems. *J. Anim. Sci.* 46:333.

Brown, H.G., S.L. Melton, M.J. Riemann, and W.R. Backus. 1979. Effects of energy intake and feed source on chemical changes and flavor of ground beef during storage. *J. Anim. Sci.* 48:338.

Brown, W.D. 1962. The concentration of myoglobin and hemoglobin in tuna flesh. *J. Food Sci.* 27:26.

Cain, W.S. 1979. To know with the nose: Keys to odor identification. *Science* 203:467.

Calkins, C.R. and S.C. Seideman. 1988. Relationships among calcium-dependent protease, cathepsins B and H, meat tenderness, and the response of muscle to aging. *J. Anim. Sci.* 66:1186.

Campion, D.R., J.D. Crouse, and M.E. Dikeman 1975. Predictive value of USDA beef quality grade factors for cooked meat palatability. *J. Food Sci.* 40:1225.

Carpenter, Z.L. 1974. Beef quality grade standards-need for modifications? *Proc. Recip. Meat Conf.* 27:122.

Civille, G.V. 1991. Food quality: Consumer acceptance and sensory attributes. *J. Food Qual.* 14:1.

Cramer, D.A., J.B. Pruett, R.M. Katting, and W.C. Schwartz. 1970. Comparing breeds of sheep. 1. Flavor differences. *Proc. West. Sec. Am. Soc. Anim. Sci.* 21:267.

Cross, H.R., Z.L. Carpenter, and G.C. Smith. 1973. Effects of intramuscular collagen and elastin on bovine muscle tenderness. *J. Food Sci.* 38:998.

Crouse, J.D., J.R. Busboom, R.A. Field, and C.L. Ferrell. 1981. The effects of breed, diet, sex, location and slaughter weight on lamb growth, carcass composition and meat flavor. *J. Anim. Sci.* 53:376.

Davis, G.W., A.B. Cole, W.R. Backus, and S.L. Melton. 1981. Effect of electrical stimulation on carcass quality and meat palatability of beef from forage- and grain-finished steers. *J. Anim. Sci.* 53:651.

Desor, J.A. and G.K. Beauchamp. 1974. The human capacity to transmit olfactory information. *Percep. Psychophysiol.* 16:551.

Dupuy, H.P., M.E. Bailey, A.J. St. Angelo, J.R. Vercellotti, and M.G. Legendre. 1987. Instrumental analyses of volatiles related to warmed-over flavor of cooked meats. In: *Warmed-Over Flavor of Meat,* A.J. St. Angelo and M.E. Bailey, eds. Academic Press, New York.

Ekeren, P.A., D.R. Smith, D.K. Lunt, and S.B. Smith. 1992. Ruminal biohydrogenation of fatty acids from high-oleate sunflower seed. *J. Anim. Sci.* 70:497.

Field, R.A. 1971. Effect of castration on meat quality and quantity. *J. Anim. Sci.* 32:849.

Fox J.B., Jr. 1987. The pigments of meat. In: *The Science of Meat and Meat Products,* 3rd ed. J.F. Price and B.S. Schweigert, eds. Food & Nutrition Press, Inc., Wesport, CT.

Hamann, D.D. and T.C. Lanier. 1986. Instrumental methods for predicting seafood sensory texture quality. *Proc. Intern. Symp, Univ. of Alska Sea Grant Prog.* Page 123. Elsevier Science Publishers B.V., Amsterdam.

Hood, R.L. and E. Allen. 1971. Influence of sex and postmortem aging on intramuscular and subcutaneous bovine lipids. *J. Food Sci.* 36:786.

Hornstein, I. and P.F. Crowe. 1960. Flavor studies on beef and pork. *J. Agri. Food Chem.* 8:494.

Hornstein, I. and A. Wasserman. 1987. Part 2-Chemistry of meat flavor. In: *The Science of Meat and Meat Products,* 3rd ed. J.F. Price and B.S. Schweigert, eds. Food & Nutrition Press, Inc., Westport, CT.

Hunt, M.C., J.C. Acton, R.C. Benedict, C.R. Calkins, D.P. Cornforth, L.E. Jeremiah, D.G. Olson, C.P. Salm, J.W. Savell, and S.D. Shivas. 1991. Guidelines for meat color evaluation. *Reciprocol Meat Conf.* 44:232

Kemp, J.D., J.M. Shelly, Jr., D.G. Ely, and J.D. Fox. 1972. Effects of castration and slaughter weight on fatness, cooking losses and palatability of lamb. *J. Anim. Sci.* 34:560.

Koohmaraie, M., J.E. Schollmeyer, and T.R. Dutson. 1986. Effect of low-calcium-requiring calcium-activated factor on myofibrils under varying pH and temperature conditions. *J. Food Sci.* 51:28.

Koohmaraie, M., S.C. Seideman, J.E. Schollmeyer, T.R. Dutson, and J.D. Crouse. 1987. Effect of postmortem storage on Ca^{++}-dependent proteases, their inhibitor and myofibril fragmentation. *Meat Sci.* 19:187.

Koohmaraie, M., A.S. Babiker, R.A. Merkel, and T.R. Dutson. 1988a. The role of Ca^{2+}-dependent proteases and lysosomal enzymes in postmortem changes in bovine skeletal muscle. *J. Food Sci.* 53:1253.

Koohmaraie, M., A.S. Babiker, A.L. Shroeder, R.A. Merkel, and T.R. Dutson. 1988b. Factors associated with the tenderness of three bovine muscles. *J. Food Sci.* 53:407.

Koohmaraie, M., A.S. Babiker, A.L. Shroeder, R.A. Merkel, and T.R. Dutson. 1988c. Acceleration of postmortem tenderization in ovine carcasses through activation of CA^{2+}-dependent proteases. *J. Food Sci.* 53:1638.

Kramlich, W.E. and A.M. Pearson. 1958. Some preliminary studies on meat flavor. *Food Res.* 23:567.

Lawless, Harry. 1991. The sense of smell in food quality and sensory evaluation. *J. Food Qual.* 14:33.

Locker, R.H. 1960. Degree of muscular contraction as a factor in tenderness of beef. *J. Food Sci.* 25:304.

Love, R.M. 1988. *The Food Fishes: Their intrinsic variation and practical implications*. Van Nostrand Reinhold Company, New York.

Love, J.D. and A.M. Pearson. 1971. Lipid oxidation in meat and meat products—A review. *J. Am. Oil Chem. Soc.* 48:547.

Mabrouk, A.F., J.K. Jarboe, and E.M. O'Conner 1967. Water-soluble flavor precursors of beef. Extraction and fractionation. *J. Agric. Food Chem.* 17:5.

Melton, S.L. 1983. Effect of forage feeding on beef flavor. *Food Technol.* 37:239.

Melton, S.L., M. Amiri, G.W. Davis, and W.R. Backus. 1982a. Flavor and selected chemical characteristics of ground beef from grass-, forage-grain- and grain-finished steers. *J. Anim. Sci.* 55:77.

Melton, S.L., J.M. Black, G.W. Davis, and W.R. Backus. 1982b. Flavor and selected chemical characteristics of ground beef from steers backgrounded on pasture and fed corn up to 140 days. *J. Food Sci.* 47:699.

Meyer, B., J. Thomas, R. Buckley, and J.W. Cole. 1960. The quality of grain-finished and grass-finished beef as affected by ripening. *Food Technol.* 14:4.

Miller, M.F., S.D. Shackelford, D.K. Hayden, and J.O. Reagan. 1990. Determination of the alteration in fatty acid profiles, sensory characteristics and carcass traits of swine fed elevated levels of monounsaturated fats in the diet. *J. Anim. Sci.* 68:1624.

Miller, R.K., J. Ophir, A.D. Whittaker, M.S. Rubio, G.C. Emesih, H. Ponnekanti, and I. Cespedes. 1993. Ultrasonic elastography to evaluate meat structure and quality. *Reciprocal Meat Conf.* Proc.46:May 1994.

Miller, R.K., L.C. Rockwell, L.A. Ray, and D.K. Lunt. 1993. Determination of the flavor attributes of cooked beef from steers fed corn or barley based diets. *J. Anim. Sci.* (submitted).

Miller, R.K., J.D. Tatum, H.R. Cross, R.A. Bowling, and R.P. Clayton. 1983. Effect of carcass maturity on collagen solubility and palatability of beef from grain-finished steers. *J. Food Sci.* 48:484.

Morgan, J.B., R.K. Miller, F.M. Mendez, D.S. Hale, and J.W. Savell. 1991. Using calcium chloride injection to improve tenderness of beef from mature cows. *J. Anim. Sci.* 69:4469.

Ophir, J., I. Cespedes, H. Ponnekanti, Y. Yazdi, and X. Li. 1991. Elastography: A quantitative method for imaging the elasticity of biological tissues. *Ultrasonic Imaging* 13(2):111.

Park, R.J., J.L. Corbeth, and E.P. Furnival. 1972. Flavour differences in meat from lambs grazed on Lucerne (Medicago sativa) or phalaris (Phalaris tuberoso) pastures. *J. Agric. Sci.* 87:47.

Parrish, F.C., Jr. 1974. Relationship of marbling to meat tenderness. Proc. Meat Ind. Res. Conf. p. 117.

Pearson, A.M. 1966. Desirability of beef-its characteristics and their measurement. *J. Anim. Sci.* 25:843.

Plsek, P.E. 1987. Defining quality and the marketing development interface. *Qual. Prog.* 20:28.

Purchas, R.W. and H.L. Davies. 1974. Palatability of beef from cattle fed maize silage and pasture. *New Zealand J. Agric. Res.* 25:183.

Reagan, J.O., J.A. Carpenter, F.T. Bauer, and R.S. Lowrey. 1977. Packaging and palatability characteristics of grass- and grass-grain fed beef. *J. Anim. Sci.* 46:716.

Rhee, K.S., T.L. Davidson, H.R. Cross, and Y.A. Ziprin. 1990. Characteristics of pork products from swine fed a high monounsaturated fat diet: Part I—Whole muscle products. *Meat Sci.* 27:329.

Richards, J.F. and B.C. Morrison. 1969. A study of fat grade, fat content and sensory properties of turkey broilers. *Can. Inst. Food Technol. J.* 2:6.

Savell, J.W. and H.R. Cross. 1988. The role of fat in the palatability of beef, pork, and lamb. In: *Designing Foods: Animal Product Options in the Marketplace*. National Academy Press, Washington, D.C.

Sawyer, F.M., A.V. Cardello and P.A. Prell. 1988. Consumer evaluation of the sensory properties of fish. *J. Food Sci.* 53:12.

Schroeder, J.W., D.A. Cramer, R.A. Bowling, and C.W. Cook. 1980. Palatability, shelf-life and chemical differences between forage-and grain-finished beef. *J. Anim. Sci.* 50:852.

Seideman, S.C., H.R. Cross, G.C. Smith, and P.R. Durland. 1984. Factors associated with fresh meat color: A review. *J. Food Qual.* 6:211.

Sink, J.D. 1979. Factors influencing the flavor of muscle foods. *J. Food Sci.* 44:1.

Skrede, T., T. Storebakken, and T. Naes. 1989. Color evaluation in raw, baked and smoked flesh of rainbow trout (*Onchorhynchus mykiss*) fed astaxathin or canthaxanthin. *J. Food Sci.* 55:1574.

Smith, G.C. and Z.L. Carpenter. 1974. Eating quality of animal products and their fat content. In: *Proc. of the Symp. on Changing the Fat Content and Composition of Anim. Products*. National Academy of Sciences, Washington, D.C.

Smith, G.C., M.I. Pike, and Z.L. Carpenter. 1974. Comparison of the palatability of goat meat and meat from four other animal species. *J. Food Sci.* 39:1145.

Solberg, T., E. Tidemann, and M. Martens. 1986. Sensory profiling of cooked, peeled and individually frozen shrimps (*Panduaus borealis*) and investigation of sensory changes during frozen storage. In: *Proc. Intern. Symp. Univ. of Alaska Sea Grant Prog*. Page 109. Elsevier Science Publishers B.V., Amsterdam.

Summers, R.L., J.D. Kemp, D.G. Ely, and J.D. Fox. 1978. Effects of weaning, feeding systems and sex of lamb on lamb characteristics and palatability. *J. Anim. Sci.* 47:622.

Tatum, D.J. 1980. Is tenderness nutritionally controlled? *Proc. Rec. Meat Conf.* 34:65.

Tatum, J.D., G.C. Smith, B.W. Berry, C.E. Murphey, F.L. Williams, and Z.L. Carpenter. 1980. Carcass characteristics, time on feed and cooked beef palatability attributes. *J. Anim. Sci.* 50:833.

USDA. 1989. Official United States standards for grades of carcass beef. Agriculture Marketing Service, USDA. Washington, D.C.

Wanderstock, J.J. and J.I. Miller. 1948. Quality and palatability of beef as affected by method of feeding and carcass grade. *Food Res.* 13:291.

Wasserman, A.E. and F. Tally. 1968. Organoleptic identification of roasted beef, veal, lamb and pork as affected by fat. *J. Food Sci.* 33:219.

Weller, M., M.W. Galgam, and M. Jacobson. 1962. Flavor and tenderness of lamb as influenced by agriculture. *J. Anim. Sci.* 21:927.

Westerling, D.B. and H.B. Hedrick. 1979. Fatty acid composition of bovine lipids as influenced by diet, sex and anatomical location and relationship to sensory characteristics. *J. Anim. Sci.* 48:1343.

Wu, J.J., D.M. Allen, M.C. Hunt, C.L. Kastner, and D.H. Kropf. 1981. Nutritional effects on beef palatability and collagen characteristics. *J. Anim. Sci.* 51(suppl. 1):71.

12

Sensory Methods to Evaluate Muscle Foods

Rhonda K. Miller

Muscle foods have properties that are related to the five senses of taste, smell, sight, feel, and sound. The four basic tastes of salty, sweet, bitter, and sour can be readily identified in different muscle foods. Further processed muscle foods, such as frankfurters or summer sausage, usually have salty or sweet tastes. Smell is a very important sensory property, as the detection of aromatics and/or odors by the olfactory nerve comprises the major components of muscle food flavor. The fishy flavor in intensely flavored fish species is an aromatic detected by the olfactory nerve. Texture or feel of food influences perceptions of acceptability. The tenderness of muscle foods has been shown to affect consumer acceptability just as the detection of mouth-coating or a residual substance, usually fat, in the mouth and throat after the consumption of high-fat meat products can influence human perception of acceptability or unacceptability.

Many factors influence the sensory characteristics of muscle foods; however, it is the identification of the sensory characteristics and the measurement of these characteristics that establish the relationship between quality characteristics and sensory properties of muscle foods. Controversy still exists between scientists on how to identify, measure, and interpret sensory information. Some scientists expound the use of consumer sensory evaluation as the purest method of measuring meaningful differences in sensory characteristics; however, the quantification of sensory attributes by trained or expert sensory panels also is supported. The methods and principles for conducting trained and consumer sensory evaluation of muscle foods will be discussed.

Evaluation of Sensory Characteristics Using Trained Sensory Panels

The use of trained humans as instruments to quantitate or identify the sensory properties of muscle foods has been extensively used in the scientific literature to understand the application of new technology on the sensory characteristics and in industry by quality control personnel to track and evaluate products on a day-to-day basis and over time. To conduct sensory evaluation, three specific areas have to be considered: (1) the environmental, product, and panel conditions that will be used for sensory testing, (2) selection and training of sensory panelists, and (3) the type and structure of tests that will be used to address sensory properties in muscle foods.

Environmental, Panel, and Product Conditions

The sensory analysis environment is extremely important as the environment can interfere or confound sensory perceptions. Removal of as many environmental factors as possible or minimizing and standardizing those factors that cannot be removed across treatments is important. This assures that the sensory response of a panelist is the result of the product characteristics, not a response confounded by the environment in which the response was evoked.

The panelist evaluation area should be separated from the food preparation area and should be free of extraneous odors, especially cooking odors. Optimally, air filtration and good ventilation systems should be used to remove odors from the evaluation area. The sensory evaluation area should contain a sample transfer mechanism where samples can be passed from the preparation area to the sensory evaluation area without transferring odors. Odors from other panelists should be minimized. For example, panelists should not use scented cosmetic or personal care products immediately prior to or during sensory evaluation. If a panelist uses tobacco products, they should be asked not to use these products on the day of evaluation as their hair and clothes carry tobacco odors. Panelists should wear lab coats that are washed in unscented detergent to minimize odors. If sinks are plumbed into the sensory evaluation area, removal of the effect of drain, pipe, or water odors should be minimized. Water used to rinse between samples should be double-distilled, deionized water, or passed through a filtration system that eliminates any flavors that could interfere in the evaluation of the product. Construction materials should be nonporous, not contain odors, be of a neutral color, off-white preferable, should not contain a pattern, and should not collect dirt or mold that would harbor odors. A constant room temperature between 22 and 24°C and a relative humidity of 45–55% should be maintained to assure comfort and a constant environment.

The noise level in the sensory evaluation area should be minimized. These sounds include noise associated with traffic flow in and out of the area, people serving samples, or any other noise from external sources. There should be no communication, visual or oral, between panelists during sensory evaluation. If one panelist makes a comment indicating dislike or like of a product, other panelists may be influenced. For example, a panelist may indicate dislike of a product (a trained panelist should never indicate like/dislike of a product, as this is a consumer response) by saying "oh, how terrible." Another panelist may have rated the product high for beefy flavor, but upon hearing the remark, the panelist may reevaluate their response and rate the product lower for beefy flavor due to the verbal influence of the other panelist.

Consistent and appropriate type of lighting should be used during sensory evaluation. Generally, 70–80 foot-candles of shadow-free light should be used. For muscle foods, red filtered lights are commonly used to reduce some visible effects due to cooking method, degree of doneness, or muscle effect. To minimize physical discomforts for panelists, comfortable seating and sufficient room per panelist should be provided.

Panelists should not be hungry at the time of evaluation, and panel evaluations should be conducted when there is a greater probability that panelists are mentally alert (subjected to a minimum amount of mental fatigue). The use of oral cleansing products or teeth brushing should be at least 1 h before panel starts. They should not have consumed food or chewed gum within 1 h of conducting oral evaluations. They should not consume liquids except water within ½ h of oral evaluation. These measures are important to minimize the effect of other foods and taste adaptation on their perception during sensory evaluation.

Unbiased presentation of samples to sensory panelists eliminates confounding effects on panelist responses. Preparation procedures of muscle foods affect the sensory attributes of the final product. Cooking method, service temperature, and humidity of the cooking apparatus, internal degree of doneness, length of cooking, time from cooking to serving, effect of holding the cooked product at a standardized temperature between cooking and serving, types of containers used to store the product, and utensils used in preparing the product can influence the sensory attributes of muscle foods. Each of these parameters must be standardized and the effect of each parameter should be evaluated before initiation of the sensory study to assure minimal influence on the sensory attributes of the product being tested. For muscle foods, panelists are usually presented a standardized size or quantity of meat and they should be given at least two serving portions per sample. For whole-muscle products, such as steaks, chops, and roasts, panelists are usually served a 1.27-cm cube, for a ground meat product, either a 0.64-, 0.42- or 0.32-cm wedge of the patty including the edges or a 1.27- or 2.54-cm square with the edges removed, and for processed meat items the product serving size is dependent on the type of product and the objectives of the test. If texture attributes are being evaluated for frankfurters, a whole frankfurter may be served to each panelist; however, if flavor attributes are being evaluated on frankfurters, a 2.54-cm section with the end pieces removed may be the serving size. The temperature of the sample when it is presented to sensory panelists can affect their sensory perception. Aromatic compounds volitalize at higher temperatures. Therefore, if temperature of the sample varies from panelist to panelist or from day to day, panelists receiving warmer samples may detect higher amounts of flavor attributes in a sample compared to when they evaluated the same product served at a colder temperature. Samples served to sensory panelists should be identified with three-digit random codes to prevent subconscious selection of a sample due to identification. Samples should be presented to sensory panelists in either glass or white glazed china to eliminate any aromatics associated with the container and to neutralize the visual attributes of the container on the sample. Plastic containers can be used if they are opaque or neutral in color and have been tested to assure that they do not impart odors. The order that samples are served to panelists should be random. In central location consumer and trained sensory evaluation tests, a strong first-order bias is present or panelists tend to rank the first sample that they evaluate either higher or lower than samples evaluated later in the order. Therefore, it is important that the order effect is either randomized across treatments or controlled so that the effect can be removed statistically.

Guidelines for cookery and sensory analysis of meat (AMSA 1978; Froning et al. 1978) should be used as standardized techniques for the sensory evaluation of muscle foods.

Selection of Sensory Panel Members

Selection of sensory panelists is critical to obtaining reliable sensory data. Panelists can be selected from internal sources within a company or program, or external panelists who have no other ties to the project or company can be recruited. The advantages to using internal panelists are that additional monetary reward is usually not provided, as being a sensory panelist becomes part of their normal responsibilities. This method reduces the direct cost of conducting sensory analysis; panelists are already on location or at work and do not have to travel to the testing site; and the pool of individuals from which the sensory panelists will be selected may be more diverse in demographics than external panelists. The disadvantages of using internal panelists are that panelists are not totally unbiased about the products as they most likely contribute to the manufacturing or are familiar with products being evaluated; work responsibilities can often interfere with sensory panel duties; and use of external panelists is directly more expensive—it can be difficult to find people who are willing to work only a few hours per day, and the selection of people may be heavily slanted toward retired people or people who work inside the home. On the other hand, external panelists have only one work commitment to the project or company which eliminates interference from other job responsibilities. As long as panelists are adequately screened and trained, biases inherent in selecting people who are available at only specific times during the day are eliminated. In my laboratory, external panelists are used. The advantage that they have only one work commitment to the program (to be a sensory panelist) which limits interference with other job related responsibilities outweighs the disadvantages previously discussed. Additionally, panel attendance is higher and turnover of panelists is lower.

After the determination of what type of panelists will be used, advertising for potential panelists in local newspapers, on bulletin boards, in company newsletters, by word of mouth, and with local organizations is conducted. As much information concerning the job requirements, total hours per week of work, and type of work should be communicated in the advertisement. Interested candidates should be given a prescreening interview where the following issues should be addressed: (1) describe in great detail what the job entails (hours per day, tasting of meat products, making decisions using a standardized scale, role of a panelist in sensory evaluation, etc.); (2) ascertain the person's availability on a daily, weekly, monthly and yearly basis (for example, an individual may be available every morning from 9 to 12 every day of the week, but they take a 2-month vacation every year; is that acceptable to you as the panel leader?), (3) determine the individual's interest (Do they like to eat different types of foods? Do they like to cook? Do they like meat? Do they like to experiment with different food ideas? Do they eat out a lot? Why would they like to be a sensory panelist?); (4) evaluate the general state of health of the individual (Do they have food allergies? Do they have frequent colds or sinus problems? Do they suffer from dental or gum diseases? Do they have frequent mouth sores? Do they take medication such as antibiotics? Are they diabetic or do they have kidney or coronary diseases? Do they have any condition that affects any of their senses of sight, smell, taste or feel? Do they miss work frequently due to illness? Do they have false teeth?); and (5) communicate the importance of the sensory program to the company or institution. During the screening process the interviewer will be trying to determine the potential panelist's reliability, interest, availability, and health. Panelists should be either disqualified or accepted for screening tests (the next step) at this point. Do not accept a panelist that does not fully fit into the requirements listed, even if you are desperate for panelists. If unacceptable or borderline individuals are accepted for further testing, issues related to why they were possibly unacceptable will resurface and could affect the efficacy of the sensory panel. For example, accepting an individual during the prescreening who shows only a mild interest in the program will many times result in a panelist who is repeatedly absent.

Screening tests are conducted to determine panelists' sensory acuity and ability to listen and follow directions. Screening of potential panelists should be conducted in four phases: (1) conduct scaling tests to determine if panelists can follow direction and make judgments; (2) evaluate sensory acuity or the individual's ability to discriminate using sniff tests and triangle tests; (3) conduct tests to determine an individual's ability to rank or rate sensory differences; and (4) conduct a personal interview to ascertain a panelist's continued interest. During the screening process one of three decisions should be made at the completion of each of the four phases listed above: (1) accept the individual for the next phase of testing; (2) reject the individual; or (3) continue testing. An individual should be accepted for all four phases before acceptance for sensory training. If a potential panelist is late or does not show up for a screening session without calling or rescheduling ahead of time, immediately reject that individual. If individuals are not responsible and timely during the screening process, these behavior patterns will continue throughout sensory testing. Examples of exercises for each of the four phases are listed in Table 12.1. Do not accept an inferior panelist even if there is concern that you will not be able to recruit enough people. It is much better to have fewer reliable, repeatable sensory panelists than to have more panelists who are not reliable and do not have the ability to discriminate.

After the selection of the sensory panelists, training begins. The overall goal of sensory panel training is to familiarize the individuals with the test procedures, to improve the individual's ability to recognize and identify sensory attributes, and to improve the individual's sensitivity to and memory for test attributes so that sensory judgments will be precise and consistent.

The number of panelists that comprise a trained panel depends on how well trained the panel is, how familiar the panelists are with the testing procedures and the product, and how small are the anticipated differences between products in the test. During training, panelists will drop out due to loss of interest, relocation, mortality, illness, and inability to grasp concepts; therefore, approximately 20% additional panelists should be trained. The first sessions of sensory training should be used to educate new panelists on the general biology of muscle foods. As panelists will be evaluating the sensorial properties, it is important that they understand what muscle fiber structure is, what connective tissue is, how the olfactory senses work, where the basic tastes are located on the tongue, etc. The subsequent sessions should familiarize panelists with the test procedures. The techniques that should be presented in these sessions are (1) how to chew a sample; (2) how to expectorate the sample versus to swallow; (3) how to rinse the mouth between each sample; and (4) familiarize them to the methods of evaluation, such as the scales to be used in testing, anchor points for the scale, the score sheet, and the terminology or lexicon to be used. During training sessions, references should be presented for each attribute that will be tested. The exercises should concentrate on one attribute each session until panelists are familiar with that attribute. Anchor points on the scale for each attribute should be presented to assure a concise understanding of what each attribute is and how to scale within that attribute. Panelists should score samples individually, then the results should be discussed as a group, the panel leader should determine the overall score for each sample, and panelists should reevaluate their sample to familiarize themselves with the score and the response in the sample. Training should proceed until consistent results are obtained *among* panelists and *by* each panelist. The degree or length of training depends on the type of panel, the number of attributes, and previous panel experience. Some panels may take up to 1 year of training if panelists have no prior experience and a large number of attributes are used. However, an existing panel may only require 2, 3, or 4

Table 12.1. Examples of Exercises Used in the Four Phases of Screening Potential Sensory Panelists

Phase of Screening	Example of Exercises to Be Used
Scaling tests	 Directions: Mark on the line following each figure proportion of the area that is shaded.  [triangle figure] None ├────────┤ All [circle figure] None ├────────┤ All [rectangle figure] None ├────────┤ All
Sensory acuity/discrimination	Sniff tests provide a good method of determining if panelists can describe or distinguish between attributes: Have panelists sniff different flavored oils that have been placed on a cotton ball or absorbant paper in a glass jar (i.e., baby food jar). Ask panelists to describe the aromatic. Examples of oils: Cedar, Cherry, Vanilla, Lemon, Lavender Triangle tests are used to evaluate a potential panelist's ability to discriminate. Examples of triangle tests (note differences between samples in a triangle test should differ by at least 2 units on the scales used in trained testing). 1. Strip loin steak cooked to 70°C and a strip loin steak cooked to 75°C and juices pressed out to get differences in juiciness. 2. Eye of round steak cooked to 70°C and a strip loin steak cooked to 70°C to get differences in tenderness. 3. Top butt steak and a strip loin steak both cooked to 70°C to get differences in flavor intensity.
Rank or rating	Examples of ranking tests: 1. Sip the solution of each cup and evaluate for sweet taste; rank the samples from least intense to most intense for sweet taste. (Solutions containing varying concentrations of sugar.) Least intense ________ ________ ________ Most intense ________
Personal interview	Examples of questions for a personal interview are as follows: A. Interest—essential Expectations—time, pay, social contact, commitment

weeks to train on the evaluation of pork after having evaluated beef using similar procedures. Training never ends. A panel leader must use verification and performance evaluation procedures to determine when panel retraining is needed.

Verification and performance evaluation of a sensory panel is an on-going, continuous process that identifies problems among individual panelists and identifies areas of further training requirements for the entire panel. The purpose of verification and performance are to determine consistency over time and to determine the ability of panelists to discriminate among samples. To accomplish this select up to nine samples that vary in sensory attributes but are within the testing parameters of the panel. Evaluate these samples in several sessions over several days. For example, each of the nine samples would be evaluated each day with the samples served in three sessions per day with three samples served per session. The nine samples would be presented to panelists over 4 days with the order of samples served randomized across sessions and order within session. The data can be analyzed by Analysis of Variance (ANOVA) in which there are nine treatments and four observations per cell or the data can be analyzed as a two-way ANOVA where session and day effects also can be studied. From the ANOVA table the F-ratio can be calculated as the Mean Square treatment divided by the Mean Square error. The larger the F-ratio, the more consistent a panelist is. A larger F-ratio means that a panelist has the ability to differentiate between samples and they evaluated the same sample similarly on each of the 4 days. Panelists should be retrained and reevaluated as needed.

The first step in evaluating the sensory properties of muscle foods is to determine the type of trained sensory panel needed to meet the objectives of the study. Two basic types of sensory panels are used in the evaluation of muscle foods—difference and descriptive attribute testing.

Difference Testing

Difference testing can be categorized into overall difference tests and attribute difference tests. Overall difference tests provide an avenue to evaluate if a sensory difference exists between samples, whereas attribute difference tests ask if a specified attribute differs between samples. Commonly used overall difference tests are triangle tests, two-out-of-five test, duo–trio test, simple difference test, "A"–"not A" test, difference-from-control test, sequential test, and similarity tests. Attribute difference tests include directional difference test, pairwise ranking test, simple ranking tests, multisample difference test—rating approach, and multisample difference test—BIB ranking test.

A triangle test involves presenting three coded samples (using three-digit random numbers) to the sensory panelist where one sample is different from the other two samples. The panelist is asked to identify the sample that is different. The number of correct responses is counted and the probability (the probability level has to be determined) that the sensory panelists can determine a difference is identified [see Meilgaard, Civille, and Carr (1991) for statistical tables].

In a two-out-of-five test, sensory panelists are presented with five coded samples and are asked to identify the two samples that are similar. As with the triangle test, the number of correct responses is counted and the probability that the two samples differ is determined [see Meilgaard, Civille, and Carr (1991) for statistical tables]. The two-out-of-five test is very effective when sensory fatigue is not an issue. With meat samples, sensory fatigue usually occurs fairly rapidly, as meat samples consist of moderate levels of flavor attributes, and lingering aftertastes can be apparent. Therefore, the two-out-of-five test is not commonly used to discriminate flavor differences between two meat samples.

The Duo–trio test provides the sensory panelists a one in two chance of identifying the correct sample. The sensory panelists are given a reference sample to evaluate. Then two coded samples are presented and the sensory panelists are asked to identify the coded sample that is the same as the reference. The correct number of replies is determined and the probability that the two samples are different is determined [see Meilgaard, Civille, and Carr (1991) for statistical tables].

To conduct a simple difference test, sensory panelists are given two samples and asked whether the samples are different or similar. Sensory panelists can be presented with either a single pair or any combination of the two pairs (1 then 2, 2 then 1, 1 then 1, or 2 then 2) for the two products being evaluated. The number of responses identified as the same and different are counted for matched and unmatched pairs, giving a total of four numbers. The placebo effect is compared to the treatment effect using χ^2 analysis. The χ^2 statistic is calculated by

$$\chi^2 = \sigma \frac{(O - E)^2}{E}$$

where O is the observed value and E is the expected number. The determined χ^2-value is compared to a χ^2table at a probability of 0.05. If the calculated value is greater than the table value, a significant difference exists between the two samples.

The difference-from-control test determines if the sample(s) are different from the control and what the magnitude of the difference is in the test sample(s) from the control. Sensory panelists are presented with a control and the test sample(s). Then, they are asked to evaluate the difference in the test sample(s) from the control using a scale. The mean is calculated from the Difference-from-Control and the blind controls. A paired *t*-test can be used to compare the means if only one sample is compared, but if two or more samples are compared to the control, Analysis of Variance should be used to test for treatment differences.

Sequential tests are a method of utilizing difference tests (such as triangle and duo–trio tests) while minimizing the number of evaluations needed to make a decision on the difference or similarity of two products. For example, utilizing the triangle test to determine if a prospective sensory panelist has the sensory acuity to be on a trained panel based on the number of correct responses was proposed by AMSA (1978). Similarity testing should be used to test if no perceivable differences exist between two samples, instead of testing if two samples are different. The triangle, duo–trio, and two-out-of-five tests can be used in similarity testing.

The directional difference test is a paired comparison test that asks if sensory panelists can determine a difference in the intensity of an attribute. The correct number of responses is counted. If the null hypothesis is that the control differs from the treatment sample, the test is one-sided. Test for significance is the same as used for the duo–trio test. A two-sided directional difference test is defined when the null hypothesis asks which sample has a lower or higher intensity of the attribute of interest or which product is preferred.

Descriptive Attribute Testing

Descriptive attribute sensory evaluation is a method of quantitating specific sensory attributes of muscle foods. A family of attributes, called a lexicon, describe the specific sensory attributes in a muscle food that can be used to evaluate the development or change in these attributes. A lexicon of descriptive attributes can be developed for any muscle food product. For example, the descriptive attributes have been defined for catfish in Table 12.2, for poultry flavors in Table 12.3, for beef-flavor attributes for warmed-over flavor in Table 12.4, and meat descriptive attributes for whole-muscle meats in Table 12.5. Therefore, descriptive attribute sensory evaluation is very useful in identifying the sensory attributes of a product and then monitoring how each attribute is altered with treatment or time. The descriptive attributes of muscle foods can be defined by a trained sensory panel through (1) ballot development sessions where a trained panel comes to a consensus as to the attributes of a muscle food, (2) established lexicons as seen in Tables 12.2 through 12.5; or (3) free profiling where individually trained panelists develop their own descriptors for a muscle food. As ballot development sessions are commonly used, this process will be described. In ballot development sessions, the trained sensory panelists identify and verbally describe the sensory attributes of the product. It is the responsibility of the panel leader to direct the discussion, identify and summarize the attributes that are provided by each panelist, and to assist in fine-tuning or directing the lexicon development. During ballot development sessions, sensory panelists should be presented with the product of interest as it will appear in the final evaluation. The product that represents the average and extremes of the product should be presented during the sessions. In my laboratory, ballot development sessions follow the outline defined in Table 12.6. The most important component

Table 12.2. Lexicon for Catfish Flavors

Descriptors	Definitions of Descriptors
Aromatics	
Nutty/Pecan	The aromatic associated with fresh pecans and other hard-shell nuts
Boiled chicken/buttery	The aromatic associated with sweet cooked chicken meat; the combination of sweet protein and some oil/fat character
Grainy/Corn/Green Vegetable	The aromatic associated with feed grains and/or corn husks: this note is an "on" characteristic at low levels but may be objectionable at higher levels or when the character is more vegetable and green than just grainlike
MIB/lagoon/bluegreen	The aromatic associated with blue-green algae growth in pond water 2-Methylisoborneol is the reference
Geosmin/musty/woody	The aromatic associated with decaying wet wood. Geosmin is the reference.
Putrid	The aromatic associated with decaying protein.
Rotten plants	The aromatic associated with decaying vegetation
Cardboard	The aromatic associated with slightly oxidized fats and oils and reminiscent of wet cardboard
Painty	The aromatic associated with linseed oil and oil based paints
Tastes	
Sweet	The taste on the tongue associated with sugars
Salty	The taste on the tongue associated with sodium ions
Feeling factors	
Astringent	The chemical feeling factor on the tongue, described as puckering/dry and associated with tannis and alum
Metallic	The chemical feeling factor on the tongue described as flat; associated with iron and copper

Source: Adapted from Johnsen, Civille, and Vercellotti (1987).

of ballot development sessions is that the panel leader adequately directs each session while maintaining an atmosphere of creativity, free discussion, and openness for the panelists. The ballot development session can proceed at the required pace, depending on the expertise and maturity of the panelists. Additionally, the panel leader can introduce a standard lexicon during ballot development for use or guidance. However, every product differs and too great a dependence on a standard lexicon can result in inadequate development of all the descriptors present in a product.

Meat Descriptive Analysis

Meat descriptive analysis has been described and standardized by AMSA (1978). For whole-muscle meat products, the major descriptive attributes are juiciness, muscle fiber tenderness, connective tissue amount, overall tenderness (the combined effect of muscle fiber tenderness and connective tissue amount),

Table 12.3. Lexicon for Poultry Flavors

Descriptors	Definitions of Descriptors
Aromatics	
Chickeny	Aromatic associated with cooked white muscle from chicken
Meaty	Aromatic associated with cooked dark muscle from chicken
Brothy	Aromatic associated with chicken stock
Liver/Organy	Aromatic associated with liver, serum or blood vessels
Browned	Aromatic associated with roasted, grilled or broiled chicken patties (not seared, blackened or burned)
Burned	Aromatic associated with excessive heating or browning (scorched, seared, charred)
Cardboard/Musty	Aromatic associated with cardboard, paper, mold, or mildew; described as nutty, stale
Warmed-over	Aromatic associated with reheated meat; not newly cooked nor rancid/painty
Rancid/Painty	Aromatic associated with oxidized fat and linseed oil
Tastes	
Sweet	Taste on the tongue associated with sucrose and sugar
Bitter	Taste on the tongue associated with bitter agents such as caffeine and quinine
Feeling factors	
Metallic	The chemical feeling factor on the tongue associated with iron and copper

Source: Adapted from Lyon (1987).

and flavor intensity. A panel is trained to discriminate for these attributes using 8-point scales as defined in Table 12.5.

Juiciness is defined as the amount of perceived juice that is released from the product during mastication. Presentation of samples that vary in juiciness such as USDA Prime strip loin steaks cooked to 65°C versus a USDA Standard strip loin steak cooked to 75°C illustrate extremes in juiciness. Muscle fiber tenderness is the ease in which the muscle fiber fragments during mastication. Connective tissue amount is the structural component of the muscle surrounding the muscle fiber that will not break down during mastication. The connective tissue component can be described as an initial popping during the first chews and the bubble-gum-like substance remaining after chewing and immediately prior to swallowing. Differentiation between muscle fiber tenderness and connective tissue amount is challenging. To illustrate the differences in these attributes, presentation of a broiled beef brisket steak and a cold-shortened or thaw-rigor steak simultaneously helps to differentiate these attributes. Overall tenderness is the average of muscle fiber tenderness and connective tissue amount when connective tissue amount is 6 or less. When connective tissue amount is a 7 or 8, overall tenderness is the same as muscle fiber tenderness.

As muscle foods from different species are evaluated using this method, flavor

Table 12.4. Lexicon for Warmed-over Flavors in Beef

Descriptors	Definitions of Descriptors
Aromatics	
Cooked beef lean	The aromatic associated with cooked beef muscle meat
Cooked beef fat	The aromatic associated with cooked beef fat
Browned	The aromatic associated with the outside of grilled or broiled beef
Serum/bloody	The aromatic associated with raw beef lean
Grainy/cowy	The aromatic associated with cow meat and/or beef with grain/feed (character is detectable)
Cardboard	The aromatic associated with slightly stale beef, refrigerated for a few days only and associated with wet cardboard and stale oils and fats
Painty	The aromatic associated with rancid oil and fat
Fishy	The aromatic associated with some rancid fats and oils
Livery/organy	The aromatic associated with beef liver and/or kidney
Tastes	
Sweet	Taste on the tongue associated with sugars
Sour	Taste on the tongue associated with acids
Salty	Taste on the tongue associated with sodium ions
Bitter	Taste on the tongue associated with bitter agents such as caffeine, quinine, etc.

Source: Adapted from Johnsen and Civille (1987).

Table 12.5. Meat Descriptive Attributes for Whole-Muscle Meat Products

Juiciness	**Muscle Fiber and Overall Tenderness**	**Connective Tissue Amount**
8 Extremely Juicy	8 Extremely Tender	8 None
7 Very Juicy	7 Very Tender	7 Practically None
6 Moderately Juicy	6 Moderately Tender	6 Traces
5 Slightly Juicy	5 Slightly Tender	5 Slight
4 Slightly Dry	4 Slightly Tough	4 Moderate
3 Moderately Dry	3 Moderately Tough	3 Slightly Abundant
2 Very Dry	2 Very Tough	2 Moderately Abundant
1 Extremely Dry	1 Extremely Tough	1 Abundant
Flavor Intensity	**Off-Flavor Characteristic**	**Off-Flavor Intensity**
8 Extremely Intense	A Acid	8 Extremely Intense
7 Very Intense	L Liver	7 Very Intense
6 Moderately Intense	M Metallic	6 Moderately Intense
5 Slightly Intense	F Fish-like	5 Slightly Intense
4 Slightly Bland	O Old (Freezer Burned)	4 Slightly Bland
3 Moderately Bland	U Rancid	3 Moderately Bland
2 Very Bland	B Bitter	2 Very Bland
1 Extremely Bland	SO Sour	1 Extremely Bland
	X Other (describe)	

Table 12.6. Sequential Steps During Ballot Development of the Descriptive Flavor Attributes for a Precooked Beef Top Round Roast Containing a New Ingredient

Ballot Session	Session Goal	Exercises Used in the Session
1	To familiarize panelists with the product being used in the study To initiate and stimulate identification of the flavor attributes in the product	1. Presentation of a baseline precooked beef top round product—stimulate discussion on the flavor attributes that can be identified
2	To continue initiation and stimulation of the flavor attributes in the product To introduce one to two products that are variations from the baseline product	1. Presentation of the baseline product. Ask panelists to describe the flavor. Begin to narrow down the descriptors into categories that are similar among panelists and to get concensus on some attributes. 2. Begin using attributes that are being defined to evaluate one or two products that are variations from the baseline. Continue to encourage and provide a free discussion of the attributes of the products.
3	Continue with identification of the flavor attributes in the product as in Session 2. Introduce the baseline product and the new ingredient in its pure form and begin identifying the flavor attributes of the ingredient.	1. Present the baseline product. Ask panelists to describe the flavor. Utilize the descriptors identified in Session 2. Continue to fine-tune these descriptors. 2. Examine one or two additional products that are variations from the baseline. Continue to discuss flavor descriptors, identifying new descriptors or altering existing descriptors. 3. Taste the new ingredient in its pure form. Begin identifying the flavor descriptors for the ingredient.
4	Repeat Session 3 until panelists are familiar and comfortable with the attributes as defined.	Same as Session 3. Examine the new ingredient in the muscle food of interest.
5	Identify references for the flavor descriptors being defined.	1. Present the baseline product. Review the current list of descriptors. 2. Introduce and examine baseline examples of each descriptor and begin anchoring the descriptor for each panelist.
6	Introduce scaling for each descriptor	1. Present the baseline product. Review the current list of descriptors. 2. Utilize references of each descriptor to continue anchoring for panelists. 3. Introduce product with the varying concentrations of the ingredient and begin scaling with the descriptors as defined. 4. Continue identification of alternate descriptors.
7	Development of a final ballot	1. Progressively continue sessions until a final ballot is defined and panelists are familiar with the descriptors and the scale.

intensity should be defined as either beef, lamb, pork, chicken, turkey, etc. The panel should be familiarized with the characteristic species flavor to assure proper use of the scale. For the identification and training of sensory panelists in the detection of off-flavor characteristics, the following compounds can provide character references. Examples of off-flavor character references are concentrated lemon juice for acid; cooked beef liver for liver, cooked beef blood or liver for metallic, cooked perch fillets or fish species having a strong fishlike aromatic, cooked meat stored for greater than 1 year that showed visible dehydration from freezer burn for old; cooked turkey or beef patties stored in a refrigerator for 2–3 days for rancid, caffeine for bitter, and soured milk for sour. After the character reference is presented to the panelists and the panelists can identify the off-flavor, the references then are presented to the panel in different dilutions to assist them in scaling within the attribute. The final step is the incorporation of the character reference in ground beef patties (ground beef patties are used as it is easy to incorporate these substances into ground beef, but other muscle foods can be used) to assist the panel in the identification of the off-flavor characteristic in combination with other muscle food flavors. Quantification of flavor intensity using the Meat Descriptive Attribute method is not as sensitive as using the Spectrum Method for Descriptive Flavor Attributes, the Tragon Quantitative Descriptive Analysis Method, or the Free Profiling Method, but the Meat Descriptive Attribute Method provides an indication of the flavor differences between treatments. If a greater understanding of changes in specific flavor attributes is needed, one of the three aforementioned methods would be the preferred sensory method.

For ground meat products, the aforementioned method of analysis can be used with the addition of ground beef attributes. Examples of other ground beef attributes are initial juiciness, final juiciness, initial beefy flavor, and final beefy flavor. If an ingredient, such as a soy product, is being added to the ground beef, identification of that attribute may or may not be added to the ballot.

For processed meat products, the aforementioned Meat Descriptive Attribute method can be used, but the ballot should be modified to reflect the characteristics of the product being tested. The Spectrum Method, the Quantitative Descriptive Analysis Method, or the Free-Choice Profiling Method are more commonly used for processed meats.

The data are analyzed using ANOVA where treatment effects are tested for each attribute on the ballot. Some multivariate analysis, such as Principle Component Analysis, can be used to determine the combined effect of each attribute on muscle food's palatability.

The Spectrum Method

The Spectrum Method is a descriptive attribute method that uses a 0–15-point universal scale where 0 = none and 15 = extremely intense to evaluate the sensory attributes of muscle foods. The intensity ratings of the universal scale are anchored so that equal distance exists between points of the scale, and intensity for a given point does not change across attributes. For example, a 2 on the universal scale equals the soda flavor in saltless saltine crackers; a 5 equals the apple flavor in Mott's apple sauce; a 7 on the universal scale equals the orange flavor in Minute Maid frozen orange juice; a 10 equals the grape flavor in Welch's grape juice; and a 12 equals the initial cinnamon flavor in Big Red Gum. The specific attributes of a product or the attributes of interest are identified either through ballot development sessions or using a standard lexicon. On the ballot the product attribute of "Other" should be included on the ballot. Panelists should be continually encouraged and given the opportunity to identify attributes that are detected in a product but are not listed on the lexicon. During training and ballot development sessions, products that represent the extremes, the average product, and all possible variations should be presented to assure inclusion of all attributes on the ballot and assure familiarization of all products used in the study by panelists. As a study proceeds, unknown attributes may become apparent. If panelists do not have the opportunity or are not looking for other attributes, important information may not be recorded. Sample ballots used to evaluate the flavor attributes of cooked beef roasts containing sodium lactate and the flavor and texture attributes of low-fat ground patties are presented (Table 12.7 and Table 12.8 respectively). These data are analyzed using ANOVA and common mean separation techniques. Multivariate analysis of variance also can be used to examine the effect of combinations of attributes on treatments.

The Quantitative Descriptive Analysis Method

The Quantitative Descriptive Analysis (QDA) Method was developed by the Tragon Corporation as a method to determine differences in sensory attributes of food products. From a large pool of trained panelists, panelists who can discriminate differences in sensory attributes of a specific product are selected. Panelists are trained as described for descriptive panelists where the product and ingredient references are used to generate descriptive terms. The panel leader acts as a facilitator, not an instructor. In the QDA sensory method, a 15-cm line scale is used to quantitate each sensory attribute. Panelists are free to develop their own approach to scoring; scoring is not anchored as in the Spectrum Method. Panelists independently evaluate each product. Panelists do not discuss data, terminology, or samples after each taste session. The only feedback on performance relative to other panel members or on any differences between samples that a panelist will receive is from the panel leader. Data are entered into a computer, analyzed statistically, and reported in graphic representation in the form of a "spider web" where a branch from the center point in the spider web represents the intensity of that attribute. An example of QDA results is presented

Table 12.7. The Ballot Used to Evaluate the Flavor Attributes of Cooked Beef Roasts Containing Sodium Lactate When Using the Spectrum Method for Quantitating Flavor Attributes

Week ______________ Name ______________________________

Day ______________ Date ______________________________

Sample Number	______	______	______	______
AROMATICS:				
Cooked Beefy/Brothy	______	______	______	______
Cooked Beef Fat	______	______	______	______
Serumy/Bloody	______	______	______	______
Grainy/Cowy	______	______	______	______
Cardboardy	______	______	______	______
Painty	______	______	______	______
Fishy	______	______	______	______
Livery/Organy	______	______	______	______
Soured (Grainy)	______	______	______	______
Medicinal	______	______	______	______
Other	______	______	______	______
FEELING FACTORS:				
Metallic	______	______	______	______
Astringent	______	______	______	______
Throat Irritation	______	______	______	______
Chemical Burn	______	______	______	______
BASIC TASTES:				
Salty	______	______	______	______
Sour	______	______	______	______
Bitter	______	______	______	______
AFTER TASTES:				
Metallic Aftertaste	______	______	______	______
Soapy Aftertaste	______	______	______	______
Other Aftertaste	______	______	______	______

in Figure 12.1. Differences in the spider web between products determine if sensory differences exist.

Free-Choice Profile Method

In the Free-Choice Profiling Method, panelists are given the freedom to develop their own sensory descriptors for a product to use the sensory scale at their preferred range. This method eliminates the concerns of uniformly anchoring panelists to a scale and assuring that all panelists are using the scale in the same manner. Panelist training entails a short session where the sensory procedures are explained. Panelists are served the samples and they describe the sensory attributes. The sensory verdicts are analyzed using the Generalized Procrustes analysis. This method adjusts the data for how the panelists used different parts

Table 12.8. The Ballot Used to Evaluate the Flavor and Texture Attributes of Low-Fat, Cooked Ground Beef Patties Using the Spectrum Method for Quantitating Texture and Flavor Attributes

Week ______________ Name ______________________________

Day ______________ Date ______________________________

Sample Number	______	______	______	______
AROMATICS:				
Cooked Beefy/Brothy	______	______	______	______
Cooked Beef Fat	______	______	______	______
Serumy/Bloody	______	______	______	______
Grainy/Cowy	______	______	______	______
Cardboardy	______	______	______	______
Painty	______	______	______	______
Fishy	______	______	______	______
Livery/Organy	______	______	______	______
Soured (Grainy)	______	______	______	______
Browned	______	______	______	______
Burnt	______	______	______	______
Other	______	______	______	______
FEELING FACTORS:				
Metallic	______	______	______	______
Astringent	______	______	______	______
BASIC TASTES:				
Salty	______	______	______	______
Sour	______	______	______	______
Bitter	______	______	______	______
Sweet	______	______	______	______
TEXTURE:				
Springiness	______	______	______	______
Cohesiveness	______	______	______	______
Juiciness	______	______	______	______
Hardness	______	______	______	______
Denseness	______	______	______	______
Fracturability	______	______	______	______

of the scale and it collates sensory terms that are similar. Sensory terms grouped as being similar are then evaluated by the sensory professional and the meaning or what was being described is determined. The advantages of this method are that sensory panelists have not been biased by training and are still considered "consumers" and the time required for training, panel performance evaluation, and retaining are almost eliminated. However, the disadvantages are that the sensory professional plays a significant role in interpreting the data and biases can be introduced into the results. Free-Choice Profiling is not commonly used in the sensory evaluation of muscle foods, but it is currently being evaluated as a viable sensory tool.

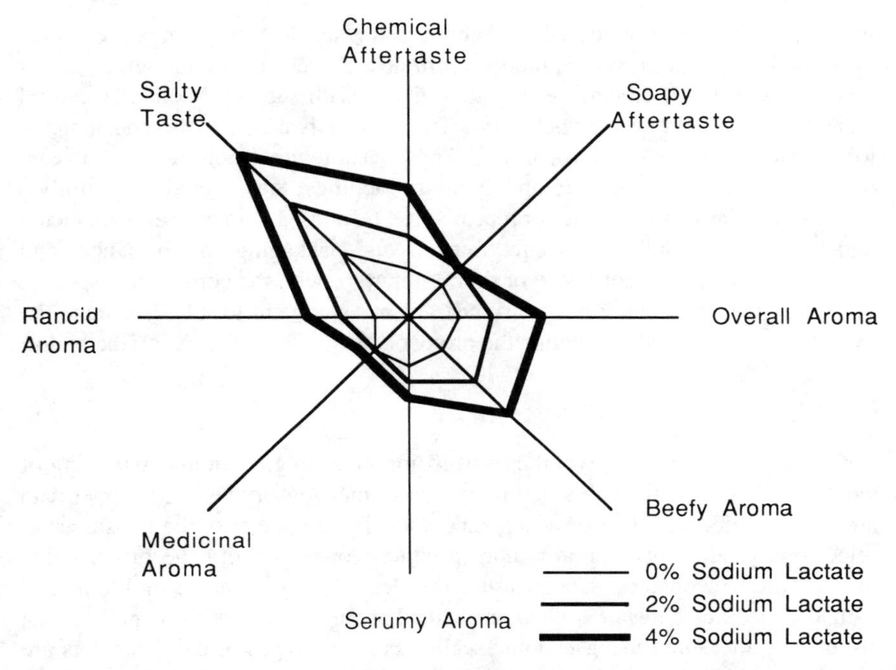

Figure 12.1. An example of a spider web graphic representation of sensory data generated from the Tragon Quantitative Descriptive Analysis Method of sensory evaluation.

Evaluation of Sensory Characteristics Using Consumer Sensory Panels

Major factors to be considered in the determination of consumer perceptions of sensory characteristics of foods: (1) consumer evaluation is not a pure marketing test, it is an evaluation of the sensory properties of the muscle food (taste, smell, sight, feel or sound); (2) do not use trained descriptive attribute terms when conducting consumer sensory tests; consumer terms can be identified in focus groups; (3) strict attention must be given to ballot design and development; it is important that the sensory test examines consumer's sensory perception and not if the consumer can properly interpret your directions or questions; and (4) strict attention must be given to the project protocol that defines product handling, preparation, and presentation: for example, the sensory test should not determine the consumer's perception of differences in cooking method between products unless that is specifically what the test is designed to address.

The first decision for consumer sensory tests is whether the test will be an in-home or central location test. Central location tests utilize a central location where the preparation and presentation of the samples are controlled. However, a central location is an artificial environment and consumers may be uncomfortable and unaccustomed to being seated in booths that could affect their sensory verdict. Central location tests can be conducted where a large room is used, such as a conference room, a community room at a church, or a community center that removes the disadvantage of unfamiliarity with sensory booths. A central location test should be conducted in a facility that is easy for the consumer to locate and that has sufficient parking. These tests require more technicians and professional time, they can tie up laboratory facilities, they exclude the family's opinion, they are less suitable for repeated use responses, and they are conducted under less real conditions where time effects, packaging, performance, and preparation issues are not tested or do not interfere with the consumer response. The next steps in designing a consumer sensory study are to, (1) determine the hypothesis to be tested, (2) define the protocol design, and (3) develop the ballot.

Hypothesis

Before defining a project hypothesis, it is important to gain an understanding of what type of data can be obtained from a consumer sensory test. Consumer data are either preference data or acceptance data. Preference asks the consumer to select one product over one or more products, or to identify the product that they prefer. Acceptance data measure the degree of acceptance or liking and then ask specifics on *what* is liked or disliked about the sample. Acceptance data are usually measured using hedonic scales. For both types of data, subjects are asked to evaluate the product based on their subjective and personal reaction to the product. The hypothesis will determine what types of questions will be used. A clearly defined hypothesis states what is to be tested in the study. Examples of hypotheses are as follows: (a) Consumers cannot detect a difference in flavor of beef or chicken fajitas with or without MSG in the seasoning mix. (b) USDA Choice and Select beef strip loin, top sirloin, and top round steaks do not differ in overall liking, flavor, or tenderness as determined by consumer sensory panelists.

Protocol Design

After the hypothesis has been clarified, the protocol design needs to be defined. The protocol design includes how the samples used in the study will be obtained, handled, packaged, and presented to consumers. An example of a protocol design is presented in Table 12.9. The protocol design for central location tests will vary from the protocol design for in-home tests.

The first item to consider in the protocol design of a central location test is the experimental controls for the subjects during the testing. The location selected to conduct the test should have low noise and limited distraction, be odor-free, comfortable, have correct and/or appropriate lighting, be temperature controlled, and the sample presentation area should be separate from the sample preparation area. The protocol design should define the sample preparation instructions in

Table 12.9. Example of a Protocol Design for a Central Location Consumer Sensory Test

Research Samples

Sample Source (Lot, Production Considerations, Production Day, Raw Materials) ______

Storage Conditions (Packaging, Temperature, Lighting) ______
Special Considerations ______
Sample Preparation
Portion size ______
Added Ingredients ______
Storage and Preparation Temperature of the Meat ______
Preparation Instructions ______
Holding Conditions (Length, Temperature, Holding Containers, Oven Model) ______
Packaging and Containers ______
Special Instructions or Considerations ______
Sample Presentation to Consumers
Total Amount of Product Needed ______
Packaging and Method/Conditions of Transportation ______
Containers and Utensils ______
Coding of Samples ______
Serving Size and Shape ______
Serving Temperature ______
Presentation Procedure ______
Order of Serving ______
Special Instructions ______

Consumers
Age ______
Sex ______
Product Usage (Light, Moderate or Heavy Users of Product) ______
Availability ______
Other ______

Source: Adapted from Civille, Munoz, and Chambers (1989).

great detail. These preparation instructions should include the amount of time required for cooking or preparation, the temperature of appliances used in meat preparation, the final cook temperature of the sample, if applicable, the materials to be used in preparation of the sample and in monitoring temperatures, and the measurements to be taken during cooking, preparation, or serving. The equipment used in the preparation of the sample should be noncontaminated and compatible. Optimally, each treatment in the study should be prepared using separate cooking materials. Roasts from different treatments should be cooked in different pans and ground beef patties should be panfried in separate electric skillets across treatments. Separate utensils should be used to handle meat from different treatments, or the utensils should be washed between uses and the washing procedure for the utensils should be defined. To assure consistent sample presentation to consumers the serving temperature, holding time between cooking and serving, and whether samples will be served monadic or simultaneous should be defined in the sample protocol. The protocol design should define the order that the samples will be served. In central location tests there is a strong first-order bias; therefore, the serving order has to be randomized or blocked. The number of samples that consumers can evaluate in a session prior to the development of taste adaptation needs to be addressed. Samples should be identified with three-digit random codes. The source of the samples should be identified and whether the samples are to be hand carried, shipped by mail, or shipped by truck to the central location. The test parameters need to be defined, such as how many samples will be in the test; how much sample variability is allowed; storage conditions; effects of storage on sample quality or presentation; the time of day that the study will be conducted; and if testing is under normal usage or extended usage conditions.

Factors related to consumer selection should also be included in the protocol design. For example, if consumers are requested not to wear perfume, brush their teeth within 1 h of evaluation, not smoke within 1 h of evaluation, not wear lipstick, etc., then it should be defined in the protocol to assure consistent communication to all consumers. The protocol design should include the detailed orientation and instructions that will be given to consumers. The orientation and instructions should not include information that would bias consumer responses such as extensive details on what the objectives of the study are or what attributes are of major importance. The orientation and instructions should include the introduction of the moderator, approximately how much time will be required to conduct the study, how many samples will be presented, and any specific instructions concerning handling or evaluation of the product.

The protocol design for in-home use tests should include the aforementioned points on detailed orientation and instructions to consumers, the sample presentation protocol, the order that samples are to be consumed in the test, the number of samples to be consumed, the source of the samples, whether the samples are to be hand carried or ship by mail or truck to a central location, and how the samples will be delivered to consumers. In addition, recipe clarity, specific additives to be used, and any equipment for preparation should be provided, when needed. The orientation and instructions used to communicate to consumers by individuals delivering the product should be standardized to assure consistent communication to all consumers. Depending on the design of the study, a last-order bias can be present in in-home use tests. Randomization of the order to which samples will be consumed is needed to remove and/or control this effect.

Ballot Design

Defining the ballot or sensory instrument is critical, as the ballot defines the independent variables that will be measured in the study. For acceptance testing, the ballot is divided into primary and secondary questions. The primary question asks the degree of liking of the product and accounts for all sensory variables that affect acceptance. Secondary questions ask acceptance of specific sensory attributes using hedonic scales, intensity ratings for a specific sensory attribute, just right questions, or attribute diagnostics to understand the reasons for preference or acceptance of a product. Open-ended questions that ask why or what was liked or disliked about a sample provide a mechanism for determining general trends (i.e., too spicy, too tough, off-flavors).

Preference tests also can be used on a consumer sensory ballot. These questions force a choice. Primary questions include the selection of one product over others or the ranking of samples. Secondary questions include open-ended questions that ask why or what was preferred about the sample selected in the primary question. When a consumer is asked to select the product that has the highest level of beefy flavor, this is an example of an attribute preference question.

Terminology development for acceptance and preference questions should not use trained panel attributes, but qualitative tests, such as focus groups or one-on-one consumer interviews, are the best way to define consumer attributes. Consumers may not understand what myofibrillar tenderness or connective tissue amount are, but they may understand what the tenderness of a meat sample is. Attributes other than those of primary interest should be included on the ballot. It is important that consumers do not know exactly what you are testing for, as, in general, people want to answer questions correctly. If they know what you are asking, they may try to give you the answer that they think you want instead of a true evaluation of their preferences or perceptions.

The type of scales used for questions on the ballot is the measurement tool for each attribute. There are many different scale types and styles (Table 12.10). There are basically three different scale styles used in consumer sensory evaluation. Hedonic scales are commonly used, as they are most similar to a consumer's initial response, they are easy to measure, but they are more susceptible to the halo effect, and it is difficult to react to negative responses. Just right scales indicate the direction for changes and are easy to understand and respond to, but data must be treated within cells and they do not indicate actual intensity, only relative intensity differences. Just right scales are bipolar and data cannot be analyzed using ANOVA. Intensity scales can be related directly to input variables and descriptive analysis; but they must be related to liking for direction, they are difficult for some consumers to use, and there is some variation in scale usage across consumers. Three scale types are commonly used in consumer sensory evaluation. Category scales define the points or boxes along a continuum and are verbally anchored either at each point, or are end and center anchored. Advantages of using category scales are that they are commonly used, they are easy to understand, and they are simple in structure. Disadvantages include that they may not have equal intervals, may not have enough points of discrimination, and it can be difficult to anchor all categories verbally. Linear scales are where

Table 12.10. Examples of the Styles and Types Scales Used in Consumer Sensory Evaluation

Scale styles

1) Hedonic scales—measures acceptance

Please indicate by placing a mark in the box your **OVERALL LIKE/DISLIKE** of this sample.

☐ ☐ ☐ ☐ ☐ ☐ ☐ ☐ ☐

EXTREMELY TENDER — NEUTRAL — EXTREMELY TOUGH

Example of a 9-point, end-anchored, hedonic scale for like and dislike with a neutral point.

2) Just right scales—have a negative and positive pole from a central just-right point.

Please indicate your evaluation of the salty flavor in the product presented.

☐ ☐ ☐ ☐ ☐ ☐ ☐

TOO LOW — JUST RIGHT — TOO HIGH

Example of a 7-point Just-Right Scale for salty flavor.

3) Intensity scales—measures the level or strength of an attribute.

Please rate how strong or intense the salty flavor is in the product presented.

☐ ☐ ☐ ☐ ☐ ☐ ☐ ☐ ☐

NONE — EXTREMELY INTENSE OR STRONG

Example of an 9-point, end-anchored intensity scale for salty flavor.

Scale types

1) Category scales

☐ ☐ ☐ ☐ ☐ ☐ ☐ ☐ ☐

EXTREMELY TENDER — NEUTRAL — EXTREMELY TOUGH

Example of an 11-point, end and central anchored category scale for tenderness.

2) Linear scales

|——————————————|

EXTREMELY TENDER — EXTREMELY TOUGH

Example of a 15 cm end-anchored linear scale for tenderness.

3) Magnitude Estimation scales.

The first sample that you evaluated had a tenderness rating of 50. Rate the second sample for tenderness in proportion to 50.

First sample ___50___
Second sample ______

Source: Adapted from Meilgaard, Civille, and Carr (1991).

a line is used and the ends are anchored to provide direction, and, depending on application, the middle point may be anchored. Advantages of linear scales are that they allow for several points of discrimination, they are simple in structure, and they are easy to understand and use. Disadvantages are that they may be used quite differently across subjects and they are difficult to tabulate. Magnitude Estimation scales are the third type of scales used. Magnitude Estimation scales are where a control is provided that is identified as "O" or a defined intensity for an attribute. The consumer is asked to evaluate the product and estimate the difference from the control and the sample. Magnitude estimation scales are not normally used in consumer sensory evaluation. Advantages of magnitude estimation scales are that they allow for several points of discrimination and may be more quantitatively related to test variables. Disadvantages are that they are more difficult for consumers to learn to use, and it can be difficult to normalize data.

After the questions and the scales have been defined, the questionnaire design should be addressed. The ballot should have a simple structure, have a defined, logical order, and clarity of instructions and terminology should be considered. The ballot should use a uniform style of terms and scales; the terms should be related to sensory properties. The sequential format should be as follows: (1) instructions; (2) primary questions; (3) secondary questions; (4) questions should follow in sequence so that the attribute will be evaluated by hedonics and then diagnostic questions; (5) open-ended questions; and (6) other questions such as preference questions.

The general instructions should include: (1) follow the instructions of the test monitor; (2) do not read ahead; (3) do not communicate with others in room; and (4) ask the test monitor quietly if you have a question. Specific instructions should include: (1) the overall test objective; (2) that you are requesting their honest opinion and their cooperation; (3) instruction on how to respond to the words; (4) how to use the scales; and (5) how and when to handle, use, apply, or taste the product. An example of a consumer ballot is presented in Table 12.11.

Selection of Consumers

The recruitment of sensory panelists is very important in a consumer sensory test. The age, geographic location (urban to rural, east/west; north/south), usage rate (whether they are light, medium, heavy users of the product), inclusion of individuals or families in the test, any specific usage types or styles, ethnic background, amount of exposure to like products, income level (usually not a big tissue), and any special problems such as allergies, dentures, or medications that could interfere in the evaluation should be included in the consumer screening questionnaire. Consumers can be recruited through phone interviews, intercepted in a central location, or group pools such as church groups or social clubs in a

Table 12.11. Example of a Consumer Sensory Ballot Used to Evaluated Cooked Beef and That Indicates the Use of Primary and Secondary Questions in italics

Primary Question—9-point, end-anchored, hedonic scale.

1. Indicate by placing a mark in the box your **OVERALL LIKE/DISLIKE** of this sample.

☐ ☐ ☐ ☐ ☐ ☐ ☐ ☐ ☐

LIKE DISLIKE

Secondary Question—9-point, end-anchored, hedonic scale.

2. Indicate by placing a mark in the box your **OVERALL LIKE/DISLIKE** for the **FLAVOR** of this sample.

☐ ☐ ☐ ☐ ☐ ☐ ☐ ☐ ☐

LIKE DISLIKE

Secondary Questions—9-point, end-anchored, intensity scales.

3. Indicate by placing a mark in the box how you feel about the **INTENSITY OF THE FLAVOR** of this sample.

☐ ☐ ☐ ☐ ☐ ☐ ☐ ☐ ☐

NONE EXTREMELY INTENSE

4. Indicate by placing a mark in the box how you rate the **INTENSITY OF THE SALTY FLAVOR** of this sample.

☐ ☐ ☐ ☐ ☐ ☐ ☐ ☐ ☐

NONE EXTREMELY INTENSE

Secondary Questions—Open-ended.

5. What did you **LIKE** about the **FLAVOR** of this sample?

____

____

6. What did you **DISLIKE** about the **FLAVOR** of this sample?

____

____

Secondary Question—9-point, end-anchored, hedonic scale.

7. Indicate by placing a mark in the box your **OVERALL LIKE/DISLIKE** for the **TEXTURE** of this sample.

☐ ☐ ☐ ☐ ☐ ☐ ☐ ☐ ☐

LIKE DISLIKE

Secondary Question—9-point, end-anchored, intensity scale.

8. Indicate by placing a mark in the box how you feel about the **LEVEL OF THE TENDERNESS** of this sample.

☐ ☐ ☐ ☐ ☐ ☐ ☐ ☐ ☐

TENDER TOUGH

Secondary Questions—Open-ended.

9. What did you **LIKE** about the **TEXTURE** of this sample?

____

____

10. What did you **DISLIKE** about the **TEXTURE** of this sample?

____

____

community can be used. A minimum of 50–100 panelists should be used in conducting consumer sensory tests, as data from consumer tests are more variable and a larger number of respondents are needed to test if differences exist in products being tested. If the product to be tested is targeted toward a specific population, such as heavy meat eaters, then a subset of that population should be selected. If strong regional differences in preference for the product are known, then consumers that represent the regional differences in preference for the product are known, then consumers that represent the regional effects should be used in the study. For example, in the National Consumer Retail Beef Study it was determined that consumers on the west coast were more concerned about the amount of fat contained in beef and were willing to sacrifice palatability (USDA Choice) for meat containing less fat (USDA Select); whereas consumers on the east coast in Philadelphia were more concerned with the palatability of the meat that they consumed (USDA Choice) and that they were willing to purchase meat with a higher fat content to assure palatability. Therefore, if you were conducting a consumer study on the acceptability and palatability of low-fat meat products, consumers from both regions should be included in the test. Consumer age should be considered, depending on the product to be tested. If meat formulas for products used in the school lunch program were being tested, children who consume the product should be the consumers in the test. It is usually not recommended to use employees or local residents for all affective testing, as these populations may not represent the consumer for which the product is targeted. However, local residents can provide a means of defining general trends and narrowing the number of variables for a study that would include regional effects and larger numbers of consumers.

Selected References

AMSA. 1978. *Guidelines of Cookery and Sensory Evaluation of Meat*. American Meat Science Association, National Live Stock and Meat Board, Chicago, IL.

ASTM. 1978. Compilation of odor and taste threshold values data. American Society for Testing and Materials, Philadelphia.

Cain, W.S. 1979. To know with the nose: Keys to odor identification. *Science* 203:467.

Civille, G.V. 1991. Food quality: Consumer acceptance and sensory attributes. *J. Food Qual*. 14:1.

Civille, G.V., A.M. Munoz, and E. Chambers, IV. 1989. *Consumer Testing: A Four Day Intensive Workshop*. Sensory Spectrum, East Hanover, NJ.

Cross, H.R., R. Moen, and M.S. Stanfield. 1978. Training and testing of judges for sensory analysis of meat quality. *Food Technol*. 32:48.

Desor, J.A. and G.K. Beauchamp. 1974. The human capacity to transmit olfactory information. *Percep. Psychophysiol*. 16:551.

Froning G.W., A.J. Maurer, K.K. Hale, and A.F. Carlin. 1978. Sensory properties of poultry meat. The Agriculture Experimental Station, University of Nebraska, Lincoln. North Central Regional Research Publ. No. 254.

Gower, J.C. 1975. Generalised Procrustes Analysis. *Psychometrika* 40:33.

Johnsen, P.B. and G.V. Civille. 1986. A standardized lexicon of meat WOF descriptors. *J. Sensory Studies* 1:99.

Johnsen, P., G.V. Civille, and J.R. Vercellotti. 1987. A lexicon of pond-raised catfish flavor descriptors. *J. Sensory Studies* 2:85.

Lawless, Harry. 1991. The sense of smell in food quality and sensory evaluation. *J. Food Qual*. 14:33.

Lyon, B.G. 1987. Development of chicken flavor descriptive attribute terms aided by multivariate statistical procedures. *J. Sensory Studies* 2:55.

Meilgaard, M., G.V. Civille, and B.T. Carr. 1991. *Sensory Evaluation Techniques*, Vol 3. CRC Press Inc., Boca Raton, FL.

Munoz, A.M., G.V. Civille, and B.T. Carr. 1992. *Sensory Evaluation in Quality Control*. Van Nostrand Reinhold, New York.

Stone, H. and J.L. Sidel. 1993. *Sensory Evaluation Practices*, 2nd ed. Academic Press, New York.

Wasserman, A.E. and F. Tally. 1968. Organoleptic identification of roasted beef, veal, lamb and pork as affected by fat. *J. Food Sci*. 33:219.

Weller, M., M.W. Galgam, and M. Jacobson. 1962. Flavor and tenderness of lamb as influenced by agriculture. *J. Anim. Sci*. 21:927.

Williams, A.A. and G.M. Arnold. 1984. A new approach to sensory analysis of foods and beverages. In: *Progress in Flavour Research, Proc. 4th Weurman Flavour Research Symp*. Elsevier, Amsterdam.

13

Aspects of Quality Assurance and Ritualistic Practices

Sam Angel

Introduction

Quality control is limited to the product only and detects faults after the handling or processing have taken place. Then action is taken to keep the percentage of rejects entering the channels of trade within certain limits. Quality control is, therefore, corrective in nature. Quality assurance, with which this chapter deals, is unlimited in scope. It covers all stages of production, including management. Quality assurance covers careful selection of the raw product; careful handling of the raw product during raising, handling, and transport; in plant or slaughterhouse control; control of storage conditions; monitoring shelf-life; and considering and evaluating consumer reaction to the product. It is preventive in nature.

Red Meats

Beef

Meat quality is a combination of attributes that makes it acceptable to consumers. Raw red meat quality is judged by its color, fat content and distribution, smell, and by the presence or absence of slime or molds. Shelf-life is limited by the development of spoilage bacteria which can cause slime, discoloration, and putrid odors. Levels of 10^6 colony forming units (CFUs) of bacteria per square centimeter of carcass surface can cause off-odors, and 10^8 cm can cause the formation of slime and discoloration on the surface. The pH of living beef muscle is around 7, and after slaughter, as rigor develops, it drops to 5.6–5.4. However if animals are exhausted before slaughter, the glycogen content of the muscle is depleted and the ultimate pH (pH_u) of the meat remains high. The color of the meat is darker than meat with a lower pH_u. Such meat is referred to as dark cutting beef or DFD which stands for dark–firm–dry meat. However at a higher pH_u, the meat is more tender and juicier. Spoilage bacteria develop to a greater extent on meat with pH above 5.8. *Pseudomonas* spp. which causes off-odors and greenish discoloration can develop on meat at around pH 6.

In order to avoid bacterial contamination, certain basic practices must be adopted by everyone handling fresh red meat. These are discussed in Chapters 9 and 14. The grading system assures that meat conforms to form and minimum quality standards, whereas the USDA's Food Safety and Inspection Service (FSIS) assures consumers that the meat is wholesome from the biological and chemical pollutants standpoint. The U.S. federal grading system for beef and poultry have been fully discussed in Chapter 7.

There are many variations on cutting up and portioning beef carcasses. In the United States today, over 80% of the beef slaughtered is sold as boxed beef. Following rigor, the carcasses are deboned and cut into primal or subprimal parts. If stored at close to 0°C (32°F), the vacuum-packed beef can keep from 1 to 2 months in very good condition.

During the long storage interval, the meat undergoes aging and becomes more tender. Retailers who receive the boxed vacuum-packed meat remove the plastic when they are ready to sell the meat. The meat acquires a fresh red color after removal from the vacuum bags. It is then cut up into retail cuts, repacked into Styrofoam trays with oxygen permeable plastic film such as polyvinyl chloride, and placed in the display cabinets. The meat has to be sold within about 3 days or it soon starts to turn brown on the surface due to the oxidation of the red oxymyoglobin pigment in the meat converting it to metmyoglobin.

For the restaurants, the cook can use the vacuum-packaged portions directly but recently, packers have also begun central prepacking of retail cuts of boxed beef for the supermarkets. The meat is packed in retail trays with oxygen-permeable film for quick sale. The central packaging of such cuts is more cost-efficient. The meat also does not have to be recut and repacked in the supermarket. The decreased handling of the meat results in less chance for cross-contamination. However, the higher oxygen content means a greater chance of rapid bacterial proliferation on the surface, and special care must be taken to maintain as low a chill temperature as possible in the display cabinet.

Boxed beef is being prepared in other countries besides the United States. In New Zealand, claims have been made that placing meat after slaughter in a chiller and bringing the temperature down to 0°C (32°F) within 24 h results in cold-shortening of the muscle and adversely affects tenderness. The cold-shortening would not occur if the meat were allowed to go through rigor before exposing to chiller temperatures close to freezing. Beef is, therefore, conditioned at 10°C (50°F) for between 18 and 24 h, then it is cut up into primal or retail cuts in a room at 10°C (50°F) and vacuum-packed. The vacuum bags of meat are then placed in cartons at the same temperature for a 60–72-h aging period and then frozen. It has been known for many years that electric stimulation of beef muscle after slaughter can bring on rigor within 1 h. The rigor is resolved shortly afterward. Electric stimulation is, therefore, also being used. The electrically stimulated muscle is placed into the chiller at 1°C (33.8°F) shortly after slaughter without it experiencing muscle fiber cold shortening. Due to the lower processing temperatures such meat is aged with a higher degree of quality assurance.

During the conditioning in New Zealand the air in the chiller is monitored not to contain more than 10 microorganisms per square foot of space. Before conditioning, 80% of the aerobic plate counts must be below 10^4 cm of surface and not above 10^5 cm of surface after conditioning. After aging, none of the meat can have a plate count above 10^6 cm of surface. Tests have shown that such beef is very tender when cooked and still hygienic.

Pork

There are two types of livestock, especially in swine: stress susceptible and stress resistant. Hormones from the adrenal gland impart physiological resistance to stress-resistant pigs. In stress-susceptible pigs, before death, the partial pressure of oxygen in the blood decreases and the heart rate increases rapidly. Following bleeding of the animal, the metabolism of the muscle quickly becomes anaerobic. Under such conditions and while body temperature is still high, rapid glycolysis can take place in stress-susceptible swine, leading to acidic muscle.

When there is suddenly no oxygen available following exsanguination, Ca^{2+} activated ATPase activity of myosin increases and Ca^{2+} uptake by the sarcoplasmic reticulum of the muscle myofibrils decreases. ATP is then depleted. This leads to a very rapid glycolysis. Lactic acid is produced, the muscle cells become acidic, and the pH falls rapidly.

As the ATP is depleted, the myosin heads become irreversibly attached to the actin filaments where the filaments overlap and the sliding movement between the filaments during muscle contraction and expansion ceases. The muscle then stiffens and enters into rigor. In the process the filaments are brought closer together laterally and the myofibrils shrink. The lateral shrinkage causes water to be squeezed out when the meat is cut in cross section. In Pale Soft Exudative (PSE) meat the shrinkage is twice as great as in normal meat and much more water is released when the meat is cut. The reason for the increased shrinkage is that in PSE meat the myosin molecule's globular heads bring the filaments closer together by 2 nm due to the heat denaturation of the proteins. Thus, a minuscule change in the myofilaments of the muscle bring about a major change in the muscle, resulting in PSE meat.

Besides PSE in pigs there exists a closely related phenomenon known as porcine stress syndrome (PSS). This occurs in animals that are very stress susceptible, and only a small number of such pigs survive stressful handling. The explanation of the phenomenon is that hormones from the adrenal medulla and the adrenal gland mediate the breakdown of glycogen and glucose, which is transported to the muscle from the liver, to lactic acid. This is followed by a rapid fall in pH, especially at elevated body and ambient temperatures. If the acidity is not controlled, it can lead to death due to generalized acidosis of the tissues.

In order to reduce stress, high temperatures should be avoided before slaughter as well as other stressor factors such as high humidity, fasting, shock, apprehension, fatigue, anoxia, and excitement. Tests are available to discern PSE and PSS in live pigs and these are being used to prevent breeding of such animals.

Pigs can also develop DFD meat due to fasting or fatigue before slaughter. DFD is associated with a relatively high pH_u of around 7. This encourages growth of spoilage bacteria and reduces shelf-life. It can cause undesirable traits such as souring in hams and a sticky consistency in bacon known as "glazing." After fatigue, the animals can be rested for a day or two before slaughter in order to build up glycogen reserves in the muscle. Sugar feeding of pigs before slaughter is also practiced to replace glycogen in the muscles and prevent pH_u. A high pH_u in pigs can have its advantages. Bacon, as a result of a high pH_u, has decreased shrink on curing and increased tenderness and juiciness when cooked.

As soon as the animal is slaughtered and bled, it is washed with cold water and immersed in approximately 60°C (140°F) water for 4–6 min, then dehaired in dehairing machines where metal or rubber scrapers remove most of the hair. The pigs should be completely bled before immersion. The scalding tank could be a source of bacterial contamination if the water is not changed frequently enough. Singeing is used to remove remaining hairs and also kills some bacteria that might adhere to the skin and hairs. Nevertheless, the singed carcass must be thoroughly cleaned before evisceration begins. All precautions during evisceration are similar to that for cattle except for certain technical differences due to the size and anatomy of the animals. Unlike cattle carcasses, the head may be left on until after chilling. In pigs as well as in cattle, following evisceration the carcasses should be washed thoroughly, especially internally. Enough water should be used to ensure a good wash, but residual water may increase the level of bacterial growth.

The storage life of pork is between 1 and 2 weeks at −1°C (30.2°F) as opposed to beef which has a storage life of over 3 weeks. However, pork is almost always cut up 24 h postmortem and much is cured.

Poultry

An important influence on the fattiness of poultry and amount of abdominal fat is the protein/energy ratio in the diet. Organoleptic tests carried out in Israel with White Rock male broilers have shown eating quality of medium-fat broiler meat is preferred over fatty or lean birds. In turkeys, vitamin E in the diet was shown to improve sensory quality.

Temperatures above 21°C (70°F) or below 15.5°C (60°F) and insufficient ventilation can cause stress to poultry in poultry houses. Stress in poultry can increase proneness to disease, slaughter bleeding time, defeathering, and affect cooked tenderness adversely. Reductions in yield and grade can result from blisters and bruises during catching and transport to the dressing plant. Breast blisters develop more in birds raised in cages than on the floor in deep litter. Bruises, especially on the breast, occur when the birds are caught and placed in large wooden or plastic cages for transport. Rough handling also results in broken bones which also cause bruised areas. Transportation to the dressing plant in the cages should be as short as possible and the birds should be kept close together inside the cages during transport. This results in less bouncing around and reduces stress. As time of transport increases, percent weight loss increases, so distances to dressing plants should not exceed 50 miles if possible. The birds should not be fed 3–4 h before slaughter. This increases glycogen content in the muscles and has a favorable effect on tenderness. A longer time without food results in stress as measured by increased glycolysis.

When the poultry reaches the plant the cages are cooled by area circulating fans pending the removal of the birds and hanging them on shackles, by the legs. Care should be taken not to injure the birds. The overhead conveying shackles move them through electric water stunners and then through rotating knives for slaughter. Following bleeding, the shackled carcasses are submerged in water at around 60°C (140°F) for scalding to aid in feather removal. To obtain a yellow bird, 52.2°C (126°F) is used. According to the Official Grades and Standards of the USDA for ready-to-cook poultry, the skin must be free of any protruding pinfeathers before a quality designation can be assigned. In the lower grades of poultry, a few scattered hairs or vestigial feathers are tolerated. These strict specifications for defeathering exist because feathers and feather follicles carry filth and are a source of bacterial contamination.

Following defeathering, the birds are eviscerated. Between 3% and 6% water can be taken up during scalding, defeathering, and washing during evisceration. Another 12% moisture can be absorbed by the carcasses during chilling and ice packing. However, 66% of the absorbed water is given up as drip and USDA regulations limit the weight increases of turkeys and chickens due to water uptake to between 4.5% and 8% for non-ice-packed and to a maximum of 12% for ice-packed poultry depending on weight of the birds. EEC and other European countries have similar weight increase regulations for poultry. Weight increases following chilling can be determined by weighing the carcasses before and after the spin chiller and determining percent drip. Excess water in the flesh can be determined by the ratio of water to protein in the flesh which in poultry varies between 3.5 to 3.75 depending on species or strain.

Chickens are usually cut up and the parts sold chilled, not frozen in the United States and other countries. Turkeys are usually sold frozen.

Seafood (Fish and Shellfish)

Stress affects the quality of the meat as it does in mammals. If fish go into rigor at a relatively high ambient temperature, such as 17°C (62.6°F) for cod, or if they are subject to rough handling or very rough seas in stormy weather, strong rigor tension within the myofibrils can weaken the connective tissues and rupture fillet muscle. Pre-rigor fish fillets tend to shrink, and if pre-rigor fish are frozen then thawed, the fillets may have mushy texture and excessive fluid losses. Filleting should be carried out after resolution of rigor by storing in coolers until rigor is resolved. If fish in full rigor are machine filleted, their texture will not be adversely affected although there will be losses in yield. Generally, the pH reduction due to lactic acid production in fish is not as low as in mammals. In cod, the decrease is from pH 7 to 6.3–6.9. In mackerel it can reach 5.8, but tuna and halibut are exceptions. There the pH_u can be 5.4–5.6. The pH_u can take 24 h in trawled fish. At pH's below 6, the muscle proteins approach their isoelectric points. Water-holding capacity decreases and frozen or cooked drip increases. The meat can become tough when cooked. In halibut, the meat resembles PSE pork. However, in halibut this condition is reached at a higher pH_u than in pork. In order to avoid or reduce this condition in halibut the fish should be kept alive for 10 h after landing. There is then only a small decrease in pH after death. This may be due to removal of lactic acid formed during the stress of catching, to the circulatory system, and to the resynthesis of glycogen in the liver.

Some fish such as herring and mackerel are not gutted after catching. If the fish were caught while feeding, proteolytic enzyme activity in the gut area can disintegrate the flesh within hours after catching and the abdomen will burst.

In fish and seafoods, with the onset of rigor, ATP is degraded to hypoxanthine (Hx) and ribose. Hx, therefore, follows autolysis in fish, and its measurement serves to indicate the degree of freshness of the fish after death. Hx production varies widely with species in the amounts and times it takes to reach maximum production. Therefore, Hx values must take into consideration the species of fish being tested.

All fish, whether caught at sea, in coastal waters, in rivers, or through aquaculture, carry bacterial loads on their skin, in their gills, and in their stomachs and intestines. Previous chapters have discussed bacteriological and toxicological aspects of fish and seafood safety and spoilage.

HACCP is a tool for quality assurance. The USDA, FSIS, and the European Common Market countries as well as other European governments have adopted or are in the process of adopting a tool called Hazard Analysis Critical Control Point (HACCP) to achieve minimal microbiological infestation and chemical contamination in meats. It is a preventive system of control. For the truth is that fresh meat is naturally practically sterile and can also be free of harmful chemicals if the necessary steps are taken to avoid contamination by contact or ingestion.

Effective control can be obtained by several "critical control points." These are places on the production line or in distribution points that are critical to maintain quality and compliance with laws and regulations. As amounts of automatic equipment are introduced into production, processing, and distribution, the degree of risk due to breakdowns or loss of control that could affect food safety increases. Therefore, the number of "critical control points" has to be increased accordingly. Three steps are necessary to achieve adequate control: identification of hazards, elimination of those hazards that are totally correctable, and control of the hazardous points that remain. The adoption of such a plan by a meat-packing plant indicates a commitment to arrive at the smallest microbial loads and chemical pollutants as is possible. This requires the full cooperation of each and every individual in the plant from top management down. It means setting up control points with written protocols all along the line, from the source of the slaughter animal and instruments and materials entering the plant up to the point of sale. Everyone along those points must be made aware of the importance of each link to the whole of the chain and know what is required and how to act in case some sanitary or other mishap occurs. Once a HACCP program is in operation, spot checks are necessary to verify that no mistakes have occurred due to automatic or monotonous repetition.

In order for the program to succeed it must be cost-effective. Every control point costs money. It is possible to set up computers at control points so that all costs can be calculated and related to management constantly. For the program to work and for management to be enthusiastic about it every animal must be accounted for as well as every part used or discarded. The information must be fed into a computer and use must be made of appropriate programs to analyze results and calculate costs of each step taken to control quality and their contribution to the profits and losses of the company.

There are new and rapid detection and testing methods for various pathogens. As the methods change, new sampling techniques, new testing procedures, or entirely new tests may be required at critical points in order to stay in control. A National Academy of Sciences (NAS) study in 1986 suggested that antemortem inspection should perhaps be varied in intensity, increasing the numbers of inspections in plants where problem types of cattle occur and where rejection is more frequent. Computers should be used to feed back information to "control points" personnel in order to expedite corrections.

In January, 1988 a new IMPS document, "Interim Quality Assurance Provisions," began to appear. This document and its future updates should be of great help to the meat industry as it will attempt to standardize quality control procedures for meat-trimming practices.

In order to increase sales, many meat packers, poultry plant operators, and fish and seafood plants are looking to foreign markets for sales. They must, therefore, be aware not only of U.S. regulations but of foreign regulations such as those of the European Common Market countries.

Since the 1960s there have been in existence international food standards. The Joint FAO/WHO Codex Alimentarius Commission comprising member nations and associate members was established by the United Nations to implement a Joint FAO Food Standards Program. Its purpose is to protect the health of consumers and to guide the preparation and promote the coordination of all food standards. The codes of practice and guidelines are of an advisory nature. The Commission has recommended international codes for fresh and processed fish, seafoods, red meats, and poultry. Several basic codes of practice pertaining to fresh and processed muscle foods are Codex International Code of Hygiene Practices for Processed Meat Products and Principles of Food Hygiene for fish and fishery products, frozen fish, canned fish, and for shrimp. An abridged version of Codex Alimentarius has recently appeared. Management should purchase these Codes of Practice and keep abreast of revisions which are made every few years.

The International Standards Organization (ISO) located in Switzerland which comprises a body of 100 countries, published the ISO 9000 series of food standards in 1985. These have been internationally acknowledged and adopted by over 50 countries. In the United States the standards are published in the ANSI/ASQC Q 90 Series European Norm (EN)2900 Series. There are five separate standards in the 9000 series, from quality system requirements for product development to final inspection and testing standards. Management should have a copy of these international standards. They can be of enormous help in establishing a HACCP system of quality assurance in plants which decide to adopt HACCP. Management and personnel in charge of the HACCP program must be aware of changes in ISO recommended standards or the common market standards and must be prepared to quickly take any necessary action to comply.

For the present, there is still a lot of grading done visually, but in order to be in total control and to be able to react and change quickly, in the future we will see more and more electronic probing with feedback systems for making corrections. If the devices can be produced at a reasonable cost, programs such as HACCP will be easier to maintain and, in the end, will be the only means to stay competitive.

Ritualistic Processes

Kosher and Halal Slaughter

Kosher or halal rules of slaughter exist for red meats or poultry. They must be slaughtered according to prescribed Jewish or Moslem religious laws. Jews and Moslems are not allowed to eat pork or meat from animals or birds of prey or meat from animals not slaughtered by a Jewish "shochet" or a Moslem male. Both cultures exclude animals that have died a natural death or were killed in an accident. The concept involved in the latter statement is health and cleanliness. Animals that die a natural death could be diseased or poisoned, and animals that have died accidentally could have shattered bones and blood clots which are forbidden to eat. Jewish dietary laws also prohibit the eating of animals that do not have split hooves or that do not chew the cud. Animals that do chew the cud and do have split hooves must not have broken or crushed limbs. (Talmud, Hul. 3:1) All such animals that were mentioned that cannot be eaten are considered tref according to Jewish dietary laws (the word originates from the Hebrew word "toref" meaning to tear, referring to animals or birds of prey), but in common language it has come to mean forbidden. Because animals must be slaughtered ritualistically according to Moslem and Jewish law they cannot be killed as by hunting with a gun or other means.

There are many anatomic abnormalities listed that prevent a slaughtered animal from receiving the kosher approval stamp following Jewish ritualistic slaughter. Carcasses of cattle or birds found to have diseased or torn lungs or lungs that adhere abnormally to the chest cavity are declared to be tref.

One form of trefa (Hebrew adverb for tref) is the nonporged hindquarter of beef or ovines. Porging means the complete removal of the sciatic nerve (*nervus ischiadicus*) from the thigh. In Hebrew the nerve is called Gid Hanasheh or the nasheh sinew, which may be the reason some people refer to porging as desinewing or deveining. The Old Testament forbids eating nonporged hindquarter meat because of the blow inflicted by the angel upon Jacob. Genesis 32:32 states: "Therefore the children of Israel eat not the gid hanasheh which is upon the halow of the thigh unto this day." Porging also requires the removal of all veins and arteries from the thighs, the cutting of the veins and arteries from the neck, and the opening and draining of the blood from the heart, so as to remove as much blood as possible. In birds, the veins and arteries must also be cut as much as possible and the heart drained of blood. Porging also requires the removal of all fat around the intestines and stomach areas. This fat is known as "helev" or sacrificial fat in Hebrew and was the fat used to perform sacrifices on the altar. It must be remembered that blood was known to carry and transmit disease and that there was no refrigeration in ancient times. Miamonides (Guide, 3:8) explained the injunction against eating "helev" in the following quotation, "The fat of the intestines is forbidden because it fattens and destroys the abdomen and creates cold and clammy blood." It should be stated here that the described laws presented in this chapter regarding the preparation of meat for eating are only part of a larger set of dietary laws regarding many foods such as milk, fish, and invertebrate foods. These laws have been regarded as so hallowed that oppressors from the ancient Syrians to modern Nazi Germany have sought to force Jews to renounce them. Many chose torture and death rather than do so. Today, in most Jewish communities porging is too costly an operation and Jews forgo eating the hindquarter meat with its many tender cuts. In Israel, very little commercially porged hindquarter meat is available. On holidays some porged lamb meat is sold. A number of religious families may make special orders of porged meat. The vast majority has to forgo this expensive treatment. Due to the high price of beef in Israel, porging the hindquarter should not add more than a few percent to the price. Porging is being considered as a commercial practice in the near future.

Jewish slaughter is known as "shechita" and the male person trained by the rabbinical authorities to carry it out is known as a "shochet." In the Moslem practice any Moslem male may carry it out if he invokes the name of Allah as he is slaughtering the animal. Kosher or halal slaughtered beef must follow the same sanitary practices as any other meat and be inspected like any other meat in the plant. Both the "shechita" and the Moslem halal slaughter must be carried out in a separate part of the plant or in a plant where no instruments or utensils have been used for hog slaughter, processing, or storage. Animals for "shechita" must be fully conscious, whereas halal in some practices today allows mild stunning. Followers of "shechita" claim the animal suffers no more in the Jewish practice than when it is stunned prior to slaughter.

The "shechita" laws were laid down thousands of years ago in the Talmud by rabbis (Talmud, Hul. 3:1; Hul. 11 a–b; Hul. 43 a) who interpreted God's laws handed to Moses on Mt. Sinai which forbade cruelty to animals. The "shechita" laws requires that the slaughter be carried out by a straight swift and steady cut across the soft flesh from front toward the back of the neck using a sharp long knife but applying no pressure so that the animal will not feel the cut. For that reason, the cut must not be performed on the standing animal with an unavoidable upward stroke but when the animal is on its back. Halel slaughter does not necessarily have this requirement, but when possible, it follows the Jewish practice in the slaughter house.

The placing of a large animal on its back for "shechita" is called "casting." The reason for the casting was to perform the "shechita" after the animal had become quiet and the front of the neck pointing upward. Years ago elaborate tying methods were invented to cause the animal to fall on its back more or less voluntarily and then the skilled "shochet" would perform the cut in the blink of an eye.

Today, in the western world there exist several types of "casting cages" that allow the turning over or casting the large animal so that "shechita" can be performed with minimum of effort or stress on the animal and under sanitary indoor conditions.

In the Weinberg Casting Pen, used in Europe and one slaughterhouse in Israel, the animal is allowed to walk into the pen and then the rounded walls and floor slowly close in and adjust to the animal's size, holding it securely. The head and the entire neck of the animal are left protruding from the front of the pen. The pen is then rotated 180°, and as soon it comes to rest, the head is secured by a collar over the upturned jaw and the "shochet" while saying an appropriate prayer performs the straight cut on the front of the neck using no pressure. The entire act from the moment the animal enters the rotating pen until "shechita" takes 30 s.

In the United States, an ASPCA slaughtering pen for cattle has been developed in the belief that casting frightens the animal. In this pen, a belly pad lifts the animal off the floor in a standing position and pressure is applied to its back. The head is lifted up and stretched by a second person and the animal is slaughtered from the side. This method may or may not be considered to satisfy the traditional laws of "shechita" and so far is only in use in the United States.

The knife used in large animal slaughter is described in great detail in the laws of "shechita." It is long, straight, and kept razor sharp and has a narrow rectangular shape with no curvatures. The weight is adequate so that no pressure need be exerted in pulling it through the soft fleshy parts of the neck. In order to ensure maximum exsanguination, the trachea and the esophagus as well as the carotid arteries, jugular veins, and the sympathetic and vagus nerves are cut in one or two swift passes. The wound then splits open and the stream of blood flows undisturbed.

As soon as the main stream of the blood flow subsides, the slaughtered animal is hoisted up by its hind legs to allow the blood to drain for several minutes. The heart forces blood through its chambers for more than 1 min after "shechita" and this enhances the bleeding process.

After that, the mechanical function of the heart stops and much less blood is pumped. Involuntary muscle contractions occur toward the end of the bleeding, forcing the remaining blood out of the blood vessels. The more blood that is expelled from the carcass, the better the quality and the longer the shelf-life of the meat, as blood deteriorates meat very quickly according to Levinger (1973b) and religious thinkers. In a recently published monograph by the FAO (1991), blood is stated to be an ideal medium for the growth of bacteria. The laws of "shechita" specify that the blood must not be gathered or used but has to be absorbed on sand or into the earth. In modern plants slaughtering kosher, the blood is absorbed on sand which is then discarded.

The FAO monograph mentioned previously as well as many people feel that because animals are fully conscious when slaughtered, kosher or halel slaughter may be less humane than slaughter following stunning. Levinger, a pharmacologist and veterinarian in Switzerland and Israel, has researched the neurophysiological aspects of ritual slaughter and has come to the following conclusions among others: The degree of exsanguination is optimal in the "shechita" method of slaughter. Within 5–6 seconds after "shechita," blood pressure drops steeply in the whole body and cannot supply body tissues. No arterial blood reaches the brain after "shechita."

The drop in pressure in the brain causes an immediate shock in the slaughtered animal and the capacity of the brain to function is very rapidly reduced especially in cortical areas. The perception of pain is greatly reduced by the clean cut. By the time a painful impulse reaches the brain, the brain is hardly capable of functioning any more. In the recumbent animal as well as in the standing animal in which the neck is stretched such as occurs in the ASPCA pen, bleeding is undisturbed. Thus, anoxia of the brain might occur even earlier. Pain perception is in the cerebral cortex and it loses its function ahead of the centers of equilibrium which are in the cerebellum.

In numerous experiments it has been proven that the animal is not aware of death and the killing instrument as such. Psychologically, the animal does not suffer pain before, during, or after "shechita." Levinger ends his conclusions by the following statement. "Any method of slaughter is cruelty to animals. In comparison to other methods of slaughter," "shechita" is at least as humane as any other method of slaughter. The same can be said of halal.

Animals smaller than cows or beef, such as sheep, can be prepared for slaughter by suspending the animal by one leg. The slaughtering is then performed by a "shochet" at a 45° angle from the side.

Chickens are slaughtered by a shochet while he bends the animal's head slightly backward. The same cut is made as for cattle, but a straight edge razor is used. It will be recalled that in kosher slaughter the trachea, esophagus, carotid artery, and jugular vein are severed. In nonkosher slaughter, only the skin, artery, and vein need be cut. In a controlled experiment in the United States some years ago it was shown that in kosher-killed chickens the number of microorganisms in the air sac was one-sixth of those in nonkosher-killed chickens.

Following the kosher slaughter of chickens, scalding is not used to ease feather removal so as not to congeal the blood even in the skin. Due to the difficulty in plucking, the birds usually go through the defeathering machines more than once, unless some other means mentioned below are employed with the agreement of the local rabbi. The multiple plucking results in skin tears, especially in Israel where strict rules against the use of other means of easing feather removal exist in the dressing plants. This is not only unsightly but the skin cannot serve as a barrier to bacterial contamination when it is torn. A syndrome known as oily bird, resulting in oily and slippery skin, can compound the difficulty in defeathering.

Several options to hot-water scalding exist besides the use of multiple mechanical defeathering. The birds could be plunged into an ice bath for 90 s. The skin stiffens and the feathers pucker up or are somewhat dislodged, increasing the ease of shearing. Most orthodox rabbis oppose this, however, because they feel the cold water could prevent blood loss during the salting step. This is the subject of ongoing debates and needs to be solved by experimentation. A third option is to use a preplucking bath containing detergent. A detergent called FSD, a product of an American company, has been shown to increase defeathering efficiency in Israel and may be in use in some dressing plants there. The detergent emulsifies the oil on the skin, allowing the rubber fingers in the machines to better grab the feathers and pluck them.

Following defeathering, the birds are eviscerated. A "mashgiach" checks the birds for any deformities or outward signs of disease and the eviscerated lungs for disease. Those found not to appear healthy or with diseased lungs are declared nonkosher and discarded.

Koshering Meat

After kosher slaughter, meat is still not fit to cook for eating purposes according to the rules of "kashrut," unless it is "koshered."

Koshering of any meat including chickens, turkeys, geese, and ducks involves three steps: hadacha:(hashraya), hamlacha, and shtifa. Hadacha (hashraya) means soaking the meat in water above 8°C (46.4°F) for ½ h. Hamlacha involves covering the meat with course salt over all surfaces, for 1 h at room temperature. The salted meat is placed on a table or on slats which allows the water and liquids in the meat, including blood, to drain away. After 1 h the meat is subjected to nipputs and shtifa which involves shaking off the salt vigorously and rinsing under running water to get rid of the surface adhering salt, then two more shtifot (plural of shtifa) in water. Cold or ice water can be used for the rinses. A "mashgiach" or overseer sent to commercial plants by the rabbinate (an official group of rabbis) is present to see that the process is carried out correctly. Koshering at home can be done without the presence of a "mashgiach."

Meat can be kept in the cooler for 72 h after slaughter before it is "koshered." However, in order for the blood not to dry on the surface such meat must be sprinkled or immersed in water from time to time. If it has not been koshered by 72 h, it cannot be koshered. Meat frozen within 72 hours after slaughter can be "koshered" after thawing within 72 h, no matter how long it has been frozen. Liver need not be salted like ordinary meat but can be considered "koshered" for eating if it is broiled. After meat is koshered, it can be treated as any other meat; for example, it can be aged by keeping it in the cooler for a certain period of time or vacuum-packed and aged.

In poultry, following evisceration the birds are rinsed and in the koshering process the first step is as mentioned above, hadacha (hashraya). At this point, bacteria remaining on the skin or flesh or occasional feather can escape into the water and cross-contaminate other birds similar to the situation in the chiller, but there the water is at least kept ice cold. There are no literature references about bacteriological testing or known standards for the hadacha (hashraya) process for either poultry or any other meats. A simulated kosher experiment versus a conventional process experiment once carried out at the University of Pennsylvania found koshered birds had a 1-day shelf-life advantage over the conventional. However the "koshering" process carried out there differed in many ways from the actual practice, and pathogenic bacteria were not tested. In Israel where upward of 130,000 tons of poultry are koshered annually in dressing plants, the process has been geared to a frozen bird. Bacterial activity is, therefore, stopped, and considerable numbers of bacteria do not survive the freezing process.

Salting for 1 h follows the water immersion. Experiments have shown that salt concentrates in the chicken skin. People who must control their salt intake should avoid eating chicken skin of koshered birds. Salting also increases rancidity in the skin.

There are no requirements for kosher slaughter or "koshering" for fish. The only requirements for fish are that they have gills and scales. Therefore shellfish, crabs, lobsters, and shrimp are excluded. Moslems also are forbidden to eat fish which do not have scales, but there is no restriction of shellfish.

In conclusion, muscle foods are a source of the best available vital nutrients in our diet. They are usually the main dishes in our meals. It is of the utmost importance that we assure a wholesome attractive and palatable supply of these foods for the consumer.

The ways to do this are as follows: establish and adhere to a HACCP program from the farm or fishing vessel through final sale; employ the latest scientific means available to monitor and maintain fresh color, composition, texture, and wholesomeness, from slaughter in the case of animals or catching in the case of fish or seafood through packaging and marketing; keep the consumers' interests uppermost in mind when promulgating grades and standards for commerce.

Kosher and halal food laws are practiced by a significant number of Jews and about a billion Moslems in the world. These rules are meant to protect slaughter animals from cruelty and suffering and man from eating unclean foods. Muscle foods have no quality for people who profess those religions unless they are produced according to the ritual rules. It behooves us as food scientists to be familiar with the kosher and halal food laws in order to be able to cater to the needs of all sections of the world population.

Selected References

Angel, S., I.M. Levinger, and Mazal Amosi. 1976. Nerves and muscles regulating feather follicle movements in chickens. *Experentia* 32:1042–1043.

Angel, S. and I.M. Levinger. 1982. Scanning electron microscope study of the feather follicle wall attachment in chicken skin. *Br. Poultry Sci.* 23:247–249.

Angel, S. and Z.G. Weinberg. 1986. Variation in salt content of broiler parts from different kosher dressing plants in Israel. *Poultry Sci.* 65:481–483.

Bailey, A.J. 1991. The eating quality of meat. *Chem. Britain* November, 1013–1016.

Bartov, I., D. Basker, and S. Angel. 1982. Effect of dietary Vitamin E on the stability and sensory quality of turkey meat. *Poultry Sci.* 4:1224–1230.

Basker, D., S. Angel, and I. Bartov. 1984. Sensory quality of chicken meat as a function of its fattiness. *J. Food Qual.* 6:253–258.

Buttkus, H. and N. Tomlinson. 1965. Some aspects of postmortem changes in fish muscle. In: *Physiology and Biochemistry of Fish Muscle,* E.J. Briskey, R.G. Cassens, and J.C. Trautmann, eds. Pages 197–201. The University of Wisconsin Press, Madison.

Cole, A.B., Jr. 1986. Retail packaging systems for fresh red meat cuts. *Reciprocal Meat Conf. Proc.* 39:106–111.

Cross, H.R. and K.E. Belk. 1992. Objective measurements of carcass and meat quality. In: *38th International Congress of Meat Science and Technology* 1:127–134.

Daumas, G. and T. Dhorne. 1992. How to normalize the methods for grading pig carcasses in the community. In: *38th International Congress of Meat Science and Technology* 5:879–882.

Dvorak, N. 1986. Producing boxed beef for a modern market. In: *5th Annual Meat Science Institute Proceedings*. Pages 139–141. University of Georgia, Atlanta.

Food and Agricultural Organization of the United Nations (FAO). 1991. Animal protection and health. In: *Guidelines for Slaughtering, Meat Cutting and Further Processing*. FAO, Rome.

Erdtsieck, B. 1975. The "Total Quality" concept. In: *The Quality of Poultry Meat, Proceedings of the Second European Symposium on Poultry Meat*. Ed. by B. Erdtsieck, Oosterbeek, The Netherlands.

Forrest, J.C. 1991. Update on current carcass evaluation techniques. *Reciprocal Meat Conf. Proc.* 44:121–127.

Huss, H.H. 1988. *Fresh Fish—Quality and Quality Changes*. Pages 43–59. FAO Rome.

Jakobsen, Mogens. 1993. Quality management—ISO 9000. *Infofish International* 2:31–36.

Jensen, L.S., I. Bartov, M.J. Beirne, Jr., J.R. Voltman, and D.L. Fletcher. 1980. Reproduction of the oily bird syndrome. *Poultry Sci.* 59:2256–2266.

Jones, S.D.M. 1989. Quantitative methods of carcass evaluation. *Reciprocal Meat Conf. Proc.* 42:57–63.

Levie, A. 1979. *The Meat Handbook*. Pages 10–41. The AVI Publishing Co., Westport, CT.

Levinger, I.M. 1966. Principles of porging. *Refuah Veterinarith* 23:164–165. (English)

Levinger, I.M. and R. Kathein, 1973. The plucking of chickens after immersion in cold water. *Br. Poultry Sci.* 14:217–219.

Levinger, I.M. 1973a. Immersion of poultry in cold water and chemical solutions prior to plucking. *Hamayan:* 38–41 (Hebrew).

Levinger, I.M. 1973b. *Trefoth, Theory and Practice*. Pages 7–28, 103–123. Institute for Agricultural Research According to the Torah, Jerusalem. (Hebrew)

Levinger, I.M. 1976a. *Medical Aspects of Shechita*. Pages 186–199. Bar Illan University, Bar Illan University. Ramat Gan, Israel.

Levinger, I.M. 1976b. *Kosher Meat Muscle Products*. Pages 145–170. Institute for Agricultural Research According to the Tora, Jerusalem.

Levinger, I.M. and S. Angel. 1976. Antagonistic muscular activity in the feather follicle. *Israel J. Med. Sci.* 12(2):178.

Levinger, I.M. and S. Angel. 1977. Effect of spinal cord transection on feather release in the slaughtered broiler. *Br. Poultry Sci.* 18:169–172.

Levinger, I.M. and S. Angel. 1984. Possible control mechanisms of feather follicle movement in the pectoral tract of the chicken. *Vet. Res. Commun.* 9:153–162.

Locker, R.H., C.L. Davey, P.M. Nottingham, D.P. Haughey, and N.H. Law, 1975 New conceptions in meat processing. *Adv. Food Res.* 21:158–222.

Mountney, G.H. 1983. *Poultry Products Technology*. Pages 32–39. The AVI Publishing Co., Westport, CT. pp 32–39.

Munk, M.L. and E. Munk. 1976. *Shechita, Religious and Historical Research on the Jewish Method of Slaughter*.

Feldheim Publishers Ltd., Jerusalem. chapter 1. Humane Slaughter, ch. III, Casting of the Animal; ch. IV, Physiological Observations During Shechita; ch. V, Anatomy of Blood Vessels Supplying the Brain in Ruminants; Blood flow to the Brain During Clamping of vessels and Severing of Carotids; ch VII, Blood Pressure After Shechita and Blood Flow Through the Vertebral Artery; ch VIII, Effects of Shechita on the Heart; Effects of Shechita on the Electroencephalograph; ch XII Shechita and Animal Psychology. pp 107–199.

Mundt, J.D., R.L. Stokes, and O.E. Goff. 1954. The skin of broilers as a barrier to bacterial invasion during processing. *Poultry Sci*. 38:799–802.

Nakai, H. 1991. Meat color evaluation. In: *Proceedings of Electronic Evaluation of Meat in Support of Value Based Marketing*. Pages 177–179. Purdue University, West Lafayette, IN.

Olson, S.C. 1988. Commercial practice in the meat and poultry industry; from boxed beef to counter ready. *ASTM Standardization News* 3:34–35.

Pearson, A.M. and R.B. Young. 1989. *Muscle and Meat Biochemistry*. Pages 426–430. Academic Press, New York. 426–430.

Price, J.F. and B.S. Schweigert, 1971. *The Science of Meat and Meat Products*. Pages 383–386. W.H. Freeman and Company, San Francisco.

Prucha, J.C. 1991. Food Safety and Inspection Service on the move to improve. *Reciprocal Meat Conf. Proc*. 44:65–67.

Purchas, R.W. 1990. An assessment of the role of pH differences in determining the relative tenderness of meat from bulls and steers. *Meat Sci*. 27:129–140.

Regenstein, J.R. and Temple Grandin. 1992. Religious slaughter and animal welfare—An introduction for animal scientists. *Reciprocal Meat Conf. Proc*. 45:155–159.

Regenstein, J.M. and Carrie E. Regenstein. 1975. An introduction to the kosher dietary laws for food scientists and food processors. *Food Technol*. 1975:89–99.

Scholtyssek, S. 1975. Influence of management factors on the meat and carcass quality. In: *The Quality of Poultry Meat, Proceedings of the Second European Symposium on Poultry Meat*. Ed. by B. Erdtsieck, Oosterbeek, The Netherlands.

Smith, B.L. 1989. *Codex Alimentarius, Abridged Version*. FAO, Rome.

Terrell, R. 1986. Update–AMI task force review of the NAS study on meat inspection. In: *28th Annual Meat Science Institute*. Pages 16–22. Food Science Division, University of Georgia, College of Agriculture and Continuing Education.

Thompson, J.E. and A.W. Kotula, 1959. Contamination of the air sac areas of chicken carcasses and its relationship to scalding and method of killing. *Poultry Sci*. 38:1437–1439.

van Middelkoop, J.H., A.R. Kuit, and A. Zegwaard, 1975. Genetic and environmental influences on broiler fat content. The Quality of Poultry Meat Proceedings of the Second European Symposium on Poultry Meat Ed. by B. Erdtsieck, Oosterbeek, The Netherlands, May 12–15, 1975, paper 47.

Weinberg, Z.G., S. Angel, and Christine Navrot, 1986. The effect of sex, age and feed on tensile strength of broiler skin. *Poultry Sci*. 65:903–906.

14

Spoilage and Preservation of Muscle Foods

Riëtte L.J.M. van Laack

Introduction

Meat preservation involves the application of measures to delay or prevent microbiological, chemical, and/or physical changes that make meat unusable as a food or that downgrade some of its quality aspects. While the various types of deteriorative changes are all significant, microbial spoilage is the most important as it often precedes other types of spoilage. For the most part, preservation measures that prevent or delay microbiological spoilage also minimize nonmicrobiological deterioration of muscle foods.

Centuries before the advent of electricity and modern refrigeration, people developed methods for preserving muscle foods. These methods included drying, smoking, sausage making, salting, and using ice or low temperatures. By the beginning of the 19th century (1810), Nicholas Appert, while not aware of the nature of the processes involved, developed canning as a means of extending the shelf-life of food products. It took some 50 years before Louis Pasteur demonstrated the role of microorganisms in spoilage and explained the effect of heating on prolonging shelf-life (pasteurization).

These preservation methods are the forerunners of today's modern processing and packaging methods—preservation methods that make it possible to have meat available to consumers year-round in any climate.

Spoilage

The usual characterization of "spoilage" is the point at which muscle foods become unfit for human consumption. Organoleptically detectable spoilage is usually attributed to microbiologically induced decomposition and putrefaction and there is seldom any doubt about the fitness of such a product for consumption.

Spoilage does not necessarily imply decomposition or putrefaction, however. Intrinsic enzymes and oxidative reactions may result in spoilage as well.

Microbiological Spoilage

Excellent overviews exist on microbial spoilage of muscle foods; those of McMeekin (1982) and Gill (1986) are recommended for further consultation on spoilage of red meats and poultry. The review of Hobbs (1983) may be consulted for information on the spoilage of fish.

As stated in Chapter 9, fresh muscle foods are an ideal medium for bacterial growth. Therefore, muscle foods are subject to rapid spoilage unless precautions are taken to retard microbial growth and activity. Microorganisms found on muscle food come from many sources. Because meat originates from a live animal, much of the microbial flora of meat product is composed of organisms found on, or in, the animal. However, processing operations, handling by workers, and plant sanitation have a great influence on the types and numbers of microorganisms found on the products. Environmental conditions and treatment of muscle foods also affect the composition of the microbial flora, the extent and rate of microbial growth, and subsequent types of spoilage that may occur.

Spoilage of Red Meats and Poultry

As meat is generally kept in chilled storage rooms, the surface microflora will be dominated by organisms capable of growing at refrigeration temperatures. Gill (1986) identifies the following species as major spoilage organisms: *Pseudomonas* and *Acinetobacter/Moraxella* which are strictly aerobic bacteria; *Alteromonas putrefaciens* and *Brochothrix thermosphacta,* both facultative anaerobes; *Lactobacillus* spp., anaerobic aerotolerant organisms; and the *Enterobacteriaceae*. The latter group encompasses a wide range of facultative anaerobes, including pathogens such as *Salmonella* and *E. coli* 0157:H7.

The spoilage microflora which develops on the surface of chilled meats is largely determined by storage conditions. If the meat is stored in air, for instance, *Pseudomonadaceae* will normally dominate. These organisms preferentially grow on small molecular compounds such as glucose. Once these compounds are exhausted, sulphurous off-odors, slime, and discoloration are produced as a result of amino acid utilization.

Anaerobic storage (under N_2 and CO_2) prevents growth of pseudomonads, thereby allowing lactobacilli and *B. thermosphacta* to grow. *Lactobacillus* spp. primarily utilize glucose and arginine, whereas *B. thermosphacta* largely relies on glucose for growth. Both genera produce off-odors and off-flavors described as "dairylike" or "cheesy" when volatile fatty acids are formed from valine and leucine. Off-odors typically become noticeable when bacterial numbers are between 7.0 and 7.5-$\log_{10}$CFU/cm^2. Detectable slime, caused by coalescence of surface colonies, production of extracellular polysaccharides, and proteolysis of muscle proteins, typically occurs when the total count approaches 7.5–8.0-$\log_{10}$CFU/cm^2.

Enterobacteriaceae usually play only a minor role in anaerobic spoilage because high densities of *Lactobacillus* spp. produce enough antimicrobial agents to suppress their growth. Under certain types of modified atmosphere packaging, however, psychrotrophic *Enterobacteriaceae* can become a significant factor in spoilage.

The following six genera of molds have been recovered from various spoilage conditions of carcasses and whole-meat cuts: *Thamnidium, Mucor,* and *Rhizopus,* all of which produce "whiskers" on beef; *Cladosporium,* which produces "black spot," *Penicillium,* which produces green patches, and *Sporotrichum,* which produces "white spot." Among the genera of yeasts that have been recovered from refrigerated spoiled meat with any regularity are *Candida, Torulopsis,* and *Rhodotorula.*

Available moisture and oxygen and pH are the primary factors determining whether bacteria, molds, or yeasts become the predominant spoilage organisms on meats. Freshly cut meats stored in a refrigerator with high humidity invariably undergo bacterial spoilage rather than fungal spoilage. Molds tend to predominate in meat spoilage when the surface is too dry or the pH is too low for bacterial growth. Due to slower growth rates, molds virtually never reach high levels on meats when conditions are conducive for bacterial growth. Generally, spoilage of unpackaged processed meat products is caused by yeasts and molds. Spoilage of packaged processed meat products is caused by gram-positive bacteria, mainly lactic acid bacteria.

Spoilage of Fish and Seafoods

In general, fish and seafoods spoil more rapidly than other muscle foods, and such spoilage is primarily bacterial in nature. Due to extremely low levels of carbohydrate and a predominantly psychrotrophic flora, proteins and other nitrogenous constituents are readily degraded, resulting in spoilage. At the time of harvesting, fish from cold waters generally have a microflora dominated by gram-negative bacteria, whereas gram-positive bacteria are predominant on fish from tropical waters. Despite this fact, spoilage of tropical fish is usually due to gram-negative bacteria. Bacteria native to fish and added adventitiously during landing and processing include *Acinetobacter, Arthrobacter, Flavobacter, Micrococcus, Moraxella,* and *Pseudomonas.* Those microorganisms primarily responsible for spoilage include psychrotrophs such as *Pseudomonas, Acinetobacter,* and *Moraxella* species.

One of several indicices characteristic of fish spoilage is trimethylamine formation. This very odoriferous amine, a major constituent of the "fishy" smell associated with spoiling seafoods, is produced by a number of bacteria capable of reducing trimethylamine oxide, a natural constituent of marine fish flesh. The formation of trimethylamine is often used as an index of quality deterioration. Trimethylamine oxide can also be degraded enzymatically during cold or frozen storage of seafoods, yielding dimethylamine and formaldehyde. Spoilage bacteria are also responsible for the formation of volatile acids, hydrogen sulfide, mercaptans, and ammonia. Proteolysis, accompanied by an increase in free amino acids and volatile sulfur compounds like hydrogen sulfide, induces adverse changes in the organoleptic quality of fish. In fact, the number of hydrogen-sulfide-producing bacteria is a better index of spoilage than is total aerobic colony count (Martin, Gray, and Pierson 1978).

Crustacean meats differ from fish in having about 0.5% carbohydrate. Nevertheless, spoilage of crustacean meats appears to be quite similar to that of fish flesh. Due to the relatively high levels of glycogen, spoilage of molluscan shellfish is basically fermentative.

Texture, Color, and Flavor Changes During Prolonged Storage

If microbial spoilage is controlled by packaging and hygiene, then product deterioration resulting from autolytic processes may emerge as the principal factor limiting storage life of chilled meat. During prolonged storage, chilled red meats become increasingly tender, and finally may lose desirable texture characteristics altogether. The autolytic activities responsible for textural changes also produce liverlike flavors that may eventually reach intensities objectionable to some consumers. Pork appears to undergo relatively slow autolytic deterioration. Jeremiah, Penney, and Gill (1992) observed that metmyoglobin precipitated from pork exudate and absorbed onto the meat surfaces, causing discoloration. Metallic and bitter notes became more intense during storage, deteriorating the flavor of samples. However, consumer rejection would have occurred due to the color changes preceding microbial spoilage and flavor deterioration.

Preservation

The general purpose of preservation is to reduce the number of microorganisms in order to extend the shelf-life of a product and to reduce or eliminate public health hazards. It should be remembered, however, that microbial spoilage is not the sole determinant of a product's shelf-life. Color changes, lipid oxidation, and loss of vitamins and other nutrients are other processes associated with the storage and, therefore, the shelf-life of a product. Preservation processes such as chilling, freezing, thermal processing, dehydration, and irradiation not only restrict (and in some cases, completely inhibit) microbial activity but also prevent or limit enzymatic, chemical, and physical reactions that would otherwise cause deteriorative changes and spoilage.

Processes, Parameters, and the "Hurdle Concept"

Preservation typically has one or two objectives: preservation of the original quality and characteristics of a product (e.g., chilled storage of fresh meat) and/or the introduction of another desirable quality (e.g., salted, cured, or acidified flavor). Although most preservation processes focus on the inhibition or elimination of microorganisms, fermentation uses the naturally present microorganisms or an added starter culture to achieve the desired shelf-life and product quality.

Table 14.1 lists a number of common preservation processes together with the parameters upon which they are based. When formulating a preserved food, or describing preservation practices, the tendency is to regard each parameter [e.g., water activity (a_w), pH, temperature] individually. The reality, though, is that microorganisms in preserved muscle foods are inhibited by a **combination** of physical parameters and additives. This means that if a particular microorganism is inhibited by an a_w of 0.88, or a pH of 5.0, or a NaCl concentration of 6%, then control of the organism in a product employing all three measures may be achieved at an a_w of 0.92, a pH of 5.5, and a NaCl concentration of 3%.

Table 14.1. Processes Used in Food Preservation and Parameters or Hurdles upon Which They Are Based

Parameters \ Processes	Heating	Chilling	Freezing	Freeze drying	Drying	Curing	Salting	Sugar addition	Acidification	Fermentation	Smoking	Oxygen removal	IMF (f)	Radiation
F (a)	X (c)	•	•	•	•	•	o	•	o	•	•	•	•	o
t (b)	• (d)	X	X	o	o	•	•	o	•	•	•	•	•	o
a_w	•	•	X	X	X	X	X	X	o	X	•	o	X	o
pH	•	•	o	o	•	•	•	X	X	X	•	•	•	o
Eh	•	•	•	•	o	•	•	•	•	•	•	X	•	o
Preservatives	•	•	o	o	•	X	•	•	•	X	X	•	X	o
Competitive flora	o (•)	o	o	o	o	•	o	o	•	X	o	•	•	o
Radiation	•	o	o	o	o	o	o	o	o	o	o	o	o	X

(a) High temperature; (b) Low temperature; (c) Main hurdle; (d) Additional hurdle; (e) Generally not important for this process; (f) Intermediate moisture foods.

Source: Leistner et al. (1981); used with permission of Academic Press.

Leistner (1992) has described the effect of superimposing several growth-limiting parameters as the "hurdle concept."

Table 14.1 lists 14 preservation processes but only 8 parameters (hurdles). Each preservation process is based on one or two main hurdles, with additional minor hurdles being required to provide the expected microbial stability. For instance, a_w is the main hurdle employed in the drying, curing, salting, freezing, and freeze-drying of muscle foods. Preservation processes such as chilling, smoking, and heating (canning) may rely on a_w as an additional hurdle. In "Black Forest"-style ham, for example, the main hurdles are water activity and preservatives (salt, nitrite, and smoke) and additional hurdles are the storage temperature, the pH, and the redox potential.

Chilling/Refrigeration

The most common method of prolonging shelf-life of meat is refrigeration, meaning storage at temperatures between −2°C and 5°C (28–41°F). Refrigeration is the single most important means of preserving fresh meats.

Until the 17th century, fresh meat was sold warm, immediately after slaughtering, or from a cool basement. Later, the use of natural ice transported from northern Europe to countries as far south as Italy was introduced, together with ice storage in insulated cellars. The first American ice machine dates from 1844, the first chill house from 1865, and the first chilled transport took place in 1869 from Chicago to Boston. The first cooling machine to use ammonia was manufactured by the Linde company in 1874.

The refrigeration of muscle foods begins shortly after slaughter or harvesting. Beef, pork, lamb, and veal carcasses are typically chilled in forced-air movement coolers. In this type of chilling process, the humidity of the air is very important. At a high humidity, evaporation (weight loss) from the carcass is limited, but the chance for bacterial growth is higher than at a low humidity. Rapid chilling, achieved by using air temperatures <0°C (32°F) and rapid air movement, can be used to dry carcass surfaces, reducing bacterial growth.

Poultry and fish are generally chilled by immersion in ice (or ice water). This rapid chilling method allows some poultry producers to slaughter, chill, cut, package, and freeze poultry in as little as 4 h.

Although refrigeration temperatures limit (and sometimes prevent) the growth of mesophilic bacteria, psychrophilic and psychrothrophic bacteria are still able to grow. Mesophiles comprise the majority of bacterial contaminants on a beef or pork carcass immediately after dressing, and only a small percentage of the flora is capable of growth at cooler temperatures. On cold-water fish, however, the majority of the microflora can grow under refrigeration. On these muscle food products, refrigeration will not result in an appreciable extension of shelf-life.

Refrigerated storage not only impedes bacterial growth but also slows the rate at which biochemical and physical changes take place in the muscle. Apart from the outright effect that temperature has, it should be remembered that temperature also influences other parameters. Temperature is related to relative humidity, and subfreezing temperatures affect not only relative humidity but also pH and possibly other parameters of microbial growth.

A remarkable consequence of rapid chilling sometimes observed in pork carcasses is the formation of reddish spots on the skin. This "discoloration" is the result of a rapid decrease in the water content of the skin and concomitant increased oxygen diffusion into the skin. Small quantities of blood remaining in the capillaries will react with the oxygen, forming oxyhemoglobin which gives the skin a bright red color. When the carcass is chilled slowly, water evaporating from the skin is replaced by a continuous supply of water from the lower layers, limiting the chance for oxygen penetration into the skin.

Freezing

Freezing is an extreme chilling procedure that results in the formation of ice crystals in the meat. In lean muscle at −5°C (23°F), approximately 85% of the water is frozen; at −30°C (−22°F), nearly 100% of the water is frozen. Due to the presence of solutes, the freezing point of muscle is approximately −2°C (28°F).

The main effect of freezing upon microorganisms is a cessation of growth. Some microbial destruction will occur during the freeze/thaw process, but freezing should not be regarded as a reliable means of destroying microorganisms other than parasites. The reader is referred to Jay (1986) for a more detailed discussion of the effect of freezing on microorganisms.

"Freezer-life"

The purpose of freezing is the preservation of the original characteristics of the product for an extended period of time. The main requirement is that the product before freezing is in an optimal condition from both microbiological and chemical standpoints. The quality of frozen meat is affected by conditions during freezing and subsequent frozen storage, and the length of the storage period. Quality changes occurring during storage at temperatures <-10°C (14°F) are unrelated to bacterial growth or metabolism and are of (bio)chemical origin (e.g., drying and oxidation).

The data in Table 14.2 illustrate that, apart from temperature, the nature of the muscle food product is a decisive factor in determining storage time. Ground product with its greater surface area and potential for oxygen penetration, for instance, is more sensitive to oxidation (rancidity) than whole pieces of meat. As pork contains more unsaturated fatty acids, it is more susceptible to oxidative rancidity than beef. Undesirable oxidation and dehydration occurring during frozen storage may be limited by proper packaging (see Chapter 18).

Table 14.2. Freezer-Life: Relationship Between Storage Time and Temperature for Different Meat Products and Fish

	Temperature		Storage time (months)
	°C	°F	
Beef	−18	0	12
Beef	−25	−13	18
Beef	−30	−22	24
Ground beef	−18	0	10
Pork	−18	0	6
Pork	−25	−13	12
Pork	−30	−22	15
Ground pork	−18	0	6
Veal	−18	0	9
Lamb	−18	0	9
Chicken	−18	0	6
Chicken	−30	−22	12
Fish	−18	0	6
Fish	−30	−22	12

Freezing Rate

The rate of freezing is of great importance, especially the track between −1°C and −4°C (30–25°F). When frozen slowly, ice crystal formation starts in the aqueous intracellular compartments. Subsequently, water in the connective tissue freezes due to a lower salt content than muscle tissue. The remaining water is more concentrated and the higher osmotic value will result in water diffusing out of the muscle cell and condensate on the outer cell ice. Eventually, the intracellular water will freeze. Under improper freezing conditions, ice crystals may grow to 150 μm (diameter) and damage the muscle cells.

When the critical temperature (−1 to −4°C; 30 to 25°F) is passed rapidly (within 80–120 min), the water in the muscle cell freezes before substantial diffusion can occur. The existence of numerous small ice crystals limits the risk for cellular deformation. Fluctuation in storage temperature may permit the small ice crystals to thaw, leading to recrystallization and the formation of large crystals. This may lead to extensive membrane damage and subsequent problems with increased rancidity and drip or purge.

Deteriorative Changes Occurring During Freezing

Oxidation of meat during frozen storage is largely due to changes in the triglyceride fraction. Susceptibility to lipid oxidation is influenced by species differences, as well as differences between tissues. For example, beef lipids are more stable than those of chicken white meat. Temperature and length of holding can alter the rate of oxidation of lipids in meat. Even at −30°C (−22°F), lipid oxidation may occur, albeit slowly. Oxidation during frozen storage is especially a problem with processed meat products as salt acts as a catalyst in oxidative reactions. Auto-oxidation proceeds faster during storage at high temperatures. The susceptibility of frozen food systems to oxidation is related to low water activity. Whereas fresh meat has an a_w of 0.99, muscle tissue frozen at $-18°C$ (0°F) has an a_w of 0.6. At water activities of 0.6–0.8, heme pigments appear to initiate lipid oxidation. Porcine myoglobin seems less stable to oxidation than myoglobin from other species. Other substances found in meat products are also known to accelerate oxidation. Lipases and lipoxydases will, especially at a lower pH, induce lipid oxidation and rancidity.

Color deterioration of frozen meat can be a serious problem. Freezing accelerates metmyoglobin formation, causing an undesirable color change. Rapidly frozen meat is lighter in color than slowly frozen meat. Localized tissue dehydration, better known as "freezer burn," may result in a lightening of dark tissue or a browning of light-colored tissues such as chicken skin. Freezer burn is an irreversible condition. It can be prevented by proper packaging.

Even with proper freezing techniques, a concentration of soluble salts within the muscle cell will occur. This will cause denaturation of proteins, and the pH of the product will change.

Freezing of muscle foods should not start before rigor has set in (see Chapter 3). The primary textural change occurring in properly frozen red meats and poultry is an increased porosity.

Freezing and frozen storage of seafoods can result in undesirable textural changes. The muscle may become tougher, drier, and lose its water-binding capacity. The overall decrease in protein extractability is attributed primarily to alterations in the myofibrillar fraction (myosin–actinomyosin binding). The nature of the reactive groups responsible for the aggregation is not well understood. Formaldehyde, a product of decomposition of trimethylamine oxide, is highly reactive with a variety of functional groups and it has been suggested that it is involved in the covalent cross-linking of proteins in frozen-stored fish. The loss of water-binding capacity is probably due to an increase in the number of cross-linkages between myofibrillar proteins. As fish lipids contain highly unsaturated fatty acids, fish muscle is very susceptible to oxidation.

Thawing

Much research has concentrated on freezing and proper procedures for freezing. The subject of thawing, however, has been largely ignored. Generally, slow thawing will result in a better product than fast thawing, meaning that thawing at refrigeration temperatures is preferable to thawing at higher temperatures. Thawing time should not be too long, as it may permit microbial growth to occur.

An important consideration in a thawed product is the amount of exudate. The exudate, containing soluble proteins, vitamins, and salts, will be increased when cell damage is extensive. Slow thawing allows for reabsorption of the fluids by the tissue.

Heating

Heating of meat and other food products has been applied since the "invention" of fire, the main purpose being to improve sensory quality. Stabilizing the product and increasing shelf-life became important in the 19th century. The French cook Appert introduced the concept of heating products so as to extend shelf-life.

Heat can be transferred from warm materials to cold materials by three mechanisms. *Conductance* is the transfer of heat from one part of a substance to another part, without movement of the parts. Conductance occurs in solid and fluid products. *Convection* is heat transfer from one location in a medium (liquid or gas) to another location by movement of the medium. *Radiation* is heat transport by electromagnetic waves (infrared light). This transport occurs between two surfaces separated by a medium that is permeable to radiation.

The efficiency of a heating procedure is influenced by the type of heat transport mechanism involved. For example, in a solid piece of meat, heating occurs via conduction; this is a slow process. In a can of minced meat in gravy, both conduction and convection are responsible for heat transport and the rate of heating is increased. Radiation is important in a process like broiling.

Thermal Destruction of Microorganisms

Most meatborne pathogens and spoilage organisms are sensitive to heat, meaning that cooking greatly reduces microbial numbers and extends the storage life of the cooked product. In addition, cooking results in a decrease in a_w, delaying microbial growth on cooked meats.

Heating temperatures up to 50°C (122°F) may result in faster growth of some organisms. At higher temperatures, however, bacteria will be destroyed. The inactivation of bacteria by heat is dependent on the following:

- The number and kind of microorganisms
- The heat resistance of these organisms
- The environment (humidity, pH, nitrite, salt, etc.).

Most non-spore-forming organisms are heat labile, meaning that they are inactivated by heating. After several minutes at 65°C (149°F), heat labile organisms like *Micrococci, Pseudomonadaceae,* and *Enterobacteriaceae* (e.g., salmonellae) die. Some species (e.g., enterococci) may survive high temperatures, however. Gram-positive bacteria tend to be more heat resistant than gram-negative bacteria. Yeasts and molds are usually quite sensitive to heat. Bacterial spores can survive very high temperatures (~ 120°C; 248°F).

Destruction of microorganisms is the consequence of several processes, such as inactivation of vital bacterial enzymes, protein denaturation, and increased membrane permeability.

Recently, there has been evidence that the temperature history of a microorganism may influence its thermotolerance. Bacteria subjected to sublethal heat treatments appear to gain resistance to temperatures that otherwise prove lethal. This phenomenon has been reported in *E. coli* and *Salmonella* spp. (Mackey and Derrick 1987; Murano and Pierson 1992) Also, it has been reported that elimination of *L. monocytogenes* was not only dependent on the ultimate temperature but also on the rate of heating; whole hams had to be heated to higher (pasteurization) temperatures than sausages to affect the same degree of thermal destruction. Due to slower heat penetration in the solid product, the potential exists for bacterial adaptation to otherwise lethal temperatures. When exposed to sublethal temperatures, bacteria form the so-called "heat shock proteins." The exact role and mode of action of heat shock proteins is, as yet, unclear. It has been suggested that heat shock proteins limit or prevent thermal denaturation of bacterial proteins, thereby providing protection against thermally induced damage. This phenomenon may have important ramifications in determining thermal processing schedules capable of destroying pathogens in muscle food products.

Based on the level of destruction of bacteria and spores, a product is said to be "pasteurized" or "sterilized." Pasteurization implies the destruction of all pathogenic bacteria, though spores and some thermoresistant spoilage organisms may survive. A commercially sterile product is free of any detectable organisms but may contain spores of thermophilic bacteria which do not germinate below 43°C (109°F). A commercially sterile product has a more or less indefinite storage life.

The fact that not all microorganisms die when a product is heated to pasteurization temperatures does not imply that a pasteurized product is unsafe or will spoil immediately. Spoilage requires more than the presence of one organism, and the necessary growth is dependent on the newly created environment (especially pH, redox potential, temperature). When pasteurized products are kept refrigerated, the chance for growth of spore-formers is minuscule.

When a product is heated in an open-kettle water bath, the maximum temperature will be $<$100°C (212°F). Under these conditions, a very long processing time would be required to attain commercial sterility. Therefore, open-kettle heating is only used for "pasteurization." To attain commercial sterility, the product is heated in a retort, a large autoclave capable of withstanding pressures of more than 3 atm (45 lb/in^2). The canned (or otherwise packaged) product is heated in pure saturated steam or in water. Alternatively, the product may be heated outside its package and then aseptically transferred to sterilized containers. This process does involve a greater risk of postprocessing contamination than in the retort procedure. However, the advantage of this approach is the ability to apply heat more uniformly, thereby shortening the heating process.

Efficacy of Heat Treatment

What is a microbiologically effective heat treatment? The primary consideration is that pathogens, a threat to public health, must be absent, or damaged to such an extent that they cannot grow. For low-acid canned foods, efficacy of the heat treatment is based on the destruction of *Clostridium botulinum*.

In order to better understand the thermal destruction of microorganisms relative to food preservation, it is necessary to define certain basic concepts associated with this technology. Due to space limitations, these terms will be explained only briefly; the reader is referred to the book by Jay (1986).

Thermal death time (TDT). This is the time required to destroy a given number of organisms at a constant, specified temperature.

***D*-value.** The decimal reduction value is the time (usually expressed in minutes) at a given temperature required to kill 90% of the bacterial cells which are present. The destruction of vegetative bacteria by heat is logarithmic and follows a first-order reaction.

***Z*-value.** The *z*-value refers to the degrees Celsius (or Fahrenheit) by which the temperature must be raised in order to obtain a 10-fold (one $\log_{10}$) increase in the death rate of bacterial cells. Mathematically, this value is equal to the reciprocal of the TDT curve (the time of heating in minutes plotted on semilog paper along the linear axis, and the number of survivors plotted along the log scale; Fig. 14.1). Whereas the *D*-value reflects the resistance of an organism to a specific temperature, the *z*-value provides information on the relative resistance of an organism at different temperatures.

***F*-value.** This value is the equivalent time, in minutes at 121°C (250°F), of all heat applied to a product, with respect to its capacity to destroy spores or vegetative cells of a particular organism. The integrated lethal value of heat received by all points in a container during processing is designated F_S or F_O, for spores or vegetative cells, respectively. The *F*-value represents a measure of the ability of a heat process to reduce the number of spores or vegetative cells of a given organism in a container.

12-D concept. The 12-D concept refers to the process lethality requirement long in effect in the canning industry. A 12-D "kill" implies that the minimum heat process should reduce the probability of survival of the most resistant *C. botulinum* spores to 10^{-12}. Because *C. botulinum* spores do not germinate and produce toxin below pH 4.6, this canning requirement is only observed in low-acid products.

In low-acid products like meat, the so-called "Botulinum-cook" or 12-D concept is usually applied. Although there are organisms/spores that are more heat

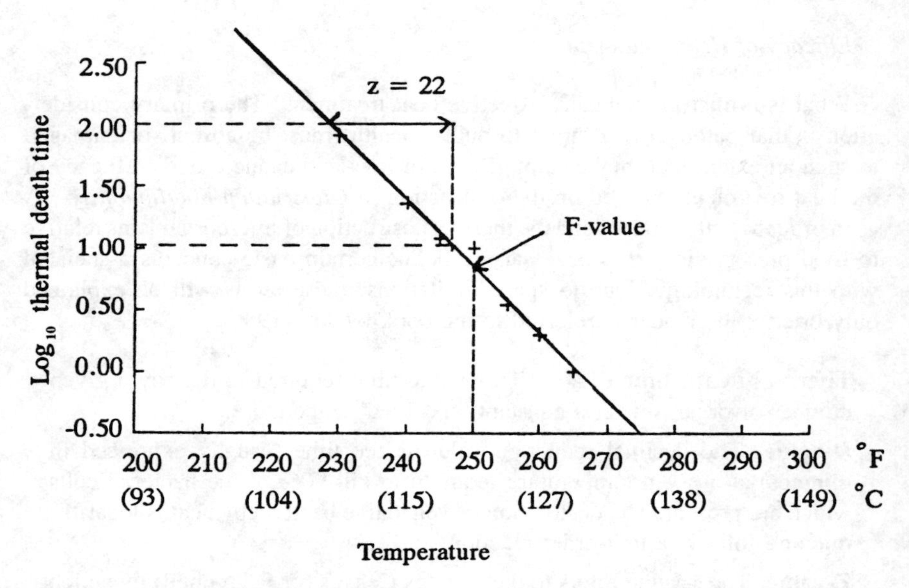

Figure 14.1. Thermal death time curve. The z-value is the increase in temperature (in °C or °F) required to obtain a ten-fold increase in the death rate of bacterial cells (e.g. a ten-fold decrease in the thermal death time). The F-value is the thermal death time at 121°C (250°F).

resistant than *C. botulinum,* they are not of public health concern. Heating in excess of the Botulinum-cook may result in unacceptable changes in other quality characteristics. In some cured meat products, a 12-D "cook" is not necessary; the number of contaminating spores is low, and the addition of NaCl and nitrite decreases the thermoresistance of the microorganisms.

If heating of canned meat products could be accomplished instantaneously, processing times could be established from the TDT curves. However, this is not the case. Thermal processing relies on a combination of time and temperature to achieve microbial destruction. In order to reach a specified end-point temperature, the product is for some time (often referred to as the "come-up" time) exposed to lower temperatures. Additional destruction of microorganisms may occur during the "come-down" time when the product is cooling to room temperature. So, thermal inactivation occurs throughout the heating process. Meat processors often speak of "equivalent time–temperature" processing schedules. For instance, in the production of roast beef at 54°C (130°F), the minimum processing time after the required temperature is reached is 121 min. A product heated to 63°C (145°F) requires no further heating (Anonymous 1993). The product heated to 54°C (130°F) retains a desirable "pink" color and possesses the same microbiological quality as that heated to a higher end-point temperature for a shorter time.

Occasionally, canned meat products will undergo spoilage, typically caused by growth of spore-formers. Spoilage of canned products is usually characterized by swollen, bulging cans due to gas production by the spore-forming microorganisms. Generally, this condition indicates underprocessing, but it may also signal a leaky can. Underprocessing may be due to an extremely high initial contamination level or to inadequate processing. Recovery of a single type of spore-forming bacteria from a swollen can is usually a good indication of underprocessing. A mixed culture, especially when it contains one or more non-heat-resistant organisms, may indicate postheating contamination, sometimes due to leaky cans and contaminated cooling water.

Physical–Chemical Changes Induced by Heat

Heating of the product results not only in bacterial inactivation but also in inactivation of enzymes; nearly all enzymes are irreversibly destroyed by a few minutes heating at 79°C (174°F). These changes can be expected to improve the quality of the product.

Heating induces protein denaturation resulting in shrinkage and loss of fluid. At 40–50°C (104–122°F), sarcoplasmic proteins lose their native structure. At 50–65°C (122–150°F), myofibrillar proteins denature, intramolecular connections are substituted by intermolecular binding, the water-binding capacity decreases and the pH increases (~ 0.3 pH unit). At ~ 65–70°C (150–158°F) similar changes occur in collagen. Between 80°C and 100°C (176–212°F), collagen slowly changes into gelatin. At still higher temperatures, proteins may decompose, resulting in the liberation of small quantities of hydrogen sulfide from sulphur-containing amino acids. During heating, carbonyl compounds are liberated and Maillard products may be formed; these are responsible for a "burned" flavor.

Lipases are destroyed by heat, whereas oxidative rancidity is accelerated by heat. Moisture further accelerates oxidative rancidity.

Warmed-over flavor (WOF) may be a problem when a cooked product is reheated before consumption. Phospholipids are thought to be primarily responsible for the development of WOF. This is attributed to disruption of the membranes during cooking and subsequent exposure of the phospholipids to oxygen. Cured meats typically do not develop WOF, as phosphates, nitrites and ascorbates inhibit the development of WOF. Some products of the Maillard reaction may have an antioxidant activity, thereby preventing development of WOF in heat-sterilized meat products (Sato, Hegarty, and Herring 1973.)

High temperatures promote the breakdown of nitrite and binding of nitrite to proteins (SH-groups). Furthermore, the nitrite may be oxidized. Nitrite losses during heating may amount to 40–50%, with additional breakdown of nitrite occurring during storage.

Microwave Preservation

Microwave heating has essentially the same effects on product shelf-life as other, more conventional forms of heating. Again, the preservation effect is dependent on the time/temperature treatment. Microwave heating has several advantages over, for instance, retort cooking:

1. Rapid provision of uniform energy throughout the product—no heat losses due to convection and conduction.
2. Microwaves have only one effect, e.g., heating.
3. The amount of heat is strictly a function of the frequency and of the dielectric loss characteristics of the food.

Despite these advantages, microwave heating for muscle food preservation has hardly been used. One of the concerns with microwave heating of products is short dwell time at the final temperature and the knowledge that cold spots may occur in the microwave (meaning that the product is not heated uniformly). Also, the ability to be heated by microwaves is dependent on the dielectric constant and loss factor of the various components of the food product. Two often-cited disadvantages of microwave cooking for meat products are that no browning or crusting of the product occurs and that product packaged in metal cans or foil pouches cannot be microwaved in the package.

Irradiation

Research on food irradiation has been ongoing for the past 40 years, making food irradiation the most scrutinized process in the history of food preservation.

Radiation may be defined as the emission and propagation of energy through space or through a material medium. The type of radiation of primary interest in food preservation is electromagnetic. The shorter wavelengths that dissipate the highest levels of energy are the most damaging to microorganisms. Those forms of radiation of interest in food preservation include microwaves, ultraviolet rays, X-rays, beta rays and gamma rays.

Energy exists in the form of waves and is defined by its wavelength. As the wavelength gets shorter, the energy of the waves increases. In general, only ionizing radiation [i.e., radiation having a wavelength of 200 nm (nanometer) or less] possesses a level of energy capable of destroying microorganisms. Ionizing radiation has such high energy levels that it can cause ejection of an orbital electron from an atom or molecule resulting in the formation of an ion. In practice, irradiation must be applied such that energy levels are insufficient to penetrate the nucleus. Therefore, the sources of radiation allowed in food processing cannot make food radioactive.

Although the term "irradiation preservation" generally refers to the use of ionizing radiation, nonionizing radiation (ultraviolet and infrared rays, microwaves) is also lethal to microorganisms and may be used for meat preservation. Because ionizing radiation destroys microorganisms without appreciably raising the temperature, the process is often termed "cold sterilization."

Following are definitions of terms used in discussing irradiation:

1. **Ultraviolet light** (260 nm). Ultraviolet (UV) light is a powerful bactericidal agent. It is nonionizing and is absorbed by proteins and nucleic acids, leading to photochemical changes that cause cell death. Due to poor penetration abilities, application of UV light is limited to food surfaces where it may catalyze oxidative changes that lead to rancidity, discolorations, and other undesirable reactions.
2. **Beta rays.** Beta rays may be defined as a stream of electrons emitted from radioactive substances. Cathode rays are similar, except that they are emitted from the cathode of an evacuated tube. These rays possess poor penetration power. Linear accelerators are among the commercial sources of cathode rays.
3. **Gamma rays.** These are electromagnetic radiations emitted from the excited nucleus of elements such as ^{60}Co and ^{137}Cs—two radionucleotides often used in food preservation due to their low cost. Gamma rays have excellent penetration power. One disadvantage of the use of radionucleotides is that the radioactive source emits rays in all directions and requires massive permanent shielding to protect workers and their environment from radiation. Other disadvantages are that radionucleotides cannot be turned "on" only when needed and the high cost and safety concerns associated with disposal.
4. **X-rays.** These rays are produced by the bombardment of heavy-metal targets with high-velocity electrons (cathode rays) within an evacuated tube. They are essentially the same as gamma rays in other respects. Advantages of electron accelerators are greater flexibility, the monodirectional characteristic, the ease with which an electron accelerator can be turned off, and the possibility to transport the radiation source without massive radiation shields. Furthermore, the efficient convertability of electron power to X-ray permits the handling of very thick products that cannot be processed by electron or gamma-ray beams.
5. **Microwaves.** When electrically neutral foods are placed in an electromagnetic field, the charged asymmetric molecules are driven first one way and then another. During this process, each asymmetric molecule attempts to align itself with the rapidly changing alternating-current field. As the molecules oscillate about their axes while attempting to go to the proper positive and negative poles, intermolecular friction is created and manifested as a heating effect. The general use of micro-

waves as a means of food preservation is limited by the heating effect that results from their use.

The basic unit of radiation is the rad, defined as being equivalent to the absorption of 100 ergs per gram of matter. The Gray (Gy) is another unit which is often used; 1 kGy equals 100 krad.

Sensitivity of Microorganisms to Irradiation

Gram-positive bacteria are more resistant to irradiation than gram-negatives. Spore-formers are, in general, more resistant than non-spore-formers. Meatborne bacteria that are most sensitive to irradiation include the pseudomonad and flavobacteria groups. Most other gram-negative bacteria are intermediate in irradiation resistance between these groups and micrococci (irradiation-resistant gram-positives). The gram-negative bacilli belonging to the genera *Moraxella* and *Acinetobacter* have been found to possess degrees of irradiation resistance higher than all other gram-negatives. Yeasts are reportedly more resistant than molds, with both groups generally being less sensitive than gram-positive bacteria. Viruses are highly irradiation resistant, whereas parasites (e.g., *Trichinella*) are very sensitive to irradiation.

Depending on the level of radiation and the degree of sterility achieved, several terms are used. **Radappertization** is equivalent to irradiation sterilization, or "commercial sterility" as it is understood in the canning industry. Irradiation needed to produce radappertized muscle foods falls in the range of 3–4 Mrad (30–40 kGy). **Radicidation** is basically equivalent to pasteurization. Specifically, it refers to a reduction in the number of specific viable non-spore-forming pathogens, other than viruses, so that none is detectable by any standard method. **Radurization** may be considered the process equivalent to pasteurization with respect to spoilage bacteria. Radurization refers to enhancement of the keeping quality of food caused by substantially reducing the numbers of viable spoilage microbes using doses in the range 0.1–0.5 Mrad (1–5 kGy). Among the organisms that survive radurization treatments are some yeasts and gram-positive bacteria of the following genera: *Micrococcus, Corynebacterium, Streptococcus, Brochothrix,* and *Lactobacillus*.

Foods subjected to radurization ultimately undergo spoilage from the surviving flora if stored at temperatures suitable for growth. The ultimate spoilage of radurized, low-temperature stored foods is invariably caused by one or more of the *Acinetobacter–Moraxella spp*. or lactic acid bacteria mentioned earlier. The most common spoilage organism in irradiation-pasteurized beef is *B. thermosphacta*. The normal spoilage flora of seafoods is so sensitive to ionizing radiation that 99% of the total flora of these products is generally destroyed by doses on the order of 0.25 Mrads (2.5 kGy).

Proteins have been shown to exert a protective effect against irradiation.

Several investigators have reported that the presence of nitrite tends to make bacterial endospores more sensitive to irradiation.

The irradiation resistance of microorganisms is greater in the absence of oxygen than in its presence. The irradiation resistance of dried cells is, in general, considerably higher than that of normal cells. Bacteria tend to be most resistant to irradiation in the lag phase, just prior to active cell division. Irradiation resistance increases at the colder temperatures and decreases at higher temperatures. Usually, muscle foods are irradiated at low temperatures in order to minimize adverse organoleptic changes.

Packaging of foods which are to be irradiated is an important consideration. Ideally, foods should be irradiated in the same package in which they are to be distributed and sold; this eliminates the potential for cross-contamination with pathogens. The choice of a packaging material depends not only on the nature of the product and its susceptibility to oxidative rancidity but also on the dosage and temperature at which the food is irradiated.

Sensory and Physical-Chemical Aspects of Irradiated Muscle Foods

Foods subjected to radappertizing doses of ionizing radiation may be expected to be as shelf-stable as commercially heat-sterilized foods. There are two differences between foods processed by these two methods that affect storage stability: radappertization does not destroy inherent enzymes that may continue to act, and some postirradiation changes may be expected to occur. Irradiation has a tenderizing effect, and irradiated meats intended for extended storage must be treated to inactivate the native tenderizing enzymes. Heat treatment prior to irradiation appears to be the best method for the prevention of undesirable enzyme-induced changes in flavor and texture.

The main drawbacks of the application of irradiation to foods are color changes and/or the production of off-odors and off-flavors. Consequently, those food products that undergo relatively minor changes in color and odor development have received the greatest amount of attention for commercial radappertization. Raw fresh meats and cured meats usually brown. Cooked fresh meat turns pink upon irradiation in the absence of oxygen. Upon exposure to air, however, the normal cooked meat color returns.

The degree of irradiation-induced flavor changes varies with the kind of meat. When fresh beef is irradiated with levels on the order of 4 Mrad (40 kGy), sulfurous odors predominate. Beyond 4–10 Mrad, an odor described as "wet dog hair" (due to S-containing carbonyl compounds) is most noticeable. Pork and chicken show little development of off-odors, even at high radiation doses. Veal and mutton are intermediate in susceptibility (Urbain 1978). By irradiating under anaerobic conditions, off-flavors and odors are somewhat minimized due to the lack of oxygen necessary to form peroxides. One of the best ways to minimize off-flavors is to irradiate at subfreezing temperatures. Lowering the temperature affects the bactericidal action of irradiation only to a very small extent. The main effect of subfreezing temperatures is to reduce or halt radiolysis and to limit the mobility of its consequent reactants.

Other than water, proteins and other nitrogenous compounds in foods appear to be the most sensitive to irradiation effects. Among the amino acids most sensitive to irradiation are the aromatic amino acids, methionine, cysteine, histidine, arginine, and tyrosine.

Irradiation of lipids and fats results in the production of carbonyls and other oxidation products such as peroxides.

Application of Irradiation

Since 1985, irradiation has been permitted for the control of *Trichinella spiralis* in pork carcasses or fresh, non-heat-processed pork cuts. The minimum dose for this purpose is 0.03 Mrad (0.3 kGy); the maximum dose is 0.1 Mrad (1 kGy).

In May 1990, the U.S. Food and Drug Administration approved the use of irradiation of poultry at doses up to 0.3 Mrad (3 kGy) for the reduction of pathogenic bacteria. This level of irradiation does not sterilize the product, and poultry treated in this way still requires refrigeration and proper handling.

Although irradiation does not eliminate spore-forming bacteria such as *C. botulinum,* neither does it imply an increased risk. When irradiated poultry was contaminated with *C. botulinum* and kept at abuse temperatures (30°C; 86°F), the natural flora multiplied and produced an off-flavor indicative of spoilage by the time the sample became toxic. Thus, enough of the normal flora survives in poultry irradiated at 0.3 Mrad (3 kGy) that *C. botulinum* would not render the product toxic before the normal signs of spoilage become evident. As an extra guarantee against the proliferation of *C. botulinum,* poultry is required to be packaged in an oxygen-permeable film (Anonymous 1993).

Since October 1992, irradiation of poultry has been applied by a commercial firm, but on a limited basis.

A great many arguments have been advanced against the use of irradiation as a means of food preservation. One argument used against irradiation is that it might be used to conceal food spoilage. This is simply not possible; a product having poor microbiological quality will still exhibit organoleptic signs of deterioration after irradiation. Also, the public appears to be confused about the difference between irradiated foods and radioactive foods. Foods treated with permitted levels of irradiation do not become radioactive.

Dehydration/Control of Water Activity

The preservation of foods by drying is based on the fact that microorganisms and enzymes need water in order to be active. Air drying of muscle foods as a means of preservation has been practiced for centuries. Low-moisture foods generally do not contain more than 25% moisture and have an $a_w < 0.6$. Another category of shelf-stable foods is those that contain between 15% and 50% moisture and have an a_w between 0.60 and 0.85—the intermediate-moisture foods.

Intermediate-Moisture Foods

Intermediate-moisture foods (IMF) are characterized by a moisture content of around 15–50% and an a_w between 0.60 and 0.85. These products are shelf-stable at ambient temperature for varying periods of time.

In preparing IMF, water may be removed either by adsorption or by desorption. In adsorption, food is first dried and then subjected to controlled rehumidification. Desorption involves placing the food in a solution of higher osmotic strength so that at equilibrium the desired a_w is reached. Other techniques used are component blending and osmotic drying. In component blending, all IMF components are weighed, blended, cooked and extruded, or otherwise combined to give the finished product the desired a_w. Osmotic drying means that foods are dehydrated by immersion in liquid humectants (glycerol, sorbitol, glycols) with an a_w lower than that of food. Water moves out of the food in an attempt to achieve osmotic equilibrium. An example of a traditional IMF product is beef jerky. A new IMF, processed by the use of humectants, is soft/moist pet food.

With the exception of cocci (lactococci, micrococci), some spore-formers, and lactobacilli, bacteria are unable to grow at the a_w of IMF products. *Staphylococcus aureus* can grow at an a_w as low as 0.85. In addition to the inhibitory effect of a low a_w, antimicrobial activity results from pH, E_h (= oxidation-reduction potential), preservatives, etc. A large number of molds are capable of growth in the 0.80 a_w range, and the shelf-life of IMF foods is generally limited to the growth of these organisms.

IMF products are not without problems, however. Lipid oxidation and Maillard browning are at their optima at the a_w and moisture content of IMF foods.

Hot-Air Drying

Cooked ground meat can be dehydrated satisfactorily by this method. The end product has a residual moisture content of approximately 5% versus 1% when freeze-dried. This moisture content is high enough to permit some deterioration to occur. In pork, hot-air drying results in rancidity unless antioxidant is added during cooking.

Hot-air drying is not suitable for uncooked meats or cooked chops and steaks; it is too slow and causes surface hardening, resulting in poor rehydration and acceptability.

Freeze-Drying

This process results in much better product than hot-air drying. Rapid freezing has been shown to produce more acceptable products than slow freezing. The major civilian use of freeze-dried meats is in dehydrated soup mixes.

After freezing, water in the form of ice is removed by sublimation (under vacuum). Conventional freeze-drying, in which the meat remains frozen throughout the drying, permits reduction of moisture to less than 2%. The end product is porous and very easy to rehydrate.

Unless heat treated before freeze-drying, freeze-dried foods retain enzymatic activity. Therefore, freeze-dried raw meat is less stable than cooked meat. Stability depends on the quality before drying, the method of drying used, the residual moisture content, the method of packaging, etc. The most prominent types of deterioration are rancidity, nonenzymatic browning, and protein denaturation. Therefore, exclusion of oxygen and moisture are requirements for a stable product.

Although some microorganisms are destroyed in the process of drying, the process per se is not lethal to microorganisms. After all, freeze-drying is one of the best known methods of preserving microorganisms. Bacterial endospores, yeasts, molds, and many bacteria will survive and can be recovered after rehydration of the product.

Curing

Meat curing is a classical example of food preservation. As a result of the introduction of refrigeration, the emphasis on curing as a preservation technique has largely been supplanted by an emphasis on the use of curing agents to impart desirable sensory characteristics to a variety of muscle foods.

The classical curing agents comprise a mixture of NaCl, sodium nitrite, and/or sodium nitrate. In the applied concentrations, most curing agents are bacteriostatic or fungistatic rather than bactericidal or fungicidal. The addition of curing salts to muscle foods creates a selective bacteriological medium in which only certain (mostly gram-positive) microorganisms are able to grow. The preservative effect of curing agents is optimal when the agents are used together.

Salt

The use of salt as a preservative is based on the fact that, at high concentrations, salt exerts a drying effect on both food (lowering the a_W) and microorganisms.

Some bacteria are inhibited by salt concentrations as low as 2%, whereas other, salt-tolerant bacteria (*Bacillus* species and micrococci), yeasts, and molds may be able to grow up to the saturation point. Halophilic microorganisms grow only in media containing high salt concentrations; they are rapidly killed when exposed to salt concentrations $<10\%$. Halophilic organisms are involved in the spoilage of natural sausage casings. They may digest the collagenous tissue to such an extent that the casings become useless. Red-pigmented halophilic bacteria are sometimes encountered as spoilage microorganisms in highly salted muscle foods.

Conventional chipped beef is an example of a meat product that is largely preserved by its high salt content. For effective preservation, the moisture content of the finished product should be 50–55% and the salt content 9–11%.

Nitrite/Nitrate

The tolerance toward nitrite varies widely among different groups of bacteria. Several explanations have been offered for the bacteriostatic properties of nitrite, namely: (1) interference with the sulphur nutrition of the organism, (2) reaction with the α-amino groups of amino acids, (3) reaction with monophenols such as tyrosine, and (4) reaction of the decomposition product nitric oxide with catalase and cytochromes. The antibotulinal activity of nitrite may be due to its inhibition of nonheme, iron–sulfur enzymes.

The bacteriostatic effect of nitrite is dependent on the pH. As pH is lowered by one unit, the bacteriostatic effect increases approximately 10-fold. Although the microorganism of greatest concern relative to nitrite inhibition is *C. botulinum,* the compound has been evaluated as an antimicrobial agent against other organisms. Nitrite and its decomposition products are unstable, especially in the presence of organic matter at low pH and at elevated temperatures.

Nitrite, in applied and acceptable concentrations (max. 120 ppm in bacon; 156 ppm in most other products), does not have an appreciable preservative effect when used alone. When added to processed muscle foods followed by substerilizing heat treatments, however, nitrite has definite antibotulinal effects. The antibotulinal effect consists of inhibition of vegetative cell growth and prevention of germination of any spores that survive the heating or smoking process. Only in combination with heat, e.g., in canned products, can nitrite prevent the outgrowth of spores. It has been suggested that the heating of meat with nitrite produced a substance or agent, the Perigo factor, which was inhibitory to *C. botulinum*. Although the Perigo factor does occur in laboratory broth systems, its involvement in meat products is somewhat dubious.

Nitrate (saltpeter) was originally used as a preservative curing compound. The presence of micrococci was necessary to convert nitrate into nitrite, and long curing periods were necessary to allow this transformation. After chemical production of nitrite became possible, the addition of nitrate became less essential. For detailed information on the use of nitrite as a food preservative, the reader is referred to the book by Cassens (1990).

The physical–chemical effects of nitrite are discussed in Chapter 5.

Sugar

Very high concentrations of sugar in a food are needed to serve as a preservative. The sugar concentrations typically used in meat curing fall far short of inhibitory concentrations.

In a wide variety of fermented sausages, sugar serves indirectly as a preservative as the result of its fermentation to lactic acid. The lowered pH in combination with the added salt and partial dehydration results in the high stability of fermented foods.

Smoke

Smoke has bacteriostatic properties, mainly due to the presence of formaldehyde and phenolic compounds. Smoke also possesses some antioxidant activity. When muscle foods are smoked in the traditional manner, the product a_W is also lowered; this may not be true when liquid smoke is applied.

Smoking is often accompanied by heat application. This combination of smoke and heat is usually effective in significantly reducing the surface bacteria population. In addition, the surface dehydration and protein coagulation seen on the outer portions of semidry and dry sausages produce a relatively effective chemical and physical barrier against microbial growth and penetration. These effects are lost when "wet" smoking is applied.

Today, only a few muscle foods (primarily ham, bacon, fish) are produced in which smoke plays an important role in preservation. In recognition of the fact that the main reason for smoking meats is to obtain a desirable flavor, liquid smokes have been developed.

High-Pressure Treatment

The use of high-pressure for food preservation was pioneered by Hite more than 90 years ago (in 1899). However, until the 1980s, only a few attempts were made to investigate the relationship between high pressure and foods.

Pressures induce a number of changes in food similar to those induced by heat. Depending on the magnitude of the pressure, proteins are denatured, enzymes are inactivated, textural properties of meat improve (Macfarlane 1985), and microorganisms are killed (Hayashi, 1989).

Pressure transmits instantaneously and evenly throughout a medium. Hayashi (1989) has proposed application of pressures less than 10,000 atm (atm) for food processing and preservation.

The capability of the pressure to kill any microorganism appears to be largely dependent on the type of microorganism and the composition of the food (Hoover et al. 1989). This may be due to membrane differences, as the cell membrane is considered to be the primary site of pressure damage. In a study by Shigehisha et al. (1991), it was shown that gram-negative bacteria were inactivated at lower pressures than gram-positive microorganisms. All tested organisms, with the exception of *B. cereus,* were killed by pressures lower than 6000 atm. At this pressure, the number of *B. cereus* only decreased 1-$\log_{10}$/g.

Pressure sensitivities of yeasts are intermediate to those of gram-positive and gram-negative organisms. The endospores of the gram-positive *Bacillus* and

Clostridium are not extensively inactivated by high pressure. Therefore, low-acid foods cannot rely on pressure treatment alone as a sterilization process.

MacFarlane (1985) reviewed the effects of high pressure on meat quality. Available data mostly concern the effect of high pressure on tenderness of pre-rigor meat. The consequences for other quality characteristics have not been investigated.

High-Electric Field Pulses

This procedure for elimination of bacteria relies on the lethal effect of strong electric fields for the inactivation of microorganisms. The external electrical field induces an electric potential over the membrane, which causes a charge separation in the cell membrane. At a critical transmembrane value ($\sim >1$ V), the repulsion between charged molecules causes the formation of pores in the cell membrane. The critical potential depends on several factors such as biological cell, cell diameter, cell-shape, and cell growth. Only when the field strength greatly exceeds the critical value will the pores become irreversible and result in cell membrane destruction. For vegetative bacterial cells, the critical electrical field strength is $\sim$ 15 kV/cm.

To date, no data have been published on the application of this process to meat or meat products.

Oscillating Magnetic Field Pulses

A procedure for inactivating microorganisms by subjecting plastic- or glass-packaged or bulk food to one or more pulses of an oscillating magnetic field is described in a patent (Anonymous 1985). The patent suggests that oscillation causes breaking of covalent bonds in certain critical molecules like DNA. Similar to the high-voltage-pulse treatment, the total exposure time is minimal, on the order of 25 μs–25 ms. It is unclear whether this process is actually effective against microorganisms without adversely affecting food quality. It is not clear whether the process is still under development or whether the idea has been abandoned (Mertens and Knorr 1992).

Intense Light Pulses

The use of intense light pulses of incoherent continuous broad spectrum light to inactivate microorganisms at the surface has been described (Anonymous 1988). The spectral distribution and energy densities of the pulsed light are chosen depending on the selected application. As food products are adversely affected by the UV bandwidth of the spectrum, the UV portion is filtered out. In this case, the inactivating effect of the intense light pulses is primarily photothermal. This implies that the application of intense-light pulses is simply a way to very quickly transfer large amounts of thermal energy to the surface of the material.

This process can be performed without significantly raising the interior temperature of the product. It has been suggested that intense-light pulses would be suitable for surface sterilization for muscle food products.

Biological Control Systems

Fermentation

There are numerous muscle foods that owe their production and quality characteristics to the activity of microorganisms. Many of these are preserved products, in that their shelf-life is considerably longer than that of the original raw materials. Moreover, all fermented foods have aroma and flavor traits resulting from the fermentation.

Traditionally, food preservation by lactic acid fermentation was an empirical process. Early sausage makers learned that fresh meat could be preserved by carefully salting and drying the products. Although various geographical regions developed unique varieties of preserved meats and sausages, the inherent stability of all products was primarily dependent on the conversion of sugar to acid by lactic acid bacteria. Salting, curing, and drying favors the growth of lactic acid bacteria over other muscle food bacteria.

Lactic acid bacteria can be divided into two groups: The homofermentatives have lactic acid as the major product of glucose fermentation, whereas the heterofermentative organisms produce equal molar amounts of lactate, carbon dioxide, and ethanol.

The majority of fermented meat products, particularly in the United States consists of dry and semidry sausages. Fermented meat products have traditionally demonstrated an extended shelf-life through a combination of reduced moisture content and lowered pH (often <5.0). Microbial cultures contribute to the shelf-life of fermented meats mainly by consistent and controlled acidification. The acid inhibits undesirable microorganisms and allows efficient dehydration. The growth of other organisms may also be suppressed by competition for nutrients, H_2O_2, antibioticlike substances, and bactericidal proteins termed bacteriocins.

Bacteriocins

Bacteriocins are antibacterial substances that can be broken down by proteases (e.g., during digestion). By definition, bacteriocins are proteinaceous macromolecules that exert a bactericidal action on susceptible bacteria, usually closely related to the producer bacterium. The ability of bacteriocins produced by lactic acid bacteria to inhibit foodborne pathogens such as *Listeria monocytogenes* make them attractive as potential "natural" food preservation agents.

The bacteriocin, nisin, from the group N streptococci has been well characterized. Nisin is bactericidal to gram-positive bacteria in general and prevents outgrowth by *Clostridium* and *Bacillus* spores. Nisin is the only bacteriocin produced by lactic acid bacteria that has realized commercial application in food processing and fermentation. Currently, it is used as a food preservative in many countries. In 1988, it was declared a GRAS compound by the FDA.

Nisin has been considered as an alternative/partial replacement for nitrite in bacon. Studies indicate that only high (uneconomic) levels of nisin achieve good control of *C. botulinum*. Reasons for the poor preservative effect include the binding of nisin to meat particles, uneven distribution, and poor solubility.

In recent years, research has focused on the antilisterial effects of bacteriocins added to meat products. It appears that these compounds have potential as preservatives. A major drawback is the presence of proteases in meat. These may inactivate the bacteriocin. Recently, Degnan and Luchansky (1992) reported the successful encapsulation of a bacteriocin within liposomes. Upon heating of the product, the liposomes break open and the bacteriocin is released into an environment in which the enzymes have been inactivated.

The addition of bacteriocin-producing lactic acid bacteria (in starter culture) may provide a biological means of preservation.

Preservation with Chemicals

Acetic Acid

Vinegar-pickled sausages, pig's feet, and herring are examples of muscle foods preserved with acetic acid.

Acetic acid possesses rather weak bacteriostatic properties. The bacteriostatic effect increases at lower pH values. Reasonably high concentrations (>3.6% for vinegar-pickled sausages) must be employed in foods in order to preserve them effectively at room temperature. The concentration is related to the presence of fermentable carbohydrates.

Propionic Acid

Molds and other highly aerobic microorganisms are efficiently inhibited by propionic acid, whereas the anaerobic microorganisms and lactic acid bacteria can tolerate relatively high acid concentrations. Propionic acid has not gained favorable acceptance in the meat industry, probably because of its undesirable effect on sensory quality.

Sorbic Acid

Hexadienoic acid, commonly known as sorbic acid, is particularly well suited as a fungistatic agent. Sorbic acid works best below pH 6.0 and is generally ineffective >pH 6.5. At concentrations on the order of 0.05%, sorbic acid inhibits molds, yeasts, and some highly aerobic bacteria. In general, the catalase-positive cocci are more sensitive than catalase-negative organisms. Resistance of the lactic acid bacteria to sorbate permits its use as a fungistat in fermented products.

Shelf-life extensions have been obtained by the use of sorbates on fresh poultry meat, vacuum-packaged poultry products, and fresh fish (Sofos and Busta 1983).

Sorbates have been studied for use in bacon in combination with nitrites. The combination of 40 ppm nitrite and 2600 ppm sorbate was proposed by the USDA in 1978. Although this combination yields bacon with similar organoleptic properties and botulinal protection as 120 ppm nitrite, the proposal was rescinded in 1979. This action was prompted by taste panel results that characterized finished bacon as having "chemical-like flavors and producing prickly mouth sensations."

Sodium Lactate

Fresh pork or beef sausage has a limited shelf-life due to the absence of antimicrobial substances such as nitrite. Sodium lactate is a neutral substance (pH 6.5–7.5) which does not significantly alter the pH of products to which it is added. Within the last few years, sodium lactate has become recognized as an effective antimicrobial agent capable of extending the shelf-life of some meat and poultry products. Very little research has been reported on the antimicrobial properties of sodium lactate. Its effectiveness is thought to be due to its ability to depress water activity, but this characteristic does not totally explain its antimicrobial activity. It has been suggested that the effect of undissociated acids such as sodium lactate was due to the ability to cross molecular membranes, become acidified inside the cell, and acidify the cell interior. Further research is needed to resolve the mechanism of inhibition.

Selected References

Anonymous. 1985. Deactivation of microorganisms by an oscillating magnetic field. World patent 85/02094. Assigned to Maxwell Laboratories. Inc. San Diego, CA.

Anonymous. 1988. Methods and apparatus for preservation of foodstuffs. World patent application 88/03369. Applied by Maxwell Laboratories, Inc., San Diego, CA.

Anonymous. 1989. Food Irradiation: A most versatile 20th century technology for tomorrow. *Food Technol.* 43:75–77.

Anonymous. 1993. Code of Federal Regulations, Title of Animals and Animal Products. Chapter III Food Safety and Inspection Service, Meat and Poultry Inspection. Department of Agriculture Part 318.

Cassens, R.G. 1990. *Nitrite-Cured Meat: A Food Safety Issue in Perspective*. Food & Nutrition Press, Trumbull, CT.

Degnan, A.J. and J.B. Luchansky. 1992. Influence of beef tallow and muscle on the antilisterial activity of pediocin AcH and liposome-encapsulated pediocin AcH. *J. Food Protect.* 55:552–554.

Gill, C.O. 1986. The control of microbial spoilage in fresh meats. In: *Advances in Meat Research, Vol. 2*. A.M. Pearson, and T.R. Dutson, eds. Pages 49–88. AVI Publishing Co., Westport, CT.

Hayashi, R. 1989. Application of high pressure to engineering food processing and preservation: Philosophy and development. In: *Engineering and Food, Vol. 2*. W.E. Spiess and H. Schubert eds. Pages 815–826. Elsevier Applied Science, London.

Hobbs, G. 1983. Microbial Spoilage of fish. In Food Microbiology Advances and Prospects, TA Roberts and FA Skinner eds. pages 217–227. Acad. Press, New York.

Hoover et al. 1989. Biological effects of high hydrostatic pressure on food microorganisms. *Food Technol*. 43:99–107.

Jay, M.J. 1986. *Modern Food Microbiology*. Van Nostrand Reinhold Company, New York.

Jeremiah, L.E., N. Penney, and C.O. Gill. 1992. The effects of prolonged storage under vacuum or CO_2 on the flavor and texture profiles of chilled pork. *Food Res. Int*. 25:9–19.

Klaenhammer, T.R. 1988. Bacteriocins of lactic acid bacteria. *Biochimie* 70:337–348.

Leistner, L. 1992. Food preservation by combined methods. *Food Res. Int*. 25:151–158.

Leistner, L., W. Rödel, and K. Krispien. 1981. Microbiology of meat and meat products in high- and intermediate-moisture ranges. In: *Water Activity: Influences on Food Quality*. L.B. Rockland, and G.F. Stewart, eds. Pages 855–920. Academic Press, New York.

Macfarlane, J.J. 1985. High pressure technology and meat quality. In: *Developments in Meat Science, Vol. 3,* R.A. Lawrie, ed. Pages 155–184. Elsevier Applied Science, London.

Mackey, M.M. and C.M. Derrick. 1987. The effect of prior heat shock in the thermoresistance of *Salmonella thompson* in foods. *Lett. Appl. Microbiol*. 5:115–119.

Martin, R.E., R.J.H. Gray, and M.D. Pierson. 1978. Quality assessment of fresh fish and the role of naturally occurring microflora. *Food Technol*. 32:188–193.

McMeekin, T.A. 1982. Microbial spoilage of meats. In: *Developments in Food Microbiology, Vol. 1,* R. Davies, ed. Pages 1–40. Applied Science Publishers, London.

Mertens, B. and D. Knorr. 1992. Developments of nonthermal processes for food preservation. *Food Technol*. 47:124–133.

Murano, E.A. and M.D. Pierson. 1992. Effect of heat shock and growth atmosphere on the heat resistance of *Escherichia coli* 0157:H7. *J. Food Protect*. 55:171–175.

Sato, K., G.R. Hegarty, and H.K. Herring. 1973. The inhibition of warmed-over flavor in cooked meats. *J. Food Sci*. 38:398–392.

Shigehisha, T., T. Ohmori, A. Saito, S. Taji, and R. Hayashi. 1991. Effects of high hydrostatic pressure on characteristics of pork slurries and inactivation of microorganisms associated with meat and meat products. *Int. J. Food Microbiol*. 12:207–216.

Sofos, J.N. and F.F. Busta. 1983. Sorbates. In: *Sorbates in Foods,* A.L. Branen and P.M. Davidson, eds. Pages 141–175. Marcel Dekker, New York.

Urbain, W.M. 1978. Meat preservation. In: *The Science of Meat and Meat Products,* J.F. Price and B.S. Schweigert, eds. Pages 402–451. Food & Nutrition Press, Westport, CT.

Urbain, W.M. 1986. *Food Irradiation*. Academic Press, Orlando, FL.

Ward, D.R. and N.J. Baj. 1988. Factors affecting microbiological quality of seafoods. *Food Technol*. 42:85–89.

15

Cookery of Muscle Foods

Anna V. A. Resurreccion

General Principles of Cookery

Cooking influences the appearance, protein, water-binding, and textural properties of meat. The cooking method markedly affects the palatability of meat, poultry, or fish.

Effect of Heat

Heating during cooking results in extensive changes in the appearance and physical properties of muscle tissue. These changes are dependent on cooking time and temperature conditions. Studies conducted on small cylinders of meat have shown that toughening occurs when the meat reaches temperatures of 60–70°C (140–158°F). This is followed by tenderization at 75.5°C (168°F). The original toughening appears to be the result of protein denaturation, whereas the tenderization is attributed to hydrolysis of collagen in the connective tissue to gelatin. The temperature at which muscle proteins denatures has been shown to occur between 50°C and 80°C (122°F and 176°F) (Stabursvik and Martens 1980). When muscle is cooked, it becomes firmer as the proteins are denatured. This process continues as the temperature is raised. At 50°C (122°F) α-actinin, the most heat-labile muscle protein, becomes insoluble. At 55°C (131°F) the light chains of myosin become insoluble; then actin becomes insoluble at 70–80°C (158–176°F). Myosin and troponin are the most heat-resistant proteins and become insoluble at 80°C (176°F). The 30,000-Da component also increases during heating, and this may occur due to heat activation of the calcium-activated protease (Hultin 1985). It appears that the connecting filaments that are situated parallel to the myofibrils are the only links in the I-band that survive heating, and the tensile strength of these elements is responsible for the residual toughness of myofibrillar proteins in cooked meat.

Denaturation of the muscle proteins decreases their water-holding capacity. Heat denaturation and surface dehydration that take place upon cooking may also cause a surface layer to form.

The toughening of the contractile proteins is counteracted by alterations in collagen. Changes in collagen begin to occur at 50°C and 60°C (122°F and 140°F) for younger and older animals, respectively. As temperature is increased, collagen's regular structure breaks (denatures) and the chains separate and fold into random structures without any residual native structure. The thermal stability of molecular collagen, as indicated by melting temperature, correlates closely with the hydroxyproline content of vertebrate collagens (Aberle and Mills 1983).

Myoglobin also undergoes denaturation during heating. The susceptibility of the heme pigment to oxidation in the denatured protein is much greater than it is in undenatured myoglobin. On heating, red meat generally turns brown due to formation of the oxidized pigment hemin (Hultin 1985); thus, the color of meat is typically used as an index for doneness (Fig. 15.1).

Heating also results in the melting of fat. This and changes in the water-holding capacity of the meat result in differences in sensory properties such as apparent juiciness and tenderness. Furthermore, the volume of the meat decreases upon cooking.

Cooking Losses

Cooking losses result from the melting of fat to form drippings, evaporation of water, and other volatiles. When meat is cooked in an open container, considerable evaporation of the water from the surface of the meat occurs. Fat losses that occur are less consistent due to the uneven distribution of fat throughout the meat. Fat that is located close to the surface will be lost faster than that found in the interior of the meat; however, the surface fat layer aids in decreasing water loss due to evaporation.

Recommended End-Point Temperatures

Suggested internal end-point temperatures are shown in Table 15.1. Most timetables for household use are based on meat at refrigeration temperature. Allowances should be made when meat is cooked from the frozen state. Frozen meat will require a cooking period one-third to one-half times longer than thawed meat because it is at a lower temperature and additional energy is needed to melt the ice crystals and to raise the temperature to that of refrigerated meat.

Dry- and Moist-Heat Cooking Methods

Muscles that are used for locomotion and provide power, such as those in the leg and shoulder areas, have more connective tissue and, in general, yield less

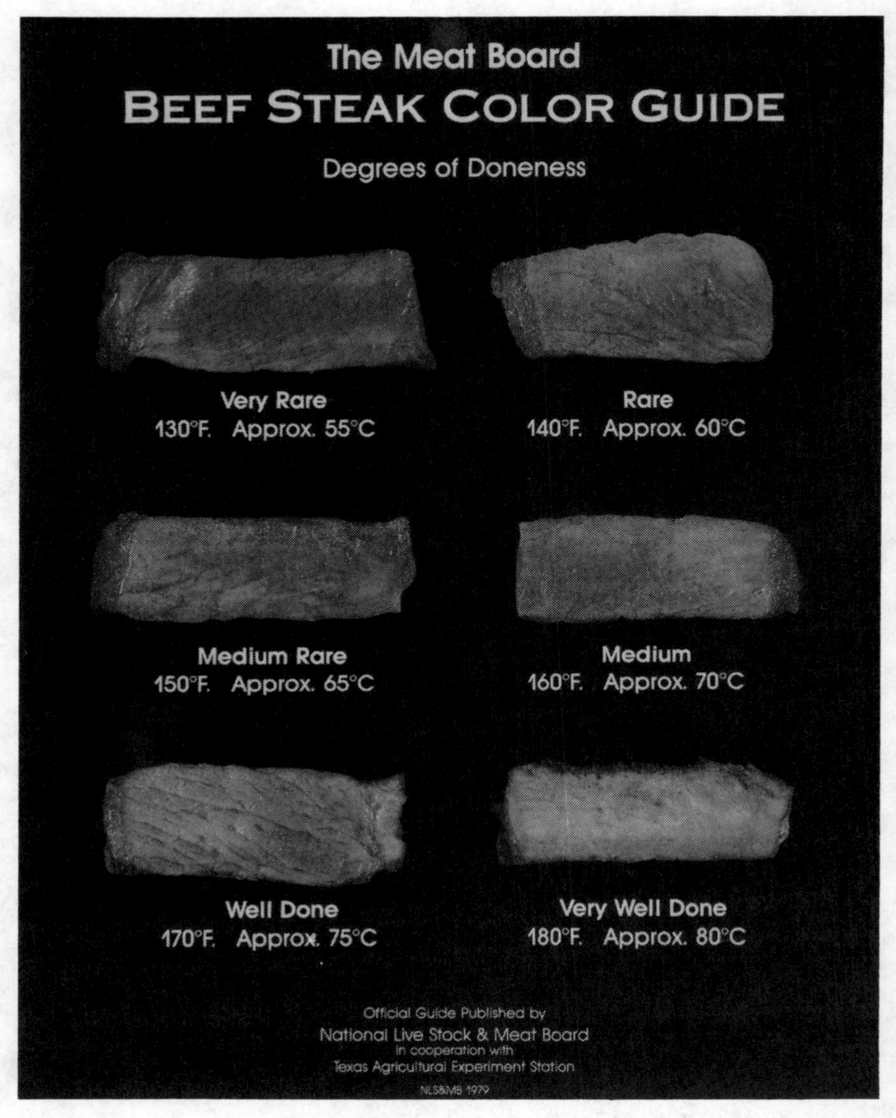

Figure 15.1. Color changes in beef cooked to varying internal end-point temperatures.

tender meat. The muscles that support, such as those in the back, rib, and loin, yield tender meat. A general guideline in selecting a cooking method is to use dry heat methods such as roasting, broiling, and panfrying to cook cuts that yield tender meat and to use moist heat methods such as braising and cooking in liquid such as in stewing and boiling for cuts that generally yield less tender meat.

Table 15.1. Suggested Internal End-Point Temperatures for Various Muscle Foods

	Degree of Doneness			
Food Item	Rare	Medium Rare	Medium	Well Done
Beef	60°C (140°F)	66°C (150°F)	71°C (160°F)	77°C (170°F)
Veal	—	—	71°C (160°F)	77°C (170°F)
Pork	—	—	71°C (160°F)	77°C (170°F)
Lamb	60°C (140°F)	66°C (150°F)	71°C (160°F)	77°C (170°F)
Chicken				
Thigh		—	—	82°C (180°F)
Breast		—	—	71°C (160°F)
Fish		—	—	80°C (175°F)

Meat, poultry, and fish are, in general, more uniformly cooked and more tender when cooked slowly. Low to moderate temperatures result in reduced toughness whatever the cooking method. For example, heating in the oven at 163°C (325°F) and simmering, rather than boiling temperatures when cooking meat, poultry, and fish in moisture are best.

Roasting

Roasting is a method by which heat is transmitted to the meat by convection in a closed preheated oven. Larger pieces of tender cuts of meat or poultry weighing 3 lbs or more are cooked uncovered and without added moisture in an oven or oven-type appliance. When fish and poultry are cooked using this method, it is referred to as baked rather than roasted. The meat or poultry is placed on a rack either in an uncovered, usually low-sided pan for even circulation of heat around the meat and to hold the food above any drippings. A beef rib roast need not be placed on a rack, unless it is boneless, because the rib bones usually form a natural rack.

When roasting is conducted in an oven, this has been described as "oven roasting." The oven door is closed and the meat is usually not turned during cooking. For steaks, that method has been referred to as "modified oven roasting" and "modified broiling."

Preheating

It is not necessary to preheat the oven. However, when preheating at a high temperature research has shown that a high starting oven temperature such as 232°C (450°F) results in greater cooking losses and less yield than when a constant temperature is used throughout the cooking period. In addition, high temperatures will likely result in more spattering and nonuniform cooking.

Procedure

When roasting, added flavorings will penetrate no more than half of an inch into the meat; therefore, it makes little difference whether seasonings are added before or after cooking. A thermocouple (or meat thermometer for household use) should be inserted into the geometric center of the meat. Place the meat or poultry in the center of the roasting pan; then place it in the center of the preheated oven at 177°C (350°F). The oven temperature may be monitored by using another thermocouple positioned near the center of the oven adjacent to the meat and should be controlled within ±5°C (±9°F). Many household-type ovens do not have sensitive controls and, if used in research, should be monitored constantly; however, models used in research are well insulated and maintain temperature to ±2°C (±3.6°F).

Suggested internal temperatures for meat, poultry, and fish are shown in Table 15.1. Meat temperature will usually rise approximately 2.8°C (5°F) but varies according to the size of the roast, during the 15–20-min recommended standing time after the roast is removed from the oven before carving.

Timetables

Timetables are guidelines to the approximate roasting time and are compiled from averages of several tests. If used, these should be in conjunction with other indices of doneness such as internal temperature and color (Fig. 15.1) of the roast.

Variation in Roasting Time

Sources of variation in the roasting time of meat result from differences in the meat such as the shape and size of the meat cut and the amount of fat and bone. Other sources of variation may result from the type of equipment and specific procedures used. The initial temperature of the roast when roasting begins and the accuracy of the equipment used are examples of such procedure and equipment variables, respectively.

Broiling

Broiling is a dry heat method that uses radiant heat to cook the meat. It is usually used for steaks, chops and patties, poultry halves, quarters, and pieces, and fish. The equipment used is an oven broiler or a special broiling appliance such as an electric broiler or an outdoor grill. When an oven broiler is used, this is referred to as "oven broiling." The heat usually radiates from one direction, so the meat must be turned during cooking. In research studies requiring the use of a special broiling appliance, investigators turned steaks frequently to maintain even browning (Berry, Smith, and Secrist 1987). In another study utilizing a special broiling unit, steaks monitored with a thermocouple were turned when an internal temperature of 40°C (104°F) was reached, then removed from the broiler at 70°C (158°F) (Crouse, Theer, and Seideman 1989). The American Meat Science Association and National Live Stock and Meat Board guidelines for cookery (Anonymous 1978) state that there is some confusion as to whether oven broiling is the same as oven roasting.

Many researchers involved in studies on meat cookery prefer to broil meat because it closely resembles the method used by consumers, whereas other researchers prefer roasting methods because the cooking conditions can be more easily controlled. Results of studies that compared roasting with broiling indicated that palatability differed little between the two methods, but cooking loss was lower for roasting than broiling. However, McCrae and Paul (1974) found cooking loss to be significantly lower in oven broiling at 204°C (400°F) compared to roasting at 163°C (325°F), both to an internal end-point temperature of 70°C (158°F) and attributed this to the relatively short cooking time required during broiling.

Procedure

To broil meat, insert the thermocouple into the geometric center of the meat, then place these on a rack or in a broiler pan in a preheated broiler or oven at 204°C (400°F). Arrange the meat and rack 8–10 cm from the heat source, varying the distance according to the initial temperature and dimensions of the meat. Broiling temperatures are generally more difficult to control than roasting temperatures and should be closely monitored. Steaks should be turned once during broiling and the meat removed when it reaches the recommended internal temperature (Fig. 15.1).

Variations in Broiling Time

Variations in broiling time for steaks, chops, and patties are mostly due to the temperature and thickness of the meat, the distance of the surface of the meat from the heat source, and the equipment and procedure used. In households, frozen meat is usually thawed before broiling. However, the distance should be adjusted when the meat is frozen and when thicker cuts are used. A rule of thumb is to broil at distance of 5 cm (2 in.) from the heat source for every 2.5 cm of thickness of steak or chop, but distance should be varied according to the broiling temperature used.

Spattering and Smoking

Spattering and smoking during broiling are caused by very high cooking temperatures. Both can be decreased by broiling at a lower temperature or increasing the distance between the meat and the heat source. Trimming excess fat from the meat will also help to minimize accumulation of rendered fat and prevent smoking.

Panbroiling

When the meat is cooked in an uncovered pan without the addition of liquid or fat, it is said to be panbroiled. Steaks and chops less than ¾–1-in. thick are usually panbroiled to brown the surface before the meat is overcooked.

Frying and Deep-Fat Frying

Comparatively thin pieces of tender meat are suitable for frying. Cuts made tender by pounding, scoring, cubing or grinding, and cooked meat are likewise suitable.

Procedure

When frying, meat is cooked, uncovered, at a moderate temperature and usually browned on both sides in a small amount of fat and turned occasionally to ensure even cooking. Deep-fat frying involves immersing and cooking the meat in hot oil at deep frying temperatures, which range from 177°C to 182°C (350–360°F). A deep-fat frying thermometer should be used to maintain oil temperature unless the meat is cooked in a thermostatically controlled fryer.

Braising

Braising is a method in which the meat is cooked slowly, in a moist atmosphere, due to added water in a closed system such as a covered pan. Traditionally, the less tender cuts are braised.

Procedure

Meat is often browned before braising. Browning is not necessary but helps to develop flavor and color. For household or institutional use, the meat may be dredged in seasoned flour before browning in a small amount of fat in a heavy skillet, then braised in an electric skillet, Dutch oven, or in a tightly covered pan on a range surface unit or oven at 165°C (329°F). In a laboratory or research setting, use a thermocouple inserted in the center of the largest muscle in the meat. In a household setting, a meat thermometer may be used to determine the end-point temperature. The meat should be placed on a rack and approximately 125 or 50 ml of water should be used for roasts, or steaks and chops, respectively. In place of water, liquids such as tomato juice or beef bouillon may be used in institutional or home cookery.

Braising was found to produce the highest drip loss, compared to oven roasting, broiling, and microwave cookery due to the added water and reduced evaporation in a covered container. Braising also resulted in the highest percent nonfat solids retained in the meat and lower percent lost into the liquid compared to roasting and boiling (McCrae and Paul 1974).

Other methods of braising include "pot roasting," cooking meat in a bag, and "fricasseeing." "Pot roasting" is a term sometimes used to refer to the braising of large pieces of meat. When cooking meat in a bag in the oven, the meat is subjected to a high moist heat environment similar to braising. The term "fricassee" may be applied to braised meats cut into smaller pieces before cooking.

Cooking in Liquid

Procedure

Boiling and stewing are methods of cooking in liquid. The procedure for stewing is very much like braising, except that enough liquid is added to cover the meat and the meat may or may not be prebrowned. Boiling is cooking meat in enough liquid to cover the meat. The liquid used is usually water and may or may not have spices added for flavor.

Microwave Cooking

In microwave cookery, meat is heated by microwave energy in a microwave oven. Meats that are particularly suited to microwave cookery are precooked and processed meats and selected meat cuts. Microwave cookery results in the greatest drip loss during cooking compared to conventional roasting and oven broiling (McCrae and Paul 1974). Howat et al. (1987) found drip loss of microwaved beef roasts to be significantly higher than when cooked in conventional and microwave–convection ovens. The uneven heating of foods during microwave cooking has likewise been recognized as a problem.

McCrae and Paul (1974) reported that greater histological changes occurred in collagenous tissue when beef was cooked by microwaves than when it was cooked in a conventional oven either by moist or dry heat methods. Gibson and Jeremiah (1988) agree with their findings that microwave cookery affects myofibrillar and stromal proteins differently than slower methods of cookery. Bodrero, Pearson, and Magee (1980) likewise stated that the shorter heating times associated with microwave cookery are probably insufficient in duration to result in significant degradation of myofibrillar or stromal proteins. In turkey muscle, however, no difference was found in fragmentation of turkey muscle cooked conventionally or by microwaves. No differences in the histological characteristics of bovine or porcine muscle cooked by dry or moist heat in a conventional or microwave oven were found by investigators who reported no differences in sensory panelist scores for collagenous tissue in beef or pork cooked by either dry or moist heat methods, in a conventional oven or in the microwave.

Procedure

To microwave meat, poultry, or fish, the food is placed in a container suitable for use in a microwave oven and covered with a suitable lid or plastic film wrap. In a study of several types of food cooked in 915- and 2450-MHz microwave appliances, those covered with a plastic film were rated better in one or more quality characteristics when compared with the same food cooked without the film cover.

Selection, Preparation, and Handling

Beef and Veal

For maximum tenderness and juiciness, beef and veal should be cooked at low to moderate temperatures. Advantages to using these temperatures are better color and flavor characteristics, and smaller cooking losses compared to cooking at high temperatures.

Dry-heat methods are used by most consumers when preparing beef. A study by Howat and Kirby (1982) on consumers' reported beef preparation practices indicated that broiling or grilling was the preparation method used by the majority or 39% of respondents, followed by stewing and braising (24%), frying (23%), and lastly, oven roasting was reported by 12%.

Selection of Cooking Method

One of the most important considerations for proper beef cookery is to know the nature of the cut to be cooked and the best cooking method to be used. However, the traditional practice of using dry-heat methods of cookery for only the tender cuts of meat is based on the assumed effect of these methods as well as on the differences in composition of the raw cuts. In steaks broiled to an internal temperature of 70°C (158°F) on an open-heart grill, McKeith et al. (1985) reported muscles from round and chuck, excluding *infraspinatus,* were scored lower in tenderness than muscles from loin and rib. The *psoas major, longissimus,* and *infraspinatus* were significantly more tender in sensory evaluation tests than seven other muscles; instrumental measures of tenderness using the Warner–Bratzler shear followed the same trend. Large ranges were observed among muscles for percent fat, sarcomere length, and total collagen.

Tender cuts of beef weighing 3 lbs or more are best cooked by dry-heat methods. These include beef rib, rib eye, top round, tip, and rump roasts. Because of its shape and tenderness, it is recommended that beef tenderloin be roasted in a 218°C (425°F) oven rather than the usual 163°C (325°F). Some large cuts such as top round, tip, and rump can usually be roasted or braised depending on the grade and tenderness of the cut. Tenderness cannot always be judged before cooking, although the USDA grade for quality is a guide. If one roasts these large cuts, it is preferable to cook them to the rare stage, then carve into thin slices to serve. Less tender cuts of beef such as flank and top round steaks may be broiled to the rare to medium stage, then carved diagonally across the muscle fibers into thin slices. These and other less tender cuts of beef can be most successfully broiled if the meat is marinated before cooking or made tender by some other method.

Meat can be tenderized by the application of a marinade that may contain weak organic acids in the form of lemon juice, tomatoes, wine or vinegar. These marinades penetrate only about 1.25 cm into the meat. Enzyme tenderizers are more effective than acid marinades. Commercial enzyme preparations in powder form or combined with seasonings are of plant origin and may include papain, bromelin, and ficin from papaya, pineapple, and figs, respectively. Overtenderizing may occur by using too much enzyme tenderizer or allowing the meat to remain in contact with the enzyme preparation for longer than recommended in the instructions. Overexposure to the enzymes at the optimal temperature for enzyme activity will result in overtenderization. Marinades and enzyme preparations are more effective on thinner cuts of meat such as steaks; their use on roasts or other thick cuts of meat will not allow sufficient penetration and, therefore, will be ineffective.

Physical methods that break or cut the muscle fibers and connective tissue may be used to tenderize meat. Such methods include chopping, pounding, or the use of special equipment that pierces the meat with multiple thin needles.

End-Point Temperatures

A range of internal endpoint temperatures is acceptable for beef cookery. The National Live Stock and Meat Board describes temperature end-points for beef and are shown in Fig. 15.1. Rare beef has a brown exterior, reddish-pink interior, and clear red juice. Beef cooked to a medium degree of doneness has a light pink interior, less juice, and lighter-colored juice than rare beef. Well-done beef is light brown throughout with slightly yellow juice.

Effect of Cooking on Beef Quality Characteristics

Studies examining the effect of final end-point temperature on palatability attributes reported decreased sensory characteristics as internal temperature increased. In a study of the effect on beef color and flavor upon roasting and broiling to internal end-point temperatures between 55°C and 85°C (131°F and 185°F), it was noted that broiled steaks were heated at a much faster rate than roasted ones. Percentage cooking losses were greater for broiled steaks (Bowers et al. 1987) and these increased as final end-point temperature increased. These losses were attributed to the shrinking of muscle fibers and shortening of sarcomeres resulting in less water-binding ability at internal temperatures above 70°C (158°F). They found that neither the final temperature nor method of heating significantly affected Instron shear and compression measurements, contrary to earlier reports indicating increases in peak shear force values as cooking temperature was increased.

A recent study on retail beef cuts by Luchak (1991) recommends that consumers should cook beef steaks for less time and to a lower degree of doneness to obtain steaks that are more tender and juicier. Flavor development and juiciness are significantly influenced by both the final end-point temperature and the method of cooking. As end-point temperature increases, the desirable aromatics such as cooked beef/brothy increase as do the undesirable notes, cardboard, and livery/organy. The serumy/bloody notes associated with beef blood decrease with increased end-point temperatures as do the metallic mouthfeel attribute and the basic tastes, sour and bitter. The muscles roasted at 149°C (300°F) compared with those broiled tended to have less of a "mouth-filling blend," more bloody, serumy, metallic, sour flavors and more juiciness (Bowers et al. 1987). Denaturation of myosin at 60°C (140°F) and sarcoplasmic protein at 67°C (153°F) may have accounted for the flavor changes.

The oxidized flavor, also referred to as warmed-over flavor (WOF), decreases as internal end-point temperature increases. The thiobarbituric acid (TBA) test is one of the most widely used methods for measuring the degree of lipid oxidation. In this procedure, malonaldehyde and other oxidation products of unsaturated lipids react with TBA to form a red complex. The TBA numbers correlate with sensory ratings by trained panelists of oxidized, rancid, and warmed-over flavors.

When the malonaldehyde content of several species was measured after cooking by conventional dry and moist heat cookery methods as an index of oxidative rancidity, beef and pork cooked for the longest periods (greater than 30 min) demonstrated the greatest increase in malonaldehyde content. Roasting for 3 h led to significant increases in malonaldehyde.

The TBA values of ground beef cooked in an autoclave were reported to decrease as internal cook temperature increased to 80, 95, 99, 100, and 110°C (176, 203, 210, 212, and 230°F). The investigators related the decrease in TBA values to Maillard-reaction products that were responsible for preventing WOF development in meat. Bailey et al. (1987) proposes three mechanisms for the prevention of WOF development by Maillard-reaction products. First, hydroperoxides react to form products that do not form free radicals. Second, free radicals formed during oxidative degradation are inactivated by transferring hydrogen atoms to hydroperoxy radicals. Lastly, heavy metal ions are complexed by the reaction products.

Gros et al. (1986) found that the oven-broiling method was found to best delay the rate of WOF development in ground beef muscles relative to microwave and combination cooking methods, which did not offer protection against WOF development after refrigeration. Oven broiling may, therefore, be the preferred method of cooking products such as beef patties which are precooked and refrigerated prior to service.

The effect of fat on the quality factors of broiled ground beef patties has been thoroughly investigated. Broiled ground beef patties containing 25% fat received higher sensory scores for tenderness and juiciness than patties containing 15% fat (Kregel, Prusa, and Hughes 1986); however, beef flavor intensity did not differ with fat content but increased with the higher internal end-point temperature. Belk et al. (1992) concluded that only minimal amounts of fat are required to maintain the desirable flavor in beef after finding that fat-trim levels ranging from 0 to 1.27 cm resulted in negligible differences. Luchak (1991) found that roasts with an external fat trim of 0.5 cm resulted in longer cooking time than roasts without external fat trim. He recommended that cuts with no external fat trim should be cooked for shorter periods of time than with external fat to avoid overcooking.

Microwave Cookery

Steaks prepared in a microwave oven result in greater cooking losses compared with steaks prepared in a convection oven or a conventional electric oven (Gibson and Jeremiah 1988). A study reported by Howat et al. (1987) on beef blade roasts cooked using moist heat conditions, by the addition of 30 g of water and covering with a plastic film cover, by microwave, microwave–convection (both at 2450 MHz, 700 W), and conventional methods found that comparable cooking losses resulted from all three. Warner–Bratzler shear values indicated no differences in samples prepared to an internal end-point temperature of 75°F (167°F) using moist heat in microwave, microwave–convection, or conventional ovens or using dry heat in a conventional oven. Sensory evaluation results stated in both studies indicated that the dry-heat conventional oven method resulted in a significantly more tender product (Howat et al. 1987) and that the eating quality of microwaved steaks was significantly compromised (Gibson and Jeremiah 1988). However, these studies included only 10 or less panelists on which to draw conclusions about perceptions of tenderness or overall palatability of the samples.

Salting

The question of when meat should be salted often occurs. If the price of meat is large, salt added prior to cooking will not penetrate the meat to a depth much greater than 1.25 cm (0.5 in.). In such a case, adding salt may result in too salty an outer layer or too salty drippings. Salt also retards the browning of meat; therefore, it is best to apply salt to steaks and chops after they are browned. In meat loaves and other ground meat dishes such as patties and meat balls, seasonings and salt are usually added to the meat before it is shaped. Sodium chloride is effective in reducing cooking losses by improving moisture retention of ground meat mixtures. Salt increases the solubility of the muscle proteins and water-binding ability (Puolanne and Terrell 1983). When cooking small pieces of meat for stews, salt may be added directly into the cooking liquid.

Buckling or Curling

Buckling or curling is due to shrinkage of the connective tissue membrane surrounding the muscle when an excessively high temperature is used, especially when broiling. This may be prevented by slashing the fat edge in several places before broiling.

Ground Beef

Ground beef is the principal beef item purchased by consumers (Berry and Hasty 1982), mainly for its leanness, price, and ease in preparation. Recently the U.S. Department of Agriculture recommended minimal holding times for minimum internal temperatures at the center of fully cooked patties. These include a 16-s minimum holding time at 69°C (155°F). A higher end-point temperature for heating ground beef allows for more evaporative loss during heating and increased cooking losses. Kregel, Prusa, and Hughes (1986) likewise found that ground beef patties heated to 77°C (170°F) had significantly higher evaporative and total cooking losses than samples heated to 71°C (160°F). Drip loss, however, was not affected by internal end-point temperature. The effect of end-point temperature on moisture content (Kregel, Prusa, and Hughes 1986) depends on the fat content of the meat.

Veal

Veal leg and shoulder cuts are fine for roasting to an internal temperature of 77°C (170°F). Veal has too little fat to broil well. Steaks and chops should be braised or panfried. Veal needs longer cooking to make it tender and palatable because it has a larger proportion of connective tissue than other meats. When cooked to the well-done stage, veal has a red-brown exterior and a gray interior color.

Pork

Cookery of pork has a major effect on its sensory and compositional characteristics. Cooking temperature and especially internal end-point temperature affect the sensory properties such as color, tenderness, flavor, and juiciness of the cooked pork (Siemens et al. 1990) as it does all meats. Pork cooked at low to moderate temperatures is generally more tender, juicy, and flavorful than pork cooked at higher temperatures.

Pork is cooked well done primarily for flavor development and to destroy *Trichinella spiralis*. Pork should be cooked until the heat penetrates into the center of the meat. To be safe, pork should be thoroughly cooked to an internal temperature of 71°C (160°F) to 77°C (170°F) corresponding to the medium to well-done stages, respectively. This provides a 13–18°C (23–33°F) safety margin because *Trichinella spiralis* is killed at 58°C (137°F). Pork is more tender and flavorful when cooked to the higher internal temperatures. A simple test for doneness of pork is to make small cuts next to the bone and into the thicker part of the meat. If the juice and meat are still pink, the meat is not done.

For roasts, the best method for ensuring doneness in a household situation is to use a meat thermometer. Research recommends an internal temperature of 77°C (170°F) for pork roasts. Pork chops may be broiled as long as cuts are no less than 1-in. thick and broiled far enough from the heat so they will be cooked to the well-done stage but not be dry by the time they are browned.

Pork center loin chops and roasts were not adversely affected by removal of fat cover or by cooking to 71°C (160°F). Chops broiled to 71°C (160°F) had greater percent protein and more moisture, but fat content was not affected (Siemens et al. 1990). Microwave reheating of conventionally precooked pork following frozen storage resulted in TBA values that were highest for pork exposed to the longest heating time.

Occasionally, some muscles of pork cooked to the well-done stage may exhibit a pink color. These reddish areas are similar to the red color found in cured meats such as corned beef and ham. In moist-heat cookery, the meat color is fixed by naturally occurring nitrates and nitrites from the water and vegetables cooked with the meat. When pork is cooked by dry heat, the presence of carbon monoxide in heating and exhaust gasses may cause the red areas to appear.

Hams and Cured Meats

Most hams are fully cooked and should have a declaration on the label. A "fully-cooked" ham has been cured, smoked to an internal temperature of approximately 66°C (150°F), and maintained at that temperature long enough to be served without further cooking. To serve "fully-cooked" hams, they need only to be warmed to 54–60°C (130–140°F) internal temperature. "Cook before eating" hams, as their name indicates, require cooking. Their texture is improved with heating to 71°C (160°F) internal temperature. Roast "cook before eating" picnics to an internal temperature of 77°C (170°F) or place in a deep kettle, cover with water and simmer, allowing 3–4 h for a 6–8 lb picnic shoulder. Roast fully cooked picnics to 60°C (140°F).

Specialty hams such as country-style, Smithfield, or Virginia hams are heavily cured hams with a distinctive flavor and texture and are "cook before eating" hams. To reduce the salty flavor, soaking and precooking in liquid is usually necessary before baking.

Lamb

Lamb has a "fell," which is a paperlike connective tissue covering. It helps to retain the shape of the meat during cooking and helps to retain moisture; thus, it should not be removed from roasts and chops purchased at the retail market.

Lamb is generally tender. Lamb cuts, with the exception of neck and shank, are tender and can be roasted, broiled, or panfried. The blade or arm chops cut from the shoulder can be either broiled or braised. A crown roast is made by using two rib roasts sewn together to form a circle or crown. Lamb rib roasts should be cooked at an oven temperature of 191°C (375°F) rather than 163°C (325°F) because of their shape and size. Lamb shanks are best braised or cooked in liquid, usually with vegetables added.

End-Point Temperatures

Lamb may be served rare, cooked to an end-point temperature of 60°C (140°F), medium, cooked to an end-point temperature of 71°C (160°F), or well done, cooked to an end-point temperature of 77°C (170°F).

Poultry

Tender chickens, turkeys, and ducks are usually roasted on a rack in a shallow uncovered pan. The body cavity can be stuffed or not, as desired. The bird is usually trussed by fastening the neck skin to the back with a skewer, turning the wing tips back of the shoulders, and tying the drumsticks to the tail. This procedure prevents the wings and legs from becoming too brown and makes the bird more attractive for serving.

An oven temperature of 160°C (320°F) is used for stuffed poultry and 200°C (392°F) for unstuffed broilers, fryers, or roasters. Poultry is usually cooked well-done but not overcooked. Poultry is done when the internal temperature is 82°C (180°F) in the thigh and 71°C (160°F) in the breast. At this temperature the leg joints move easily and the meat of the drumstick is soft when pressed.

When a small turkey is being roasted, the internal temperature in the breast should be 77°C (170°F). Turkey toms were cooked to internal temperatures of 85°C (185°F) and 95°C (203°F) corresponding to the optimum and overcooked stages of doneness, respectively (Poste et al. 1987). These final internal temperatures did not have an effect on drip or jelly losses, although increased evaporation and fat losses were observed in samples cooked to the higher end-point temperature. Sensory ratings of the cooked turkey appearance were not significantly different, but the turkeys cooked at 95°C (203°F) were significantly less juicy and less tender than those cooked to 85°C (185°F).

Broiling

Usually the smallest of the broiler–fryers are selected for broiling. A popular size is approximately 1 kg (2.25 lbs), although young chickens of any weight can be broiled. Depending on the size, the chickens are cut in half lengthwise or quartered and placed skin side down in a broiler pan or other shallow pan, one layer deep. The use of a rack is optional. The chicken is placed in the broiler so that the highest part is 12–15 cm away from the heat. Pieces are turned as they brown.

Frying

Young chicken of any weight and cut in any way desired can be fried, with or without breading. Enough fat, about 1 cm deep is placed in a frying pan or in an electric skillet and heated until the recommended temperature is reached. The pieces of chicken are browned in the fat, turning as necessary.

Moist-Heat Cookery

Older poultry such as mature hens are often cooked by moist-heat methods. Such poultry can be fricasseed, steamed, or simmered in water.

Microwave Cookery

Poultry can also be microwaved, but care must be taken to ensure that the poultry is thoroughly cooked. The use of plastic film cover during microwave cooking in 915- and 2450-MHz appliances resulted in a significant reduction of total cooking losses in chicken but not in drip loss. Furthermore, sensory comparisons of appearance, color, texture, flavor, and overall preference were rated better by 50% of the panel when chicken was cooked with the plastic cover, but moistness was not rated better. Turkey that was covered with plastic wrap, then microwaved, was rated better than the uncovered microwaved control on all attributes evaluated.

Reheated Poultry

In a study of the development of warmed-over flavor of reheated precooked broiler breast, thigh, and skin, Ang and Lyon (1990) found that TBA values and trained sensory panelist ratings of cardboard, rancid/painty, and warmed-over flavor were higher in day-old samples compared with the control. They did not conduct tests to determine whether typical consumers are able to differentiate these flavors found in day-old samples from those found in the control.

Fish

Fish may be baked, broiled, fried, microwaved, poached, or broiled. Baked fish is prepared in a preheated oven at 204°C (400°F) to an internal temperature of 80°C (176°F). Broiling may be accomplished by cooking the fish 2 in. from the heat source until well done, then turned and broiled on the other side to an internal temperature of 75–90°C (167–194°F). Fried fish can be cooked using vegetable shortening or cooking oil. The fish may be dredged in flour and fried until browned on each side and cooked to an internal temperature of 75–90°C (167–194°F). In some instances, fish may be poached by placing it on a rack and immersing this in water at 95°C (203°F). The fish is covered while cooking and heated to an internal temperature of 80–90°C (176–194°F). Fish may also be boiled in water to an internal temperature of 80–90°C (176–194°F).

Effect of Cookery Method on Fish Quality

In a study to determine the effects of several methods of preparing striped bass on their sensory properties, Armbruster et al. (1987) found that panelists rated the color, odor, flavor, tenderness, juiciness, and overall acceptance of fish higher when prepared by using dry-heat methods such as baking, broiling, and frying in comparison with those cooked with moist-heat methods such as microwaving, poaching, and boiling.

The quality of fried battered fish portions depends on several factors. The thicker the portions, the longer the cooking time, and the more frying fat is absorbed. The volume of the fish portions placed in the fryer also affects cooking time if the volume of the fish causes a rapid drop in frying temperature, as much as 15–30°C (27–54°F), thereby resulting in an increased cooking time because of the slower heat penetration rate.

Several factors determine the amount of fat absorbed by the battered fish portions, such as the temperature of the frying oil, properties and composition of the batter, and postpreparation holding time. Crude lipid of the batter decreased as frying temperature increased. The longer the cooking time, the more fat was absorbed by the batter coating. Flick et al. (1989) found that the crude lipid content of battered fish after frying did not increase or decrease with frying temperature and was not different regardless of the length of the cooking period. This observation may be attributed to differences in initial fish lipid content or to the effects of the higher temperature on rapid vaporization of moisture in the fish to form steam and force the fat out of the batter coating and fish. The frying oil itself can have a significant effect on the flavor of the food and may affect the acceptability of the food. The acceptability of deep-fried, battered fish will depend on the greasiness and flavor of the finished product.

Precooked, Restructured, and Miscellaneous Products

Recently, technological advances and processing combinations have been used to develop and restructure products which resemble intact cuts of meat in appearance and taste. Sensory data indicated that tenderness and juiciness of restructured roasts did not differ from that in intact rounds (Liu et al. 1990). The palatability of restructured steak with fat contents of 20% or less were broiled to an end-point temperature of 70°C (158°F) and showed no differences in objective and subjective measures of palatability. The effect of cooking method on restructured beef steaks was studied by Penfield et al. (1989), who compared microwave cookery with broiling in a conventional oven, both to an internal end-point temperature of 73°C (163.4°F). Microwave cooking required less time and energy, and cooking losses and penetration hardness were greater for the microwaved steaks. The microwaved steaks also appeared more well done, were harder, less moist, and more resistant to chewing.

Considerations in Cookery of Muscle Foods

Monitoring Temperature Changes

Poorly regulated household-type ovens are sometimes responsible for overcooked meats. In such cases, the oven temperature should be checked with an accurate oven thermometer when meat seems to brown too quickly during roasting or if the roast is dry and crusty.

Broiling temperature can be regulated either by changing the distance between the meat and the source of the heat, or, in some cases, by changing the thermostat setting. Whether the oven door is left open or closed also influences broiling time and temperature. Because broiling equipment varies greatly in construction and heating capacity, it is best to follow the manufacturer's suggestions for the distance to place meat from the source of heat and the position of the door during broiling.

Temperature Control

Temperature control is very important in meat cookery, especially for research purposes. Thermocouples and a recording device provide a precise system for measuring and recording internal temperature. If thermocouples are not available, chemical thermometers may be used. Household thermometers do not have the accuracy required for research.

A thermocouple with a wire diameter less than 0.02 cm is recommended. When using a thermocouple, use a metal probe to insert it into the geometric center of the steak, chop, or roast and make sure it does not touch either bone or fat. Do not use a metal thermocouple in the microwave oven.

Effect of Freezing

Consumers prefer to purchase fresh meat; yet most of this meat purchased is later frozen at home. Freezing may result in profound changes in muscle foods, which have the potential for reducing quality characteristics such as juiciness and tenderness. During freezing, ice crystallization and ionic shifts of minerals can result in loss of muscle protein solubility (Reid 1983), causing decreases in water-holding capacity (WHC) and increases in thaw drip and cook loss. Overall acceptability scores for fresh pork loin chops were scored higher than chops frozen for 1–4 weeks. Brewer and Harbers (1991) investigated the effect of packaging on the physical and sensory characteristics of frozen ground pork stored for 13, 26, and 39 weeks and found changes with time regardless of packaging material. Vacuum-packaged pork maintained physical and sensory characteristics for the longest period and changed least over 39 weeks of storage. Saran wrap, aluminum foil, and a Saran–foil combination slowed the rate of change due to freezing and were often not significantly different from each other over time. The implications of these findings are that ground pork purchased at retail in commercial polyvinyl-chloride (PVC) packaging should be repackaged prior to freezing to maintain "fresh" pork characteristics. As vacuum-packaging is not a viable option for most consumers, the use of Saran wrap, aluminum foil, or a Saran–foil combination would be equally effective. Based on sensory data, frozen storage of ground pork should be limited to 26 weeks even with vacuum-packaging if "fresh" pork sensory and physical characteristics are to be maintained (Brewer and Harbers 1991).

Effect of Initial Temperature

Thawing procedures should be very carefully controlled to prevent temperature abuse. Beef loin steaks thawed to a high (10.5°C/50.9°F) initial temperature produced higher tenderness and juiciness scores, less cooking loss, shorter cooking times, and a more rare degree of doneness than steaks thawed to a lower (4.1°C/39.4°F) initial temperature (Berry and Leddy 1990). However, the higher initial temperature produced lower beef flavor intensity in steaks, from the higher marbling, than the lower initial temperature. Therefore, with more highly marbled steaks, some additional broiling time may be necessary to develop beef flavor, if higher internal temperatures were used prior to broiling.

Reheating Precooked Meat and Poultry

Meats can be cooked in advance and reheated prior to serving. However, optimum flavor is found in freshly cooked meats compared to precooked meats loosely wrapped in aluminum foil and refrigerated over 4 days (White, Resurreccion, and Lillard 1988). Reheated meat that has oxidized and developed a warmed-over flavor resulted in reduced intensity of fresh cooked beef flavors (cooked beef lean, cooked beef fat, browned, serum/bloody, grainy/cowy), and cardboard and oxidized flavors were noted by the descriptive flavor panel used in these tests.

To maintain the freshly cooked flavor, the cooked meat must be vacuum-packaged or very tightly wrapped to prevent off-flavor development. Reheating of vacuum-packaged precooked beef in a microwave oven will not result in development of warmed-over flavor if the product is refrigerated for up to 21 days, the suggested shelf-life of the product (Resurreccion and Reynolds 1990). Moreover, studies indicate that precooked vacuum-packaged beef can be held at refrigeration temperatures for up to 42 days without loss of palatability as determined by consumers. Jones, Carr, and McKeith (1987) found that in vacuum-packaged precooked pork held for up to 28 days, TBA reactive products did not change over time, indicating that oxidative rancidity and/or warmed-over flavor was not a problem under these conditions. When Stites et al. (1989) developed a precooked beef roast, they found that the product could be stored in excess of 28 days and retain its palatability attributes. Roast beef slices reheated by microwaves after 2 days of storage (4°C/39.2°F) compared favorably with conventionally reheated samples. Experienced judges indicated significantly less intense warmed-over aroma in microwave than in conventionally heated samples, but the warmed-over flavor or TBA numbers of the meat reheated by the two methods were not different. In a similar study, by Boles and Parrish (1990), on a vacuum-packaged precooked microwave-reheated pork product stored for 7, 14, and 28 days, storage time had no effect on juiciness, off-flavor, or overall palatability. The TBA numbers of the phosphate-treated products were above 1.0, which is the published threshold at which trained panelists can detect the warmed-over flavor but their untrained panelists were unable to detect a difference in flavor. Their finding supports reports by White, Resurreccion, and Lillard (1988) that consumer panelists who are untrained were unable to detect significant differences in the flavor of reheated, precooked, stored meat when the TBA number was 6.3 mg/kg or less.

Acknowledgment

The author wishes to thank Ms. Jye-Yin Liao for her technical assistance in all aspects of the preparation of this chapter.

References

Aberle, E.D. and E.W. Mills. 1983. Recent advances in collagen biochemistry. *Reciprocal Meat Conf. Proc.* 36:125–130.

Ang, C.Y.W. and B.G. Lyon. 1990. Evaluations of warmed-over flavor during chill storage of cooked broiler breast, thigh and skin by chemical, instrumental and sensory methods. S. Food Sci. 55(3):664–648.

Armbruster, G., K.G. Gerow, W.H. Gutenmann, C.B. Littman, and D.L. Lisk. 1987. The effects of several methods of fish preparation on residues of polychlorinated biphenyls and sensory characteristics in striped bass. *J. Food Safety* 8:235–243.

Bailey, M.E., S.Y. Shin-Lee, H.P. Dupuy, A.J. St. Angelo, and J.R. Vercellotti. 1987. Inhibition of warmed-over flavor by Maillard reaction products. In: *Warmed-over Flavor of Meat,* A.J. St. Angelo and M.E. Bailey, eds. Page 237. Academic Press, New York.

Belk, K.E., R.K. Miller, L.L. Evans, S.P. Liu, and G.R. Acuff. 1992. Flavor attributes and microbial levels of fresh beef roasts cooked with varying foodservice methodologies. *J. Muscle Foods* 4:321–337.

Berry, B.W. and K.F. Leddy. 1990. Influence of steak temperature at the beginning of broiling on palatability, shear and cooking properties of beef from steaks differing in marbling. *J. Food Serv. Syst.* 5:287–298.

Berry, B.W. and R.W. Hasty. 1982. Influence of demographic factors on consumer purchasing patterns and preferences for ground beef. *J. Consumer Studies Home Econ.* 6:351–361.

Berry, B.W., J.J. Smith, and J.L. Secrist. 1987. Effects of flake size on textural and cooking properties of restructured beef and pork steaks. *J. Food Sci.* 52:558–563.

Bodrero, K.G., A.M. Pearson, and W.T. Magee. 1980. Optimum cooking times for flavor development and evaluation of flavor qualies of beef cooked by microwave and conventional methods. *J. Food Sci.* 45:613–616.

Boles, J.A. and F.C. Parrish, Jr. 1990. Sensory and chemical characteristics of precooked microwave-reheatable pork roasts. *J. Food Sci.* 55:618–620.

Bowers, J.A., J.A. Craig, D.H. Kropf, and T.J. Tucker. 1987. Flavor, color, and other characteristics of beef longissimus muscle heated to seven internal temperatures between 55° and 85°C. *J. Food Sci.* 52:533–536.

Brewer, M.S. and C.A.Z. Harbers. 1991. Effect of packaging on physical and sensory characteristics of ground pork in long-term frozen storage. *J. Food Sci.* 56:627–631.

Crouse, J.D., L.K. Theer, and S.C. Seideman. 1989. The measurement of shear force by core location in longissimus dorsi beef steaks from four tenderness groups. *J. Food Qual.* 11:341–347.

Flick, G.J., Y.-Y. Gwo, R.L. Ory, W.L. Baran, R.J. Sasiela, J. Boling, C.H. Vinnett, R.E. Martin, and G.C. Arganosa. 1989. Effects of cooking conditions and post-preparation procedures on the quality of battered fish portions. *J. Food Qual.* 12:227–242.

Gibson L.L. and L.E. Jeremiah. 1988. The effect of cookery method on beef palatability and cooking properties. *J. Consumer Studies Home Econ.* 12:313–319.

Gros, J.N., P.M. Howat, M.T. Younathan, A.M. Saxton, and K.W. McMillin. 1986. Warmed-over flavor development in beef patties prepared by three dry heat methods. *J. Food Sci.* 51:1152–1155.

Hearne, L.E., M.P. Penfield, and G.E. Goertz. 1978. Heating effects of bovine semitendinosus: shear, muscle fiber measurements, and cooking losses. *J. Food Sci.* 43:10–12, 21.

Howat, P.M., J.N. Gros, K.W. McMillin, A.M. Saxton, and F. Hoskins. 1987. A comparison of beef blade roasts cooked by microwave, microwave-convection and conventional oven. *J. Microwave Power Electromagn. Energy* 22(2):95–98.

Howat, P.M. and A.L. Kirby. 1982. How consumers choose beef cuts. *Louisiana Agric.* 26(2):8–10.

Hultin, H.O. 1985. Characteristics of muscle tissue. In: *Food Chemistry,* O.R.Y. Fenneman, ed. Page 7. Marcel Dekker, Inc., New York.

Johnston, M.B. and R.E. Baldwin. 1980. Influence of microwave reheating on selected quality factors of roast beef. *J. Food Sci.* 45:1460–14621.

Jones, W.L., T.R. Carr, and F.K. McKeith. 1987. Palatability and storage characteristics of precooked pork roasts. *J. Food Sci.* 52:279–281, 285.

Kregel, K.K., K.J. Prusa, and K.V. Hughes, 1986. Cholesterol content and sensory analysis of ground beef as influenced by fat level, heating and storage. *J. Food Sci.* 51:1162–1190.

Liu, C.W., D.L. Huffman, W.R. Egbert, and M.N. Liu. 1990. Effects of trimming and added connective tissue on compositional, physical and sensory properties of restructured, precooked beef roasts. *J. Food Sci.* 55:1258–1263.

Luchak, G.L. 1991. The determination of cooking characteristics of USDA Choice versus USDA Select retail beef cuts. Page 44, Master's thesis. Texas A&M University. College Station, TX.

Martens, H., E. Stabursvik, and M. Martens. 1982. Texture and colour changes in meat during cooking related to thermal denaturation of muscle proteins. *J. Texture Studies* 13:291.

McCrae, S.E. and P.C. Paul. 1974. Rate of heating as it affects the solubilization of beef muscle collagen. *J. Food Sci.* 39:18–21.

McKeith, F.K., D.L. DeVol, R.S. Miles, P.J. Bechtel, and T.R. Carr. 1985. Chemical and sensory properties of thirteen major beef muscles. *J. Food Sci.* 45:869–872.

Penfield, M.P., C.A. Costello, M.A. McNeil, and M.J. Riemann. 1989. Effects of fat level and cooking methods on physical and sensory characteristics of restructured beef steaks. *J. Food Qual.* 11:349–356.

Poste, L.M., E.T. Moran, Jr., G. Butler, and V. Agar, 1987. Subjective and objective evaluation of meat from turkeys differing in basting and final internal cooking temperature. *Can. Inst. Food Sci. Technol.* 53(2):383–386.

Puolanne, E.J. and R.N. Terrell. 1983. Effects of salt levels in prerigor blends and cooked sausages on water binding, released fat and pH. *J. Food Sci.* 48:1022–1024.

Reid, D.S. 1983. Fundamental physicochemical aspect of freezing. *J. Food Technol.* 37(4):76.

Resurreccion, A.V.A. and A.E. Reynolds, Jr. 1990. Consumer responses to vacuum packaged precooked beef roasts held for various storage periods. Annual Meeting of the Institute of Food Technologists. June 17–20, Anaheim, CA., p. 137.

Siemens, A.L., H. Heymann, H.B. Hedrick, M.A. Karrasch, M. Ellersieck, and M.K. Eggeman. 1990. Effects of external fat cover, bone removal and endpoint cooking temperature on sensory attributes and composition of pork center loin chops and roasts. *J. Sensory Studies* 4:179–188.

Stabursvik, E. and H. Martens. 1980. Thermal denaturation of proteins in post-rigor muscle tissue as studied by differential scanning colorimetry. *J. Sci. Food Agric.* 31:1034–1042.

Stites, C.R., F.K. McKeith, P.J. Bechtel, and T.R. Carr. 1989. Palatability and storage characteristics of precooked beef roasts. *J. Food Sci.* 54:3–6.

White, F.D., A.V.A. Resurreccion, and D.A. Lillard. 1988. Effect of warmed-over flavor on consumer acceptance and purchase of precooked top round steaks. *J. Food Sci.* 53:1251–1252, 1257.

Selected References

Anonymous. 1978. *Guidelines for Cookery and Sensory Evaluation of Meat.* American Meat Science Association and National Live Stock and Meat Board, Chicago, IL.

Armbruster, G. and C. Haeffele. 1975. Quality of foods after cooking in 915 Mhz and 2450 Mhz microwave appliances using plastic film covers. *J. Food Sci.* 40:721–723.

Bannister, M.A., D.L. Harrison, A.D. Dayton, D.H. Kropf, and H.J. Tuma. 1971. Effects of a cryogenic and three home freezing methods on selected characteristics of pork loin chops. *J. Food Sci.* 36:951–954.

Bouton, P.E. and P.V. Harris. 1981. Changes in the tenderness of meat cooked at 50°–65°C. *J. Food Sci.* 46:475–478.

Bowers, J.A. and M.C. Heier. 1970, Microwave cooked turkey: heating patterns, eating quality and histological appearance. *Microwave Energy Applic. Newletter* 3(6):3.

Chambers, E., O.A. Cowan, and D.L. Harrison. 1982. Histological characteristics of beef and pork cooked by dry or moist heat in a conventional or microwave oven. *J. Food Sci.* 47:1936–1939, 1947.

Davis, G.W., G.C. Smith, Z.L. Carpenter, T.R. Dutson, and H.R. Cross. 1979. Tenderness variation among beef steaks from carcasses of the same USDA quality grade. *J. Animal Sci.* 49(1):103–114.

Huang, W.H. and B.E. Green, 1978. Effect of cooking method on TBA numbers of stored beef. *J. Food Sci.* 43:1201–1203, 1209.

Hearne, L.E., M.P. Penfield, and G.E. Goertz. 1978. Heating effects of bovine semitendinosus: shear, muscle fiber measurements, and cooking loses. *J. Food Sci.* 43:10–12, 21.

Igene, J.O., K. Yamanchi, A.M. Pearson, J.I. Gary, and S.D. Aust. 1985. Evaluation of 2-thiobarbituric acid reactive substrates (TBRS) in relation to warmed-over flavor (WOF) development in cooked chicken. *J. Agric. Food Chem.* 33:364–367.

Jeremiah, L.E. 1980. Effect of frozen storage and protective wrap upon cooking losses, palatability, and rancidity of fresh and cured pork cuts. *J. Food Sci.* 45:187–196.

Johnson, P.B. and G.V. Civille, 1986. A standardized lexicon of meat WOF descriptors. *J. Sensory Studies* 1:99–104.

Johnston, M.B. and R.E. Baldwin. 1980. Influence of microwave reheating on selected quality factors of roast beef. *J. Food Sci.* 45:1460–1462.

Kendall, P.A., D.L. Harrison, and A.D. Dayton. 1974. Quality attributes of ground beef on the retail market. *J. Food Sci.* 39:610–614.

Marriott, N.G., S.K. Phelps, C.A. Costello, and P.P. Graham. 1988. Restructured beef with fat variation. *J. Food Qual.* 11:53–61.

Martens, H., E. Stabursvik, and M. Martens. 1982. Texture and colour changes in meat during cooking related to thermal denaturation of muscle proteins. *J. Texture Studies* 13:291.

Melton, S.L. 1983. Methodology for following lipid oxidation in muscle foods. *Food Technol.* 37(7):105–111, 116.

Miller, A.J., S.A. Ackerman, and X. Palumbo. 1980. Effect of frozen storage on functionality of meat for processing. *J. Food Sci.* 45:1466–1471.

Penner, K.K. and J.A. Bowers. 1973. Flavor and chemical characteristics of conventionally and microwave reheated pork. *J. Food Sci.* 38:553–555.

Schmidt, J.G., E.A. Kline, and F.C. Parrish, Jr. 1970. Effect of carcass maturity and internal temperature on bovine longissimus attributes. *J. Animal Sci.* 31:861–865.

Siu, G.M. and H.H. Draper. 1978. A survey of the malonaldehyde content of retail meats and fish. *J. Food Sci.* 43:1147–1149.

St. Angelo, A.J., J.R. Vercellotti, M.G. Legendre, C.H. Vinnett, J.W. Kuan, C. James, Jr., and H.P. Dupuy. 1987. Chemical and instrumental analysis of warmed-over flavor in beef. *J. Food Sci.* 52:1163–1168.

Tim, M. and B.M. Watts. 1958. The protection of cooked meats with phosphates. *Food Techn.* 12(1): 240–243.

Zipser, M.W. and B.M. Watts. 1961. Lipid oxidation in heat-sterilized beef. *Food Technol.* 15(10):445–447.

16

Nutritional Value of Muscle Foods

J. Samuel Godber

Introduction

The concept of nutritional value has undergone considerable modification over the years. The old view that nutritional value was merely a statement of the nutritional content of a food has given way to a new view that also takes into account the digestibility and utilization of nutrient components in food and diets. These additional considerations have been loosely termed "nutrient bioavailability." Also, concern for potential adverse nutritional aspects of food products, such as contribution to heart disease or cancer, has been included in the concept of nutritional value. The modern muscle food scientist must be aware of the full scope of factors that should be considered when contemplating and discussing the nutritional value of muscle foods.

The intent of this chapter is to provide a broad overview of the factors that must be considered to understand the complex concept of nutritional value. Special attention will be given to unique nutritional properties of muscle foods and to how these properties provide important nutritional benefits. Additionally, a brief discussion of certain criticisms of muscle foods will be provided with the intent of clarifying the current state of our knowledge with regard to these controversies. The final section of the chapter will provide a unified discussion of the value of muscle foods in healthful diets.

Nutritional Value of Food

Nutritional properties of food have been appraised in a number of different ways. The most simplistic approach to assessing the nutritional worth of a food would be to state the abundance of various nutrients contained within the food. Numerous publications exist that utilize this approach to nutritional properties of food, the most comprehensive of which is the series of USDA publications referred to as Handbook 8. This approach is also frequently utilized by the food industry when making nutritional claims about food products and/or food supplements for marketing purposes. Unfortunately, in many cases, the amount of a nutrient present in a food may not accurately reflect its nutritive value. In addition to consideration of nutrient abundance in food and diets, biological factors that affect the way that the body utilizes nutrients must be taken into account. In other words, foods may contain nutrients in forms that the body cannot utilize. This has led, in recent years, to an emphasis on nutrient bioavailability. However, terms such as "bioavailability" or "availability" are used somewhat loosely to describe a number of often ill-defined measurements of nutritive value. In addition to the terms "availability" and "bioavailability," terms such as "digestibility" and "absorbability" have also been used to describe the way nutrients are used by our bodies. An understanding of the concept of nutrient bioavailability is essential in the study of nutritive value of foods.

The Metabolic Process

In order to understand the role of biological factors in utilization of nutrients and the concept of bioavailability, it is necessary to briefly summarize the metabolic processing of nutrients in general terms. Metabolism is the sum of the processes by which a particular substance is handled in the living body. For a specific nutrient, this would entail all of the events that take place from the time the nutrient is ingested up to the utilization of the nutrient in a biological function. A diagram of the metabolic process with nutritionally important events is given in Fig. 16.1.

The metabolic process begins the moment that food is taken into the mouth. The initial steps in the process are referred to collectively as digestion, the total process by which food is broken down to the molecular level for incorporation into the body via the digestive tract. The process of chewing, or mastication, breaks large food particles into smaller units that expose more surface area and are more easily mixed with digestive secretions. The saliva in our mouth represents the first digestive secretion, as it serves several digestive functions. In addition to the obvious function of lubrication, saliva also contains an enzyme, ptyalin, that begins breaking down starch.

Once the food has been masticated, it passes through the esophagus into the stomach. Here it mixes with gastric secretions to form a semifluid mixture called chyme, which potentiates a number of functions. Gastric secretions are highly acidic (pH of <1), which lowers the pH of the chyme to around 2 and initiates the difficult process of breaking food down into its chemical components. These secretions also contain enzymes, the most important of which is pepsin, an acid protease that aids in the breakdown of proteinaceous material. The action of pepsin is especially important in the digestion of muscle foods because it is able

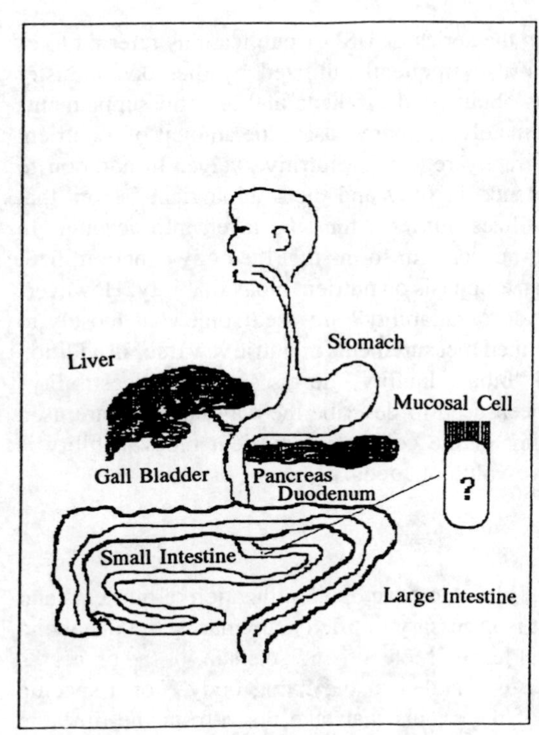

Mastication:
-breaks up large particles

Salivary:
-lubrication
-starch degradation

Gastric:
-acid
-pepsin; collagen
-intrinsic factors

Duodenum:
-primary site of digestion
-pancreatic enzymes
-bicarbonate from pancreas
-biliary secretions for fat absorption

Small intestine:
-absorption of nutrients

Mucosal Cell:
-site of nutrient absorption and transfer to circulatory system, many specifics still largely unknown.

Large Intestine:
-primary site of absorption for several nutrients, e.g. Vitamin K.

Figure 16.1. Diagram of the metabolic process.

to break down collagen, a protein unique to muscle foods. Other factors are also introduced in the stomach that are important to the digestive process. One of these is a compound referred to as intrinsic factor, which is critical to the utilization of vitamin B-12. It also has been theorized that a similar compound may be important to the absorption of iron and possibly other trace elements.

The chyme mixture leaves the stomach via the pyloric sphincter and enters the duodenum portion of the small intestine. This process is controlled by nervous and hormonal mechanisms that are dependent on gastric volume and food composition. Meat in the diet stimulates the release of the hormone gastrin, which, in turn, causes increased gastric secretion and enhances the activity of the pyloric pump. However, fat tends to slow the rate of gastric emptying; thus, the relative effect of meat on gastric emptying will depend on its fat content.

It is in the duodenum that perhaps the most significant events in digestion occur. The first event that takes place in the duodenum is the release of biliary and pancreatic secretions that are involved in an array of functions. Biliary secretions contain emulsifying agents (bile salts) that are essential to the absorption of lipid and lipid-soluble nutrients such as the fat-soluble vitamins. Pancreatic secretions contain buffers, most important of which is bicarbonate, that cause the pH of the digesta to increase from the acid range established in the stomach toward the neutral range. This is necessary to protect the intestinal mucosa from the acidic chyme and because pancreatic enzymes work best in the neutral pH range. Unfortunately, many essential dietary minerals are less soluble at neutral pH and are less well absorbed. The implications of reduced solubility to mineral bioavailability will be discussed later.

Pancreatic secretions contain numerous digestive enzymes. These include proteinases such as trypsin and chymotrypsin as well as amylases and lipases. Thus, enzymes that break down all classes of chemical compounds are present in pancreatic secretions and facilitate the initial absorption of nutrients into intestinal mucosal cells.

Absorption takes place generally by one of three processes. The most common process of absorption is active transport, for which a carrier and energy source are required for absorption. The other two processes include passive transport, which is a simple diffusion process, and facilitated passive transport, which requires a carrier but does not require energy. The mode of transport can have a tremendous effect on utilization of nutrients. For many nutrients, binding sites exist on the outer portion of the mucosal cell (brush border). Often times, several nutrients have common binding sites and/or cotransport systems, which impact utilization (i.e., nutrients may have to compete for binding sites and carrier molecules).

Proteins are generally absorbed by an active transport process in the form of amino acids or possibly dipeptides and sometimes even as intact protein (although this is by a different mechanism called pinocytosis). Carbohydrates are absorbed as monosaccharides or disaccharides with an active transport mechanism, except for fructose, which is absorbed by facilitated passive transport. Lipid absorption takes place as a diffusion process involving micelles formed by bile salts. Lipids are broken down to glycerol, monoglycerides, and free fatty acids in the GI tract, absorbed into the intestinal cell where they are resynthesized into triglycerides for transport in the lymph system as chylomicrons. It is possible, however, for short-chain fatty acids, such as those found in dairy products, to be incorporated directly into the portal blood. Minerals generally are absorbed by active transport. Complex cotransport systems exist for the macrominerals sodium and potassium. Calcium absorption is controlled by parathyroid hormone and vitamin D in response to the body's needs. Trace elements such as zinc and iron are absorbed as complexes with organic material because they are normally insoluble at intestinal pH. This is especially relevant for muscle foods because they provide highly available organic complexes. Absorption processes of vitamins are variable, ranging from diffusion processes for fat-soluble vitamins to complex carrier-mediated processes for vitamin B-12. Uptake of the fat-soluble vitamin K is dependent on microbial activity in the large intestine.

Once absorption has taken place, numerous other processes may occur in the intestinal mucosal cell prior to nutrient transport into the circulatory system. For the most part, however, the mucosal cell is still a "black box," the processes of which are the subject of intense study. It is known that further hydrolysis of chemical components can take place within the mucosal cells because peptidases, disaccharidase, and lipase have been isolated from mucosal cells. Also, carrier molecule transfers and even restructuring of molecular character within the mucosal cell is known to take place. It is believed that many of the processes that occur within the mucosal cell affect the utilization of nutrients; this is especially true in the case of trace elements.

Finally, nutrients pass from the intestinal mucosal cell into the circulatory system where transport takes place to appropriate distal sites within the body for incorporation into biochemical mechanisms. Nutrients may also be placed into storage for future utilization.

Definition of Bioavailability

With this brief overview of the digestive process in mind, it is now possible to define the term bioavailability. A nutrient is bioavailable if it is present in the food, is properly digested into a form that is absorbable, and, once absorbed, is utilized in a biologically meaningful way, that is, is either incorporated into a biochemical pathway or is stored for future use. In other words, a nutrient must pass through all of the physiological processes previously mentioned and achieve ultimate utilization before it can be deemed to have nutrient value. It is not enough merely to be present in the food.

A host of factors can prevent utilization of nutrients. These factors may be incurred at any point in the metabolic process. Furthermore, environmental, physiological, and chemical occurrences may be involved. The first category, referred to as extrinsic factors, relates to conditions that exist in the food or as a component of a diet. The latter two, referred to as intrinsic factors, relate to chemical aspects of the nutrient or to physiological conditions present in the organism that impact bioavailability. The influence in both cases may have either a positive or negative effect. Examples of extrinsic and intrinsic factors that affect nutrient bioavailability are given in Table 16.1.

Nutrient Density

The concept of nutrient density is also an important consideration when appraising the nutritional value of food products. A simplistic definition of nutrient density would be "concentration of nutrients per unit of energy (calories)." Because there are many nutrients, it is difficult to quantify nutrient density on the basis of all nutrients present. Nevertheless, nutrient density has been assessed in terms such as "Index of Food Quality" and "Nutrient Quotient" in which an index is established using selected nutrients. At present, however, nutrient density is not widely

Table 16.1. Factors That Influence Nutrient Bioavailability

Factor	Example	Effect
Extrinsic		
Antidigestion agent	Protease inhibitors	Reduces biological value of protein by preventing enzymatic hydrolysis of peptides.
Chelating agent	Ascorbic acid, phytic acid, EDTA	May reduce or improve bioavailability of minerals depending on ligand binding strength relative to that of musocal absorptive sites.
Competitive inhibitors	Minerals with similar chemical properties	Reduces biological value of nutrient by competing for absorptive sites
Intrinsic		
Chemical form of nutrient	Nonheme versus heme iron	Bioavailability dependent on chemical properties of the nutrient
Nutritional status	Iron deficiency	Deficiency state may increase the bioavailability of deficient nutrient
Digestive factors	Intrinsic factor for vitamin B-12	Endogenously produced factors required for proper utilization of nutrients
Digestive disease states	Achlorhydria	Disease conditions that impact digestive or other metabolic mechanisms related to nutrient uptake and metabolism

applied in making nutritional recommendations. This situation is unfortunate in the case of muscle foods because, as will be discussed later, muscle foods are highly nutrient dense.

Nutritional Properties of Muscle Foods

As has been summarized in the previous sections, the nutritional properties of any given food are dependent on a complex set of factors and interrelationships. Nutritional value of a food is related to the quantity and utilization of nutrients and the potential of the food to facilitate or impede nutrient utilization. The following subsection provides an overview of the nutritional properties of muscle foods, i.e., red meat, poultry, and seafood.

Abundance of Nutrients in Muscle Foods

The first aspect of nutritional value is, of course, the type and amount of nutrients present in a food. In general, there are five major classes of nutrients present in all foods: protein, lipid, carbohydrate, vitamins, and minerals. Technically, all chemical components found in food can be classified into one of these categories, or as in the case of many vitamins, may fall into more than one classification.

The recently approved Nutrition Labeling and Education Act of 1990 has inspired new food labeling requirements. One of the new features is the concept of % Daily Value. Percent Daily Value is based on the USRDAs (Recommended Dietary Allowance) published by the National Academy of Sciences (Washington, D.C.) for vitamins and minerals. For total fat, saturated fat and cholesterol, Daily Value is based on the suggested maximum amount of those nutrients for a diet at a given level of calories. Food manufacturers will be required to provide % Daily Values for diets that contain either 2000 or 2500 calories but must point out that % Daily Value will change as total calories change. A diet of 2000 calories must be used for the calculation.

Protein

Proteins are large molecular weight polymers of amino acids bonded together by peptide linkages. They comprise the structural unit of all muscle cells and connective tissue. Other proteins include enzymes, certain hormones, and carrier molecules such as lipoproteins. Our bodies require protein for growth and maintenance of proteinaceous tissues. Muscle foods have high-quality protein because they have amino acid profiles that match or exceed our body's requirements for amino acids.

There are approximately 20 amino acids that compose the protein in muscle foods, as is also the case for the proteins in our bodies. Of the 20 amino acids, 9 are considered to be essential. A nutrient is considered essential if the body cannot synthesize it. Thus, essential amino acids must be present in the diet to meet the body's need. The body's requirement for each essential amino acid is different, so a food with high-quality protein is one that provides adequate total protein and appropriate amounts of essential amino acids. The World Health Organization (WHO) has estimated an ideal amino acid profile that will meet our body's requirements for amino acids. The National Academy of Sciences (NAS) in this country has established a similar ideal profile that includes histidine, which has only recently been recognized as an essential amino acid in both infants and adults. In Table 16.2, a comparison of amino acid profiles of several types of muscle and nonmuscle foods relative to the WHO and NAS profiles is given. In nearly all cases, muscle foods meet or exceed the estimated ideal concentration for all essential amino acids. This is also true for other nonmuscle animal foods, for example, milk and eggs. The cereals, for example, rice and wheat, have low concentrations of lysine. Other nonanimal foods invariably lack sufficient concentrations of one or more of the amino acids.

Adult males (25 years and older) need approximately 63 g of protein per day according to the most recent USRDAs. Females and younger males require less protein, but when calculating USRDAs, the population group with the highest requirement is used (excluding pregnant or lactating women) as the standard. This figure is based on research that establishes the approximate daily requirements for

Table 16.2. Comparison of Essential Amino Acid Profiles of Selected Muscle Foods to Selected Nonmuscle Foods, and to Food and Agriculture/World Health Organization (FAO/WHO) and National Academy of Science (NAS) Ideal Profiles

	Amino Acids (g/100 g N)								
Food Product	His	Ile	Leu	Lys	Met + Cys	Phy + Tyr	Thr	Trp	Val
FAO/WHO	–	25	44	34	22	38	25	6	31
NAS	11	26	44	32	16	46	22	7	30
Beef	21	28	49	52	23	45	27	7	30
Pork	26	27	49	59	21	44	27	7	30
Lamb	20	30	49	55	24	46	27	7	34
Veal	23	31	50	52	22	45	27	6	35
Chicken	18	31	45	51	25	44	26	7	30
Turkey	19	31	49	57	24	48	27	7	33
Tuna	18	29	51	57	25	46	27	7	32
Shrimp	13	30	50	54	25	47	25	9	29
Oysters	12	27	44	47	22	42	27	7	27
Milk	17	39	62	51	22	62	27	9	43
Egg	15	39	55	42	35	61	31	10	45
Peanut	14	19	35	20	14	51	19	5	23
Soy flour	15	28	47	38	17	51	25	8	29
Wheat flour	13	20	39	13	22	46	16	7	23
White rice	14	26	49	21	26	52	21	7	36
Broccoli	11	23	28	30	11	31	19	6	27
Corn	17	25	68	27	18	53	25	5	36
Apple (+skin)	10	26	40	40	16	30	23	7	30
Oranges	12	17	15	31	20	31	10	6	27

Note: Standard three letter abbreviations were used for amino acids.

protein due to growth or maintenance, and accounts for differences in amino acid composition between proteins. Also, it builds in a safety factor by increasing the recommended allowance by two standard deviations from the average requirement of the population group. The requirement will be affected by age, sex, level of physical activity, and other factors that relate to protein usage such as wound healing, and so on. The most recent edition of the USRDAs differentiates the requirements based on sex and age.

The protein content of most muscle foods on a wet weight basis is between 15% and 35%. This figure will change due to moisture and lipid content. When muscle foods are considered on a raw weight basis as purchased at the store, the protein content is generally less than 20%. However, few people eat muscle foods raw, and most trim away visible fat in red meat species and discard the skin in poultry products. This is especially true with today's health-conscious consumer. Therefore, muscle foods, as consumed, have much higher protein contents. A 85-g (3-oz) serving of muscle food generally provides more than 20 g of protein, or between 30% and 45% of the *Daily Value* for protein, as can be seen in Table 16.3. Thus, muscle foods not only possess a nearly ideal amino acid profile but provide a significant portion of the daily requirement for protein in a relatively small serving. This is not the case with cereals, vegetables, and fruits, although legumes such as peanuts and soybeans do provide high-protein contents, as do milk and eggs (Table 16.3).

Lipid

Lipid includes a diverse group of substances that are loosely characterized as being soluble in organic solvents. In other words, lipids tend to be highly nonpolar molecules that are soluble in nonpolar solvents such as hexane. The most common food lipids include the glycerides (glycerol esterified with one, two, or three fatty acids), phospholipids (glycerol esterified with two fatty acids and one phosphate-containing compound such as phosphatidyl choline) and sterols (compounds with a squalene-derived structure) as can be seen in Fig. 16.2. The basic component of most food lipids is the fatty acid, which possesses a carbon

Table 16.3. Protein Content of Selected Muscle and Nonmuscle Foods.[a]

Food	Grams in Typical Serving[b]	Protein/100 g (g)	Protein/ serving (g)	% Daily Value	INQ
Beef, top round, braised	85	36.12	30.71	49	7.41
Pork, loin, roasted	85	27.84	23.66	38	4.68
Lamb, leg, roasted	85	28.30	24.05	38	6.44
Veal, leg, broiled	85	36.16	30.74	49	7.34
Chicken, breast, roasted	85	31.02	26.68	42	8.17
Turkey, light meat, roasted	85	29.90	25.42	40	8.38
Tuna, light, canned in water	85	25.51	21.68	34	8.84
Shrimp, cooked, moist heat	85	20.91	17.77	28	9.19
Oysters, cooked, moist heat	85	14.12	12.00	19	4.48
Milk, whole	244	3.29	8.03	13	2.27
Egg, whole	50	12.14	6.07	10	3.31
Peanut butter	32	24.59	7.87	13	1.80
Soy flour	42	34.80	14.62	23	3.41
Wheat flour	42	10.33	6.40	10	0.67
Rice, white, cooked	102	2.69	2.74	4	0.90
Broccoli, boiled	78	2.97	2.32	4	4.52
Corn, boiled	77	3.32	2.56	4	1.36
Apple, with skin, raw	138	0.19	0.27	0	0.12
Orange, raw	131	0.94	1.23	2	0.73

[a] Data obtained from USDA Handbook 8-1–8-21, Washington, DC

[b] Typical serving for all muscle foods is 3 oz; for milk 1 fluid cup; for peanut butter 2 oz; for soy, wheat, rice and broccoli 1/2 cup; for egg, corn, apple, and orange whole product.

Note: INQ is Index of Nutritional Quality; INQ = mg nutrient/100 g divided by USRDA divided by the dividend of calories/100 g divided by 2700.

COMMON LIPID STRUCTURES

SATURATED FATTY ACID: Stearic acid; C18:0

CH3-CH2-CH2-CH2-CH2-CH2-CH2-CH2-CH2-CH2-CH2-CH2-CH2-CH2-CH2-CH2-CH2-COOH

MONO-UNSATURATED FATTY ACID: Oleic acid; C18:1

CH3-CH2-CH2-CH2-CH2-CH2-CH2-CH2-CH=CH-CH2-CH2-CH2-CH2-CH2-CH2-CH2-COOH

GLYCEROL	MONOGLYCERIDE	DIGLYCERIDE	TRIGLYCERIDE	PHOSPHOLIPID
H_2C-OH	H_2C-OH	H_2C-OH	H_2C-OCOR'	H_2C-OCOR'
HC-OH	HC-OH	HC-OCOR'	HC-OCOR''	HC-OCOR''
H_2C-OH	H_2C-OCOR'	H_2C-OCOR''	H_2C-OCOR'''	H_2C-OPO_3X

SQUALENE:

CHOLESTEROL:

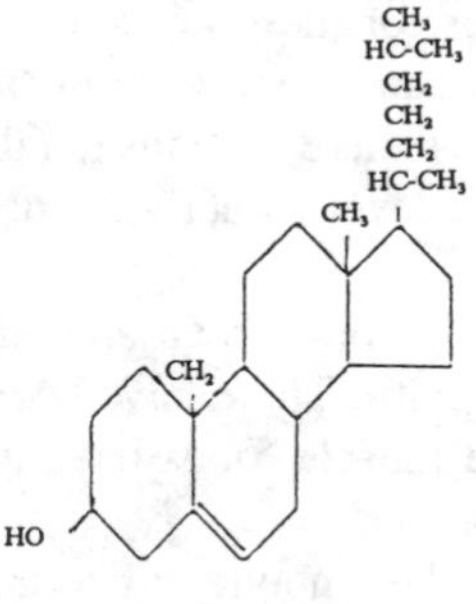

CH_3
HC-CH_3
CH_2
CH_2
CH_2
HC-CH_3
CH_3
CH_3
HO

R', R'', R''' are fatty acid hydrocarbon backbones

X is usually an alcoholic compound; most commonly choline (lecithin) and ethanolamine (cephalin)

Figure 16.2 Structural characteristics of major lipid components.

backbone with a terminal carboxyl group. The carbon backbone can be saturated (fully hydrogenated) or unsaturated (having double bonds) and extends from 2 to 24 carbons in length. Generally, food fatty acids have even numbers of carbon atoms and are between 12 and 20 carbon units in length. Fatty acids that are fully hydrogenated are referred to as saturated fatty acids, those with one double bond as monounsaturated fatty acids, and those with two or more double bonds as polyunsaturated fatty acids (PUFA). Seafoods possess the unique quality of having relatively high proportions of fatty acids with carbon chain lengths of 20 to 24 with 5 or 6 double bonds. These fatty acids are referred to as highly unsaturated fatty acids (HUFA).

The relative amount of lipid in muscle foods is perhaps the most variable aspect of nutrient abundance (Table 16.4). Seafoods typically have lower levels of lipid, with many shellfish species having a lipid content of less than 2%, although the so-called "fatty fishes" such as salmon or mackerel may have lipid contents as high as 30%. Red meats tend to have higher levels of lipid, generally greater than 5% with a normal range between 10% and 40%. Poultry, when skin is removed, falls between seafoods and red meats in lipid content.

Red meat food products have historically been considered high-fat foods, but in recent years, a great deal of emphasis has been placed on reducing fat content. This has been accomplished by innovative breeding and feeding practices in the production of meat animals and through increased trimming of visible fat, in the processing of muscle food products (Table 16.5). Also, the importance of considering the actual lean edible portion has been emphasized. When the total raw carcass is used to depict fat content, it is generally above 20%. However, when a lean edible portion of cooked product is considered, the percentage is lowered to around 10% and in some cases even less. Percent Daily Value (Table 16.4) for fat, saturated fat and cholesterol indicates that muscle foods would easily fit into dietary guidelines for these nutrients.

Fat content of muscle foods can vary due to many factors other than species and processing. Within species there is considerable variation depending on age, breeding, feeding history, level of activity, use of feed additives such as repartitioning agents, and so on.

Health aspects of dietary lipids have been among the most highly discussed topics of food, nutrition, and health. There have been both positive and negative sides of these discussions, and muscle foods have tended to be at the center of the controversy for many of them.

From a negative standpoint, the amount and composition of fat, especially in the red meats, has been a central issue. Saturated fatty acids, which are in higher concentrations in red meats, have been found to be associated with heart disease. Perhaps the strongest criticisms of muscle foods in general is that they contain cholesterol, which may contribute to plaque buildup in blood vessels. These issues are highly complex, however, and will be discussed at length later in this chapter.

From a positive standpoint, fat is the most concentrated form of energy (9 kcal/g versus 4 kcal/g for carbohydrate and protein). When processing foods for use in areas with widespread caloric deficiencies, fats are an important functional ingredient. However, in industrialized countries, high-fat intake contributes to excess energy consumption that causes major health problems.

The relative amounts of the different forms of fatty acids present in foods must also be considered. There is evidence that the HUFA found in many seafoods

Table 16.4. Lipid Content of Selected Muscle and Nonmuscle Foods[a]

Food	g/serve	g lipid/ 100 g	g lipid/ serving	% daily value	g saturated lipid/serving	% daily value	Cholesterol mg/serving	% daily value
Beef, top round, braised	85	5.80	4.93	8	1.78	9	76	25
Pork, loin, roasted	85	15.29	13.00	20	4.57	23	81	27
Lamb, leg, roasted	85	7.74	6.58	10	2.31	12	76	25
Veal, leg, broiled	85	6.33	5.38	8	2.19	11	114	38
Chicken, breast, roasted	85	3.57	3.07	5	0.92	5	73	24
Turkey, light meat, roasted	85	3.22	2.74	4	1.01	5	59	20
Tuna, light, canned in water	85	0.82	0.70	1	0.23	1	25	8
Shrimp, cooked, moist heat	85	1.08	0.92	1	0.28	1	166	55
Oysters, cooked, moist heat	85	4.95	4.21	6	0.51	3	93	31
Milk, whole	244	3.34	8.03	12	5.02	25	33	11
Egg, whole, hard boiled	50	11.15	5.58	9	1.66	8	274	91
Peanut butter, smooth	32	49.98	15.99	25	3.12	16	0	0
Soy flour	42	21.86	9.18	14	1.43	7	0	0
Wheat flour	42	0.98	0.41	1	<0.1	0	0	0
Rice, white, cooked	102	0.28	0.29	0	<0.1	0	0	0
Broccoli, boiled	78	0.28	0.22	0	<0.1	0	0	0
Corn, boiled	77	1.28	0.98	2	0.18	1	0	0
Apple, with skin, raw	138	0.36	0.49	1	<0.1	0	0	0
Orange, raw	131	0.12	0.16	0	<0.1	0	0	0

[a]Data obtained from USDA Handbook 8-1–8-21, Washington, DC.

[b]Typical serving for all muscle foods is 3 oz., for milk is 1 fluid cup, for peanut butter is 2 oz., for soy, wheat, rice, broccoli is 1/2 cup, for egg, corn, apple and orange is whole product.

Table 16.5. Influence of Production and Processing of Selected Muscle Foods[a]

Production or Process Parameter	Variable Tested	Lipid %	Sat/ UnSat	Cholesterol mg/100 g
Trimming	Untrimmed beef Sirloin steak	20.00		
	Trimmed beef Sirloin steak	14.61		
	Untrimmed beef Round steak	18.00		
	Trimmed beef Round steak	11.07		
Altering dietary fat	Frankfurter from pig fed control		34.8/63.9	
	Frankfurter from pig fed 20% canola		16.6/82.5	
	Loin eye from pig fed control		35.9/64.1	
	Loin eye from pig fed 12% HOSO[a]		29.1/70.9	
Create analogue products	Natural shrimp	1.08		195.0
	Imitation shrimp	0.25–1.80		30.0–45.0
	Natural king crab	1.54		53.0
	Imitation king crab	0.20–1.30		4.0–20.0

Source: Adapted from Beitz (1992).

[a]HOSO is high oleic safflower oil.

have a beneficial effect in heart disease and other chronic disorders. Recent research suggests that monounsaturated fatty acids, such as oleic acid found abundantly in muscle foods, may have a beneficial effect as well. Also, stearic acid, the 18-carbon saturated fatty acid that makes up a significant portion of the saturated fat in muscle foods, may not be as detrimental as other saturated fats with regard to heart disease.

Fat also provides much of the flavor that we associate with food products. This is important from a nutritional standpoint because foods must be palatable for people to eat them. Muscle foods have always been considered highly palatable and desirable components of the diet, largely due to their flavorful fat content.

Finally, fats contain several essential nutrients such as the essential fatty acids linoleic and linolenic acid, and fat-soluble vitamins A, E, and K.

Carbohydrate

Carbohydrate represents a diverse group of compounds known as saccharides. The most common food saccharide is sucrose (or table sugar), a disaccharide composed of one each of the monosaccharides fructose and glucose. The most common polysaccharide is starch, composed of glucose molecules linked together primarily by alpha 1–4 linkages, with a smaller number of alpha 1–6 linkages (the branching linkage). Cellulose, the most common nondigestible saccharide, is also made up entirely of glucose molecules but with beta 1–4 linkages, for which our bodies lack a digestive enzyme.

Almost all carbohydrate compounds are derived from plant sources. Muscle foods are low in carbohydrate. The only naturally occurring carbohydrate found in muscle food products is glycogen, or animal starch. Glycogen is also composed of glucose-linked alpha 1–4 but has a higher proportion of alpha 1–6 linkages than plant starch.

Glycogen serves as an energy reserve in muscle tissues and is used especially during anaerobic metabolism. The small energy reserve represented by glycogen in muscle tissue is expended during the slaughtering process when the muscle cells undergo glycolytic metabolism due to the loss of the blood supply. Therefore, little glycogen remains in most muscle food products. The major exceptions are certain species of shellfish such as mussels and oysters, which may contain in excess of 5% glycogen.

There are several types of processing that increase the relative carbohydrate content of muscle food products. Perhaps the most well-known use of carbohydrate in processing muscle foods is the inclusion of sucrose or glucose in curing pork products. "Sugar-cured" is a well-known claim in country-style ham and bacon products of the southern portion of the United States. Granted, in this application, sugar makes up a relatively small proportion of the curing mixture and the accumulation of sugar in the muscle would be expected to be low. On the other hand, the manufacture of surimi, a frozen, washed fish mince, utilizes fairly high levels of carbohydrate as cryoprotectants.

Surimi is a semipurified preparation of fish myofibril protein that has been stabilized for frozen storage. In the original usage by the Japanese, the freshly prepared paste was formed and cooked immediately, yielding a product called kamaboko. With the advent of seafood analogue products that are formulated from the semipurified fish mince, greater demand for the paste necessitated storage. Frozen storage, however, reduced the functional properties of the protein necessary for analogue applications. Carbohydrates such as sucrose, glucose, sorbitol, and even starch have been used to protect against freeze denaturation of myofibril protein. The normal percentage of carbohydrate that is added is around 8%. Thus, in these products, carbohydrate is a significant component from a nutritional standpoint.

Vitamins

The category of vitamins encompasses a wide range of compounds with different chemical forms and biological functions, all of which are essential nutrients. Vitamins function primarily as cofactors in biochemical pathways. They are divided into two categories based on solubility, that is, fat soluble and water soluble. A nomenclature based on alphabetical designations originally described vitamins as they were being discovered before their chemical identity was known. The letter system is now obsolete but survives for some vitamins, e.g. vitamin A, for simplicity.

Muscle foods are a fair to good source of most water-soluble vitamins (Table 16.6). Vitamin B-12 is exceptional because it is found primarily in animal products. Pork products are one of the best sources of thiamin. Muscle foods are not as good a source of fat-soluble vitamins as are vegetable sources, although fish liver oil has high levels of vitamins A and E, which is partially responsible for its use as a dietary supplement.

Minerals

Minerals comprise the inorganic portion of food products but most often reside in organic complexes. Minerals are divided into the categories of macrominerals and microminerals (or trace minerals) according to their relative abundance in foods. Because the body cannot synthesize minerals, those required for biological functions are considered essential nutrients. Determining the essentiality of minerals is one of the more recent endeavors of nutritional scientists and certain trace minerals such as boron, nickel, and silicon are currently being evaluated for essentiality. Most minerals possess biological function by virtue of being a prosthetic inclusion in enzymes or as cofactors to enzyme systems. However, it is difficult to determine their exact role in many of these organic complexes.

Muscle foods are a good sources of minerals (Table 16.7). This is not surprising, as the biologically important minerals in human nutrition would be most likely present in sufficient quantities in the flesh of animals that possess similar mineral-dependent biological systems.

Certain seafoods, e.g., oysters, are known for high levels of zinc. Zinc now has been found to be related to a large number of enzyme systems, although its discovery as an essential nutrient occurred in males with delayed reproductive development. Because including zinc in their diet improved reproductive development, a general misconception has been created that zinc, and oysters by association, improve male reproductive performance.

Although calcium is usually not high in muscle foods, processed seafood products such as canned sardines and salmon are excellent sources of calcium because their bones, softened by the canning process, are generally consumed with the flesh. Processed meat products that utilize mechanically deboned flesh would also be sources of calcium because of the inadvertent inclusion of small bone fragments.

The mineral of greatest nutritional significance in muscle foods is iron. Iron is a prosthetic element in the heme-containing pigments myoglobin and hemoglobin, which serve the essential role of transporting oxygen within the body. As such,

Table 16.6. Comparison of Selected Vitamins in Muscle and Nonmuscle Foods[a]

Food Product	Grams in Typical Serving[b]	Thiamin				Riboflavin				Niacin				Folic Acid				Vitamin B-12			
		mg/ 100 g	mg/ serve	% Daily Value	INQ	mg/ 100 g	mg/ serve	% Daily Value	INQ	mg/ 100 g	mg/ serve	% Daily Value	INQ	mg/ 100 g	mg/ serv	% Daily Value	INQ	mg/ 100 g	mg/ serve	% Daily Value	INQ
Beef, top round, braised	85	0.070	0.060	4	0.60	0.250	0.220	13	1.90	3.81	3.24	22	2.59	9	8	4	0.58	2.70	2.29	115	17.46
Pork, loin, roasted	85	0.971	0.825	55	6.85	0.421	0.358	21	2.62	5.95	5.06	34	3.31	5	5	3	0.26	1.08	0.92	46	5.72
Lamb, leg, roasted	85	0.110	0.090	6	1.05	0.290	0.250	15	2.44	6.34	5.39	36	4.78	23	20	10	1.64	2.64	2.24	112	18.93
Veal, leg, broiled	85	0.060	0.050	3	0.51	0.350	0.300	18	2.63	10.56	8.98	60	7.11	18	15	8	1.15	1.17	1.00	50	7.49
Chicken, breast, roasted	85	0.070	0.060	4	0.77	0.110	0.100	6	1.07	13.71	11.79	79	11.98	4	3	2	0.33	0.34	0.29	15	2.82
Turkey, light meat, roasted	85	0.060	0.050	3	0.70	0.130	0.110	7	1.35	6.84	5.81	39	6.36	6	5	3	0.53	0.37	0.32	16	3.26
Tuna, light, canned in water	85	0.032	0.027	2	0.46	0.074	0.063	4	0.94	13.28	11.29	75	15.25	4	3	2	0.43	2.99	2.54	127	32.64
Shrimp, cooked, moist heat	85	0.031	0.026	2	0.57	0.032	0.027	2	0.52	2.59	2.20	15	3.77	4	3	2	0.55	1.49	1.27	64	20.63
Oysters, cooked, moist heat	85	—	—	—	—	0.332	0.282	17	3.90	2.49	2.12	14	2.62	1 8	15	8	1.80	38.27	32.53	> 1000	382.70
Milk, whole	244	0.038	0.093	6	1.10	0.162	0.395	23	4.16	0.08	0.21	1	0.18	5	12	6	1.09	0.36	0.87	44	7.86
Egg, whole	50	0.074	0.037	3	0.84	0.286	0.143	8	2.89	0.06	0.03	0	0.05	49	24	12	4.21	1.32	0.66	33	11.35
Peanut butter	32	0.136	0.044	3	0.41	0.099	0.032	2	0.26	13.09	4.19	28	3.18	78	25	13	1.80	0	0	0	0.00
Soy flour	42	0.412	0.173	12	1.69	0.941	0.395	23	3.41	3.29	1.38	9	1.06	227	96	48	7.01	0	0	0	0.00
Wheat flour	42	0.785*	0.487*	33	3.22	0.494*	0.306*	18	1.79	5.90[c]	3.66[c]	24	1.91	26	16	8	0.80	0	0	0	0.00
Rice, white, cooked	102	0.163*	0.166*	11	2.29	0.013	0.014	1	0.16	1.48[c]	1.51[c]	10	1.64	3	3	2	0.31	0	0	0	0.00
Broccoli, boiled	78	0.082	0.064	4	5.25	0.207	0.161	10	11.71	0.76	0.59	4	3.84	68	53	27	32.69	0	0	0	0.00
Corn, boiled	77	0.215	0.166	11	3.71	0.072	0.055	3	1.09	1.61	1.24	8	2.19	46	36	18	5.95	0	0	0	0.00
Apple, with skin, raw	138	0.017	0.023	2	0.47	0.014	0.019	1	0.34	0.08	0.11	1	0.17	3	4	2	0.63	0	0	0	0.00
Orange, raw	131	0.087	0.114	8	2.84	0.040	0.052	3	1.15	0.28	0.37	3	0.72	30	40	20	7.35	0	0	0	0.00

[a] Data obtained from USDA Handbook 8-1–8-21, Washington, DC.

[b] Typical serving for all muscle foods is 3 oz; for milk 1 fluid cup; for peanut butter 2 oz; for soy, wheat, rice and broccoli 1/2 cup; for egg, corn, apple, and orange whole product.

[c] Enriched

Note: INQ is index of nutritional quality; INQ = mg nutrient/100 g divided by USRDA divided by the dividend of calories/100 g divided by 2700.

Table 16.7. Comparison of Selected Minerals in Muscle and Nonmuscle Foods[a]

Food Product	Grams in Typical Serving[b]	Phosphorus				Manganese[c]				Iron[d]				Zinc				Copper			
		mg/ 100 g	mg/ serve	% Daily Value	INQ	mg/ 100 g	mg/ serve	% Daily Value	INQ	mg/ 100 g	mg/ serve	% Daily Value	INQ	mg/ 100 g	mg/ serve	% Daily Value	INQ	mg/ 100 g	mg/ serve	% Daily Value	INQ
Beef, top round, braised	85	226	192	24	3.65	0.050	0.042	2	0.32	3.32	2.82	19	2.86	4.56	3.88	26	3.93	0.124	0.105	7	1.06
Pork, loin, roasted	85	279	237	30	3.69	0.010	0.009	1	0.05	0.93	0.79	5	0.66	2.92	2.48	17	2.06	0.098	0.083	6	0.69
Lamb, leg, roasted	85	206	175	22	3.70	0.028	0.024	1	0.20	2.12	1.81	12	2.03	4.94	4.20	28	4.73	0.120	0.102	7	1.14
Veal, leg, broiled	85	249	212	27	3.99	0.039	0.034	2	0.25	1.32	1.12	8	1.13	3.96	3.37	23	3.38	0.140	0.119	8	1.19
Chicken, breast, roasted	85	228	196	25	4.72	0.017	0.015	1	0.14	1.04	0.89	6	1.15	1.00	0.86	6	1.10	0.049	0.042	3	0.54
Turkey, light meat, roasted	85	219	186	23	4.83	0.020	0.017	1	0.17	1.35	1.15	8	1.59	2.04	1.73	12	2.40	0.042	0.036	2	0.49
Tuna, light, canned in water	85	163	139	17	4.44	0.011	0.009	1	0.12	1.53	1.30	9	2.22	0.77	0.65	4	1.12	0.051	0.044	3	0.74
Shrimp, cooked, moist heat	85	137	116	15	4.77	0.034	0.029	2	0.47	3.09	2.62	18	5.74	1.56	1.33	9	2.90	0.193	0.164	11	3.58
Oysters, cooked, moist heat	85	278	236	30	6.95	—	—	—	—	13.40	11.40	76	17.87	181.9	154.6	> 1000	242.53	8.922	7.584	506	118.96
Milk, whole	244	93	228	29	5.06	—	—	—	—	0.05	0.12	1	0.14	0.38	0.93	6	1.10	—	—	—	—
Egg, whole	50	180	90	11	3.87	—	—	—	—	2.09	1.04	7	2.40	1.44	0.72	5	1.65	—	—	—	—
Peanut butter	32	323	103	13	1.87	1.536	0.492	25	3.55	1.67	0.53	4	0.51	2.51	0.80	5	0.77	0.556	0.178	12	1.71
Soy flour	42	476	200	25	3.68	2.077	0.872	44	6.41	5.82	2.44	16	2.40	3.58	1.50	10	1.47	2.221	0.933	62	9.14
Wheat flour	42	108	67	8	0.83	0.682	0.423	21	2.09	4.64[e]	2.88[e]	19	1.90	0.70	0.44	3	0.29	0.144	0.089	6	0.59
Rice, white, cooked	102	47	47	6	1.24	0.467	0.477	24	4.92	1.10	1.12	8	1.55	0.46	0.47	3	0.65	0.063	0.065	4	0.88
Broccoli, boiled	78	48	37	5	5.77	0.245	0.191	10	11.77	1.15	0.89	6	7.38	0.15	0.12	1	0.96	0.069	0.054	4	4.42
Corn, boiled	77	103	79	10	3.35	0.194	0.149	8	2.52	0.61	0.47	3	1.06	0.48	0.37	3	0.83	0.053	0.041	3	0.91
Apple, with skin, raw	138	7	10	1	0.37	0.045	0.062	3	0.94	0.18	0.25	2	0.51	0.04	0.05	0	0.11	0.041	0.057	4	1.15
Orange, raw	131	14	18	2	0.86	0.025	0.033	2	0.61	0.10	0.13	1	0.33	0.07	0.09	1	0.23	0.045	0.059	4	1.471

[a] Data obtained from USDA Handbook 8-1–8-21, Washington, D.C.

[b] Typical serving for all muscle foods is 3 oz; for milk 1 fluid cup; for peanut butter 2 oz; for soy, wheat, rice, and broccoli 1/2 cup; for egg, corn, apple, and orange whole product.

[c] No USRDA; upper limit of recommended safe and adequate used to calculate % USRDA and INQ.

[d] USRDA for females age 11–50 used to calculate % USRDA and INQ.

[e] Enriched.

it would follow that muscle foods, with their abundance of these pigments, would be a good source of iron. Perhaps more importantly, it has been found that iron in the heme form is more bioavailable than nonheme iron, as will be discussed in the next section. Comparisons of mineral contents in selected muscle foods with non-muscle foods is given in Table 16.7.

Bioavailability of Nutrients in Muscle Foods

The bioavailability of nutrients in diets that contain muscle foods is a highly important positive nutritional consideration. It is believed that muscle foods possess intrinsic factors, most of which have not been entirely delineated, that improve the bioavailability of a variety of nutrients. The best established intrinsic effect of muscle foods on a nutrient involves heme iron.

It has been found that there are two pools of dietary iron, heme and nonheme, each with its own absorptive pathway. Heme iron is absorbed in the heme complex and broken down into elemental iron inside the intestinal mucosal cell by an enzyme, porphyrin oxidase. The exact mechanism for the absorption of nonheme iron is still unknown, but it is believed that an active transport system exists when iron is present in small concentrations. The significance of this two-pool concept is that the different forms of iron are affected by other dietary factors differently. Factors that are known to reduce the absorption of nonheme iron, such as phytic acid, do not have the same effect on heme iron. Thus, heme iron is much more highly bioavailable than nonheme iron.

Other studies indicate that trace minerals such as zinc and copper are more bioavailable from muscle foods than from plant foods. Indirect evidence has been presented that muscle foods may even promote the bioavailability of calcium and or magnesium due to an apparent effect on bone mineralization. All of these potential effects are likely related to the role of muscle foods in maintaining nutrients in a more absorbable form throughout the digestive process. This may simply be due to a protective effect against adverse dietary factors.

It has also been suggested that certain other nutrients are more readily available from animal sources than from plant sources, including vitamin B-6, folic acid, and niacin. The reasons for these differences are not known. It is possible that factors such as solubility or absorptive site compatibility suggested for trace elements are also important in the absorptive mechanism of these vitamins. Much more research is needed to clarify the exact value of muscle foods in promoting the bioavailability of nutrients.

The Meat Factor

The possibility that muscle foods promote the bioavailability of nutrients in nonmuscle foods when the two are eaten together has been clearly established in the case of nonheme iron. Not only is heme iron in muscle foods more highly bioavailable than nonheme iron, but there is also conclusive evidence that nonheme iron in foods that are eaten with meat has higher bioavailability. This phenomenon has been referred to as the "meat factor." Thus far, it has only been seen with iron, although it has been speculated to exist with other nutrients, especially other trace elements. Also, the exact mechanism by which the effect comes about is not known.

Numerous theories have been proposed to explain the meat factor. One of the simplest explanations is that the protein or amino acid content of muscle foods is responsible. However, other animal food products with similar protein contents, such as eggs and milk, do not produce a similar effect. Also, a mixture of amino acids similar to the amino acid content of meat does not exhibit the same degree of bioavailability enhancement. The possibility that the meat factor is due to the content of nucleoproteins in meat has also been found to lack merit. The two most often cited and likely explanations for the meat factor are related to the effect of muscle foods during the digestive process. Primarily, it has been found that low-molecular-weight digestion products produced by protein degradation protect iron from factors that would otherwise reduce solubility and, thus, absorption. Second, it is believed that during the digestion of muscle foods, pH changes favor iron solubility. Iron, like many transition elements, is more soluble at lower pH. The digestion of muscle foods stimulates gastric acid secretion, produces buffering capacity, and maintains a lower pH for a longer period of time in the duodenum, where the majority of trace-element absorption takes place.

The possibility that the meat factor exists for nutrients other than iron is predicated on several biochemical and physiological features of the digestive mechanism. Nutrients that share similar characteristics, for example, the multivalence of transition elements, are thought to have similar metabolic pathways. In this sense, copper and possibly zinc may be affected in a manner similar to iron. These minerals have common absorptive sites and carriers during the absorption process, so it is possible that they may possess a meat-factor-like effect with regard to bioavailability. It is also likely that digestion products of muscle food degradation contribute to improved absorption of a variety of nutrients. Support for this possibility resides in the teleological development of man as an omnivore. It has been found that carnivorous animals utilize iron from meat sources much better than from plant sources, whereas the opposite is true for herbivores. Biochemical and physiological mechanisms may have developed that are specific to the type of diet that is consumed. As an omnivore, man's digestive mechanism would be suited to a mixture of plant and muscle foods.

Nutrient Density of Muscle Foods

Nutrient density in muscle foods is high. This is related to the fact that muscle foods have compositions that are similar to the tissues of our bodies, which means that there are high levels of essential nutrients per unit of weight. Plant foods on the other hand tend to be high in carbohydrate, which only provides energy, and thus the relative abundance of essential nutrients per unit of energy is low. Foods or constituents of food that contain only calories, such as table sugar, are sometimes referred to as "empty" calories.

In Tables 16.3, 16.6, and 16.7 the Index of Nutritional Quality (INQ) has been included for each nutrient. A specified quantity of the food, e.g., 100 g, was used to calculate a ratio of the percentage of the USRDA that is achieved for each nutrient to the percentage of a typical caloric intake, e.g., 2700 calories. If a food contributes a greater percentage of the nutrient than calories, the ratio will be above 1, which is considered a nutrient-dense food for that nutrient.

For almost all nutrients, the percentage of the USRDA provided by muscle foods is higher than the percentage of calories. This is an indication of high nutrient density. Plant foods have high INQ for vitamins A and C. It would appear that many plant foods have relatively high INQs for a large number of nutrients, but they also tend to contribute fewer calories per unit of weight. Low concentration of a nutrient relative to weight must also be considered. As a general rule it is necessary to take both the percent of the USRDA provided by a typical serving and the INQ into account when appraising the relative nutritional value of a food.

Protein INQs are quite high for most muscle foods, as are most minerals, especially in shellfish. Manganese and calcium are not generally abundant in muscle foods, whereas magnesium and copper have only marginally good abundance. Vitamin B-12 has the highest INQ in muscle foods among the vitamins, and thiamin INQ is high for pork products, although this is not the case for other muscle foods. Folate is the only water-soluble vitamin for which the majority of muscle foods have relatively low INQs, and muscle foods are not a good source of fat-soluble vitamins.

Effect of Processing on Nutrients in Muscle Foods

A variety of processing procedures is used in the manufacture of muscle food products, each of which affects nutrient quality. Also, production practices influence nutrient quality.

Technically, processing begins at slaughter or harvest. The processing steps involved with conversion of muscle to meat are considered primary processing. Conversion of meat into meat products is referred to as secondary processing. It is mainly secondary processing that impacts nutritional quality, although it is conceivable that certain occurrences in primary processing may affect nutrient quality. For example, it has been speculated that delayed bleeding could increase the proportion of heme iron in muscle tissue. On the other hand, highly exudative meat, such as the case of pale, soft, exudative (PSE) pork, a condition related to animal stress during the slaughter process, could reduce the quantity of water-soluble nutrients.

Secondary processing may simply involve fabrication of wholesale or retail cuts from carcasses or may include extensive processing such as restructuring and cooking. The degree of impact each processing step has on nutritional quality is variable.

During fabrication of wholesale or retail cuts of meat, external and internal fat is removed. This would increase the nutrient density of many nutrients because of the empty calories represented by external fat. When muscle foods are cooked, certain nutrients could be lost, either due to heat degradation or through drip losses. During freezing and thawing, water-soluble nutrients may be lost due to thaw exudation. Oxidation of lipid during long-term storage of muscle foods results in the loss of nutrients susceptible to oxidative degradation such as fat-soluble vitamins A and E and certain water-soluble vitamins including thiamin and folic acid.

It may be possible to influence the nutritional quality of muscle food products that undergo restructuring processes such as boneless ham and restructured roast beef. During restructuring, excess fat is removed and additives are included that could influence nutritional quality. Adding antioxidants could reduce the effect of oxidation on susceptible nutrients. On the other hand, use of phosphates and certain nondigestible polysaccharides, such as carrageenan or alginate, may reduce the bioavailability of trace elements.

Adverse Health Claims Against Meat in the Diet: Scientific Clarification

Consideration of nutritional value of foods also should include an assessment of possible adverse nutritional properties. In the case of muscle foods, this consideration has been emphasized to a considerable extent in recent years, to the point that a skewed representation of the nutritional value of muscle foods has emerged.

The Role of Muscle Foods in Heart Disease

Coronary heart disease (CHD) is the most widespread chronic disease confronting modern humankind. It has been associated primarily with affluent nations. One of the principal dietary differences that occurs as a population becomes more affluent is an increased consumption of animal foods. Much of the evidence that relates heart disease to diet is based on epidemiological (or large-scale population) studies. Thus, highly positive correlations between diets high in animal foods and heart disease have been established. However, it should be noted that epidemiological studies of Eskimo populations have indicated that diets high in muscle food of marine origin have lower rates of CHD. Thus, the diet–heart disease issue is complicated and requires careful study.

Before it is possible to discuss the exact role of muscle foods in CHD, it is necessary to briefly summarize the etiology of CHD. Damage to and deterioration of the coronary blood vessels are primary features of CHD. This process is referred to as atherosclerosis, which is characterized by a constriction of the blood vessel caused by an accumulation of sclerotic tissue. This tissue arises as a result of damage to the lining of the vessel followed by an immune-type phagocytic response. During the immune response, lipid-rich cellular artifacts are generated that contain high levels of cholesterol. Ultimately, these materials become fibrous and calcified, causing narrowing of the blood vessel. Considerable debate as to the mechanism of the original injury has ensued, and a picture of the entire process remains the subject of much research.

Muscle foods have been linked to CHD in two ways. The most obvious association is related to the fact that plaque is composed of high levels of cholesterol, and animal foods are the only dietary source of cholesterol. The second association focuses on the quantity and composition of fat in muscle foods. Diets high in fat have been implicated in the onset of CHD, as have lipids that are high in saturated fatty acids. Certain muscle foods, e.g., red meats, have the reputation of being high in both total fat and saturated fatty acids.

Cholesterol is an essential biochemical that is the precursor to many hormones, is a cellular constituent of membranes, and is the precursor to bile salts needed for proper digestion of lipid and lipid-soluble vitamins. Every cell in our body requires cholesterol for proper function and our body manufactures a much greater proportion of cholesterol than is consumed in the diet. Indeed, it is now widely believed that abnormal circulating cholesterol levels are more a function of genetics or excess dietary lipid and/or energy than of dietary cholesterol.

With regard to the contribution by muscle foods to lipid intake, a closer look at the data reveals several interesting facts. For example, red meats do not contribute as much dietary fat as previously believed because the data upon which associations were first made were based on total carcass weight rather than actual weight of muscle consumed. Also, it has been established that consumers trim a large portion of the external fat prior to consumption, which would reduce the relative proportion of fat in muscle foods considerably.

The belief that saturated fat was primarily responsible for CHD was based on studies in which saturated fat was provided by tropical fats. Certain tropical fats, e.g., coconut and palm oil, have a high proportion of their total lipid as saturated fatty acids with medium-chain-length carbon backbones. The lipid profile of most muscle foods on the other hand is lower in total saturated fat than these tropical fats and has a higher proportion of longer-chain fatty acids. Also, evidence indicates that several fatty acids found predominately in muscle foods, e.g., oleic and stearic, are not highly associated with CHD. The HUFAs found in marine lipids have been classified as agents that lower the risk of CHD. Modern approaches to production and processing of muscle foods are reducing fat content and even modifying the composition toward more beneficial types of fatty acids (Table 16.5).

One other aspect of muscle foods that has been implicated in the etiology of CHD is the high salt level of certain processed muscle food products. High levels of sodium contributed mainly by salt (NaCl) have been associated with hypertension, or abnormally high arterial pressures. Hypertension is related to CHD because of its potential to cause injury to vessel walls, the first step in the development of atherosclerosis.

The role of NaCl, or in actuality the sodium ion, in hypertension and the general recommendation for widespread reduction in NaCl consumption is extremely controversial. The population appears to be divided into two main categories with regard to the effect of sodium on hypertension, primarily based on genetic susceptibility. The majority of the population has a genetic protection against developing hypertension. These individuals can consume relatively high levels of sodium in their diets without risk of developing hypertension. Between 5% and 20% of the adult population is at risk of having essential hypertension, so named because its cause is unknown. In these individuals, consumption of high levels of sodium has been shown to promote hypertension.

The problem is that it is difficult to differentiate between these two segments of the population, i.e., those who are sensitive to sodium and those who are not, especially in light of the fact that the symptoms of hypertension are not outwardly apparent. Certain segments of the medical community believe that it would be prudent to recommend population-wide reductions in sodium consumption, even though more than 80% of the population is not endangered by higher levels of sodium intake, because of the often unknowing 20% that is at risk. This is a difficult situation because so much of our processed food supply is manufactured with NaCl or other sodium-containing compounds as functional ingredients. In many cases, the safety of the food product is dependent on the presence of the sodium-containing compounds as water activity lowering or antimicrobial agents.

Processed muscle food products are just one type of food that contains sodium, yet they are often cited as a major source of dietary sodium. Bakery products and snack foods contribute greatly to dietary sodium consumption.

Nevertheless, the muscle food processing industry is making a concerted effort to reduce sodium use in the manufacturing process by using functional substitutes or innovative processing procedures.

The Role of Muscle Foods in Cancer

Cancer is considered the next most deadly disease confronting modern humankind and probably instills a higher degree of fear and anxiety due to public perception of debilitating effects and poor prognosis. Cancer is the term used to describe diseases characterized by uncontrolled, abnormal cellular growth. Cancer may occur in virtually any cell in the human body given the appropriate set of circumstances. Because cancers are established and generated long before symptoms are manifested, it has been difficult to establish exact causes.

Simplistically, cancer arises when a causative agent, referred to as a carcinogen, invades a normal cell and creates a potential for cancer to occur, followed at some later time by a promoter agent that causes the cancer to develop. Certain cancers have been fairly well characterized as to the most likely carcinogens and promoter agents, but in most cases, the exact circumstances that produce cancers of various types are not well understood.

Muscle foods have been mentioned specifically in certain types of cancer and indirectly as a major source of fat in cases where dietary fat is believed to be highly correlated to occurrence. A type of cancer that is often associated with consumption of muscle foods is cancer of the colon and/or lower GI tract. The reasoning for this association has been that muscle foods increase fecal transit times and produce higher levels of toxic compounds through fermentative reactions. The scientific evidence for this association, however, is lacking.

A cancer often cited as being related to fat consumption is breast cancer. Because certain muscle foods have the reputation of being higher in fat, they have been implicated as a potential causative dietary factor in this disease. Newer theories suggest that cancers related to fat consumption may be more dependent on the type of fat consumed, rather than being dependent solely on total dietary fat. Unsaturated fat is more susceptible to lipid oxidation, a reaction that produces highly reactive intermediate organic compounds called free radicals. The so-called antioxidant defense mechanism of our body defends against free radicals. In fact, free radical reactions have been highly related to possible cancer-causing agents. Lipids of plant origin, especially the highly purified vegetable oils so prevalent in developed countries, are very high in unsaturated fat. It is equally possible that increased consumption of this type of lipid, as has occurred over the last 50 years, may make a population more susceptible to cancer.

A great deal more research will be needed to understand the exact nature of both cancer and CHD, so it would not be prudent to espouse general dietary recommendations for abstinence from any food group. To do so would produce two highly undesirable effects. The first would be a false sense of security, and the second would be a potentially harmful change in dietary practice. A false sense of security could lead to reduced diligence in monitoring early warning signs for these diseases and/or obtaining regular medical screening tests for them. A major change in dietary practice, such as eliminating muscle foods from one's diet, could lead to nutrient deficiencies and increased susceptibility to diseases, even cancer. A principal component of our body's defense against cancer is the immune system, which, in turn, is highly dependent on adequate nutritional status. If nutritional status declines, it is entirely possible that the body becomes more susceptible to chronic diseases.

Integration of Meat and Meat Products into Healthful Diets

The willingness of consumers to use particular food products is going to be more and more dependent on a product's nutritional and health value or, at least, consumers' perception of those values. Muscle foods should be considered an important component of healthful diets for a host of reasons.

The body must be nourished properly to be healthy and to stay healthy. A vast array of nutrients are needed to properly nourish our bodies. Many nutrients are considered essential because our bodies cannot manufacture them in great enough quantity to meet our needs. Muscle foods provide an abundance of nutrients, generally at higher concentrations than most other foods relative to caloric content. Most of the essential nutrients are present in muscle foods. Furthermore, muscle foods provide nutrients in a form that potentiates greater bioavailability than nutrients from other food sources. Indeed, in diets that lack muscle foods, great care must be taken to ensure that adequate levels of essential nutrients are present and bioavailable.

Perhaps of even greater significance is the fact that muscle foods appear to make nutrients from other foods more utilizable or bioavailable when eaten together. This may be due to the fact that man evolved as an omnivore in which biochemical mechanisms were developed that favored diets with mixtures of plant and animal foods. The exact nature of this phenomenon is not known, nor is its full nutritional impact, but it is clear that its potential existence places even more relevance on the role of muscle foods in healthful diets.

Finally, the potential adverse effect of muscle foods on health must be put into proper perspective. Mark Twain once said that "Water taken in moderation never harmed anyone." Obviously, this wry remark is directed at advocates of moderate behavior, but moderation should certainly be the mainstream of current thinking in nutrition and health. It is true that evidence exists, albeit mostly by association, that certain degenerative diseases of modern humankind may be related in part to consumption of muscle foods. However, the irony to the relationship is that those populations susceptible to higher levels of these degenerative diseases have longer life expectancies, which, in turn, could be attributed in part to superior nutritional status. Other factors, such as increased consumption of processed foods and certain vegetable oils, could also be associated with this higher incidence of degenerative disease. The etiology of degenerative diseases is so complicated and controversial that it has been difficult to determine precisely the causative factors.

When all is taken into account, it would appear that including muscle foods in the diet is highly favorable when the good is weighed against the bad. That is not to say that precautions should not be taken to reduce the potential adverse effects of fat in muscle and other foods by limiting consumption to prudent levels. Thus, with deference to Mark Twain's views, moderation is the best advice with regard to dietary habits aimed at preventing or delaying the onset of degenerative diseases.

Input is needed to further clarify the role of muscle foods in healthful diets. Does the meat factor occur in nutrients other than iron? What are the mechanisms for improved bioavailability of nutrients? What levels of muscle foods provide optimal benefits? Past lack of attention to healthful aspects of muscle foods by the muscle foods industry as a whole has allowed the benefits of muscle foods to be overshadowed by controversial claims of negative effects. Current efforts in this area have been greatly expanded, and it is imperative that the role of muscle foods in healthful diets be emphasized and updated.

Selected References

Beitz, D.C., et al. 1992. *Food Fats and Health*. Task Force Report No. 118. Council for Agricultural Science and Technology, Ames, IA.

Bodwell, C.E. and B.A. Anderson. 1986. Nutritional composition and value of meat and meat products. In: *Muscle as Food*. P.J. Bechtel, ed. Academic Press, New York.

Godber, J.S. 1990. Nutrient bioavailability in humans and experimental animals. *J. Food Qual*. 13:21–36.

Guyton, A.C. 1981. *Textbook of Medical Physiology,* 6th ed. W.B. Saunders Co., Philadelphia.

Krause, M.V. and L.K. Mahan. 1984. *Food Nutrition and Diet Therapy,* 7th ed. W.B. Saunders Co., Philadelphia.

Miller, D.D. 1991. Nutritional quality of foods and diets. In: *Encyclopedia of Human Biology Volume 5*. Academic Press, New York.

National Research Council. 1989. *Recommended Dietary Allowances,* 10th ed. National Academy Press, Washington, DC.

Nettleton, J.A. 1985. *Seafood Nutrition: Facts, Issues and Marketing of Nutrition in Fish and Shellfish*. Osprey Books, Huntington, NY.

U.S. Department of Agriculture. 1986–1991. Composition of foods: Raw, processed, prepared, *Agriculture Handbook 8,* revised. Vols. 8-1–8-21. U.S. Department of Agriculture, Washington, DC.

Wardlaw, G.M. and P.M. Insel. 1990. *Perspectives in Nutrition*. Times Mirror/Mosby College Publishing, St. Louis, MO.

Zapsalis, C. and R.A. Beck. 1985. *Food Chemistry and Nutritional Biochemistry*. John Wiley and Sons, New York.

17

Product Development

Larry W. Hand

Introduction

> *There are three rules of new product success. Unfortunately, no one has any idea what they are.*
>
> Phil Fletcher, President and CEO, ConAgra. (1989)

Phil Fletcher's statement illustrates the many frustrations involved with product development and ultimately product success. New products are a main source of corporate growth. The objective of this chapter is to present an overview of product development in muscle food products from the perspective of a food technologist. A complete overview of product development would include various marketing, production, sales, and management discussions. These topics will not be described, but resources are available in the list of references.

Product development is not a technologist's individual responsibility. It is necessary, regardless of the size of the company, for those responsible for product development to function as a team within the company in order to achieve product success. In the team concept, technologists work closely with marketing and business people on specific projects. In this way, the appropriate people in the company are available to make the necessary decisions, thus expediting the product development process. The ultimate objective of product development is to generate products that perform at or better than predetermined levels and, thus, generate profits for the company. Target sales levels and company objectives may range from obtaining the largest market share, growing a new category of production, or increasing volume.

Product development may take many forms depending on the corporate philosophy, current economy, competition, company size, technical resources, and market. Regardless of the system used, there are a few basic steps involved in product development that are not system-specific.

Product development for muscle foods is not typically high-tech. Therefore, speed and marketplace timing are more critical than a particular system or development strategy. One of the myths associated with the development of products is that large organizations with large research and development budgets achieve higher new product success ratios than smaller companies (Bujake 1972). Small companies tend to be entrepreneurial and flexible in nature and limited resources which may allow them to make fewer mistakes, so top management becomes more closely involved with product development. Large corporations require such a large return on investment that they sometimes overlook the small segments that later develop into large markets. There is a trend for larger manufacturers to move to regional marketing strategies, where small processors are already prospering. Small firms tend to invent new products out of necessity and closeness to the consumer, whereas large firms tend to imitate the small firm successes but on a larger scale. *Venture groups* are a means of enabling large corporations to incorporate this entrepreneurial spirit. The "product champion" system, assigning one individual the responsibility for a product, is another method for encouraging entrepreneurial spirit and increasing motivation within the corporation.

It is often difficult for the Research and Development (R&D) departments to develop new products due to their limited resources of personnel, time, and equipment. New products are not the only responsibility of the R&D department. The many demands of a company on the R&D resources include product and process improvement and technical service. Due to the daily demands on resources and technical support to improve and maintain the company's existing product lines, long-term objectives are often pushed aside. To overcome this dilemma, separate product development groups that function outside the system of R&D can be formed. This may take the form of a separate group including members of quality control, marketing, and production along with R&D with a different reporting structure than their original departments. This is usually a temporary, product-specific group, and the members return to their original function at the termination of the project.

Understanding the Business

The Risks of Product Development

The rate of success from products was outlined by Suess (1984) and is shown in Table 17.1. As a rule of thumb, 10% of all new products succeed past consumer testing; only 10% of these successful products make it to market. Only 5.4% of products on a grocery store's shelves 25 years ago are still there (Scheringer 1989). With these great risks it is surprising anyone attempts new product development. However, successful products return large dividends in growth for the company.

Table 17.1. Fate of 100 New Products

Fail during development	95
Fail in test markets	2
Fail after national market roll out	1
Succeed	2

Source: Adapted from Suess (1984).

The Product Life Cycle

The typical product life cycle of introduction, growth, maturity, and decline is discussed by Urban and Hauser (1980). In maturity or decline, an organization must (1) expand the product line and, thus, extend the life cycle, (2) improve the product to make it superior, or (3) develop a new product in another category to maintain revenue. This life-cycle improvement is an ongoing, continual process. As the market for new products becomes more competitive, product life cycles shorten and companies are under increased pressure to maximize the value and number of new products introduced.

Marketing Strategies

The primary responsibility of the technologist is to develop the type of product that the company desires and write the *specifications* that will ensure the consistent production of that product. The speed that a product moves through the life cycle is dependent on many things from the quality and consistency of the product to the ever-changing whims of the consumer. Many of these are areas over which the technologist does not have control.

Product development can take many forms and definitions. Many of these are generated from the marketing strategies of the company. R&D has little value by itself; it has to be coupled with the market. The innovative firms are not necessarily the ones that produce the best technological output but the ones that know what is marketable (Mansfield 1976). Some companies are technically driven, whereas others are marketing driven. Those that are technically driven typically develop the product and then sales and marketing are challenged to sell and market the product. Conversely, a marketing-driven company will challenge technology to develop a product to meet the consumer's need.

A *new product* is one that is unique and puts the company into a new business area, usually as a result of new technology, innovation, or unique consumer demand. Because of its uniqueness, the new product is the first onto the market in that particular category. This "first to market" strategy is usually based on strong research and development, technical leadership, and willingness to take risks of the company. This type of product development usually requires greater resources and long-term planning because the time line is much longer than many other areas of product development. In fact, Best (1990) reported on R&D Budget allocations and found that 35.9% of the budgets were spent on new technology and product development and 27.6% on line extensions and existing technology development.

Line extensions are products that are new varieties of existing products. Examples of line extensions include new sizes, shapes, weights, slicing thickness, flavors, and packages. The majority of product development activities are to create line extensions. These products are to increase current *shelf presence* or retail sales space and expand a particular market category. However, this type of product development carries as its greatest threat "*cannibalism,*" or the sacrificing of sales of one product for the sales of the new variety. When properly executed, an expanded market category will increase the overall business through minimized cannibalism, higher profit margins, and/or increased market shares at the expense of the competition. An advantage of this type of product development is that the risk is reduced because developmental costs and effort are small. A disadvantage of this strategy is that competitors can regain their lost market share with similar products in a relatively short time.

An *improved product* is a product that is reformulated or modified to meet the needs of the business to become the best cost producer. Companies use this strategy when new ingredients become available, or ingredients become cost-effective, or cost savings can occur by replacing ingredients. This marketing strategy is based on modifying products to fit the needs of particular customers in mature markets.

A *competitor match* is when products are developed to compete with specific competitors currently in the marketplace. This "follow the leader" strategy is based on strong developmental resources and the ability to act quickly as the market begins its growth phase. This also is an area of "Me-Too" competition to protect existing product lines and requires superior manufacturing efficiency and cost control.

Swientek (1992) pointed out that the amount of time necessary to get a new product to market is one of the most critical parameters for new product success. Getting a new product to market first is a tremendous advantage. A truly new product is viewed as unique and it establishes consumer expectations. To succeed, the second and third entries into a category must often be better than the original and deliver a competitive point-of-difference to the consumer or more simply, be considerably less expensive.

Steps in Product Development

The basic steps of product development are as follows:

1. Idea generation
2. Concept development

3. Feasibility
4. Technical development
5. Precommercialization
6. Product introduction and commercialization.
7. Maintenance

In a step-by-step scenario of product development, an idea is generated from some source; the marketing and technology groups meet with management to pursue the idea; the technologist develops several *prototype* models; the business group decides if it is feasible; the prototype is then produced in larger numbers; *test marketed*; a sales plan is developed by the marketing group; the product is introduced (rolled out) into the market; product manufacturing improvements are continuously made. Throughout this process, the decision to continue the product development activity is repeatedly being made. Of course, the ultimate "go or no-go" decision is made by the consumer. Several step-by-step approaches are available to organize the product development process. However, for a successful product to be marketed, the entire organization from management to line worker needs to be involved in the implementation phases.

Product development is a time-sensitive task. There are many ways to keep on target and speed up the process. Frequently, simultaneous activities take place, such as package development while the final ingredients are being tested. To assist this process, flowcharts are used to keep track of the progress of several different simultaneous activities and deadlines. These charts detail each task involved in the development of the product and the time involved, along with the interaction of each of these tasks.

Idea Generation

The development of ideas for new or improved products is not an exact science nor does it follow a defined path. Theoretically, ideas can be generated by anyone, and the product development person must be aware of this and be alert to all sources of information for possible new product trends, ideas, or concepts. Ideas come from many different sources, from the CEO to the dock worker; however, the ideas with the best results come from consumers. Kraft General Foods chairman and CEO Michael Miles (1991) stated that "product success starts with the consumer."

As ideas can come from many sources, technologists need to stay abreast of research and patents, not only in their own area, but also in auxiliary areas in food science, engineering, etc. The technologist also needs to stay current with consumer trends. The biggest mistake a technologist and company can make is to not listen to, and ignore, the consumer.

Recognition of demand is a more frequent factor in successful innovation than recognition of technical potential (Marquis 1969). One of the ways to tap consumers is through the use of consumer *focus groups* that discuss a variety of issues and may present an actual product to be developed or identify critical needs, wants, or attributes for a product. Another aspect, depending on the type of business philosophy that exists in a company, is to review all the competitors' products in a certain category and examine for possible improvements, obvious missing varieties, or meeting recent consumer trends.

The idea for a product or product target has to take into consideration volume and profit requirements and expectations, consumer trends and needs, technical innovations, and preliminary economic analysis. These ideas are then screened in the marketplace through market research. Two common uses of consumer data gathering at this point in the product development process are the use of focus groups and the use of *consumer intercepts*. Focus groups bring a group of consumers of known traits (buying history, demographics, etc.) together and present concepts of products and solicit their opinions. Focus groups are qualitative in nature, whereas consumer intercepts are more quantitative. One problem with this method is that the consumers may be very supportive of a product concept or idea, but they have not paid for that product and may not buy the product in the marketplace due to price or other constraints. The intercept market research system solicits consumer input through the use of a survey either through the mail or by personal visitation such as in shopping malls. One of the risks associated with intercepts is that the survey does not usually allow the flexibility to obtain more thorough definitions of answers or opinions that can occur in the focus groups.

Bujake (1972) explored the myth that a technological breakthrough is needed for new product success. What is really needed is a marketing breakthrough. This breakthrough occurs with the identification of a real, unfulfilled consumer need or untapped market and the development of a product concept to satisfy that need. Any company can produce a product that a consumer may try one time. Real product successes come in repeat purchases. Company management must commit the resources necessary to develop an idea, as an idea will not make a successful product by itself. An idea can only develop into a successful product when the product fills a consumer need and the company has put the appropriate resources in terms of people, time, marketing, advertising, and distribution behind the product.

Company philosophy or currently available resources may hamper the product development process. The technologist must take into consideration that some great ideas cannot be developed due to the company philosophy or limited resources. Some products just do not fit with the company image. One way to avoid these problems is to review the company's current product line and ask the following questions:

1. Are there needs that current products are not satisfying?
2. Is there a problem or consumer dissatisfaction with existing products?

The answers to these questions concerning the company's current line of products may lead to further restructuring and the development of new or improved products.

Concept Development

The concept and prototype development phase is one of the more creative stages for the technologist. From the concept, the technologist will develop a bench-top prototype by hand or by using small existing equipment. Usually, certain parameters are given to the technologist from marketing, production, and/or the business functions of the company. The most critical parameter is the cost restrictions that may be invoked. These restrictions may force the technologist to be more creative in developing the prototypes, because it usually limits the resources available to produce the prototype. The development of the prototype usually proceeds from recipe development, manufacturing, and forming on the bench top, and then to a presentation to the company at the prototype review. Often, several different prototypes are generated to help further refine the idea. The prototype development is also an opportunity for the technologist to screen some of the limitations involved due to technology available in relation to the product desired along with production feasibility. The technologist must analyze similar competitive products currently on the market including chemical, physical, and microbiological attributes. Included in the prototype development is the development of preliminary packaging needs and production feasibility. In examining the production feasibility, one must examine the possible equipment needs and constraints that may be associated with actual volume production. Along with the prototype development, the technologist also develops the inputs to quality standards and costs. Once the prototype is developed it is screened within the company among other technologists and management. Sometimes, a more formalized group such as a sensory *taste panel* of personnel or consumers is used in this screening process. Consumer feedback at this time can prove invaluable in telling the technologist that the product is meeting the projected needs. The prototype(s) can also be presented to consumers for their evaluation either in a controlled "tasting" or a take-home trial. With the information from these tests, the business and marketing functions must reevaluate the preliminary economic analysis. For the technologists, this includes another product cost revision. At this step it is necessary to make a decision to proceed or stop the project, before major developmental funds are committed.

Feasibility

The feasibility phase is an ongoing process to determine if the product idea can become a marketable product that accommodates the company's goals. The first aspect is an assessment of the technology needed to manufacture the product. If the technology is currently available within the company, the developmental time lines are much shorter. If the technology must be developed or purchased, the developmental time lines are greatly increased and are often outside the technologist's control. If the product requires expanding into new technology areas, the company must be committed to the project. Sufficient time and capital must be provided for equipment installation and start up, the education of employees, and the development of quality control and maintenance programs.

In developing new products, companies must recognize capability limitations in manufacturing, distribution, and sales. These limitations should not be barriers to innovation but should not be ignored either. One of the biggest mistakes some companies encounter is to develop and introduce new products in order to fill available production capacity. This can be a misplaced focus if consumer needs are not fulfilled by the new product.

Frequently, contract manufacturing (co-packing) becomes a good option for new product introduction. Contract manufacturing arrangements range from totally producing and shipping the product to producing and/or packaging only. These arrangements are one method to overcome both technology shortfalls and production capabilities of the company. The primary advantages for a company to use contract manufacturing is that the company does not have to use precious production space for a new untested product. Also, the ability to utilize the expertise of the contract manufacturing company reduces the risk for a company introducing products outside their area of production experience. A co-packer can produce a product until the sales volumes are large enough to warrant the incorporation of the production into the company's current production facilities or building additional space. The greatest disadvantages are the possible increased cost of the product and exposing confidential processes and procedures. Depending on the contract arrangement, the company may be locked into a production lease of *X* amount of volume for a certain amount of time regardless of the life of the product. The contract manufacturer option provides an additional layer of management with which the technologist must coordinate details.

The marketing and business areas are responsible for developing the projected costs and projected volumes for the new product. The technologist must provide information for the development of the raw material and production associated costs. Estimated market share and projected volumes must be accurately forecast to determine if the new product will be a profitable venture for the company. When developing totally new products or new names, the legal aspect needs additional attention. The evaluation of possible trademark or copyright infringements is invaluable to protect the company. When using new technologies or adaptations, patents must be examined for possible infringement.

The company management must carefully evaluate all the risks and benefits involved in this enterprise. Those risks include the possible cannibalization of other products in the company's lines, if the product is an acceptable fit for the types of products currently manufactured, what type of market share or volume is available and, above all, if the new product will be a profitable venture.

Technical Development

Concept Formulation

For the technologist, the development of a successful new product formulation is sometimes a quick task. Most often, there is a considerable investment in time and effort in the development of the particular product nuances and flavors. The formulation develops initially from the prototype concept. Careful recording of ingredients added in the development phase will be beneficial later when the product is converted from the prototype to production. After the formula is developed in the laboratory, it must be converted to measurable items for production. For example, a clove of garlic, one onion, one green pepper, and one chicken breast all must be converted to a weight basis so that production can accurately manufacture the product. Also, regulated ingredients, such as sodium nitrite and sodium phosphates, must be controlled from the initial concept phase throughout development so that these ingredients will be in compliance with governmental regulations once actual production begins.

Ingredients

The specific ingredients used are a part of the overall quality of a product. The current trend in muscle product development is to use muscle foods as an ingredient, rather than a main item, such as stir-fry beef or chicken. This requires increased knowledge and research in the areas of interaction of different foods and their properties. The technologist must determine what holding temperatures are critical, what ingredient shelf-lives are, and the specific parameters of the added ingredients that impact their performance in the new product. For example, if the fresh onions needed will spoil or discolor rapidly in the company's current refrigerated storage facilities, one may consider dehydrated onions or an alternative ingredient. Another consideration to review is the possible seasonality of ingredients or raw materials. The identification and approval of possible *primary* and *secondary ingredient suppliers* must also take place to assure product production continuity. Aspects of supplier approval include quality, costs, delivery schedules, specifications, letters of guarantee, and volume projections.

Suppliers

Suppliers are playing an increasingly large role in product development for food companies both large and small. The use of their expertise and advice in the development process can save valuable product development time. With increased costs associated with product development, suppliers and companies are forging cooperative alliances to develop new products. This may take the form of assisting the company technologist minimally with providing trial ingredients, recommending optimal levels of ingredients, to the greater extreme of prototype through product development. This association between supplier and company benefits both entities, as the supplier will benefit with increased business, and the company with decreased developmental costs and possibly less staff and overhead. On the downside, the alliance may not be proprietary, allowing the supplier the right to sell their ideas to other manufacturers.

Costs

Cost projections are required at all points in the development process and are critical to the development process. Typically, the initial cost projections will be altered several times during the development process. Changes in the concept may reduce the price, or alternative sources for ingredients will reduce prices. In certain types of product development, especially in the "Me-Too" strategies, cost may be the ultimate measure. For example, the product to be developed, while delivering certain quality attributes, must only cost *X* amount. These competitive cost levels often hamper the product development phase. From a more optimistic viewpoint, this forces the technologist to be more creative in the development of the product and working with ingredient suppliers. For example, it is difficult to develop a chicken breast product at a retail price point that is below the current raw material cost before ingredients, cooking losses, packaging, labor, and distribution costs are included in the calculation. When this occurs, the technologist must talk with the marketing and business functions to consider alternatives in product concept or possibly reconsider whether the company should become involved in the manufacture and marketing of this type of product concept. To be a good product development food technologist, one must understand cost accounting and financial project planning (marketing).

Shelf-Life

Shelf-life information should be obtained as the product is developed. Often, the product concept identifies a targeted shelf-life. Shelf-life testing should be accomplished on plant-manufactured product. Products that are manufactured in a pilot plant or by hand are not a true representation of the product and may lead to erroneous information. Preliminary studies may be accomplished on the pilot-plant-manufactured product; however, this information is only of general use. Shelf-life testing typically examines the spoilage (microbial, chemical, and sensory) of a product over time at specific storage temperatures. With increased emphasis on food safety, shelf-life testing often includes abuse testing at higher temperatures so that information is available about the potential risk involved in product abuse.

The product formulation is scrutinized not only for cost and specific ingredients but, more importantly, flavor and other quality attributes such as color and texture. The use of a trained taste panel is necessary to differentiate between these quality attributes. Taste panel optimization consists of developing test products of varying levels of a particular ingredient (spice) or attribute (texture) and using the information generated by the panel to evaluate and create a better quality product, often at a lower cost. Further explanation of the use of taste panels is found in Chapter 12.

Nutritional Properties

The nutritional properties of the products need to be determined as soon as possible once formulations are finalized. This may include the use of nutritional databases along with product analysis. Nutritional analyses need to be conducted, as this information has to be included on all retail food labels. Specific nutritional attributes may be part of the guidelines given by the marketing function in the concept phase of the product development. The technologist will then develop the product around these requirements.

Packaging

Packaging of the product is another element where the technologist must work closely with packaging personnel and/or suppliers. The specific parameters of the concept help to dictate the type of packaging material that is necessary. For instance, will the use of a cellophane overwrap with an outer paperboard box be adequate, or is the packaging concept for this product going to include special films, oxygen scavengers, modified atmosphere, or vacuum? More importantly, does the company have the equipment necessary to handle this type of film and package? The *lead times* necessary for the production, purchase, and installation of this additional equipment must be calculated into the developmental time of the product. Along with package development is the development of the label. From the technologist's standpoint this may be as simple as developing the *ingredient statement* and nutritional information for the label to complete label development which includes color decisions, product representation, or photography. Of course, the approval of the appropriate governmental agencies (USDA, FDA) along with all necessary internal management is necessary for packaging materials and label graphics. In developing packaging, the technologist is primarily concerned with the product name, ingredient statement, nutritional analysis, cooking or reconstitution instructions, product size, and net weight. The packaging group is usually responsible for the development of specifications for label and case size, preprinted or plain labels, special legal wording, box size, number per case, preprinted or plain boxes, interleaves, and quantities of each needed.

Documentation

After management has agreed with the continuance of the product, the technologist must develop preliminary manufacturing instructions for the *plant tests*. Once the plant tests are conducted, more formalized standard operating procedures are developed to ensure the consistent manufacture of the product. Critical production steps must be identified and a process control program be provided.

The development of the proper documentation involved with any new product is vitally important to the future success of the product. Documentation for the technologist involves developing specifications on ingredients, standard operating procedures, quality control specifications, hazard analysis critical control points (HACCP), and government regulations. Without proper documentation, the repeat purchaser will receive a different product than the one they were expecting from the initial trial. Almost any company can produce a product that someone will try initially, but repeat purchases are necessary for success. Detailed information on how to use standard operating procedures and quality standards is covered in Chapter 13.

Standard operating procedures are designed to ensure the product is manufactured in a similar manner each time it is produced, whether that is in different plants, shifts, or lines or using different ingredient suppliers. A basic rule of thumb is to take nothing for granted, and document all procedures involved in the process. The technologist needs to pay particular attention to those variables that would directly impact the quality and safety of the product, for example, temperature and composition of the raw materials, cooking temperatures, packaging systems, and storage temperatures. Initially, the standards are defined in broad terms and then refined as the manufacturing department becomes more familiar with the product. Finished product quality standards are the final product specifications and are used to define what is acceptable for a product in terms of quality attributes such as size, weight, color, defects, and microbial quality. Ingredient specifications can be developed with the suppliers (internal and external) as to what is acceptable in terms of the raw material ingredients. These specifications are necessary for the purchasing agents in securing ingredients from different suppliers.

Throughout the technical development phase, it is important to consider that the concept may change, due to market conditions and consumer feedback. Constant changes and reconfiguration of the product are considered a normal part of developing a successful product.

Precommercialization

Plant Test

The plant test is conducted to assist the company in developing the parameters necessary for production of the product on a regular basis. The plant test generates a tremendous amount of information that confirms the product performance. Yield estimates are updated, costs are further refined especially labor and machinery hours, and product is submitted for shelf-life testing. Sufficient product is also produced to perform a *shipping test*. This test allows the technologist and packaging group the opportunity to examine package performance and identify any handling and distribution problems that may arise. These tests provide the information necessary for management to make a decision to proceed or abort further development of the product.

Test Marketing

Test marketing is much more involved than a plant test and requires a great deal of team planning. The test market requires product for sale in a controlled sales area. The specific sales area is chosen for a variety of reasons, such as current brand strength or lack of strength, competitors' similar products, population demographics, and type of company sales force. Urban and Hauser (1980) pointed out several advantages and disadvantages of test marketing. The three major advantages of conducting a test market include (1) risk reduction, (2) improvement information, and (3) production and distribution improvement. In considering risk, the company may lose 1 million dollars in a test market failure, whereas a national roll out failure may result in a 10 million dollar mistake. Besides the economic losses, new product failures lower sales morale, disrupt marketing channel relationships, erode consumer confidence, and compromise brand loyalty. Product improvement information generated by the use of a test market includes information, not only how to improve the specific quality parameters of the product but also how ads, promotion, and price manipulate product sales. Test marketing provides the opportunity to discover difficulties in the production and marketing channels that need to be remedied. The disadvantages of performing a test market are that risk reduction and strategic improvement may not be justified especially with a high initial manufacturing start-up cost. The test market also may be considered a disadvantage as it takes valuable time and the competitors have time to match the product before a national roll out can occur. Test marketing results must be carefully reviewed because competitors can manipulate the results of a test market with excessive advertising of their products or, for poor products, decreasing advertising to encourage sales of a possible failure. These promotion strategy changes can give erroneous results of true performance of long-term success of the products and may result in incorrect decisions on the product being made.

Another possible misinterpretation of the test market situation was pointed out by Daniels (1990). In many cases, the products that are being tested are significantly superior to the products that will be manufactured and distributed. Unfortunately, the researchers conducting these studies may be misinterpreting the data and believe that they are testing if the new standard product, not the "perfect" product, is better than the competition. Because the "perfect" product cannot be produced and delivered, the researchers' interpretation can be misleading and may lead to incorrect conclusions.

Product Introduction and Commercialization

Either in the initial plant test or the start-up for test marketing, adjustments will need to be made in the production of the product. The procedures that worked wonderfully in the laboratory do not always work when scaled up in plant production. Small things such as mixing times or packaging speeds can amount to major problems in the actual production of a product. A product that was mixed for 10 minutes in the small lab mixer may require much less or much more time to result in the correct texture required for the product to which management has agreed. The technologist has to be able to make decisions and improvements or adjustments to the preliminary procedures in order to make a product similar to the laboratory prototype. This is one area in which having a pilot plant facility available can decrease the number of variables before the actual plant test. A pilot plant test facility allows technologists to make less expensive mistakes. The cost of a 50-lb mistake is much smaller than a 5000-lb mistake made in a plant situation. Pilot-plant testing results in fewer production-related problems, reducing plant test costs and reducing start-up time. This also helps to maintain better production facility and technologist working relationships.

The distribution of the new product is also examined during the plant start-up/test marketing phase. The distribution of a product is a step that can determine the success or failure of a new product. Critical issues of storage conditions and shelf-life need to be determined so that no surprises arise. Distribution in terms of company fit is also an important item. It is difficult for a distribution and sales force who are used to handling all frozen product to have to market a refrigerated product. Entering new distribution channels requires a company and sales force education process.

Product Maintenance

Optimization

In the rush of developing a new product and meeting deadlines, all possible product improvements have not been examined. During the maintenance phase of a product life, certain economies of scale arise. Because production volumes are more certain than they were initially, adjustments can be made in process, distribution, packaging, and ingredients. During this period, the competition has time to retrench and combat the product developed for the market.

The optimization phase allows the technologist to reformulate to take advantage of certain aspects of the product performance that have occurred in the initial product roll out phase. All aspects of product manufacture must be examined for possible improvements, including ingredients, process, packaging, quality, and, most importantly, profit levels, and consideration must be given to line extensions. The technologist often begins removing him/herself from the front line as other functions of the company start doing their jobs. However, the development technologist should keep abreast of product success and regularly monitor the product to ensure the product is delivering required attributes. One way in which to monitor the product changes over time is a practice of measuring against a "gold" standard or bench mark. This basically is comparing existing products with products that are made with the original formulation specifications. Information generated from this technique enables the improved product from being altered so much that it does not resemble the original goal or product desired.

Costs

Product development costs have been discussed throughout the technical development phase, but the awareness of costs and their obvious role in product profit and ultimate success are critical. Awareness of costs is vital for product success. However, in a survey, Best (1989) reported that 24% of the R&D executives were not involved with cost-justifying projects, 20% used best guess estimates, and 24% indicated that financial analysis was not used in project evaluation in their companies. Because of the high costs associated with product development and the high risks involved to the company, trying something out, consumer testing, and seeing what happens is no longer possible. The trade costs must also be critically analyzed. More products take up more shelf space. The retailer's profitability often suffers, and they make up for the losses with *slotting allowances* and *failure fees*. Along with these costs for new products, predicting demand for new products is much harder than for established products. Forecasting errors are often enormous. Overestimating may lead to excess product that must be sold at a loss or destroyed. Underestimating may lead to frustrated consumers and distributors who may not return to purchase the product again, because the product was not on the shelf. These costs are in addition to the costs normally associated with new products, such as new equipment, ingredients, and the time and effort spent in developing products. The successful company must be willing to rationalize unsuccessful new products. Rules for volume and profit thresholds need to be established and should be followed. With more items requiring marketing support, it is crucial to understand how different products respond to different kinds of promotional spending.

Summary

Product development is a high-risk, expensive side of the muscle foods business. The benefits derived from successful new products are necessary to maintain the longevity of the company. The technologist involved in product development soon realizes that although he may have appreciable control over the specific attributes of how a product is developed, the ultimate success is still based on the entire company working together as a team.

Product Development Checklist

STEP I—Idea Generation

1) Determine food product target(s) based on:
    a) Volume requirements/expectations
    b) Profit requirements/expectations
    c) Consumer trends and needs
    d) Technical innovations
    3) Preliminary economic analysis
2) Screen ideas in the market.
    a) Focus Group Method
    b) Intercept Method
3) Develop the product concept and market positioning statements.

STEP II—Concept Evaluation

1) Identify, develop, or prepare a prototype to physically represent the idea.
2) Determine technical and production feasibility.
3) Develop preliminary packaging needs and feasibility.
4) Revise preliminary economic analysis.
5) Make a decision to proceed or stop the project.

STEP III—Prototype Product Development

1) Design a "Bench" formula using commercial ingredients in batch sizes less than 50 lbs.
2) Conduct Bench sample testing evaluation and comparison to the prototype.
3) Institute formula modifications based on results of the previous step and conduct further product evaluations to finalize formula.
4) Identify primary and alternative ingredient suppliers.
5) Develop revised formula cost information for bench sample.
6) Repeat previous steps as necessary to meet original plan or change the plan.
7) Prepare preliminary ingredient statements and manufacturing instructions.
8) Conduct a plant test.
9) Prepare or update product yield estimates.
10) Revise costing.
11) Conduct shipping test.
12) Make a decision to proceed or stop the project.

STEP IV—Package and Process Development (Can Run Concurrently With Product Development Phase)

1) Determine operating requirements such as new equipment, facility expansion, equipment, layout, etc.
2) Determine packaging options.
3) Select and prepare prototype package and label design.
4) Determine packaging, label lead times, specifications with suppliers, and test packaging.
5) Finalize packaging specifications.
6) Finalize label design.
7) Finalize photography or other art work.
8) Obtain FDA/USDA approval (if appropriate).
9) Submit sales estimates for equipment and packaging purchase.
10) Finalize packaging/label specifications.
11) Make decision to proceed or stop the project.
12) Procure packaging and equipment.

STEP V—Pre-commercialization

1) Finalize marketing plans.
2) Develop test market plan and finalize return on investment analysis.
3) Determine final cost/pricing/financial plan.
4) Make decision to proceed or stop the project.
5) Develop creative introduction materials and advertising strategy.
6) Develop production plan and schedule.
7) Finalize quality control documents.

STEP VI—Product Test Market

1) Establish test market objectives and develop support programs.
2) Conduct preliminary production run.
3) Conduct test market.
4) Analyze test market results.
5) Make decision to proceed or terminate the project.

STEP VII—Product Introduction

1) Prepare sales forecast.
2) Release product to production scheduling.
3) Start full production.
4) Introduce and present new product to the sales force.
5) Distribute the product.
6) Sell.

STEP VII—Commercialization

1) Provide sales support materials during initial sales introduction.
2) Review project status, track initial product placement success.

3) Make decision to proceed or stop the project.
4) Review original plan and objectives and revise if necessary.
5) Regularly monitor production run product to ensure product is delivering required attributes.

Adapted from Tybor and Reynolds (1989)

Selected References

Best, D. 1989. R&D revs up for the '90s. *Prepared Foods,* 59:70–72, 74, 76.

Best, D. 1990. Food Industry remains bullish on R&D. Prepared Foods. 160:62–63, 65–66, 68.

Bujake, J.E., Jr. 1972. Ten myths about new product development. *Res. Management.* 15:33–43.

Daniels, R.W. 1990. Ignorance is Bliss. But it may be expensive! Presented at 11th Annual Marketing Research Conference, American Marketing Association. In *Audits International/Monthly*. Audits International, Highland Park, IL.

Earle, M.D. and A.M. Anderson (eds.). 1985. *Product and Process Development in the Food Industry*. Harwood, New York.

Fletcher, P. 1989. In "Gorman conference previews foods for a new millennium." Dornblaser, L., J. Friedrick, and R. Matthews, *Prepared Foods* 159:14.

Graf, E. and I.S. Saguy (eds). 1991. *Food Product Development From Concept to Marketplace*. Van Nostrand Reinhold, New York.

Kinnear, T.C. and J.R. Taylor. 1979. *Marketing Research: An Applied Approach.* McGraw–Hill, New York.

McVicker, R. 1991. In "Food Technology in the 2000's. Anonymous. *Food Processing*. 52:66–68.

Mansfield, E. 1976. Business Week 2435:59. [In *Design and Marketing of New Products*. G.L. Urban and J.R. Hauser, eds. 1980. Page 24. Prentice-Hall. Englewood Cliffs, NJ].

Marquis, D. 1969. [In *Design and Marketing of New Products*. G.L. Urban and J.R. Hauser, eds. Page 29. 1980. Prentice-Hall. Englewood Cliffs, NJ.].

Marquis, D. 1969. The anatomy of successful information, Innovation 1:28–37 [In *Design and Marketing of New Products*. G.L. Urban and J.R. Hauser eds. 1980 p. 29 Prentice-Hall. Englewood Cliffs, N.J.]

Miles, M. 1991. Going for the big numbers. *Prepared Foods* 161:30.

Scheringer, J. 1989. 25 years of new products. *Prepared Foods,* New Products Annual, 159:31.

Schwartz, W.C. 1986. Bringing a new product to market. Presented at 28th Annual Meat Science Institute. University of Georgia, Athens. March 16–19.

Suess, G.G. 1984. Developing a new meat product. *Reciprocal Meat Conf. Proc*. 37:58–65.

Swientek, R.J. 1992. Building speed in new product development. *Food Processing* 53:26–32.

Tybor, P.T. and E. Reynolds. 1989. *Food Product Development—Making Profits Grow*. Bulletin 1024. Cooperative Extension Service, The University of Georgia.

Urban, G.L. and J.R. Hauser. 1980. *Design and Marketing of New Products*. Prentice-Hall. Englewood Cliffs, NJ.

18

Packaging Muscle Foods

Joseph H. Hotchkiss

Introduction

"Muscle foods" or meats, poultry, and fish comprise a broad category of food products that require several different types of packaging. Muscle foods such as fresh red meats require the presence of oxygen (O_2) to maintain consumer appeal, whereas others such as cured meats are rapidly degraded in the presence of O_2. Each requires a different type of packaging.

Two decisions must be made when selecting muscle food packaging: the shape or form, and the material. Selection will depend on product factors such as color stability, storage conditions, microbial condition, preservatives, and degree of processing. Market factors such as distribution time/shelf-life, package size and cost, premarket pricing, and brand labeling need consideration. Fresh muscle foods have a shorter shelf-life than preserved. Processed products require more sophisticated and expensive packaging because they will be stored at higher temperatures for longer periods than refrigerated products.

General Functions of Muscle Food Packaging

The most fundamental function is to contain and unitize the product into sizes and cuts required for the market. This may mean single-serving packages for retail or large packages for food service. Packaging must have both high strength and integrity. Thermally processed and nitrite-cured meats often require more physical protection compared to fresh meats. Products that are packaged at the same location as they are sold, such as fresh meats, do not require such a high degree of integrity or strength in their packaging. Sanitation and consumer appeal are of primary concern for these products.

Muscle food packaging must be a barrier between the product and the environment in order to protect and preserve. The degree of barrier required depends on the type of meat. Meats which are stored at room temperature for months (e.g., commercially sterile meats) must be protected from the effects of O_2. Packaging for such products must be a high barrier to O_2 ingress and loss of water. Frozen muscle foods must also be protected against water loss. Packaging must also provide a barrier against biological, chemical, and physical agents that would detract from quality or safety. Components of the package that might transfer from the package to the product during storage must not be toxic. In the United States, the Food and Drug Administration has specific requirements on the amount and type of material that may be extracted from a plastic that is in contact with foods. Users of plastic packaging must get confirmation from suppliers that their materials meet these requirements.

Lastly, packaging must allow the meat product to be produced and distributed efficiently and economically. Costly and inefficient packages will not be successful. Packaging is a significant factor in the low cost of food in much of the developed world.

Materials Used in Muscle Food Packaging

The types and forms of materials used to package muscle foods include fiber based (paper and paperboard), glass, and metal. Nearly all meat packaging incorporates plastics either as coatings, linings, or in the construction of films, bags, or plastic cans. Table 18.1 lists several of the most common plastic materials used to package muscle foods.

The term "plastic" refers to a broad group of materials with different mechanical, physical, and chemical properties. Plastic is analogous to "metal," which also describes a broad group of materials. Polymers are made by linking together small molecules known as "monomers" into very large molecules. There are only 20 or so basic monomers used to make polymers for food packaging, but there are many different ways of linking them together. Polymers formed by linking two or more different monomers together are known as copolymers as opposed to homopolymers which are made from only one type of monomer. The reason for this is to achieve properties that are a combination of the two materials. For example, linking vinyl chloride and ethylene monomers into a copolymer gives a material that has a relatively high moisture barrier and low O_2 barrier. This material is used to package fresh red meats. Combinations of properties can also be achieved by combining layers of different materials either by lamination or by coextrusion. Coextrusion is a method by which two or more layers of different polymers are formed into a single film by simultaneously manufacturing both layers. For example, laminating nylon to polyethylene will give a material that is puncture resistant, a good O_2 barrier due to the nylon, a good moisture barrier, and is able to be heat sealed as a result of using polyethylene.

Many polymer properties influence the quality and shelf-life of packaged meats, but of primary importance is the resistance that the film offers to the

Table 18.1. Selected Polymers Used for Packaging Muscle Foods

Polymer	Use	GTR(O_2)[a]	WVTR[b]	Comment
Ionomer	Heat-seal layer	226–484	1.3–2.1	Resists seal contamination
Nylon (uncoated)	Films, thermoformed trays	2.6	24–26	Also used as bone-guards
Nylon[c] (PVdC coated)	Films, thermoformed trays	0.5	0.2	
PET[c] (uncoated)	Films, trays	5.0	1.3	Good clarity
PET[c] (PVdC[c] coated)	Film	0.2	0.6	
LDPE[c]	Bags, poultry wraps	250–840	1.2	Low-cost, low-gas barrier
LLDPE[c]	Heat-seal layer	250–840	1.2	Good clarity
EVA[c]–LDPE[c] copolymer	Seal layer, films, wraps	515–645	3.9	Heat shrinkable
pp[c] (non-oriented)	Semirigid containers	84–415	0.5–0.65	
PVC[c]	Fresh meat wrap	5–1500	<4.0	GTR depends on plasticization
PVdC[c]	Barrier Layer	0.08–1.7	0.05–0.3	Barrier less affected by moisture

Source: Adapted from *Packaging Encyclopedia, 1988* The Cahners Publishing, Newton, MA.

[a] Gas Transmission Rate. cc/0.001 in./100 in.^{2}/24 h @ atm, 73°F, 0% relative humidity.

[b] Water Vapor Transmission Rate, gm/24 h/100 in.^{2} @ 90% relative humidity.

[c] PVdC, polyvinylidenechloride; PET, polyethylene terephthalate (polyester); LDPE, low-density polyethylene; EVA, ethylene vinyl acetate; LLDPE, linear low-density polyethylene; PP, polypropylene; PVC, polyvinyl chloride

transfer of gases (e.g., O_2) and vapors (e.g., water/odors) through the material. This resistance is known as the barrier property of the film. Gases and vapors pass through plastics by a process known as "permeation." All common polymers are permeable to O_2 and water vapor, but the rate (i.e., the amount of gas per unit time and unit area) differs over three orders of magnitude for low- to-high-barrier materials (Table 18.2). If O_2 ingress (or water loss) is responsible for the deterioration of a meat product, then products packaged in a high-barrier material will have a longer shelf-life than those in low-barrier packages.

Thermal properties of plastics are important. Most bags and trays used for meats are sealed by fusing two pieces together by the application heat. In many cases, meats are packaged in skin-tight packages. This is achieved by evacuating the air from inside the package, heat sealing the package closed, then giving the package a heat treatment of a few seconds. Films that have been stretched during manufacture will shrink to tightly fit the product when heat is applied. Many meat products are thermally processed after packaging. In these cases, the poly-

Table 18.2. Typical Constructions for Packaging Used for Muscle Foods

Product	Package Form	Materials[a]	Comments
Fresh red meats	Overwrap films	Plasticized stretch PVC. EVA/LDPE stretch and shrink.	Treated with antifog agents
	Trays	PS or Pulp	Foamed and clear trays
	Drip pads	Polyolefin-covered cellulose	Polyolefin prevents sticking
	Chub (tube) packs	EVA/nylon/EVA with or without EVOH EVA/PVdC/EVA LDPE	Ground beef, course and fine grind Heat-processed pork sausage Cold-processed pork sausage
Fresh fish	Trays and overwrap	PS, pulp trays, plasticized PVC wraps	Packaging usually similar to red meats
Fresh poultry	Overwrap films	EVA/LDPE and PVC stretch/shrink	Often preprinted
	Trays	EPS and pulp	
	Bags	LDPE	Often clip tied
Frozen poultry	Shrink bags	EVA and LLDPE	Often clip tied
Frozen fish	Paperboard cartons	Polyolefin-coated solid bleached sulfate fiberboard	Requires good moisture resistance
	Shrink film overwrap	LDPE or EVA	Barriers such as PVdC or EVOH sometimes added as coextrusions or laminants
	Paper overwrap	Waxed bleached paper	Poor low-temperature stability
Modified Atmosphere Packaged meats	Thermoformed trays	PVC/EVOH, or PS/PVdC, or Nylon with added barrier	EVOH or PVdC barriers. Tie layers may have antifog additives, often preprinted
	Lidding Stock	PET or biaxially orientated nylon with barrier layer (PVdc), ionomer heat seal layer	
Processed meats	Thermoform film trays	Nylon with ionomer or EVA seal layer. PVdC or EVOH barrier to prevent discoloration	Franks, luncheon meats
	Rigid plastic for carded meats	HIPS or PET rigid base	PVdC or EVOH added for barrier
Wholesale and distribution meats	Shrink bags	EVA/PVdC/EVA shrink bags	Polyolefin bone-guards to prevent film puncture
	Corrugated boxes	Kraft corrugated	"Boxed beef"

[a] See Table 18.1 for abbreviations

mer must be able to withstand the temperatures used in processing, up to 100°C (212°F) for pasteurized products and 121°C (250°F) for retorted products.

Packaging materials used for foods are carefully regulated in the United States by the Food and Drug Administration and for meats and meat products, by the U.S. Department of Agriculture. The main concerns are safety and economic protection. The materials from which packaging is made may contain substances that could cause off-flavors or off-colors or could be of toxicological concern. Polymer processing additives or solvents from printing inks are examples. For this reason, all food-contact packaging materials must be specifically approved for food use. Written assurance that the packaging materials meet current regulations should be acquired from the materials supplier prior to use.

Levels of Muscle Food Packaging

Meat packaging can be divided into four levels or types based on type of product and the means of distribution:

1. **Retail packaging**: Retail meat packaging must protect and unitize the product into sizes that can be used by consumers and must present the product in a manner that is appealing. For red meats, this means maintaining the bright red color of oxygenated myoglobin in the meat. Prepricing and brand labeling by the meat packer are sometimes part of retail packaging, particularly for poultry, but for red meats, the retailer most often does the packaging and pricing. In fact, red meats are virtually the only food commodity in which the last steps in manufacturing are left to the retailer. Fish and poultry are most often prepackaged by the manufacturer.
2. **Hotel, Restaurant, Institutional (HRI) Packaging**: Meats for the HRI trade need not have an appealing appearance because the consumer does not see the packaging. Institutional packaging, in most cases, must unitize larger quantities with less variability between individual servings. Labor is a major cost in restaurants and institutional feeding, and packaging of unitized portions is intended to help reduce costs. Institutions have also become concerned about the costs of disposal of used packaging.
3. **Distribution Packaging**: Most fish and fresh red meats are distributed in wholesale lots which are not suitable for consumer use. Wholesale cuts must be processed into consumer cuts and repackaged at retail. Distribution packaging must provide protection and sufficient shelf-life for distribution. Recently, larger wholesale-type cuts such as tenderloins or other beef primal cuts are being offered for sale at retail in vacuum packages. Poultry is often distributed prepackaged in consumer-sized packages. These are sometimes prelabeled and prepriced.

4. **Processed Product Packaging**: Processed products such as cured meats (e.g., hams, luncheon meats, hot dogs) are expected to have a long shelf-life and require a high degree of protection. Materials that have a high resistance to the permeation of O_2 and water vapor are used. Higher-barrier materials are usually more expensive because of the type of plastics used and/or the increased thickness. The packaging must also unitize product. Other consumer convenience features such as reclosure and the ability to be heated in a microwave oven are an added benefit. Frozen meats typically have not required the use of high-barrier materials.

Fresh Red Meat Packaging

Package Requirements

Red meats are, in most cases, packaged twice prior to sale; once for wholesale distribution as a primal or subprimal cut, and a second time as a consumer package after further breakdown into suitable sizes for retail display and consumer use. This final step is conducted in the retail store because the "case-life" (i.e., time that a product can be stored in the retail display case) is only 2 or 3 days in typical consumer packaging, which is too short for distribution from the packaging plant to the retail store.

Wholesale packaging of fresh meats must provide sufficient shelf-life for distribution and storage. Longer shelf-life is achieved by reducing the O_2 tension in the package by vacuum-packaging and providing a high barrier to O_2 and water vapor transfer. Reduction in O_2 permeation rate causes three changes in meats that influence shelf-life. First is a color change from bright cherry-red (oxygenated myoglobin) to a dark purple color (unoxygenated myoglobin). Maintaining the bright-red color is not of concern as long as the meat will "bloom" when exposed to air. Reduction in O_2 ingress also reduces the growth rate of spoilage microorganisms and lipid oxidation. Wholesale cuts are typically vacuum-packaged in flexible high-barrier polymer bags made from coextrusion films. These bags are shrunk to fit the shape of the cut of meat using a heat tunnel or hot-water dip. Bags must be resistant to puncture so that sharp bones do not penetrate the film and release the vacuum. Larger cuts of meat sometimes have protective covers placed over sharp bones to prevent puncture. Barrier bags are fitted with highly puncture-resistant polyolefin patches laminated to the bag at key locations for the same purpose. These materials are referred to as "bone guards" (Fig. 18.1).

Reduction in the O_2 surrounding meats can increase shelf-life. The most important is a reduction in the growth rate of aerobic psychrophilic (i.e., cold tolerant) microorganisms which produce slime, off-odors, and discoloration. Microorganisms such as *Pseudomonas* and certain molds are capable of growth in meats under normal refrigeration temperatures. However, these organisms

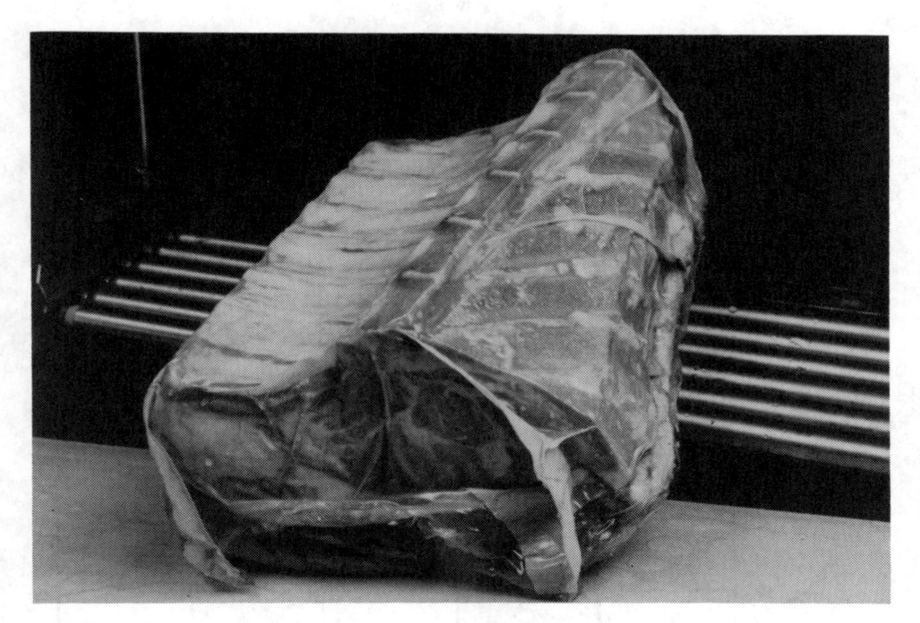

Figure 18.1. Fresh beef loin packaged in boneguard barrier bag. (Courtesy of Cryovac Division of W. R. Grace & Co.)

require O_2 for growth. Vacuum-packaging reduces the O_2 tension sufficiently to inhibit growth of these organisms and extends shelf-life beyond what would be expected from unpackaged hanging carcasses or primal cuts packaged in low-barrier films. There is a shift in the type of microflora from aerobic to facultative anaerobes. Proper chilling and refrigeration during distribution are also important to maintaining the extended shelf-life of vacuum-packaged meats. Vacuum-packaging has other benefits; it prevents desiccation (water loss) during storage and distribution, minimizes purge, prevents contamination, reduces lipid oxidation, unitizes cuts, and reduces distribution weight loss. Reduction to primal and subprimal cuts is conducted in a centralized facility where efficiencies of scale can be realized. A larger portion of the waste can be left at the processor, reducing shipping costs and maximizing its value as a rendered product. Box beef also gives increased flexibility in the types of cuts delivered to an individual store and reduces the in-store costs.

The requirements for retail packaging are different than wholesale packaging. The primary requirement is proper O_2 tension so that myoglobin is oxygenated to its bright red "bloom." At low O_2 pressures, myoglobin is purple. If the environment is too oxidative, the ferrous (Fe^{2+}) iron in the oxymyoglobin can be oxidized to ferric (Fe^{3+}) metmyoglobin iron which is brown (Fig. 18.2).

Other requirements for retail packaging include transparency on at least one side, control of drip, retardation of water loss to reduce drying and weight loss,

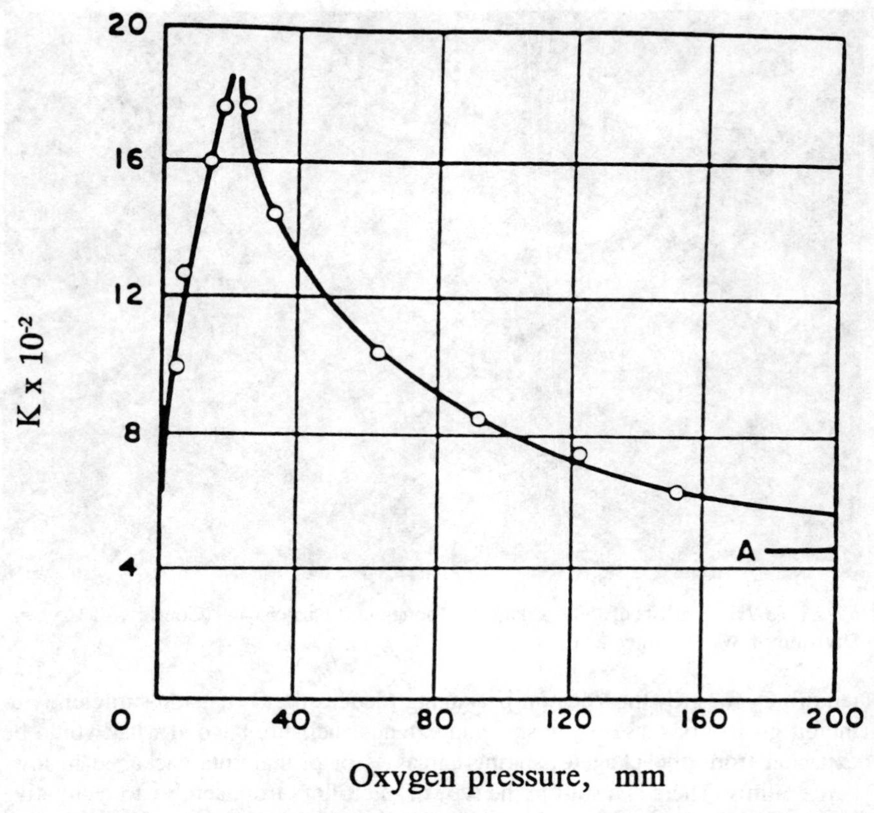

Figure 18.2. The relationship between oxygen pressure and the rate of brown oxidized myoglobin. A = value at atmospheric oxygen pressure. The highest rate of brown pigment formation occurs at a pressure of 20 mm Hg. [From *Meat Science* (Lawrie 1985) with permission].

puncture resistance, and use of antifog additives so the product remains visible. The packaging must prevent microbial contamination of the meats.

Packaging Materials and Forms

Wholesale primal and subprimal meats are vacuum-packaged in multilayered bags at a pressure of approximately 0.1 atm (Fig. 18.3). The bag is heat-sealed and passed through a hot-water bath (approximately 90°C (195°F)) to shrink the film around the meat. Table 18.2 lists some typical constructions for films used for wholesale red meats. Among the most commonly used films are heat-shrinkable bags made of EVA/PVdC/EVA coextrusions or laminants. (See Table 18.1 for abbreviations.) Other constructions include nylon or EVOH. In some

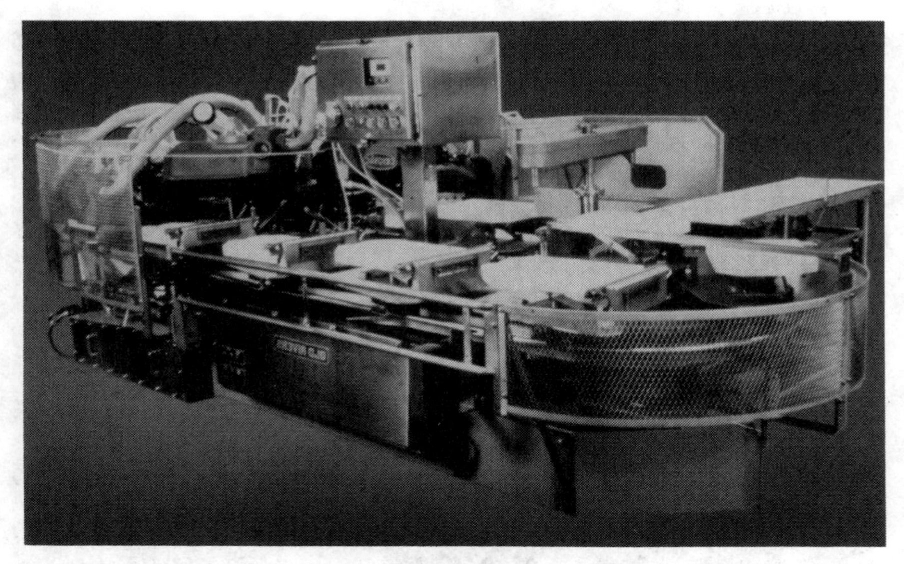

Figure 18.3. A Cryovac 8300 Rotary Vacuum Chamber Machine used to vacuum package fresh and processed meats in barrier shrink bags. (Courtesy of Cryovac Division of W. R. Grace & Co.).

cases, nonshrink bags are used. Bags are unitized into 20–50-lb cartons made of corrugated paperboard that may have an additive in order to improve wet strength in high-moisture environments (Fig. 18.4). Ground meats are often packaged in "chubs," which are large tubes that are filled with ground meat and both ends tied or clipped closed.

The growth in self-serve meat counters in the 1950s required the use of new consumer-sized packaging for retail display. Plasticized PVC is nearly universally used in retail packaging. PVC has an O_2 transmission rate which allows sufficient O_2 in the package for myoglobin to be oxygenated, yet has sufficient resistance to moisture to inhibit drying out. The film is treated with antifog agents which reduce the surface tension of water so that condensation droplets do not form on the inside. This allows the product to remain visible in the refrigerated case. Several types of trays have been used for retail. The pulp tray has been nearly replaced by expanded polystyrene (EPS), orientated polypropylene (OPP), or clear polystyrene which have the advantage of not sticking to meats, not absorbing moisture and weakening, and of being more sanitary. Pulp trays have increased in use in recent years due to environmental concerns. Drip is commonly controlled by inserting a "drip pad" inside the package to absorb purge or drip. This pad contains absorbent cellulose with a thin polyolefin polymer layer on one side. The polyolefin layer keeps the absorbent cellulose pad from sticking to the meat.

Meat has a shorter shelf-life in PVC packaging than vacuum-packaged meats.

Figure 18.4. Boxed beef packaged in barrier bags for wholesale distribution. (Courtesy of Cryovac Division of W. R. Grace & Co.)

This raises the question of why vacuum-packaging is not used for consumer packages. Vacuum-packaging by definition reduces the O_2 pressure surrounding meats and, hence, changes bright-red oxymyoglobin to purple myoglobin. Although this change has no adverse effect on the meat, consumers judge quality of red meats, in large part, by the bright-red oxymyoglobin color. The bright-red oxymyoglobin is considered to be a measure of freshness and so vacuum-packaged meats have not been successful at the retail level even though the overall quality of the vacuum-packaged meat might be higher.

Newer Technologies

Distribution of vacuum-packaged primal and subprimal meats has the disadvantage of requiring processing and repackaging in the retail store. Packaging twice, first for distribution and then for retail display, requires an excessive amount of material and creates solid waste. Stores are required to have a trained staff and facility to unpack wholesale cuts, reduce them to consumer cuts, and repackage by hand. Several centralized packaging technologies have been developed to allow muscle foods to be distributed in consumer packages that have been unitized into shipping cartons. These "case-ready" packages are removed from the shipping carton and placed directly in the display case without further processing (Fig. 18.5).

Figure 18.5. "Case-Ready" or "Counter-Ready" fresh lamb modified atmosphere packaged in a master pack for extended shelf-life. (Courtesy of CVP Systems, Inc.)

The most widely studied technology for case ready meats is modified atmosphere packaging (MAP). In MAP, consumer-sized cuts are packaged in high barrier containers from which the air has been removed either by vacuum or flushing and replaced by a gas mixture that is capable of extending the shelf-life. For red meats such as beef or pork, the most commonly used gas mixtures have been up to 50–80% O_2, 20–50% CO_2, and often nitrogen. The O_2 maintains the bloom and the nitrogen, which is inert, helps maintain package shape by countering the tendency of the package to form a vacuum as the CO_2 is absorbed by the meat. The CO_2 is biostatic. In general, gram-negative psychrophilic microorganisms are inhibited by CO_2 at levels greater than 10%. The inhibitory effect of CO_2 appears not to be due to a change in the pH of the meat as a result of the formation of carbonic acid but to a more direct effect on the microorganisms. Discoloration has been a problem with some modified atmosphere packaging systems; hence, the use of high O_2 atmospheres to maintain desirable color. In at least one process, pork is treated with a combination of ascorbic acid (vitamin C), phosphates, and citrate as a dip or spay in order to help maintain color and appearance.

There are two basic design and materials requirements for MAP. First, the package must contain sufficient headspace to act as a reservoir for the gas(es). The *amount* of each gas available is more important than the percentage in the mixture. The amount will be a function of not only the percent in the mixture but also the size of the headspace. Second, the barrier properties of the container affect how long the gas mixture will remain in the package. The container must

Figure 18.6. O_2 absorbing packet added to sliced bologna package compared to control package.

have a sufficient barrier to inhibit the loss of the modified atmosphere, particularly CO_2. This requires high-barrier plastics which adds cost. The loss of CO_2 by permeation accounts for the observation that the common practice of adding solid CO_2 when grinding meats does not increase shelf-life when packaged in low-barrier packaging.

MAP is also used as a distribution method. In this form, several conventional consumer packages are placed in bulk containers containing a high CO_2 and low O_2 atmosphere. The trays are then removed from this container and allowed to "bloom" in the retail case. This slows deterioration during shipment, but the effect of the modified atmosphere packaging is lost when the shipping container is opened and the product displayed.

Recently, the incorporation of packets or sachets containing materials that alter the atmosphere surrounding a packaged food has been introduced. These packets act by absorbing O_2 from the headspace or by releasing CO_2. The O_2 absorbers are iron compounds that react with atmospheric O_2 in the presence of moisture to form iron oxides (i.e., rust) thus inhibiting oxidation. This idea seems particularly applicable to processed meats where even small amounts of O_2 are detrimental (Fig. 18.6).

Fresh Poultry Packaging

Package Requirements

The requirements for poultry packaging are that it prevent cross-contamination and that it help provide sufficient shelf-life for distribution. Other factors such as cost, brand labeling, drip control, and transparency are also important. Unlike red meats, the poultry industry has moved rapidly to centralized packaging. Centralized poultry packaging is technically simpler because poultry meat is less sensitive to color changes and 0_2 tension is less critical. Poultry does not require aging as does red meat. The major mode of deterioration of fresh poultry is microbiological. Growth of microorganisms causes slime and undesirable odors in fresh poultry in as little as 6–8 days. "Deep chilled" (i.e., crust frozen) poultry lasts 10–14 days at 7°C (45°F). Shelf-life of 18 or more days has been reported for MAP poultry.

With the exception of whole turkeys and cooked and processed products, the market for frozen poultry has remained small. Therefore, packaging for frozen products remains a relatively minor portion of poultry sales.

Packaging Materials and Forms

Fresh poultry is packaged and distributed in one of four ways. The oldest is the "wet shipper." Whole carcasses are packed in shaved ice (approximately 2:3, w:w, ice:chicken) in wax- or plastic-coated fiberboard cartons, wire-wound crates, or plastic totes. This is inefficient because the ice must be made, transported, and disposed of, the cartons are not reusable, and the carcasses must be repackaged. The move by consumers from whole birds to parts and precut carcasses has hastened the decline in wet shippers. The major advantage of wet shippers is that carcasses do not lose weight during shipment and can actually gain weight.

Carcasses are also shipped in "dry shippers" in which whole birds or parts are placed in coated corrugated fiberboard or plastic totes without ice. Loss of moisture by evaporation or drip during shipment is a concern if carcasses are not individually packaged. For this reason, the carcasses are most often prebagged in polyolefin bags to prevent moisture loss. However, this system may not always achieve desired shelf-life.

The third system is known as "deep chilled," or "crust-frozen pack." Whole birds or parts are packaged in overwrapped trays. The packages are rapidly passed through a −40°C (−40°F) or lower chill tunnel which freezes the surface to a depth of approximately ¼ in. The product then thaws (i.e, equilibrates) during distribution at approximately 2.2°C (28°F). This low-temperature treatment reduces microbiological activity and extends shelf-life. The overwrap film is usually a stretch PVC or stretch/shrink polyolefin such as polyethylene (PE) and the trays are foamed polystyrene. The film is often preprinted for brand labeling.

In the last decade, modified atmosphere shipment of poultry in "master packs" has become widespread. This system makes use of the antimicrobial properties of CO_2. Parts or whole carcasses are packaged in consumer packages made of O_2 permeable films such as LDPE or PVC. Groups of packages are unitized in 25–50-lb groups in corrugated cartons lined with large heavy-walled plastic bags.

The air in the bag is evacuated and replaced with CO_2 or mixtures of CO_2 and O_2 and/or nitrogen. The bag is hermetically sealed and the carton closed. In commercial practice, the shelf-life can be extended from the normal 10–14 days up to 18–21 days. The disadvantage is in the cost of the large bag and gases. However, these costs must be compared to the cost of the extra refrigeration required for the "chill pack" systems.

Both fresh and frozen whole turkeys are bagged, evacuated, heat sealed, and sent through a short-duration hot-water bath to shrink the bag tightly around the carcass. The bags are made of materials that are puncture resistant and have a sufficient moisture barrier to prevent desiccation during storage. Corrugated shipping cartons that have been treated to increase wet strength are used.

Packaging Processed and Value-Added Poultry Products

Precooked and processed poultry products have been successful. Many of these products are imitations of red-meat-based products such as frankfurters and ham and may be packaged in the same manner. However, several unique cooked or processed products have also been introduced. These products range from breaded and fried patties and fully cooked cut parts to whole cooked chickens and dinners (Fig. 18.7). These products are generally designed for direct and rapid reheating so the packaging must be compatible with microwave-oven use. The product must also have a sufficient shelf-life for long-distance distribution, storage, and retail display, usually a minimum of 3–4 weeks. In several cases, MAP has been applied to these products. Often, gas mixtures containing reduced O_2 ($<1\%$) and elevated CO_2 are used. Nitrogen is added to inhibit package collapse as the $C0_2$ is absorbed. One manufacturer has used a foamed polystyrene tray that contains an EVOH barrier layer. The trays are lidded with a lamination of EVOH–nylon–PE. The PE provides good heat seals, and the EVOH acts as a barrier. Films must contain antifog additives to provide clarity.

Fresh Fish and Seafood Packaging

Package Requirements

Fresh fish is highly perishable due to rapid chemical and microbiological changes. Short shelf-life requires shipment as soon after harvest as possible. Some seafood such as lobster have such short shelf-lives that they must be shipped alive. The major packaging requirements are inhibition of spoilage, unitization, and prevention of contamination.

Frozen fish has considerably longer shelf-life than fresh but is susceptible to desiccation and undesirable chemical changes. Packaging for frozen fish has the additional requirement of inhibiting moisture loss. Fish also undergoes undesirable changes during frozen storage resulting from the inherent chemical nature

Figure 18.7. Breaded and cooked poultry products packaged in modified atmosphere and high-barrier trays and overwrapping.

of the tissue. Many fish contain trimethylamineoxide which is enzymically converted to trimethylamine, even when stored frozen. Trimethylamine is responsible for much of the "fishy" odor associated with fish. Enzymic hydrolysis of trimethylamineoxide also results in the formation of formaldehyde which cross-links proteins and toughens fish during frozen storage.

Packaging Materials and Forms

Traditionally, fish intended to be sold without having been frozen are packed in shaved ice, often after evisceration. Frequently, the fish and ice are packed in reusable and returnable polyolefin or wooden tubs. The fish are then sold without further packaging or sometimes after overwrapping in a manner similar to poultry or red meats. This method has the same disadvantages as ice packing poultry.

Newer Packaging Technologies—Modified Atmosphere

Several attempts to increase the shelf-life of fresh fish have been made. The most technically successful have been MAP and vacuum-packaging. Dozens of scientific articles have documented that careful selection of the atmosphere surrounding refrigerated fish and seafood can inhibit the growth of spoilage microorganisms. In most applications, fresh fish are packaged in high-barrier polymer trays with a flexible film sealed over the tray, usually in consumer-sized portions. The O_2 content of the package is reduced to $<5\%$, and a mixture of 40–80% nitrogen and 20–60% CO_2 introduced. Wholesale distribution systems have also been developed. In its most common form, MAP of fresh fish consists of trays made of PVdC–LDPE, EVA–PVdC, or PVC–LDPE laminants. EVOH or PVdC serve as the barriers.

MAP of fresh fish has not been widely practiced in the United States. This is due, in part, to concerns over safety. Fish from marine estuaries are often contaminated with *Clostridium botulinum* Type E. This bacteria can produce a lethal toxin under anaerobic (i.e., without O_2) environments. Some types of *C. botulinum* that are associated with fish are capable of producing toxin at temperatures as low as 3.3°C (38°F). This concern has led the U.S. National Fisheries Institute to issue guidelines for MAP of fish:

1. Only high-quality fresh fish should be used for MAP.
2. Fish should be treated with a sanitizing solution.
3. Only relatively high O_2 transmission rate films should be used.
4. The gases should be at least 70% CO_2 with the balance N_2 and O_2.
5. Product should be chilled to 2.2°C (28°F) and shipped as soon as possible.
6. Products should be maintained under "good refrigeration."

Frozen Meats, Fish, and Poultry Packaging

Package Requirements

The requirements for packaging frozen muscle foods are different than fresh products principally because microbiological deterioration is no longer a primary concern. Frozen foods are expected to be stored for months depending on the temperature and the degree of protection offered by the packaging. Fish have a shorter frozen shelf-life than meat or poultry.

Two major packaging-related types of deterioration occur during frozen storage. The first is desiccation due to the sublimation of water from the product. This water loss adversely affects texture and color and is commonly termed "freezer burn." Frozen-meat packaging should have a sufficient water vapor barrier to inhibit any noticeable loss of water during frozen storage.

The second packaging-related deterioration results from oxidation. Oxidation occurs at frozen food storage temperatures as well as higher temperatures. Shelf-

Figure 18.8. Vacuum skin packaged frozen tuna fillets.

life of frozen meats will be inversely proportional to the rate of O_2 migration into the package.

Frozen-meat packaging must also be made of puncture-resistant material. Sharp corners and bones in meats must be prevented from punching holes. Often "bone-guards" which are heavy puncture-resistant films are placed over sharp bones to prevent puncture. The packaging materials must also not become brittle at freezer temperatures.

Packaging Materials

Frozen poultry products are often packaged in heat-shrinkable polyolefin bags with metal clip closures or heat seals. Frozen uncooked fish have been traditionally packaged in coated and bleached paperboard cartons that are overwrapped with wax- or PE-treated paper to inhibit moisture loss. Wax has the disadvantage of cracking at freezer temperatures, allowing O_2 and water vapor to migrate and, thus, hastening deterioration. More recently, frozen fish has been successfully marketed in vacuum skin packaging in which the fish is placed on a semirigid plastic tray and plastic "skin" formed around the frozen fish (Fig. 18.8).

Heat-sealed plastic films have been increasingly used to package frozen meats. Polyolefin films such as LDPE have the advantage of low cost but are only moderate O_2 barriers. Coextrusions or laminants of LDPE or EVA with barrier materials such as PVdC, nylon, and/or EVOH are often used to prevent oxidation and desiccation. These films may be used as either shrink wraps for cartons or as bags and skin packaging.

Processed Meat Packaging

Package Requirements

Processed products include nitrite-cured meats, poultry, and fish that are sold in the refrigerated case. Processed products include canned and dried meats, both of which are stored at ambient temperatures. Processing partially or completely inhibits the growth of microorganisms in these products. Canned and dried meats may be stored at room temperature for periods of months to years. Nitrite-cured products are not microbiologically stable and are usually stored at refrigeration temperatures. The mechanisms of deterioration are loss of water, oxidation of lipids and pigments, or, in some cases, microbiological deterioration. Oxidation results in the development of rancid off-flavors and/or changes in color and appearance. Oxidation of the nitrosohemochrome that gives nitrite-cured meats their pink color can be catalyzed by lights such as those used in display cases. Packaging can reduce this type of oxidation by reducing the amount of O_2 available to react with the pigments and by reducing the transmission of UV light.

Barrier properties are used to control both microbiological and oxidative forms of deterioration. Inhibition of O_2 ingress inhibits the growth of psychrophilic (i.e., able to grow at refrigeration temperatures) microorganisms which are mostly aerobic and inhibits chemical oxidation.

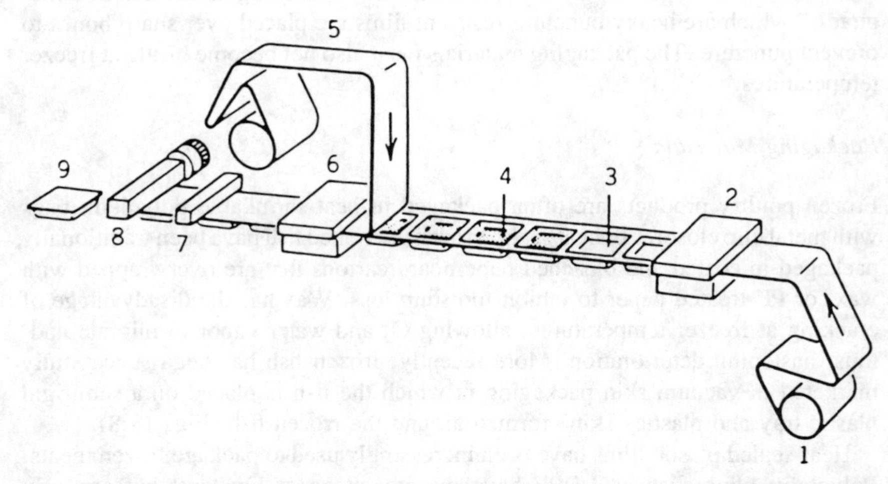

Figure 18.9. Thermoform–fill–seal (t-f-s) machine and process commonly used for processed and cured meats. 1—lower forming web; 2—thermoforming die; 3—formed unfilled pocket; 4—filled pocket; 5—upper or lidding web; 6—vacuum/gas flush/sealing die; 7—knife for cross-machine package cutoff; 8—rotary knives for machine direction cutoff; 9—finished package. [From *Packaging Encyclopedia* (Bakker 1986) with permission.]

In addition to strength and high barrier, polymer films used for processed meats must have excellent heat-sealing properties. The sealing area is often contaminated with oil and fat from the meat. Some films do not form strong seals in the presence of contamination, whereas others, particularly the ionomer-based films, are less affected. The films must also resist cracking during refrigerated distribution and storage, yet be easy to open. In some cases, reclosure is a useful feature. Luncheon meats are often packaged in plastic "trays" or "bubbles" which are then sealed with a thick layer of film that forms a lid. These meats are often hung from pegs in refrigerated cases for retail display.

Packaging Materials

Processed meats are typically packaged in heat-shrinkable and heat-sealable bags of either EVA/PVdC/EVA or Nylon/EVOH/ionomer coextrusions. Films containing carboxylic acid salts of metals such as zinc or sodium, known as ionomers, are particularly good heat-sealing layers because they are not affected by small amounts of fat in the seal area.

Processed meats are often packaged on horizontal form–fill–seal machines. Typically these machines thermoform a pocket or cavity in a continuous film web (Fig. 18.9). The pocket is filled with product and a top film is applied over the filled cavity. The package is evacuated and the top film heat sealed to the lower cavity.

Figure 18.10. Beef rib eye packaged in barrier cook-in-bags being moved into steam cabinet. (Courtesy of Cryovac Division of W. R. Grace & Co.)

Either nylon or PET-based film with a coextrusion or lamination of a heat-sealable film such as ionomor or EVA is applied. For products needing a longer shelf-life, a high-barrier material such as PVdC or EVOH can be included.

Dried meats are stored at room temperature and require considerable protection against O_2 and moisture. Packages require a high-barrier material such as PVdC or EVOH. In some cases, products are packaged in materials containing aluminum foil or in glass jars. Oxygen-absorbing packets have been used.

Some ground or processed meats such as sausage and liverwurst are packaged in "chub" packs. These are tubelike packages in which the ends are clipped closed or sealed closed. Those products that have a short shelf-life can be

Figure 18.11. CVP A-200 snorkel-type vacuum/gas flush machine being used to package meats in master packs for extended shelf-life distribution. (Courtesy of CVP Systems, Inc.)

packaged in a lower-cost material such as LDPE, whereas processed meats, sausage, or hot-filled meats require higher-barrier materials such as PVdC.

A newer technology for prepared meats is the use of cook-in-the-bag. These are restructured deli-type processed meats such as turkey breast or ham or roast beef which are cooked in heat-shrink packages rather than open heating (Fig. 18.10). The product is distributed in the package and opened at retail for slicing. This system has the advantage of eliminating significant weight loss during cooking and processing.

Future Trends in Muscle Food Packaging: Centralized Packing

As with other parts of the food industry, meats, poultry, and fish have undergone changes. One difference between red meat and poultry is centralized packaging. These are called "case ready" products (Fig. 18.11). Poultry has the advantage of being centrally processed into many products that appeal to consumers. Poultry is brand-labeled which allows advertising, marketing, and promotion. There are economics of scale when large amounts of product are processed in an assembly line fashion as is common in the poultry industry. Less waste must be shipped and can be rendered at the plant. Lastly, centralized packaging gives companies incentive to control quality and consistency. Without centralized packing and brand labeling, there is little reason for companies to develop better products. The red meat industry must depend on the retailer to produce cuts that are appealing to consumers and to market their product.

The problem with centralized red meat packaging is insufficient shelf-life. There are at least three current technologies that could overcome this problem. The first and simplest is vacuum-packaging. Meats that are vacuum-packaged in high-barrier films have sufficient shelf-life (2–3 weeks) for centralized packaging. Unfortunately, consumers are not used to the purple color and are unwilling to purchase vacuum-packaged meats. The second technology is to group and vacuum-package consumer cuts. These "tray ready" cuts are then repackaged at the retail level. This eliminates the need for further cutting at retail but does not allow for brand labeling and requires repackaging at retail. The third technology is to prepack retail cuts and distribute them in unitized groups under modified atmospheres or other technologies intended to increase shelf-life. The disadvantage is in the cost of packaging materials and the process.

Selected References

Bakker, Marilyn (editor-in-chief). 1986. *The Wiley Encyclopedia of Packaging Technology*. John Wiley & Sons, New York.

Benning, Calvin J. 1983. *Plastic Films for Packaging: Technology, Applications, and Process Economics*. Technomic Publishing Company, Lancaster, PA.

Brody, A.L. (ed.). 1989. *Controlled/Modified Atmosphere/Vacuum Packaging of Foods*. Food & Nutrition Press, Trumbull, CT.

Brown, W.E. 1992. In: *Plastics in Food Packaging: Properties, Design, and Fabrication*. H.A. Hughes, ed. Marcel Dekker, New York.

Hanlon, J.F. 1992. *Handbook of Packaging Engineering, Second Edition*. Technomic Publishing Company, Lancaster, PA.

Hotchkiss, J.H. 1990. Advances in and aspects of modified atmosphere packaging in fresh red meats. In: *Proceedings, 42nd Annual Reciprocal Meat Conference of the American Meat Science Association*. Pages 31–40. National Live Stock and Meat Board, Chicago, IL.

Jenkins, Wilmer A. and James P. Harrington. 1991. *Packaging Foods with Plastics*. Technomic Publishing Company, Lancaster, PA.

Lawrie, R.A. 1985. *Meat Science,* 4th ed. Pergamon Press, Oxford.

Paine, Frank A. and Heather Y. Paine. 1983. *A Handbook of Food Packaging*. Leonard Hill, Glasgow.

Robertson, Gordon L. 1993. *Food Packaging*. Marcel Dekker, New York.

Takashi, Kadoya (ed.). 1990. *Food Packaging*. Academic Press, San Diego, CA:

19

Methods and Economics of Meat Merchandising

Blaine B. Breidenstein

Introduction

The economics of meat merchandising could not be discussed without acknowledgment of the common economic terms, supply and demand. Most of the comments will be made on demand with the assumption that the agricultural system can, within a short time, produce or supply what the consumer demands. The National Live Stock and Meat Board published a report "Demand Strategies" Special Report: The Meat Consumer Today (1990). This report was designed to provide a comprehensive analysis of the modern consumer of meat and look at implications for marketers. They came to the following conclusions.

First, we must recognize that consumers have changed over the last 25 years. What are these changes? Changing household structure has resulted in greater emphasis on convenience. An aging population has different needs than younger consumers. There is increasing awareness and concern about health issues; thus the nutritional value of meats has become important in consumer buying decisions. Microwave technology and other advances in food processing capabilities have increased preparation options. Most consumers are more price conscious, thereby reducing the demand for more expensive meat products. Changing lifestyles have resulted in the moving away from meals consisting of meat and potatoes to those made up of entrees such as seafood and a variety of ethnic dishes.

Factors driving demand are variety, convenience, health, and price. The desire for variety has resulted in increased consumption of nonmeat entrees and new ways to prepare meats. Changing-life styles and population structure has played a major role in the emerging of convenience as an important factor. Single-member or single-parent households, the percentage of working women and the two-income families, a drop in leisure time, the increased use of microwave ovens, and the impressive growth of fast-food sales are some of the commonly cited factors responsible for the increasing importance of convenience. Health issues, many of which center around fat and cholesterol, are driving the industry to provide new products. Price is especially important to families with lower incomes and has been a factor in the increased consumption of poultry products.

The information presented in this chapter will describe an existing system of meat distribution and merchandising that will change from necessity. Some of the issues presented in "Demand Strategies" are already influencing this system. Those institutions involved in the production, processing, distribution, and marketing of meat products must constantly review their product and services. They must always remember that the consumer makes the final choice and the consumer has more options to choose from than ever before. Also, most decisions made by these institutions must be based on good economic information that allows them to make a profit. Without profit, a business institution fails regardless of the quality of the product or the service provided.

Sources of Meat Products (Muscle Foods)

Muscle foods are derived from a wide range of living creatures. Historically, meat has been thought of as edible tissue from domestic livestock, namely, beef, lamb, pork, and veal. Poultry generally has been derived from chicken and turkey with limited quantities from duck, geese, and other fowl. Fish and seafood have been harvested from large bodies of water or rivers with aquaculture supplying limited quantities. Exotic or game animals and birds have included deer, elk, caribou, bison, moose, bear, pheasant, and ostrich. These animals have for the most part been undomesticated and living under natural environmental conditions.

Major changes are occurring in the production and utilization of all these groups. Animal health, nutrition, selection for specific growth and muscle characteristics, introduction of biotechnology, and other factors are constantly improving the production of animals and animal products. Improved material handling systems, refrigeration, sanitation, and processing technology and procedures continue to supply an array of muscle food products, both traditional and new from the many sources of muscle foods. Distribution and marketing systems are adjusting to meet the needs of the producer of muscle food products and the many varied needs and desires of today's consumer.

All these muscle foods compete in the marketplace for the consumer dollars spent on food and food-related items. If we consider the four factors driving demand (variety, convenience, health, and price), some of the changes in muscle food product consumption in the last 20 years are better understood. Poultry products have gained in consumption, whereas the red meats have declined or remained constant (Table 19.1). Price per edible unit has played a major role in this increased utilization (Table 19.2).

Table 19.1. Per Capita Disappearance of Meat, Poultry, and Fish

	Beef		Veal		Pork		Lamb/Mutt		Poultry	Fish
Year	Carc	Retl	Carc	Retl	Carc	Retl	Carc	Retl	Retl	Retl
1960	85.7	63.4	6.2	5.0	78.3	59.1	4.7	4.2	34.1	10.3
1965	101.0	74.7	5.4	4.3	68.1	51.8	3.7	3.3	41.2	10.8
1970	114.4	84.6	3.0	2.4	73.1	56.0	3.3	2.9	48.7	11.8
1975	119.2	88.2	4.2	3.3	55.9	43.0	2.0	1.8	48.9	12.2
1980	103.4	76.6	1.8	1.5	74.0	57.3	1.5	1.4	61.0	12.8
1985	106.7	78.9	2.2	1.8	66.3	51.7	1.6	1.4	70.3	14.4
1990	96.1	67.8	1.3	1.1	64.1	49.8	1.7	1.5	90.3	15.5

Sources: U.S. Department of Agriculture and American Meat Institute. 1991. *Meat Facts*.

Carc = Carcass weight basis.

Retl = Retail weight basis.

Table 19.2. U.S. Average Retail Meat Prices, 1981–1990 in Dollars per Pound

	1981	1990	Increase per lb.	% Increase from 1981
Beef (Choice grade)	2.35	2.81	0.46	19.6
Sirloin steak	2.99	3.67	0.38	22.7
Round steak	2.86	3.32	0.46	16.1
Round roast	2.63	2.93	0.30	11.4
Rib roast	3.03	4.49	1.46	48.2
Chuck roast	1.82	2.09	0.27	14.8
Ground chuck	1.80	1.97	0.17	9.4
Fresh pork	1.52	2.13	0.61	40.1
Pork Chops	2.13	3.27	1.14	53.5
Sliced bacon	1.67	2.13	0.46	27.5
Smoked picnic	1.04	1.28	0.24	23.1
Canned ham	2.46	2.77	0.33	12.6
Fresh pork sausage	1.61	2.35	0.74	46.0
Frankfurters	1.77	2.29	0.52	29.4
Bologna	2.11	2.51	0.40	19.0
Poultry				
Whole broilers	0.74	0.90	0.16	21.6
Frozen hen turkeys	0.98	0.99	0.01	1.0

Sources: U.S. Department of Labor, Bureau of Labor Statistics, and American Meat Institute 1991 Meat Facts.

Just as important are the changes in products offered to the consumer both in variety and convenience. Combine these factors with the view that poultry is a healthier food and we now have many of the reasons for poultry's increased favor as a muscle food. Fish and seafood have also gained in popularity as a food. Producers of these products offer a wide selection of products to the consumer that are competitive in cost per serving and are quickly and easily prepared in the home. Fish products are considered the most healthy of all muscle foods by some nutritionists and are a major source of the omega-3 fatty acids. The rapid development of aquaculture has contributed much to the supply of fish and seafood and had a major impact on the price of finished edible products. Animals historically considered as wild have become more plentiful. Animals such as deer, elk, caribou, bison, ostriches, llama, and others are being raised on farms and ranches. Most of these produce muscle foods that are low in fat and considered healthier than beef or pork. Quantities produced are small compared to the other sources, but these specialty products are gaining in popularity.

As poultry, seafood, and, to a very limited extent, game animals have gained popularity, products from the traditional red meats have been losing their position in the market. Aggressive marketing programs have reduced the rate of decline in consumption of beef and pork. The wasteful production practices, especially in the case of beef, are slowly changing. Animals are being selected and grown that produce lean carcasses with acceptable quality. The beef industry cannot afford to continue the practice of overfinishing cattle and then trim fat that has limited value from the meat. Overfinished cattle produce muscle food with an excess of nontrimmable fat. Pork producers have made some adjustment in the lean-to-fat ratio in their animals. An active marketing campaign to inform consumers of pork as the "other white meat" and its nutritional, health-related attributes has been effective. The saturated–unsaturated fatty acid ratio in pork is considered more desirable than that found in beef.

Role of Local, State, and Federal Governments

Meat inspection and grading is covered in another section and the reader should refer to that section for detailed information. Local city, county, and state ordinances and regulations are not uniform in their monitoring of marketing of perishable food items. The operators of meat slaughter, processing, retail, and foodservice businesses need to be familiar with and follow federal, state, and local regulations. Likewise, state regulations vary from state to state with some states allowing fairly free movement within the state of products produced in state-inspected plants. Others are more restrictive of products and have limits on volume or dollar sales. The monitoring by local and state governments in food processing establishments, retail markets, and foodservice and institutional outlets is directed heavily toward sanitation and public health issues with some emphasis on weights and measures.

The federal government's role is exercised primarily at the slaughter and processing level. Federal meat inspection covers meat wholesomeness, safety, and accurate labeling. All labels on federally inspected meat products must be approved by the USDA and contain information that is appropriate for that specific product. Nutritional labeling is scheduled to become mandatory on most food items including meat by May 1994. Although meat and meat products are inspected and controlled by the United States Department of Agriculture (USDA) under the Food Safety and Inspection Service (FSIS), the nutritional regulations guidelines will coincide with those imposed on other foods by the Food and Drug Administration (FDA).

FSIS requires mandatory labeling of most meat and poultry products except raw, single-ingredient products (e.g., chicken breasts). The label information will be based on products as packaged rather than as prepared. Nutrition information as consumed may also be presented, provided no ingredients are added. The agency encourages voluntary labeling of raw, single-ingredient products, with the information presented as packaged or consumed. As an alternative to providing information on the labels, manufacturers of raw, single-ingredient meat and poultry products may provide nutrition information through point-of-purchase materials such as posters, brochures, and videos. Labels on meat and poultry will require prior approval by FSIS (Mermelstein, 1993).

The Compliance Division of FSIS monitors businesses engaged in interstate food marketing and distribution. Their program operates between the processor and the consumer as a "second line of defense" against unfit products in marketing channels. Their aim is to prevent fraudulent or illegal practices once the product has left the processing plant. Compliance officers check for uninspected meat or poultry, counterfeit inspection stamps, inaccurate labels, and contamination or spoilage of product. Meat and poultry product handlers are periodically reviewed by compliance officers.

Slaughter, Processing, and Distribution Systems

Prior to 1960, beef cattle were slaughtered and shipped as carcasses or primal cuts to retail outlets or companies supplying the hotel, restaurants, or institutional trade. Carcasses and cuts were generally shipped including the bone and most of the fat. Retailers or purveyors processed the carcasses and cuts into the desired products. Considerable amounts of bone and fat were accumulated in the preparation of the carcasses and cuts into the final products. Low-quality beef cattle were (and still are today) slaughtered, boned, and sold as boneless beef for grinding and incorporated into ground beef or used in sausage manufacturing.

Around 1960, boxed beef was introduced into the marketplace. Beef carcasses today are broken into primal or subprimal units, vacuum-packaged, and shipped in boxes to retailers or purveyors for cutting into the final products (Table 19.3). Beef carcasses weighing less than 725 lbs are used for the retail trade (Cutting Method A & B). Carcasses over 725 lbs are generally made into boneless cuts and are directed to the Hotel, Restaurant, and Institutional (HRI) markets (Cutting Method C). Market studies have shown that 80–85% of the fed beef (high quality) is now marketed as boxed beef.

Table 19.3. Beef Carcass Conversion to Vacuum-Packed Boxed Beef Cuts as Percent of Chilled Carcass Weight

IMPS#	Product Name	Cutting Method[a] A	B	C
	Primal or Subprimal Cuts			
158A	Round, diamond cut	19.42	19.42	
167A	Round, knuckled, peeled			3.07
169	Round, top (inside)			5.55
170	Round, bottom (gooseneck)			7.76
172	Loin, full loin, diamond cut	14.47	14.47	
180	Loin, strip loin, short-cut, boneless			3.61
184	Loin, top sirloin butt			3.51
189A	Loin, full tenderloin, side muscle on, defatted			1.42
107	Rib, oven-prepared	7.34		
109	Beef rib, roast ready		5.80	
112A	Rib, ribeye roll			3.25
113	Chuck, square cut (2 pieces)	26.70		
115	Arm chuck, boneless (3 way)		24.42	24.42
	Miscellaneous Cuts Credit Items			
	Special lean trim		0.34	0.90
120	Brisket, boneless, deckle off	2.11	2.11	1.88
121E	Plate, skirt steak (diaphragm), skinned, outer	0.38	0.38	0.38
121D	Plate, skirt steak (transmersus abdom.), inner	0.52	0.52	0.55
123A	Rib, short plate, short ribs, trimmed	0.80	0.80	
124	Rib, back rib			0.99
185B	Beef Loin, bottom sirloin butt, ball tip			0.46
185C	Beef Loin, Bottom Sirloin Butt, tri-tip (Triangle)			0.59
193	Beef Flank Steak	0.42	0.42	0.41
	Plate, short plate, boneless	6.06	6.06	6.55
	Lean trim 85/15	3.15		1.52
	50/50 trim	7.94	8.67	11.68
	Fat, bone, and shrink	10.69	16.59	21.50

Source: Personal files.

[a] Cutting Methods: A—Bone-in round, loin, rib, and round cuts from carcasses weighing less than 725 lbs. B.—Bone-in round and loin, reduced bone in rib and boneless chuck. Generally these are from carcasses weighing less than 725 lbs. C.—All products boneless, round, and loins separated into smaller units. Usually these are from carcasses over 725 lbs.

With boxed beef, much of the fat and bone that did move to the retailer or purveyor is kept at the slaughter-processing facility where its quality is higher and, therefore, has greater value as by-products. The boxes of vacuum-packaged beef products in the plant move by a mechanical belt system to a scaling station where a computer-printed label (bar code) is attached to the box. This label identifies the box, product, weight, grade, processing date, etc. Eventually a "Uniform Product Code" system will be used for all meat products. This will allow retailers or other warehouse systems to scan product into and out of their warehouses. Assuming the container box is identified with a number, the warehouse using this computer system can follow the movement of the boxes to specific stores or hotel restaurant, and institutional (HRI) units as well as maintain a current warehouse computerized inventory system.

Boxed cuts of fresh pork are commonly gas flushed with carbon dioxide (or with carbon dioxide making up most of the gas), sealed in plastic bags, and shipped to retail or purveyors' outlets. The gas flushing extends the shelf-life of the fresh pork items. The pork cuts are processed into retail cuts or cuts for the HRI trade.

Chicken from larger poultry slaughterers packaged as individual birds in plastic bags may be vacuum-packaged. If birds are loosely packaged in plastic bags, the master carton may be gas flushed to extend shelf-life. Much of the chicken is cut centrally into various cuts, tray packed, sealed in plastic, and shipped in master cartons using some CO_2 (either snow or pellets) to keep the temperature in the -3.3 to -2.2°C (26–28°F) range. Some whole birds are sold in ice-packed containers, but the number shipped this way is diminishing.

Turkey, which was a seasonal item sold mostly during the Thanksgiving–Christmas season, now has year-round sales appeal. Fresh turkey parts as well as whole birds are available throughout the year. Frozen turkeys still make up a high proportion of the turkey sales and are also available throughout the year.

Poultry items, both chicken and turkey, are viewed with favor by the health-conscious consumer. This has stimulated the poultry industry to become very active in product development. Many new and different chicken and turkey items are available in the retail and HRI trade. Chicken products have become a favorite of many fast-food consumers.

Veal and lamb products are available in vacuum-packaged systems. Some processors do make and package individual retail or HRI cuts, but vacuum-packaged cuts are generally made into retail cuts at the store or HRI cuts by a purveyor.

Processed meats and sausages are packaged in consumer-sized packages by the processors. Products are sliced or portioned and packaged in varying sizes to accommodate individual consumer needs. Most processed meats are sold as a "branded" product. The "branded" product identifies the processor's registered name and is used to advertise and promote a product by that processor.

Further processing (central processing) of fresh meat into case-ready meat is used by the poultry industry more than the beef, pork, lamb, or veal processors. Poultry products are more adaptable to packaging systems that reduce moisture loss and oxidative changes than beef, pork, lamb, or veal. Films with very low moisture vapor and gas transmission rates can be used. Centrally processed prepackaged retail poultry items can be kept at lower temperatures during storage and shipping (slightly frozen) with less purge accumulation in the packages. The lower temperature inhibits or delays microbial growth. Poultry products maintain a more normal color when vacuum-packaged. Beef and pork products have a different color under vacuum than when they are packaged in oxygen-permeable films that allow the conversion of myoglobin to oxymyoglobin. Myoglobin, especially in beef, is the dominant pigment (dark purple color) when packaged under vacuum. The consumer has been conditioned to want the characteristic fresh (bright-red) color and has not shown a real desire to accept the darker colored vacuum-packaged fresh beef.

Central Processing Case-Ready Meats

Centralized processing and packaging of case-ready meats would result in reduced labor costs, more efficient use of equipment and facilities, automated packaging systems, lower packaging cost, and improved by-product utilization. Careful inventory control and good product flow through the marketing system are essential. Central processing eventually will become the normal method of preparing products for both the HRI and retail trade. The long-term economic appeal for case-ready meat is strong, as it offers a number of ways to reduce cost of operation (William Bishop Consulting, Ltd, 1988). Retail marketing programs and pricing systems (explained under retail pricing) used today are major deterrents in establishing successful central processing programs of case-ready meats.

Central processing of red meats into case-ready meat products requires rigid sanitation programs. Temperature in the processing area should be less than 14.4°C (40°F). The case-ready products should have a temperature of 14.4°C (<40°F) when processed. These products need to be maintained at a uniform temperature during storage, shipping, and marketing at −1.6 to −1.1°C (29–30°F). Packaging needs to be closely monitored and controlled. Modified or controlled atmosphere packaging, used in some European markets, increases the packaging bulk significantly over vacuum-packaged product. The variety of fresh retail meat items offered in retail markets today in the United States presents a major challenge to a central processor. Shelf-life limitations on fresh case-ready items combined with unpredictable sales patterns are two additional factors. Today's retailer processes, cuts, and packages fresh red meat products in their back rooms as sales dictate. A central processing system will require improved sales projections by retailers.

Distribution Centers

Distribution centers were maintained by most of the large packing companies into the 1970s. These were called branch houses. Some of these are still being used today and continue to service small retail companies and some purveyors that supply the HRI business. During the 1960s and 1970s, the large retail companies built and began to operate their own warehouse-distribution centers. Other private warehouse systems were developed to service different geographic areas of the United States, Store-door drop deliveries by packers are uncommon today. A central warehouse system, providing full service to a group of retail stores or HRI businesses, is much more efficient than the old store-door delivery system.

Retail Store Types

Super Stores are large retail stores that may offer their customers the opportunity to buy hard line (auto supplies and hardware) and soft line (clothing, purses, shoes, and jewelry) items, drug prescriptions, and other services as well as food. Warehouse supermarket stores are common in the larger cities, offering limited service and every-day lower prices. There are traditional supermarkets providing products and services as they did in the 1960s, 1970s, and 1980s.

These types of stores have many similarities in their retail meat business. Most of the stores cut, wrap, and price fresh meat items in a cutting area. These items are then moved to the self-service display cases for customer sales. Many of these stores have delicatessen sales areas where luncheon meat and cheese items are sliced at the customer's request. Most of them have electronic scanning systems at customer checkout stations, so the cutting room scaling system is equipped with a computer bar coding system to facilitate the checkout procedure. The information incorporated into a bar code system is decided by the group using the system. Uniform Product Codes (UPC) are being developed and adapted to bring about improved buying, shipping, and receiving of products. The computerized scanning system provides a method for capturing marketing information on each type of item sold. This information, properly processed, can give current point-of-sales information that is beneficial in scheduling labor, production of products to sales, sales planning and purchasing of items, and materials for sale. Prior to this, much of the planning was done based on historical data that were grouped into categories such as meats, deli, frozen foods, etc. In addition, yield information can be captured on such items as boxed beef, fresh pork, lamb, and other materials that are converted to retail products in the meat cutting area.

Warehouse supermarket stores, because of their size and variety of products offered, may do some preparation and processing of a limited line of sausage-type products within their store. These may be fresh or cooked. These products can provide an opportunity to add value to materials that are generated in preparing products for retail sales. These stores, as do most food retail stores and local food service companies (doing business within a state), come under the county, city, and state regulations for food-related businesses.

Supermarket Stores

Supermarket stores, depending on their size and the market area they service, have varying sizes of service delis and fresh fish and fresh meat service cases that display a variety of beef, pork, lamb, veal, chicken, and turkey products. The attendants are qualified butchers that can make special cuts and provide cooking and serving advice and recommendations to the customer. Large self-service cases are provided for fresh-packaged meat items, cured and smoked meats, frozen meats, fish, self-service deli, and some precooked prepackaged meats. New products appear regularly that are a combination of vegetables, gravies, and meat products that are ready to eat, oven-ready, or microwaveable. These products may be advertised to appeal to the health-conscious consumer and are definitely in the convenience category. Many are prepared and packaged as a single serving and are designed to provide a nutritious meal to the one-person homemaker. The number of consumers in this category continues to increase as our society ages and more people remain unmarried until reaching their late twenties and early thirties.

Warehouse Club Stores

Warehouse club stores (stores that require membership cards) provide a variety of meat products to their customers. The philosophy behind warehouse clubs is volume. They buy large quantities and sell at low prices. Historically, they bought prepackaged meat products, generally in a frozen form. Today, many of the meat products are prepared and packaged at the club store and sold in large family packages. One warehouse club group buys beef primal cuts vacuum-packaged in family-style multipacks from a major beef supplier rather than preparing these cuts themselves. These multipack items are priced in the $8–$12 range and provide customers with a quality product at a good price.

Some studies show the club stores selling 12%–13% of the grocery items today. This is expected to continue and reach 30% in the future. The rapid expansion of these clubs means more opportunity for meat and poultry companies (Nunes, 1992).

Butcher Shops, Locker Plants, and Specialty Meat Shops

Butcher shops, locker plants, and specialty meat shops are common merchandisers of meat products. In many parts of the world, butcher shops are the most common meat market offering uncut, unwrapped fresh meat. Some may have glass-fronted display cases. Customers make the selection after which sales personnel cut, weigh, package, and price the items. Butcher shops or meat shops that offer a variety of fresh, processed, or frozen meat products are common in mid-sized to large cities in the United States.

These products may be branded and prepackaged by major meat processors or prepared on the premises and carry the butcher shop's own brand and inspection number. Fresh meat products may be made from vacuum-packaged boxed beef, pork, veal, or lamb products. Some of these butcher shops offer both retail and wholesale products. If under state inspection, many of them are purveyors of meat items to the local HRI trade of the city where they are located. They may be federally inspected and offer a variety of products both for retail and wholesale markets. These shops appeal to the local trade in their locality, providing the customer with an array of fresh, frozen, or precooked products as well as information, suggestions, and recommendations for cookery and serving.

State-inspected locker plants became common in rural communities and small villages, towns, and cities after World War II in the United States. These establishments initially provided slaughter, processing, packaging, and frozen storage of beef, pork, and lamb products to the local community. Changes in transportation, rural electrification, and the popularity of home freezers, migration of rural people to larger cities, and the shopping patterns of the people serviced have resulted in some of these local locker plants becoming federal-inspected facilities. Many have home retail and wholesale outlets similar to those described as butcher shops. Some deliver finished products to customers by United Parcel Service and other rapid deliver systems using catalogues and printed brochures to advertise and market their products. Others have established route sales delivering meat products to retail stores, restaurants, and other institutional groups located in rural areas and small cities. Although some have successfully made the transition from small state-inspected locker plants to federally inspected plants, many, because of limited finances and other reasons, continue to operate as state-inspected plants.

Convenience Stores and Fast-Food Markets

Convenience stores and fast-food markets are important meat merchandising outlets. Shelf-stable snack foods are commonly displayed and offered for sale. Some self-service deli-type items are available also. Fast-food establishments and the U.S. consumers love affair with hamburgers served by them has resulted in a major growth area for the merchandising of meat products. Initially, hamburgers were the main meat item. The companies managing these establishments soon recognized the need to offer a variety of products ranging from hamburgers, roast beef, hot dogs, steak sandwiches, chicken fillets, fish sandwiches, and numerous other meat and nonmeat products to their customers. Today, these companies are concerned with nutrition, health, and environmental pollution as well as convenience and fast service.

Specialty Food Shops

Specialty food shops frequently found in large shopping malls offer attractively packaged items. Dry sausages, country-cured hams, jerky, and other nonrefrigerated meat products are merchandised through these outlets. These shelf-stable nonrefrigerated meat items are attractive gift items that can be transported considerable distances through the U.S. Postal System without danger of spoilage or shipping damage.

Retail Merchandizing

Purchasing Fresh Meat

Vacuum-packaged boxed beef has had a great impact on marketing of beef to retailers and food service. Prior to boxed beef, much of the beef was sold as carcass beef. Carcasses were cut into the forequarters, hindquarters, or primal cuts such as the arm chuck, brisket, rib, plate, loin, round, and flank. Prior to the development of the national interstate road system in the United States, much of this product moved to distant markets by refrigerated rail cars. Most of the large slaughter facilities were located close to large rail systems. As the interstate road network developed, the transportation system changed to over the road truck deliveries. The slaughter facilities relocated from the large city locations to rural areas where feedlots were located. During this reallocation process, single-level slaughter facilities were built to handle beef only. The slaughter plants were highly mechanized and could slaughter 200–400 head of cattle per hour. Highly mechanized beef cut areas developed to disassemble the carcass into the boxed beef cuts as we know them today. The purchaser of boxed beef cuts no longer had to buy carcass components they did not want to merchandise. The cuts going into the vacuum-packaging program are offered as bone-in or boneless. USDA quality grades Select, Choice, or Prime are available. Yield grade 3 or better are generally offered as standard for the trade; however, yield grade 2s may be available for a price premium. The cuts may be available in selected weight ranges, and varying levels of exterior fat thicknesses can be specified. Pieces per box are uniform from a supplier but can vary from supplier to supplier.

Market reporting systems have had difficulty adjusting to the boxed beef program. When much of the beef was sold as carcass beef, packers had 5–7 days to ship the product so slaughter numbers gave reliable information on the movement of beef through the marketing system. The vacuum-packaged boxed beef product, packaged on today's high-vacuum-packaging systems, has a much greater shelf-life than carcass beef. Vacuum-packaged bone-in beef cuts with good temperature control will hold up well for 18–21 days. Boneless beef products can be held for 30 days or more. When vacuum-packaged boxed beef with bone-in exceeds 12–15 days post-vacuum-packaging, retail shelf-life starts to be shortened. Boneless beef cuts may start to show shortened acceptable retail shelf-life time after 24–25 days post-vacuum-packaging. Slaughter numbers are not as highly correlated to beef sales, and consequently, market reporting systems are searching for reliable sources of information.

Much of the pricing for boxed beef is on a cattle formula base. Subprimal cuts are cut using well-described anatomical features of the carcass as described in the "Institutional Meat Purchase Specifications" (IMPS) developed and maintained by the Agricultural Marketing Service (AMS) of the United States Department of Agriculture (USDA). The packer uses average yield information in converting the carcass into a complete set of primal or subprimal cuts and the components of the carcass designated as credits (Table 19.3). A value is determined for the credits which is credited toward the cost of the cattle. Also, the weight of the credit components is calculated and then subtracted from the carcass weight. The difference between the carcass weight and the credit components is the total weight of the subprimals resulting from that particular carcass cutting method. The packer adds a processing and packaging cost to the residual carcass cost. Average processing cost of $6.75/cwt for 550/700-lb cattle and $7.75/cwt. for over 700-lb cattle is added. (Tracking the Market, *National Provisioner,* 31 August 1992, p. 27). This total cost less the value of the credits divided by the carcass weight less credit component weight is the average cost for the subprimals. The market has assigned a value for the subprimals so individual subprimals can be purchased in carload or less than carload lots. Subprimal units are generally more expensive when purchased as subprimal units than as a portion of a full carcass unit. Full carcass units reduce inventory imbalances at the packer level because all subprimal units move uniformly. Imbalances in inventory may require offering some subprimal units at discount prices.

Merchandising of beef by retailers has changed as boxed beef sales have increased. Prior to boxed beef, when full carcass or quarters were featured, retailers had difficulty in getting all products to sell uniformly. As a result, some items accumulated, resulting in excessive shrinks and some deterioration in quality. Full carcass or quarter sales required a high labor input. Retail stores, in many cases, have reduced their labor force to 50–60% of what it was prior to the acceptance of boxed beef.

Promotional sales today are held for a relatively short time and margins are kept at a higher level compared to the historical carcass, forequarter, or hind quarter retail promotional sales. The end result has been a reduction in beef promotional sales tonnage and higher margin returns to the retailer. Prior to boxed beef's popularity during the 1960s and early 1970s, retailers sold beef with a 16–20% gross margin and would get only 8–13% gross margin on heavily promoted items. During the 1980s, the margins had increased to 28–35% on regular sales and promotional items sold for 20–30% gross margins.

Beef consumption has declined in the United States because of the consumer concern with healthy food issues associated with beef and the selling of beef at higher margins. The higher margins were the result of acceptance of boxed beef (a higher cost product than carcass beef) by retailers. Other muscle food retail margins have remained fairly steady during these years.

Beef checkoff dollars in the late 1980s were used to promote beef by advertising lean beef as a healthy nutritious food. These activities have been credited with bringing about the steady beef consumption patterns the last 2–3 years.

The "Institutional Meat Purchase Specifications" are available as 10 documents; General Requirements; Fresh Beef—Series 100; Fresh Lamb and Mutton—Series

200; Fresh Veal and Calf—Series 300; Fresh Pork—Series 400; Cured, Cured and Smoked, and Fully Cooked Pork Products—Series 500; Cured, Dried, and Smoked Beef Products—Series 600; Edible By-Products—Series 700; Sausage Products—Series 800; and Quality Assurance Provisions. These specifications are recommended for use by any company procuring meat products. For assurance that procured items comply with the detailed requirements, the USDA, through its Meat Grading and Certification branch, provides a voluntary Meat Certification Service. Because the service is voluntary, a user fee is charged. This fee is the same as the Beef Grading fees. Purchasers desiring this service should contact:

USDA, AMS, Livestock and Seed Division
Meat Grading and Certification Branch
Room 2628 South, P.O. Box 96456
Washington, DC 20090-6456

The National Association of Meat Purveyors has, since 1961, printed numerous revisions of the Meat Buyer's Guide. Information about this can be obtained by contacting

National Association of Meat Purveyors
8365-B Greensboro Drive
McLean, Virginia 22102
(703) 827-5754

Packaging Options, Functions and Materials

When the subprimal vacuum-packaged meats arrive at the store, they are fabricated into consumer portions. These consumer-sized cuts are generally placed into polystyrene foam trays or in some cases pulpboard or clear polystyrene trays. The trays are then overwrapped with stretchable polyvinyl chloride film having a low water vapor transmission rate and high oxygen permeability [8,000–12,000 cc/m^2, 24 h, 1 atm, 23°C (73°F)] to allow maximum oxymyoglobin development or bloom (Cole, A.B., Jr., 1986). Consumer packages are then priced, labeled, and placed into retail cases 0–4°C (32–40°F). Typical retail shelf-life of tray-overwrapped fresh retail meat is from 3–5 days, dependent on storage conditions, size and shape of cut, ground or unground, and meat quality. Spoilage in this 3–5-day period is generally from oxidation causing the development of metmyoglobin. Aerobic microorganisms coming from dirty knives, cutting tables, and meat-cutters' hands can accelerate metmyoglobin development. *Pseudomonas* have been identified as a dominant microflora found in PVC-overwrapped meats. A retail outlet that cuts product to sales should have minimal product that requires reworking or repackaging. Reworking generally downgrades the product into low-value ground beef or stew meats.

The shelf-life of meat can be greatly extended by vacuum-packaging meat in flexible materials of low gas permeability. A primary purpose of vacuum packaging is to reduce the oxygen in the package, effectively eliminating the growth of aerobic bacteria. Anaerobic bacteria, especially the *Lactobacilli* species, are dominant in vacuum-packaged cuts and suppress the growth of many undesirable spoilage bacteria. Continued respiration of the meat and growth of aerobic microorganisms use up the residual oxygen. Carbon dioxide is produced during this time helping to inhibit undesirable spoilage, as well as bacterial growth.

A decision must be made by the retailer or processor to determine the muscle color (oxygenated or nonoxygenated) to be marketed. Systems include those involving modified atmospheres high in oxygen, which can enhance and improve the bright cherry-red beef color, and vacuum-packaging systems, which exclude oxygen giving the fresh meat product a purplish-red color.

In 1972, a patent was assigned for vacuum skin packaging which used a heated dome to heat the packaging material to a semimolten state. No product distortion occurs from this process, as the film conforms to the exact shape of the product. These are nonoxygenated systems giving the fresh meat product a purplish-red color. Advantages of these systems are improved shelf-life from high-oxygen-barrier materials, improved product appearance, and shelf-life similar to normal vacuum-packaging (Cole, A.B., Jr., 1986).

Modified atmospheres will provide the normal bloomed color of meat for 8–14 days. After this period, deleterious oxidative rancidity and microbial and color changes diminish the acceptability of fresh retail meat. Vacuum-packaging can practically double the shelf-life of fresh retail meat compared to modified atmosphere packaging.

Modified atmosphere packaging systems were commercialized in the 1970s in Europe. The marketing system in Europe, with short transporting distances for packaged meat, is more adaptable to this system than the long transporting distances from processor to retailer in the United States. The Atmospack concept uses a unique modified gas atmosphere system coupled with conventional thermoforming roll-stock equipment and low-oxygen-permeability materials. Typical thermoforming vacuuming machines have gas flush capabilities. After vacuumizing, the specific gas mixtures are injected into the package and the package is hermetically sealed. The meat products packaged in these systems with high (50–80%) oxygen concentrations maintain a very desirable red color for 8–14 days. Carbon dioxide gas (20–50%) inhibits microbiological growth.

The modified atmosphere pack is not well suited for freezing, as surface dehydration develops. Normal vacuum-packaging and vacuum skin packaging perform well in freezer storage of fresh meats with no surface dehydration because of the reduced air spaces in the packages resulting from close film-to-meat contact.

A master pack, containing overwrapped trayed products (stretch wrap, high oxygen permeable, flexible packaging film), is placed in a large bag or pouch that is vacuumized. This is then filled with a modified gas atmosphere and sealed.

The modified atmosphere gas migrates into the overwrapped tray and provides the necessary bacteriostatic effect. At store level, the overwrapped tray packages are placed directly into retail display cases. The oxygen in the atmosphere then migrates into the overwrapped tray, allowing the meat product to bloom.

Laminated thermoforming machinery can be used for retail packaging systems to produce fresh retail meat, vacuum-packaged cuts. This system allows processors and retail chains to prepackage fresh retail meat and market their products through distribution systems similar to boxed beef channels.

Packaging costs of the various packaging systems are not easily described or determined. The trays overwrapped with stretchable polyvinyl chloride film are most commonly used with red meats and are lowest in cost. Vacuum-packaging and modified atmosphere packages are intermediate in price, and vacuum skin packaging is slightly more costly. These systems need to be evaluated not only on initial cost of materials, equipment, and labor but also on consumer acceptance and product integrity.

Retail Meat Cuts and Identification

Guidelines commonly used by the meat industry to make wholesale and retail cuts are the following: Separate tender portions of the carcass from the less-tender portions; separate lean areas from the portions having greater amounts of fat; and separate thicker, more heavily muscled portions of the carcass from the thin-muscled areas.

Wholesale or primal cutting separates the legs, made up of large locomotion-muscle systems, from the back, composed of large support-muscle systems from the thinner body-wall sections of the carcass. Wholesale cuts have been defined as large subdivisions of the carcass that have been traded in volume by segments of the meat industry. Relatively rigid standardization of wholesale cuts has been established for many years to provide efficient communication between buyers and sellers.

Retail cuts are subdivisions of wholesale cuts or carcasses that are sold to the consumer in a ready-to-cook or ready-to-eat form. Consumers have learned to identify cuts by differences in size, color, fat characteristics, and shape of bones present in bone-in products. Beef cuts are large and have a cherry-red color and a white firm fat. Pork cuts are intermediate in size, are grayish-pink in color, and have softer fat. Lamb cuts are small in size, are dark pink or light red in color, and have a hard white fat. Veal cuts are almost devoid of fat and pink in color. Poultry packages vary from light skin color to a yellow tinge to the skin. Poultry muscle has a light pink color. Fish is variable in color depending on the species. Packages should be free of fluid and muscle should have a bright rather than a dull color.

Bone-in retail cuts can be classified into one of seven types based on muscle/bone shape and size relationships. Loin cuts often contain a section of the tenderloin muscle (*psoas major*) in addition to the *loin eye*. The *lumbar* vertebra has the typical T-bone configuration formed by the lateral process, the body, and dorsal process of the vertebra. Rib cuts have the rib eye (*longissimus*) but no tenderloin and may contain a section of rib costal bone. A cross section of the hind leg (round) has the top (*semimembranosus*), bottom (*biceps femoris*), eye (*semitendinosus*), and sirloin tip (*quadriceps*) muscles that are easily distinguished from the arm cuts which have more small muscles. Flat or irregular bones are present in the hip cuts composed of a small number of relatively large, parallel muscles compared to blade cuts which display numerous small nonparallel muscles. Belly or plate cuts can be recognized by the alternating layers of fat and lean and rib bones.

Poultry carcasses are traditionally referred to as parts. Breasts and wings are composed of mostly white muscles, whereas drumsticks and thighs consist mainly of dark muscles. Backs and necks are used largely as raw material for processed meat. Parts presented in retail stores usually have skin intact, but breasts are often sold with skin removed.

Fish are sold fresh in a variety of forms. Whole or round fish are as they come from the water; drawn fish have been eviscerated; dressed fish have viscera, head, tail, and fins removed; steaks are cross-sectional slices of large fish; fillets are the sides of fish cut away from the backbone; butterfly fillets are the two single fillets held together by the belly skin; and portions or sticks are cut from the frozen block and may or may not be coated with batter or breading. The shells of crabs, oysters, and scallops usually are removed before presentation to consumers. Clams may be offered with or without shells; examples of shellstock are steamers, cherrystones, quahogs, etc. Shrimp may be merchandised with only heads removed or with heads and shells removed.

Retail meat cuts from beef, pork, lamb and veal have been known by a variety of names. Many names had regional significance and were recognized in restricted geographic areas. More than 1000 names were used to identify the approximately 300 fresh retail cuts of these species. Because standardization of cutting and nomenclature of retail cuts is essential for good retailing, an Industrial Cooperative Meat Identification Standards Committee was formed in 1972. Representatives from packers and processors, different sizes of grocery and supermarket operations, and livestock producers represented by the National Live Stock and Meat Board produced the "Uniform Retail Meat Identity Standards." Copies can by obtained from the National Live Stock and Meat Board, 444 North Michigan Avenue, Chicago, IL 60611.

Labels

Retail meat labels vary from simple to complex with respect to the amount of information. Depending on geographic location, some information items are required by law and others are optional with the retailer. The Uniform Retail

Meat Identity Standards meat cut names contains three parts: species, wholesale cut, and retail cut.

Retail meat weights must equal the specified weight on the label. Some consumers use the package weight when purchasing specific quantities for specific needs.

One of the most important items of information is the total cost of the package of meat. Many purchases are made on the basis of budgetary expenditures for meat for a specific meal or time span. Consequently, a variety of package sizes and unit costs within similar products is essential.

The date on which the meat is packaged or a pull date should appear on the label for the benefit of both the consumer and the retailer. The date is an index of freshness on which many consumers rely.

Brief cooking instructions are helpful to many customers. Labels or printed materials accompanying products may suggest appropriate cooking methods or provide recipes.

Some kinds of meat are promoted and merchandised on the basis of U.S. Department of Agriculture (USDA) grade. Some consumers rely on brand names supplied by the slaughterer or processor. Most fresh poultry and some other products carry brand names. Branded products often attract or gain consumer loyalty that does not develop when meat is sold merely as a commodity. Brand loyalty is something developed over time. Consistent taste, fat level, color, and quality are factors that relate to brand loyalty.

Retail Pricing Policy

Retail pricing policy has considerable influence on consumer choices. Because grocery people generally manage supermarket companies, they are familiar with products that are inherently not readily adaptable to a unit sales or unit pricing like the meat industry. Their pricing system has been a factor influencing the reduced demand or consumption of red meat and especially beef products and an increase in consumption of poultry. This can be demonstrated by following the pricing policy of vacuum-packed/boxed beef compared to carcass beef.

When the beef industry was still delivering carcass beef to the supermarket, the gross margin expected was 25%. Gross margin percentage is calculated dividing the difference between selling price/value and purchase cost by the selling price/value. As bone-in boxed cattle units came into the supermarket, they expected 25% gross margin, and boneless beef was expected to return a gross margin of 25%.

This means that the retailer on carcass beef wants 25% gross margin on the selling price with $1.00 per pound carcass cost, an average selling price of $1.33/lb. or 33 cents per pound margin.

The retailer using the basic bone-in boxed beef cattle unit also wants 25% gross margin on the selling price, but now because he is using a value-added product that requires 30% less labor, the cost is $1.34 per pound. His gross profit now brings him $0.45 per pound with an average selling price of $1.78/lb.

The retailer using boneless boxed beef products has reduced his labor requirements by 60% to produce case-ready products compared to carcass beef. His gross margin goal is still 25%, but now his cost is $2.00 per pound, producing a $0.66 per pound margin or an average selling price per pound of $2.66.

This pricing policy places a disadvantage on boneless middle meat cuts: lip-on rib eyes, for example. Lip-on rib eyes costing the retailer $3.50 per pound would have a selling price of about $5.00 per pound.

This example used a 25% gross margin objective to calculate and show the changes in average selling price per pound. When vacuum-packed products are converted to retail items, usually limited trimming material for stewing or grinding is generated along with fat and bone. These low-value portions of the vacuum-packaged item result in an increased selling price for the more desirable roasts, steaks, or chops. This is true because of the targeted average selling price or gross margin expected by the retailer.

Retailers establish the gross margin percentage they need for their meat departments. Gross margin objectives are attained by using different gross margin percentages for beef, pork, poultry, lamb, processed meat, fish and, in some cases, the deli, depending on their individual accounting systems. These range from 25% to 45% and usually average about 33%. Nevertheless, the objective of this discussion was to illustrate what happens in a pricing system when percent gross margin is applied in pricing of products. If gross margin percent is the pricing mechanism, the selling price per pound increased at a phenomenal rate as the wholesale cost per pound increases. Because beef, pork, and lamb have a higher wholesale cost per pound than chicken or turkey, the profit or margin in actual dollars per pound is higher for the beef, pork, and lamb. This means that beef, pork, and lamb are contributing more dollars per pound of sales to the meat department's earnings than is chicken or turkey.

What can be done to help this situation? The retailer could establish pricing of all meat items based on his production costs. This means the meat department would require some alterations in standard retail accounting policy and systems. The retailer needs to know the combined cost of operations and the amount of dollar profit desired along with the average weekly tonnage. Using this information, the retailer can find the cents per pound margin needed to achieve the dollar margin desired. The end result, using a cents per pound rather than a percent margin pricing system, would be a more equitable pricing system and undoubtedly improved consumption of red meat, because a major factor influencing their sales is price (Vesta 1984).

Display Systems

Meat display systems are located in the back area of retail food stores or along the edges of the store product display areas. Retailers vary in their preference of location, but the rear location is the most common. Many aisles of grocery products on display end near the meat display case. The customers' shopping traffic pattern takes them into the meat marketing area by this arrangement.

The refrigerated display space is an expensive area in a store. Consequently, meat items compete with each other and with other refrigerated foods for this space. Retail display cases are laid out with careful planning to entice the customer to purchase meat products. The layout may reflect the buying preferences of the customers that shop in that store, as well as considering the profit potential along with sales promotion items. Large retail companies prepare case layout plans for their stores. Layouts normally are not altered much from one week to the next because customers do not like major changes in layout.

Generally, retailers place fresh meat cuts of one species in a common display area. This practice allows customers to locate desired items more readily, as many customers are species oriented in their meat purchases. The order species are found in retail cases varies from one geographic area of the country to another. Species order may be influenced by consumption patterns within areas that, in turn, can be the result of economic situations, ethnic and religious considerations, as well as other factors influencing consumption of specific meat products.

Display Lighting

The appearance of meat is dependent on the type of light under which it is displayed. Florescent lights are commonly used rather than incandescent because they produce less heat than the incandescent lights. Meat display areas usually are well lighted and have the potential of producing a considerable amount of heat, which has an undesirable effect on fresh meat color. The increased temperature causes the oxymyoglobin to change to metmyoglobin, bacteria can grow more quickly contributing to the color change, and moisture may condense on the inner surface of the packaging film. A specific type of florescent lighting is important because of differences in color balance or retention, i.e., the degree to which the spectral energy of the light coincides with spectral reflectance of meat. Sources that provide good color balance have about 20–35% of their emission in the red range. Higher strength in the red range could cause the meat to appear misleadingly red. Light intensity is an important consideration and a range of 800–1100 lux is optimum (Judge et al. 1989).

Display Temperatures

Display cases should be closely monitored for temperature, and defrost cycles should be established to have minimal effect on the product on display. The freezing point of fresh meat is about −2°C (28°F) and temperature should be maintained close to that level. Low and uniform display temperatures suppress bacterial growth, enzyme activity, color change, oxidation, dehydration, and weep. Retail display cases can vary in temperature within a case or between cases. Short defrost cycles are desirable and should allow no more than a −16°C (3°F) increase in product temperature. Retail meat cases should maintain a rack temperature of no less than −1°C (30°F) and not more than 4°C (40°F) during the defrost cycle. Defrost cycles should be no more than 20 min long. The air circulation pattern should not be inhibited and product should be placed within the refrigerated area of the display case. Excessive loading of a case with product causes the temperature of product to increase and results in a reduction of shelf-life. At no time should the product be placed in the display case such that it is out of the refrigeration zone.

Marketing Frozen Meats

Frozen meat marketing has what would appear to be advantages over fresh meat marketing and merchandising. Many of the problems associated with fresh meat merchandising would be eliminated by distribution in the frozen form. However, frozen meat sales represent a small segment of retail meat merchandising. Some large meat processors invested large sums of money and time in frozen meat processing systems. The programs were discontinued because of lack of consumer acceptance of fresh frozen meat products. Consumers would rather buy fresh meat in the conventional packages and place it in their freezer at home than buy the same red meat cut frozen. This is true even though home freezers are less able to freeze the meat properly than commercial freezing systems. Consumers feel more comfortable choosing fresh meat and have more confidence in its quality as a fresh product than in the same item frozen. Several fish products are readily accepted by the consumer as well as some poultry items in the frozen packaged form. Product appearance of frozen red meat cuts along with the cost may be factors. Frozen prepared meats are generally highly trimmed in preparation. This makes the cost per serving high because the retailer uses a pricing system similar to fresh meat.

Packaging materials used for frozen meats have barrier properties similar to those used in fresh meat vacuum-packaging because prevention of moisture loss and oxidation are essential to prevent freezer burn. Freezer burn is the loss of moisture from the meat (dehydration). Ice crystals form in wrapped product between the meat and wrapping material. Quality deteriorates as freezer burn develops. Skin tight wraps are required to avoid accumulation of ice crystals in the packages. Packaging material must be pliable, durable, and puncture resistant at freezer temperatures.

Frozen meats need to be sold within certain time limits to maintain high quality. Freezer storage time varies depending on the type of meat that is in the package as well as the freezer temperature. Meat containing salt and other seasonings should be stored shorter times than meat without them.

Frozen meat cases like fresh meat cases need to be carefully monitored for temperature. Ice buildup in the case should be kept to a minimum. A stable temperature of −18°C (0°F) or lower is recommended to avoid ice crystal formation and excessive drip loss from thawing.

Processed Meats

Processed meats are supplied to the store in bulk form or are prepared for consumers in retail packages at the processor level. Bulk products are generally sold in the service deli. The bulk products are generally cooked ready-to-eat meat in cylindrical, rectangular, or loaf form. Customers choose the product, slice thickness, and the quantity. The deli service person slices the product, wraps, and prices the meat for the customer.

Processed meat prepared for the self-service deli area usually is packaged in systems that require two types of laminated film. One film is a forming film which, when heated, forms a cavity into which the sliced meat is placed. A nonforming film is placed over the cavity and product, vacuum is drawn, and the package is sealed. In some cases, nitrogen gas is infused into the package. With good sanitation and proper refrigeration, meat packaged by this method remains wholesome for 60 days or longer. Meat should not be in the display area for more than 2–3 weeks.

Retention of color in cured meats depends on the absence of oxygen or the exclusion of light. Cured meat pigment fades under strong fluorescent light if oxygen is present. Packaging requirements for cured meat necessitates the use of materials that are oxygen or light impermeable and moisture proof. Most cured products are packaged in vacuum packages. Some are displayed in nonvacuum opaque material such as cardboard with a limited portion of the product visible through a window. Some are placed in packages that are both vacuumized and opaque (Judge et al. 1989).

Package sizes are influenced by the manner in which the processed meat is consumed. Relatively small packages are customary, as these are usually used for snacks or lunches. The products deteriorate fairly rapidly after opening and spoilage may occur. Small packages are consistent with eating habits and needs of many customers. Packaging costs per unit of weight for these items increases as the unit weight decreases.

Processed meats packaged by the processor for consumer sales must conform to the USDA FSIS labeling requirements. Labels of this nature are USDA approved for specific products and must have an ingredient listing, "keep refrigerated" statement, product name, the USDA plant inspection legend, and net weight of the package. After July 1, 1994, packages will be required to have nutritional information.

Most prepackaged processed meat items today have a postprocessing shelf-life in excess of 60 days. Sell-by dates may appear on the packages. Prepackaged processed meat items generally give the retailer a gross margin of 40% or more.

Foodservice

Traditionally foodservice has been referred to as Hotel, Restaurant, and Institutional (HRI) types of service. These were establishments serving food not consumed in the home. Today, supermarkets in their service delis offer prepared foods that fit the foodservice designation.

Hotels

Foodservices in hotels usually are components of their hospitality and numerous amenities. These are designed for customer service and convenience and are capable of providing meal service throughout the day. Most hotels have facilities capable of serving banquets and can serve a wide range of meat items that satisfy the needs of specific conventions or banquet attendees.

Restaurants

Traditional foodservices for restaurants include dining rooms, cafeterias, snack bars, fast-food outlets, and street vendors. Each of these types of foodservice require meat products that in many cases are prepared specifically for that type of restaurant sales. Consequently, their menu choices, speed of service, costs, and service styles are different. Quality and cost of the meat served are important factors, and a restaurant's success is highly dependent on customer satisfaction.

Institutions

Institutional dining facilities are those associated with schools, hospitals, residence halls, government agencies, airlines, industrial plants, or offices. Customer satisfaction in many institutions is not as important as it is in the case of hotel/restaurant foodservice programs. Therefore, a wider range of quality of meat is used in institutional foodservice organizations. Menu planning is important in institutions, depending on the consumers' specific needs. Also, foods served in these facilities are consumed repeatedly by some consumers.

Supermarket Carry-Out Items

Carry-out items from supermarket delis are gaining in popularity. Some food specialists estimate that 25% of the food eaten at home in the mid-1990s will be prepared outside of the home. Sales are expected to increase from $7 billion in 1987 to $15 billion in 1994 (Zayas 1991). Retailers have indicated an interest in aligning themselves with outside suppliers that provide products equal to what is being offered in their delis today. Regional suppliers will have an opportunity to be a supplier and participate in today's growing deli, fresh prepared food business.

Meat Acquisition

Menu development is the first step of acquisition for foodservice. Reputation of eating establishments is related to careful menu planning, food preparation, and serving. Foodservice operators are looking for quality, consistency, and uniformity in the products they purchase along with product value. New products are being developed to meet marketing trends. Some of the trends and products are:

Ethnic—for example, Mexican, Cajun, and Chinese

Healthful—leaner, less calories, lower cholesterol

Natural—natural in, chemicals out

Finger Foods—snack foods, meat nuggets, steak, pork, chicken, etc.

Convenience Foods—microwaveable entrees prepared, portion controlled for food and plate cost

Packaging—extended shelf-life, cooking bags, smaller box sizes, portion packs for vending, meat slices

Fresh Versus Frozen—frozen fish concept or fresh product gas flushed

Value-Added Products—brand identity, new processes, formulas such as entrees, bread battered items.

Portional control meats offer some advantages to an operator compared to on-site preparation:

Convenience—storage and handling of large cuts is eliminated

Cost savings—no skilled meat cutter is required and waste is almost nonexistent

Planning—consistent supply of exact portions, so menu items can be compared for cost

Reserve supply—steaks and other items can be thawed quickly at all times.

Meat Orders

Quantities of meat products ordered by food service establishments must be based on projected business volume. Historical serving and use information should be invaluable. Numerous factors must be considered such as number of customers, serving size, and cooking loss. Cooking loss of fresh meat usually ranges from 25% to 33% and is influenced by fat content, water-binding capacity, degree of doneness, cooking method, and temperature, among other things.

Delivery schedules should be closely aligned with serving schedules, quantities of meat required, storage space, and product shelf-life. Frequent deliveries are beneficial and generally assure product freshness but add cost to the product, which must be passed on to the consumer.

Receiving and Storing Meat

When meat is received, it should be examined to see that it meets purchase specifications. The USDA grading service will certify that products meet (IMPS) standards for a fee. Delivery information should include appropriate information such as product identity, count, total weight by item, grade, shipping date, and delivery date.

Ideally, product temperature of refrigerated products should be checked upon delivery. Delivery temperature of less than 5°C (40°F) is desirable for good storage life. Frozen product boxes should be carefully inspected to determine if good temperature control has been maintained (−17°C; <0°F).

Meat products need to be stored in clean refrigerated storage areas with temperature maintained at less than 2°C (36°F). Some air circulation is desirable to maintain uniform temperatures and high humidity (60–80% RH) reduces dehydration of exposed meat surfaces. Meat, poultry and fish need to be stored away (in separate coolers) from fruits, vegetables, and other foods, and from each other. Carefully controlled inventory rotation (FIFO = first in/first out policy) is important in both frozen and fresh meat products to ensure consistent freshness and flavor.

In-Unit Processing

Processing within a foodservice unit is fairly common. Usually only special products are made. To process meat requires a refrigerated area and equipment along with skilled labor. Primal and subprimal cuts can be purchased to provide the foodservice items desired. Limited uses for by-products, processing costs, and inappropriate yields may make this exercise uneconomical.

Institutional Meat Purchasing Specifications (IMPS)

The USDA Federal Grading Service has developed specifications for many fresh and processed items purchased and used by foodservice units. The federal government is a major customer for institutional meats, so these specifications are used in all government meat procurement. These specifications are known as the Institutional Meat Purchase Specifications (IMPS).

The IMPS were referred to in the retail section and are the specifications most vacuum-packaged boxed beef fabricators follow in their carcass-breaking operations. Some modifications or alterations of these can be negotiated, but these are the base from which boxed beef products are priced and marketed. IMPS are available from the Superintendent of Documents, U.S. Government Printing Office, Washington, DC 20402. The National Association of Meat

Purveyors has published an illustrated reference for these standardized products known as *The Meat Buyer's Guide*. The color prints in this publication present the characteristic colors of the muscle for each species. Measurements and anatomical position of knife and saw cuts to break carcasses into the IMPS identified cut are well illustrated. A Food Service cut section is shown for each species. The first Meat Buyer's Guide was printed in 1961 and the latest in 1988.

Quality Grades

The USDA has conducted a voluntary beef quality grade system since 1927. A USDA quality grade stamp will appear on the surface of a carcass or primal cut when this service is used. Quality grade information is not widely used to merchandise foodservice meat. However, when used for foodservice meat products, beef quality grades are more important than quality grades for lamb, veal, and poultry. Fresh pork quality grade designations are not used, but a bright greyish-pink color and muscle structure are important. The color is associated with freshness and the muscle structure to animal maturity and tenderness.

Hotels, restaurants, and institutions specializing in high-quality beef will purchase either intact beef cuts or foodservice cuts derived from the USDA Prime, Choice, or Select grade. Utility and Commercial grade are used by some foodservice groups, but the steaks should be tenderized by some method. Choice or Select grade beef is commonly used in foodservice because these grades are acceptable to most consumers in palatability and leanness, and are more economical than those from the Prime grade. Prime grade or very high Choice grade are used routinely by some of the best restaurants and hotels. Demand for the Prime grade has been reduced as the consumers have become conscious of fat in their diet. Even so, some customers continue to buy the Prime grade rib and loin cuts to get the best in beef flavor, juiciness, and tenderness. Foodservice operators usually purchase Choice grades of lamb and veal and grade A poultry.

Fat Limitations

Purveyors trim fat from meat cuts based on buyer specifications. Roasts and subprimal meat cuts are usually purchased on an average fat thickness with maximum thickness limitations. The IMPS standards describe these relationships. Steaks specifications have the same fat trim standards as roasts. If no specifications on fat limitations are given, they must not exceed an average of 1.25 cm (½ in) in thickness and cannot exceed 1.87 cm (¾ in). Surface fat on chops, cutlets, and fillets, unless otherwise specified by purchaser, must not exceed an average of 0.64 cm (¼ in) in thickness, and the thickness at any one point must not be more than 0.95 cm (⅜ in). Fat content of patties is specified as average percent of raw weight. The IMPS standards for ground beef have been lowered to not exceed 22% fat.

The purchaser of ground beef may specify any fat content providing it is between 10% and 30%. Considerations of fat level specified must include cost (labor and weight loss as fat), functions of fat during cooking (basting action and protection from dehydration), nutritive effects (caloric content) and taste, flavor, and juiciness, especially with ground beef.

Refrigeration State

Foodservice meat is available as fresh or frozen cuts. Meat properly frozen, stored, and rotated is similar in quality to unfrozen fresh meat. Fresh meat, because of its limited shelf-life, must be purchased more often and in smaller quantities. A major advantage of frozen portion-cut meats is that inventories and microbiological quantity are easier to control. Freezer facilities must be available and maintained in good working condition. If the meat products are not cooked from a frozen state, time and refrigerated space must be available to allow for proper thawing procedures. Cooking steaks and chops from the frozen state requires more time than cooking similar products unfrozen. Some thin or small products such as patties and nuggets can be successfully cooked from the frozen state because the cooking time is relatively short.

Size of Cut

The Meat Buyer's Guide shows weight ranges designated as Range A, B, C, and D. Weight ranges are shown for carcasses, primals, and subprimals for all the IMPS cuts of beef, pork, lamb, and veal. Uniformity of size along with yields and meat cost may be predicted along with determining cooking time and serving size.

Size of portion cut meats (steaks and chops) is controlled to a large extent by the cutting procedure, whereas roast size is determined more by carcass size. Because muscles vary in size, the purchaser may specify weight or thickness of the cut, but not both. Standardized weight results in variation in thickness. Thickness variations are acceptable in preparing steaks and chops where individual pieces can be carefully attended to, and variations in doneness is beneficial. Standardized thickness results in variation in weight. Uniform product thickness requires fairly consistent cooking time to reach desired doneness. Variations in weight (or in this case size) may be acceptable when customers can select their chop or steak as is done in buffets.

Chicken and broiler parts and quarter sizes are controlled by weight. Weight tolerances are very low and most suppliers offer a variety of weights.

Meat patties are controlled by both thickness and weight. Weight tolerances are based on a 4.54 kg (10-lb) unit and given as the number of patties over or under the specified number. For instance, 114 g (4-oz) patties should have 40 patties per 4.54 kg (10-lb) unit and can contain 39–41 patties and must be within the weight tolerance.

Aged Beef

Very few purveyors or steak houses offer beef aged by dry aging because of the cost associated with the moisture loss and external trimming of dehydrated surface areas. Dry aging of high quality (Choice or Prime) beef ribs or loins is done in a cooler at 4–5°C (32–40°F), 85–90% humidity, and for 25–50 days. Aging in vacuum packages provides the tenderization effect of dry aging, but the flavor development is not the same. Aged boxed beef under vacuum greatly reduces the cost by retaining the moisture in the meat, has little or no trim loss, and provides a flavor acceptable to most consumers.

Processed Meat

Some smoked meat and sausages are included in the IMPS. Because of the large number of meat products that fall under this category, standardization is very difficult. Foodservice is offered the same products as retail. However, the presentation/packaging of the product is quite different.

Most of the cured and smoked products are made from pork. Ham and bacon are popular pork items with Canadian-style bacon and smoked pork chops found on some foodservice menus. Corned beef is popular in some areas of the country and smoked turkey has gained popularity. Smoked and cured meat items contain salt and nitrite and have a greater shelf-life than fresh meat. Most are unfrozen so they are readily available to use in foodservice operations for quick customer service.

Breakfast sausages in natural or collagen or artificial casings are a favorite sausage product in foodservice. These usually contain fresh meat (generally pork) containing salt and other seasoning. Shelf-life is relatively short, 5–7 days, so inventory has to be replenished often and carefully controlled.

Precooked and Restructured Meat

Some foodservice establishments successfully use precooked frozen entrees. These products save time, reduce labor in the foodservice establishment, and make menu costs more predictable. Many of these items require only short heating periods before serving. Battered and breaded fish and chicken fillets or nuggets are the items most favored. Sliced, restructured roast beef and toppings for pizza and tacos along with some meat stews, pies, rolls, and roasts are other precooked products requiring short heating times to serve.

Precooked refrigerated convenience food use has increased in both retail and foodservice establishments. Examples of this class of foods include precooked BBQ meats, precooked roast beef, refrigerated stews, and pasta-style entrees, as well as food packaged by the "sous-vide" process (cooking and rapid chilling following vacuum-packaging). "Sous-vide" means under vacuum. Some characteristics of these foods are the following: (1) Further cooking is unnecessary. It may be reheated for aesthetic purposes. These foods are not sterile so they should be consumed in less than 3 weeks after preparation: (2) Products should be stored below 3°C (38°F). Low-temperature microorganisms may develop during prolonged storage: (3) Most precooked convenience foods are low acid foods (pH >4.6) and are considered to be of higher risk, as they do not rely on pH as a means to help control growth of microorganisms: (4) Water activity is relatively high and is not a substantial factor in inhibiting microbial growth.

Some quality and safety requirements are the following: (1) The components used to prepare the food products must have proper and desirable characteristics: (2) The cooking process must be sufficient to inactivate any bacterial pathogens that might be present: (3) Precooked food products must be lowered to an acceptable storage temperature quickly after cooking: (4) After cooking and cooling, the food must be handled properly to avoid contamination: (5) Packaging must take place with careful sanitation precautions: (6) Storage and distribution of packaged foods must include proper refrigeration and delivery within safe time limits (Beauchemin 1990).

Selected References

Beauchemin, Micheline. 1990. Sous-Vide technology. *Reciprocal Meat Conf. Proc.* 43:103.

Cole, A.B., Jr. 1986. Retail packaging systems for fresh red meat cuts. *Reciprocal Meat Conf. Proc.* 39:106.

"Demand Strategies" Special Report. The Meat Consumer Today. National Live Stock and Meat Board. 1990. National Live Stock and Meat Board, Chicago, IL.

Institutional Meat Purchase Specifications. USDA, AMS, Livestock and Seed Division, Grading and Certification Branch, Washington, DC.

Judge, M.D., E.D. Aberle, J.C. Forrest, H.B. Hedrick, and R.A. Merkel. 1989. *Principles of Meat Science. Retail Merchandising.* Kendall/Hunt Publishing Company, Dubuque, Iowa.

Meat Buyers Guide. National Association of Meat Purveyors, McLean, VA.

Meat Facts. 1992. American Meat Institute, Washington, DC.

Mermelstein, N.H. 1993. A new era in food labeling. *Food Technol.* 47(2):81.

Nunes, K. 1992 .Warehouse clubs now a niche in the retail market. *Meat and Poultry* 38:9, 30.

Tracking the markets. Boxed beef equivalent values. *National Provisioner,* 31 August 1992, p. 27.

Uniform Retail Meat Identity Standards. National Live Stock and Meat Board, Chicago, IL.

Vesta, R. 1984. Value incentives reflected through the marketing system. *Reciprocal Meat Conf. Proc*. 37:123.

William Bishop Consulting, Ltd. 1988. *A Study of the Economics of Case-Ready Meats*. American Meat Institute, Washington, DC.

Zayas, A. 1991. Consumers go ga ga for supermarket carry-out. *National Provisioner,* 2 September 1991, p. 12.

20

Biotechnology for Muscle Food Enhancement

Morse B. Solomon

Introduction

Biotechnology is the implementation of biological sciences for the improvement of technology. Use of science for the improvement of muscle foods has involved natural selection of dominant traits, selection of preferred traits by crossbreeding, the use of endogenous and exogenous growth factors, and ultimately gene manipulation to produce desirable changes in meat/carcass quality and yield.

Until recently, improvements in the quality of meat products that reached the market place were largely the result of postharvest technology. Extensive postharvest efforts have been implemented to improve or to control the tenderness, flavor, and juiciness. Tenderness, flavor, and juiciness are the sensory attributes that make meat products palatable.

However, consumers are currently not only interested in the quality (palatability) but are also concerned with the nutritional value, safety, and wholesomeness of the meat they consume. The public is inundated with warnings about the health risks of consuming certain types or classes of foods. The Surgeon General's Report on Diet and Health, which was released in 1988, stated that "the primary priority for dietary change is the recommendation to reduce intake of total fats, especially saturated fat, because of their relationship to the development of several chronic disease conditions" (USDHHS 1988).

A wide range of biotechnology strategies for altering the balance between lean and adipose tissue growth and deposition in meat-producing animals is available. These include genetic selection and management (production) strategies. More recently, the confirmation of the growth-promoting and nutrient repartitioning effects of somatotropin, somatomedin, β-adrenergic agonists, immunization of animals against target circulating hormones or releasing factors, and gene manipulation techniques have given rise to a technological revolution for altering growth and development in meat-producing animals.

Genetic Selection and Management Strategies

The main genetic alteration during the past 30 years has been to decrease carcass fatness and increase lean tissue deposition. These alterations have been via growth rate and mature size of meat animals particularly through genotypic and sex manipulation.

Breeds

Animal breeding relies on the selection of the most appropriate stock for the next generation. Animal breeding employs a wide gene pool but suffers in the length of time required for cross-breeding and the evaluation of the resultant meat quality. The common process for acquiring new genetic variation or genes is by cross-breeding with lines, strains, or breeds that possess desirable traits imprinted on the gene(s). Exotic breeds were introduced into the original breeding stock to combine superior traits of the parent breeds to maximize the benefit in the crossbred carcass.

Genetic selection trials have shown that it is possible to change the composition of animals within any breed. The major disadvantage is that the process is long term and slow. However, the benefit is that once genetic changes have been made, they are permanent in the population and can present a desirable cost : benefit ratio. The present unique and diverse genetic pool in this country offers excellent opportunities for sources of potentially valuable genetic material.

Sex Condition

The history of castration is probably as old as the history of domestication of animals by man wherein he sought to fulfill his requirements for meat, animal products, and draft power (Turton 1969). The original reason for castrating was probably to render the male more manageable and to enable males to be grazed along with mature females without indiscriminate breeding (Crighton 1980). Thus, its use was one of the first actions taken to manage the genetic pool in animal populations at the farm. This practice was probably reinforced by the observation that the castrated male had a larger deposition of fat than its intact counterpart. This procedure was particularly important at a time when a large amount of fat was a highly desirable feature of the carcass. However, because of current market demands for leaner meat and the necessity for intensification, the traditional practice of castrating males has been questioned.

Much interest surrounds the use of young intact males in modern meat production systems. Such interest stems from the fact that males gain weight more rapidly, utilize feed more efficiently, and produce higher-yielding carcasses with less fat and more red meat than castrates and females. These advantages have been associated with the presence of circulating testosterone levels. In fact, if only half of the cattle finished in the U.S. feedlots were finished as bulls rather than steers, a savings of millions of dollars in feed costs would result because approximately 15% less feed would be required per unit gain by bulls.

Nevertheless, increased production efficiency obtained through the use of intact males has often been offset by management problems, particularly with animal behavior. Furthermore, meat production from intact males has often encountered strong resistance from packers because intact males produce carcasses that are less tender and have lower quality grades, darker lean color, and coarser textured lean. All of these factors result in lower consumer acceptance at the retail level. Packers also have indicated that hides are more difficult to remove from intact males.

Growth-Promoting Agents

Growth-promoting agents are substances that enhance the growth rate of animals without being used to provide nutrients for growth such as nutrient partitioning agents. Growth-promoting agents are anabolic; that is, they produce more body tissues and thereby result in more rapid animal growth. Growth-promoting agents cause changes in carcass composition, mature weight, and efficiency of growth. Many substances qualify as "growth-promoting agents" despite their varied origin and chemical nature. Growth-promoting agents influence growth in three ways: (1) by stimulating feed intake (appetite stimulants) and thereby increasing the supply of nutrients available for growth, (2) by altering the efficiency of the digestive process resulting in improved supply and/or balance of consumed nutrients, and (3) by altering how animals utilize or partition absorbed nutrients for specific growth processes.

Appetite Stimulants

Many growth-promoting agents influence voluntary feed intake; however, responses in voluntary intake are generally small and inconsistent. Stimulating appetite (intake) could result in improved growth rate and the production of edible protein (lean tissue). Scientists are investigating the use of drugs to control appetite. If successful substances are developed, it may be possible to enhance meat production in the future by increasing or decreasing an animal's appetite.

Antibacterial Growth Promoters

A wide variety of antibiotics (antibacterial agents) enhance growth in primarily nonruminant animals. Some common antibiotics that are used to remedy clinical infections are successful at promoting growth. Included in these common antibiotics are penicillins, tetracyclines, bacitracin, avoparcin, and virginiamycin. These antibiotics are active against gram-positive bacteria and, in most instances, are not retained in the tissues of the animal. As much as a 20% improvement in growth rate and feed efficiency have been observed with the administration of antibacterial agents to nonruminants.

Rumen Modifiers

A special class of antibiotics whose principal site of action is the reticulorumen of ruminants has been investigated. These include monensin (Rumensin®, Romensin®) and lasalocid (Bovatec®) which are classified as ionophores. Rumen additives are generally administered to growing ruminant animals. The type of diet fed to the animal receiving rumen modifiers has an effect on the outcome. Voluntary intake is depressed when rumen modifiers are included in concentrate diets, whereas little depression in intake is observed when included in forage-based diets. Improvements in growth rate and productivity are generally less than 10%. The mode of action of rumen modifiers is poorly understood but is thought to be a result of altering the metabolism (digestive process) of rumen microflora. They act against gram-positive bacteria in the rumen, causing a shift in patterns of volatile fatty acid production, improved digestive efficiency, reduced bloat, and production of methane and hydrogen.

Steroids and Related Substances

Castration of males, which removes their primary source of androgens (testosterone), is the oldest method of manipulating growth by nonnutritional mechanisms. A number of steroid-related growth-promoting agents that are either available for commercial use or have been extensively researched exist. These are listed in Table 20.1. These agents are most effective in castrated ruminant males or females and are not effective in pigs. Steroid-related growth-promoting agents generally increase live weight gain and improve carcass lean : fat ratios in ruminants but are not as effective in swine. Their mode of action is unclear, but

Table 20.1. Steroid-Related Growth-Promoting Agents

Class	Active Agent(s)	Trade Name
Natural estrogens	17β-estradiol	Compudose®
Estrogen analogues	Diethylstilbesterol	DES
	Zeranol (resorcyclic acid lactone)	Ralgro®
Natural androgens	Testosterone	
Androgen analogues	Trenbolone acetate	Finaplix®
Natural progestogens	Progesterone	
Combined products	Trenbolone acetate + estradiol	Revalor®
	Testosterone + estradiol	Implix BF®
	Testosterone propionate + estradiol benzoate	Synovex-H®
	Progesterone + estradiol	Implix BF®
	Progesterone + estradiol benzoate	Synovex-S®
	Zeranol + trenbolone acetate	Forplix®

many consider a direct effect on muscle cells. Growth-promoting agents, such as zeranol and trenbolone acetate, appear to have little, if any, anabolic effect in turkeys. The growth response of chickens to androgens is ambiguous, with both increases and decreases being reported. Androgens appear to be androgenic but not anabolic when administered to chickens.

Endogenous Somatotropin

A group of peptide hormones, e.g., somatotropin (ST), growth hormone-releasing factor (GRF), somatostatin, insulinlike growth factor-I (IGF-I), insulin, and thyrotrophic hormone, work in harmony to regulate and coordinate the metabolic pathways responsible for tissue formation and development. Even though relationships between growth and circulating levels of some of these peptide hormones have often produced conflicting results, the majority of data indicates that the genetic capacity for growth is related to increased circulating levels of somatotropin and IGF-I in livestock.

The anterior pituitary secretes three hormones (ST, prolactin, and a thyroid-stimulating hormone) that influence growth and carcass composition. Somatotropin, often called growth hormone (GH), is the most notable with commercial growth-promoting potential. Somatotropin is a small, single-chain polypeptide, made up of 191 amino acids, secreted by the pars distalis of the pituitary's adenohypophysis. The structure of somatotropin varies among species. Release of ST is stimulated by a growth-hormone-releasing factor (GRF or GHRH) produced in the hypothalamus (Hardy 1981). Understanding how to control the production of these hormones within the meat animal is a long-term goal of scientists.

Exogenous Somatotropin

Porcine

There is a growing database supporting the use of pituitary-derived porcine somatotropin (pST) as an agent to improve efficiency of growth and carcass composition in swine. Turman and Andrews (1955) and Machlin (1972) were the first to demonstrate that daily exogenous administration (injection) of highly purified pST dramatically altered nutrient use, resulting in improved growth rate and feed conversion of growing–finishing pigs. Pigs injected with pST had less (35%) fat and more (8%) protein.

However, their original observations were of little practical significance because purification of porcine ST from pituitary glands was not economical. A single dose required 25–100 pituitary glands. More recently, the development of recombinant deoxyribonucleic acid (DNA) technology has provided a mechanism for large-scale production of somatotropin. The gene for ST protein is inserted into a laboratory strain of *Escherichia coli* which can be grown on a large scale and from which ST can be purified and concentrated for use. There is a growing database supporting the use of recombinantly derived pST (rpST). No significant differences in the effectiveness of pST versus rpST have been observed.

With greater emphasis on lean tissue deposition and less lipid, the optimal genetic potential for protein deposition of an animal is a very important concept in that this potential, or ceiling, defines the protein requirement of the animal. In defining the optimal/genetic potential for protein deposition, ST is used as a tool to maximize genetic potential for protein accretion. Administration of pST to growing pigs elicits a pleiotropic response that results in altered nutrient partitioning. In studies with growing pigs, significant improvements of 40% in average daily gain and 30% in feed conversions can be achieved by administration of pST. Research has also shown that the effect of pST is enhanced by good management and nutritional practices (Campbell et al. 1990; Caperna et al. 1990). Furthermore, a 60% reduction in carcass fat and a 70% increase in carcass protein content can be attained. The magnitude of response has varied in the various studies performed since the initial, classical studies by Turman and Andrews (1955) and Machlin (1972). Different interpretations in response have been attributed to differences in experimental designs. These include initial and final weight of pigs, length of study, genotype, sex, dose of ST, nutritional conditions, and time of injection.

Daily injections of pST has been the method of choice for pST administration. Porcine ST must be administered by injection because it is a protein and would be inactivated by digestive enzymes if given orally. The response from pST administration is not related to the site of injection nor to the depth of injection. Optimal benefit would be realized by a delivery system that mimicked a daily surge of pST.

Mechanisms of pST action have been reviewed and discussed in numerous reviews. Clearly, pST affects many metabolic pathways that influence the flow of nutrients among various tissues of the body. Many have concluded that the mechanism by which pST decreases fat content in pigs is via the inhibition of lipogenesis (Table 20.2). These changes in metabolism and cell proliferation lead to the alteration of carcass composition via the reduction of nutrients normally destined to be deposited as lipid to other tissues. Studies have demonstrated that regardless of the stage of growth, pigs respond to exogenous pST at all ages. However, rapidly growing animals do not seem to benefit from exogenous pST as much as do finishing animals with respect to fat alterations. This is not surprising because pigs that are growing rapidly and are more efficient are not producing much fat.

Meat Tenderness

Administration of pST represents a technology with a promise for packers and retailers to offer leaner pork products. However, the administration of pST to

Table 20.2. Biological Effects of Somatotropin on Adipose and Skeletal Muscle Tissue

Tissue	Effect	Physiological Process Affected
Adipose tissue	↓ [a]	Glucose uptake
	↓	Glucose oxidation
	↓	Lipid synthesis
	↓	Lipogenic enzyme activity
	↓	Insulin stimulation of glucose metabolism
	↑ [b]	Basal lipolysis
	↑	Catecholamine-stimulated lipolysis
	↑	Ability of insulin to inhibit lipolysis
		Insulin binding unaffected
		Somatotropin binding unaffected
Skeletal muscle (growth)	↓	Protein degradation
	↑	Protein synthesis
	↑	Satellite cell proliferation

[a] ↓ = decreases.

[b] ↑ = increases.

barrows has been shown to reduce tenderness by as much as 39% when compared to controls. It is difficult to determine whether tenderness differences will be perceived by the consumer; however, it should be noted that most of the shear-force values were within the shear-force values associated with normal pork products. In a recent study by Solomon et al. (1994a), time postmortem of sampling muscle from pST-treated pigs for subsequent shear-force analysis had a significant effect on tenderness. Differences in shear-force tenderness between pST and control pigs were virtually eliminated when loin chops were removed from the carcass and frozen within 1.5 h postmortem compared to controls (frozen 5 days postmortem). Some of the inconsistencies reported in the literature for shear force and tenderness as a result of pST administration may be a result of inconsistencies in the time that the meat sample is removed and the time that the sample is frozen.

Minimal observable differences in processing yields, color retention, or composition of products from control and pST-treated pigs have been observed. From the wealth of literature, it appears that pST affects carcass composition and not quality (other than possibly tenderness). However, pale, soft, exudative muscles (PSE) have been observed in a couple of pST studies (discussed in Muscle Morphology subsection).

Muscle Morphology

The consensus is that pST exerts a hypertrophic response on carcass muscles that can be seen at both the cellular level (fiber types) and with the naked eye (carcass conformation and loin–eye area). Most of the research demonstrates that muscle fiber area increased in size with the use of pST. However, a rate-limiting factor in the hypertrophic response of muscles to pST administration was the level of dietary protein used (Solomon et al. 1994a) in combination with pST administration. This confirms that the beneficial effect of pST is dependent on good management and nutritional practices. Porcine ST administration has little effect on the (percentage) distribution of muscle fiber types. Solomon et al., in extensive studies on pST administration to pigs, reported numerous observations of giant, hypertrophied muscle fibers as result of pST treatment. In two pST studies (Solomon, Campbell, and Steele 1990b; Solomon et al. 1991), pST-treated pigs exhibited PSE muscle (30% and 62% incidence, respectively) which is much higher than the 12% that is reported as normal occurrence by packers. One explanation offered for the increased incidence of PSE in these studies was a seasonal effect. The experiments were conducted during the summer with average temperatures ranging above 35°C for the duration of the experiments. However, one cannot discount the possibility that the occurrence of PSE in these two studies and the absence of PSE in other studies could be due to genetic differences among pigs used in the different studies.

Carcass Composition

Lipid composition studies have demonstrated that the lipid content from pST-treated pigs was as much as 27% less in the lean tissue compared to controls. The administration of pST resulted in lean tissue containing as much as 40% less saturated fatty acids (SFA), 37% less monounsaturated fatty acids (MUFA), and no difference in polyunsaturated fatty acids (PUFA) compared to controls (Solomon, Caperna, and Steele 1990a). Cholesterol content of lean from pigs receiving pST was greater than controls. The decrease in total SFA and MUFA, and virtually no change in PUFA, support the conclusion that the mechanism by which pST decreases carcass fat content in pigs is by the inhibition of lipogenesis. A decrease in fat synthesis would lead to a decrease in the production of SFA and MUFA with little effect (change) in the amount of PUFA. The majority of PUFA in pig tissues are the result of dietary fatty acids linoleic and linolenic and are not synthesized. However, the possibility of an increased turnover of storage lipids, at the level of triacylglycerol synthesis or hydrolysis, exists (Clark, Wander, and Hu 1992).

Use of pST is similar to cattle implants and anabolic agents used to enhance growth in that all cause increases in growth and more efficient utilization of feed. However, they are different in chemical structure. Porcine ST is a protein that is not active orally and is readily digested like any protein. Because pST is a protein and is broken down in the gastrointestinal tract, human ingestion of pST would present no dangers because digestive processes would inactivate the protein and provide no residues. Digestion would break the protein down into its component amino acids, making it available for normal metabolic processes. This is the argument for the acceptance of the use of pST in pigs.

Bovine and Ovine

There is evidence that exogenous bovine and ovine ST improves efficiency of growth (~20%) and lean-to-fat ratio (~40%) in ruminant animals. The database is less extensive than it is for swine; however, the conclusion that ST treatment improves efficiency of gain and reduces fat deposition in cattle and sheep is supported.

Poultry

Daily injection of broiler chickens with exogenous chicken somatotropin (cST) does not stimulate growth rate or improve feed efficiency. The endocrine regulation of growth in chickens appears to be distinctly different from that described for domestic mammals. Administration of exogenous cST exerts a strong lipogenic rather than lipolytic action in rapidly growing broiler chickens that results in carcass from cST-treated birds containing more fat than controls. No effect of ST on turkey growth has been reported.

Growth-Hormone-Releasing Factor

Somatocrinin, often called growth-hormone-releasing factor (GRF) or growth-hormone-releasing hormone (GHRH), is a peptide hormone belonging to the glucagon family of the gut. Effects on growth of the administration or manipulation of GHRH are likely to be due to direct effects on the secretion of ST. Exogenous GRF administration increases concentrations of ST in serum of meat animals (Moseley et al. 1985). Long-term administration of GRF has been shown to stimulate growth in the rat and in man by increasing both the secretion and synthesis of pituitary ST (Clark et al. 1986).

Somatostatin

Somatostatin, another hormone produced by the hypothalamus, acts directly on the adenohypophysis of the pituitary gland to inhibit ST release (Hardy 1981). Endogenous ST secretion could be enhanced by GRF agonists or somatostatin antagonists. Enkephalins, also hypothalamic peptides, stimulate ST release and it is likely that enkephalin agonists could also be used to enhance endogenous secretion (Convey 1988).

Although somatostatin was originally purified from the hypothalamus, it is now recognized that many cells and tissues throughout the body secrete this peptide. One of these networks releases somatostatin to affect ST release (Tannenbaum 1985). Several studies [reviewed by Spencer (1986)] used somatostatin antibodies to attempt to neutralize circulating levels of somatostatin in blood and promote growth by increasing blood ST concentrations. Although this technique appeared to be an attractive possibility for manipulating growth, immunization against somatostatin did not consistently stimulate whole body growth. Lack of a response may be because of the multitude of sources (cells and tissues) that secrete somatostatin. Immunization against somatostatin has led to other strategies based on immunization techniques that may neutralize or amplify hormonal signals via receptors for hormones.

A somatostatin antagonist that completely blocks the inhibitory effects of somatostatin on ST release was evaluated by Spencer and Hallett (1985). This antagonist improved the rate of gain (7.5%) in rats when infused at a rate of 36 μg/d for 2 weeks. No carcass data were presented. Recently, Solomon et al. (1993) investigated the effect of the thiol agent cysteamine (mercaptoethylamine) on carcass composition of growing turkeys. They observed a 50% reduction in carcass fat when turkeys were fed cysteamine in the diet. As much as a 66% reduction in carcass fat was found when birds were intubated with cysteamine. These findings confirmed the possible increase in growth hormone concentration and decrease in somatostatin secretion that, in turn, would result in more lean and less fat accretion.

β-*Adrenergic Agonists*

The recent discovery of β-adrenergic agonists, which are chemical analogues of epinephrine, norepinephrine, and catecholamines, is a promising development in growth promotion applications of animals. Included in this class of compounds are the following: clenbuterol, cimeterol, ractopamine, L-644,969 and L-640,033, and isoproterenol. β-adrenergic agonists are orally active and, thus, may be administered in the feed. Many are chemically stable and extremely potent, making successful development of implants for cattle, sheep, and swine likely. β-agonists improved live weight gain (15%) and feed conversion efficiency (15%) in meat-producing animals in research trials. Carcass protein (muscle) content is substantially increased (25%) while carcass fat is decreased (30%). These changes were observed in intact and castrated males and females, unlike anabolic steroids, which are sex dependent. They represent an altered pattern of metabolism such that nutrients are directed or partitioned away from adipose (fat) tissue and directed toward lean (muscle) tissue. For this reason, the term "nutrient repartitioning agents" is commonly applied to adrenergic agonists. Mechanisms (Table 20.3) by which β-agonists influence growth have been reviewed throughout the literature.

Changes in carcass lean content is primarily due to hypertrophy (increased cell size), rather than hyperplasia (increased cell number). β-agonists appear to exert their effects on skeletal muscle by reducing degradation rate without altering the rate of protein synthesis. Muscle tissue is continually being degraded and resynthesized in animals. Nitrogen retention in skeletal muscle is increased by β-agonists, yet decreased in skin/hide and visceral organs. Hypertrophy of cross-sectional areas of skeletal muscle fibers is typical of all red-meat-producing

Table 20.3. Biological Effects of β-Adrenergic Agonists on Adipose and Skeletal Muscle Tissue

Tissue	Effect	Physiological Process Affected
Adipose tissue	↓[a]	Glucose uptake
		Glucose oxidation unaffected
	↓	Lipid synthesis
	↓	Lipogenic enzyme activity
	?[b]	Insulin stimulation of glucose metabolism
	↑[c]	Basal lipolysis
	↑	Catecholamine-stimulated lipolysis
	↑	Insulin inhibits stimulation of lipolysis
		Insulin binding unaffected
		Somatotropin binding unaffected
Skeletal muscle (growth)	↓	Protein degradation
	↑	Protein synthesis
	↑	Satellite cell proliferation

[a]↓ = decreases.

[b]? = questionable.

[c]↑ = increases.

animals treated with β-agonists. Stimulating increased rates of lipolysis (fat breakdown) and decreased rates of lipogenesis (fat synthesis) reflect decreased carcass fat deposition (Muir 1988). Additionally, the supply of energy available for fat synthesis may be reduced in treated animals, both because an increased proportion of dietary energy is used for protein synthesis and because the β-agonists elicit a general increase in metabolic rate.

A critical factor in the usefulness of β-agonists is likely to be the degree to which their effects are retained after withdrawal, as a recommended withdrawal period is likely to be required. Growth rates decline following withdrawal. Carcass effects are not affected as rapidly but the advantages are not sustained after prolonged periods of withdrawal.

Porcine

In the majority of studies, daily gain was not significantly increased by feeding of β-agonists. In fact, some studies showed a depression in growth rate. The repartitioning effects of β-agonists appear to increase with dose rate in pigs and result in a decrease in carcass fat. Minimal differences in response to sex (gilt versus barrow versus boar) have been observed. Minor increases in meat toughness have been reported for pigs treated with β-agonists. In a recent review, Warriss (1989) concluded that β-agonists fed to pigs do not promote pale, soft, exudative (PSE) meat but might lead to a greater propensity for dark, firm, dry meat. β-agonists appear to have no effect on water-holding capacity and ultimate muscle pH.

Bovine and Ovine

Bulls, steers, and heifers as well as rams, wethers, and ewes have been studied in different trials as to their response to treatment with β-agonists. Responses have ranged from no response to an improvement of 48% in growth rate and feed conversion. Beef carcasses treated with β-agonists during the finishing phase contain less fat and more lean than controls. β-agonists are very effective in repartitioning utilizable energy away from fat deposition and toward protein accretion in rams, wethers, and ewes. Ruminants seem to be quite susceptible to meat toughening when treated with β-agonists.

Hamby et al. (1986) were among the first to report that toughening of meat (lamb) occurs in β-agonist-treated animals. Shear force in the longissimus muscle increased 114% in treated lambs. Wang and Beermann (1988) demonstrated as much as a 70% reduction of micromolar calcium-dependent proteinase (μM CDP) activity in skeletal muscle of lambs fed a β-agonist for 3 or 6 weeks. This μM CDP reduction may explain the dose-dependent increase in shearforce values of treated lambs. (Beermann et al. 1989)

Poultry

The addition of β-agonists to the diets of growing chickens has improved growth rate (~ 5%), lean tissue yield, and reduced carcass fat content (~ 12%). However, these compounds reduce tenderness in poultry muscle. β-agonists are only recommended during the growing/finishing period. Optimal dose level and duration of feeding have not been established. Sufficient evidence exists that feeding β-agonists only for the finishing period is more effective than feeding for longer periods. Effects on carcass composition tended to be greater in females than in males.

Transgenic Animals

In the past decade, development of recombinant DNA technology has enabled scientists to isolate single genes, analyze and modify their nucleotide structures, make copies of these isolated genes, and transfer copies into the genomes of livestock species. Such direct manipulation of genetic composition is referred to as "genetic engineering," and the term "transgenic animal" denotes an animal whose genome contains recombinant DNA.

Scientists are struggling with practical problems involved in transferring genes from one animal to another. Scientists inserting genes from a variety of different species, for example, the pig, have encountered significant problems. One unresolved major difficulty is the inability to insert genes precisely into animals' DNA, the building block of genetics. Instead, the gene is inserted into the nucleus of a cell with the hope that the gene lands in an appropriate location. In addition, many pigs develop severe health problems, e.g., ulcers, pneumonia, arthritis, cardiomegaly, dermatitis, and renal disease. Another major area of public concern is the potential impact these animals, once released, might have on the environment. Will the engineered animal outcompete native animals in the environment for available food sources. The public has also expressed concern regarding the potential effect bioengineered genes might have on the consumer if the transgenically altered products are used for human food, particularly when the gene product is a hormone or bioactive peptide. Although biotechnology can increase agricultural productivity, produce greater profits, lower food costs, and improve the competitiveness of U.S. agriculture in world markets, there are real and perceived adverse socioeconomic effects.

Steps involved in gene transfer for the engineering of animals include the following: (1) isolation of the gene that produces a described protein, (2) removal of an embryo from an animal, (3) injecting the gene into the embryo (pronucleus of a single-celled embryo) so that it becomes part of the embryo's DNA, and (4) implanting the genetically altered embryo into a female host (surrogate). Once the genetically engineered animal is born, it, as well as future generations, should carry the new gene. As many as 25% of transgenic embryos implanted into host mothers result in the birth of transgenic animals.

Although this technology has great potential as a source of new medicines and foods, real and perceived risks need to be addressed. One minor risk is the possibility that the desired trait will not be expressed even though it has been inserted into the genome of the host animal. Furthermore, although a gene may be inserted and expressed by the host animal, there can be difficulty in turning the gene on and off at the appropriate time at the appropriate tissue. Although the gene or trait is engineered and expressed by the host, the potential always exists for loss of this trait in succeeding generations, prompting the suggestion that multigenerational studies must be performed to fully understand the relative value of a transgenically modified animal. More complex is the potential impact the transferred gene or its products will have on the general well-being of the host animal. The successful use of genetic engineering to enhance carcass composition and efficiency of meat production in livestock depends on many factors. These include identification, isolation, and modification of useful genes or groups of genes that influence meat quality and quantity. Control of the time and level of expression of the inserted genes in transgenic animals so that their health status is either improved or not diminished affects the successful insertion of these genes into the genome.

Few single genes have been identified that have major effects on carcass composition. A national effort to map the genes of meat animals is underway.

In cattle, the double-muscle gene is responsible for muscle hypertrophy and enhanced lean tissue deposition. In pigs, the halothane sensitivity gene (Hal) is associated with increased yield of lean meat and porcine stress syndrome. Pigs homozygous for "Hal" are susceptible to stress and have a high incidence of pale, soft, exudative (PSE) meat. These genes offer considerable potential for investigation of carcass composition in meat-producing animals. However, except for the Hal gene, which has been identified as a single mutation in the ryanudine receptor gene, the specific product of each gene remains to be identified.

The transfer of cloned genes into the mouse genome in 1980 (Brinster et al. 1985) was immediately recognized as an important scientific achievement. However, the subsequent creation of the "super" mouse by the transfer of a rat somatotropin gene provided the convincing evidence that demonstrated the potential offered by gene transfer (Palmiter et al. 1982). Recently at Ohio University (Athens) transgenic mice carrying a modified version of the bovine somatotropin gene that originally created the "super" mouse was found to produce "mini" mice (approximately half the size as the controls). Modifying a somatotropin gene and incorporating it into mouse DNA that, in turn, prevents stimulating growth lends itself as a powerful tool for probing the hormone's function. This suggests that somatotropin does more than promote growth.

Swine with Growth-Related Transgenes

A number of transgenic swine contain various ST transgenes (Pursel et al. 1993). Production of excess ST in transgenic animals caused multiple physiological affects but did not result in "giantism" as was expected based on the earlier production of "super" mice as described by Palmiter et al. (1982). However, transgenic pigs that have excess ST levels exhibited numerous unique carcass traits. Reduced carcass fat, alteration of muscle fibers, thickening of the skin, enlargement of bones, and redistribution of major carcass components occurred in transgenic pigs. Some of these effects are similar to those observed after daily injections of pST, whereas others are considerably different. Possibly, these differences are the consequence of the continual presence of excess ST in the transgenics, whereas injections of pST provide a daily pulse of excess ST.

Carcass Composition and Meat Tenderness

Carcass fat was dramatically reduced in transgenic pigs that expressed a bST transgene at five different live weights (Solomon et al. 1994b). This difference in fat became greater among transgenic and nontransgenic littermates as the pigs approached market weight. Total cholesterol content of ground carcass tissue was not different between bST transgenic pigs and sibling controls. Analysis of fatty acids showed that carcasses of bST transgenic pigs consistently contained less (as much as 85%) total SFA than sibling control pigs at each body weight.

These differences in SFA were primarily a result of reductions in palmitic, stearic, and myristic acid. Carcasses from bST transgenic pigs contained less total MUFA and PUFA fatty acids (as much as 87% and 67%, respectively) than sibling control pigs. Carcasses of transgenic pigs had near the optimum ratio of 1 : 1 : 1 for SFA : MUFA : PUFA as recommended by NRC (1988).

When carcasses were separated into the four primal (pork) cuts, the hams of the bST transgenic pigs were significantly larger and the loins were significantly smaller than those of the sibling control pigs. The intramuscular fat for each primal cut (lean portion only) showed large differences between bST transgenic pigs and the controls. In spite of these dramatic reductions of fat throughout all primal cuts, evaluation of tenderness by shear-force determination indicated there were no significant differences between the two groups of pigs for the longissimus (loin) muscle.

Muscle Morphology

Morphological evaluation of bST transgenic-pig skeletal muscles revealed bST transgenic pigs had fewer red (βR) fibers and more intermediate (αR) fibers than control pigs. The population of white (αW) fibers was similar; however, the classical porcine fiber arrangement with βR fibers grouped in clusters surrounded by αR and αW fibers was less evident in the transgenic muscle. Hypertrophied (giant) fibers, that were identified in pST-injected pigs, were not present in bST transgenic pigs. The shift in the percentage of βR fibers to αR fibers in the bST transgenic pigs has not been identified in pigs that have received daily injections of pST. In bST transgenic pigs, only βR fibers appear to increase in size. Even though bST transgenic pigs were highly stress sensitive, there were no signs of pale, soft, exudative meat.

Fish with Growth-Related Transgenes

Transgenic carp, carrying the growth hormone gene from rainbow trout, were developed in order to increase the production of faster growing, more nutritious fish. The genetically altered carp grew 20–40% faster than nontransgenic carp.

Bovine and Ovine with Growth-Related Transgenes

To date, transgenic cattle produced by microinjection of DNA into pronuclei is inefficient and extremely costly, in large part due to the cost of maintaining numerous pregnancies to term. Numerous pregnancies result in nontransgenic progeny. The success rate in both bovine and ovine is significantly less than that for swine. No carcass data are available for transgenic bovine or ovine.

Selected References

Beermann, D.H., S.Y. Wang, G. Armbruster, H.W. Dickson, E.L. Rickes, and L.G. Larson. 1989. Influences of β-agonist L-655,971 and electrical stimulation on postmortem muscle metabolism and tenderness in lambs. *Recip. Meat Conf.* 42:54 (Abstr).

Brinster, R.L., H.Y. Chen, M.E. Trumbauer, M.K. Yagle, and R.D. Palmiter. 1985. Factors affecting the efficiency of introducing foreign DNA into mice by microinjecting eggs. *Proc. Natl. Acad. Sci. USA* 82:4438–4444.

Campbell, R.G., R.J. Johnson, R.H. King, M.R. Taverner, and D.J. Meisinger. 1990. Interaction of dietary protein content and exogenous porcine growth hormone administration on protein and lipid accretion rates in growing pigs. *J. Animal Sci.* 68:3217–3225.

Caperna, T.J., N.C. Steele, D.R. Komarek, J.P. McMurtry, R.W. Rosebrough, M.B. Solomon, and A.D. Mitchell. 1990. Influence of dietary protein and recombinant porcine somatotropin administration in young pigs' growth, body composition and hormone status. *J. Animal Sci.* 68:4243–4252.

Clark, R.G., G. Chambers, J. Lewin, and I.C.A.F. Robinson. 1986. Automated repetitive microsampling of blood: growth hormone profiles in conscious male rats. *J. Endocrinol.* 111:27–35.

Clark, S.L., R.C. Wander, and C.Y. Hu. 1992. The effect of porcine somatotropin supplementation in pigs on the lipid profile of subcutaneous and intermuscular adipose tissue and longissimus muscle. *J. Animal Sci.* 70:3435–3441.

Convey, E.M. 1988. Strategies to increase meat yield and reduce fat/cholesterol. In: *Proc. of 6th Biennial Symp. American Academy Vet. Pharm. and Therapeutics on Animal Drugs and Food Safety*, pp. 27–32.

Crighton, D.B. 1980. Endocrinology of meat production. In: *Developments in Meat Science—1*. R.A. Lawrie, ed. Pages 1–36. Applied Science Publishers Ltd., London.

Hamby, P.L., J.R. Stouffer, and S.B. Smith. 1986. Muscle metabolism and real-time ultrasound measurement of muscle and subcutaneous adipose tissue growth in lambs fed diets containing a β-agonist. *J. Animal Sci.* 63:1410–1421.

Hardy, R.N. 1981. *Endocrine Physiology*. Pages 80–89. Edward Arnold Ltd., London.

Machlin, L.J. 1972. Effect of porcine growth hormone on growth and carcass composition of the pig. *J. Animal Sci.* 35:794–800.

Moseley, W.M., L.F. Krabill, A.R. Friedman, and R.F. Olsen. 1985. Administration of synthetic human pancreatic growth hormone-releasing factor for five days sustains raised serum concentrations of growth hormone in steers. *J. Endocrinol.* 104:433–439.

Muir, L.A. 1988. Effects of beta-adrenergic agonists on growth and carcass characteristics of animals. In: *Designing Foods*. Pages 184–193. National Research Council, Washington, DC.

National Research Council. 1988. *Designing Foods. Animal Product Options in the Market Place*. National Academy Press, Washington, DC.

Palmiter, R.D., R.L. Brinster, R.E. Hammer, M.E. Trumbauer, M.G. Rosenfeld, N.C. Birnberg, and R.M. Evans. 1982. Dramatic growth of mice that develop from eggs microinjected with metallothionein-growth hormone fusion genes. *Nature* 300:611.

Pursel, V.G., and C.E. Rexroad, Jr. 1993. Status of research with transgenic farm animals. *J. Animal Sci.* 71 (suppl. 3): 10–19.

Solomon, M.B., T.J. Caperna, and N.C. Steele. 1990a. Lipid composition of muscle and adipose tissue from pigs treated with exogenous porcine somatotropin. *J. Animal Sci.* 68 (suppl. 1):217.

Solomon, M.B., R.G. Campbell, and N.C. Steele. 1990b. Effect of sex and exogenous porcine somatotropin on longissimus muscle fiber characteristics of growing pigs. *J. Animal Sci.* 68:1176–1182.

Solomon, M.B., R.G. Campbell, N.C. Steele, and T.J. Caperna. 1991. Effects of exogenous porcine somatotropin administration between 30 and 60 kilograms on longissimus muscle fiber morphology and meat tenderness of pigs grown to 90 kilograms. *J. Animal Sci.* 69:641–645.

Solomon, M.B., T.J. Caperna, R.J. Mroz, and N.C. Steele. 1994a. Influence of dietary protein and recombinant porcine somatotropin administration in young pigs: III. Muscle fiber morphology and meat tenderness. *J. Animal Sci.* 72:615–621.

Solomon, M.B., K. Maruyama, R.J. Mroz, and B. Russell. 1993. Effect of cysteamine administration on body composition in growing turkeys. In *Proc. 39th International Congress of Meat Science and Technology* Pages 121–128.

Solomon, M.B., V.G. Pursel, E.W. Paroczay, D.J. Bolt, R.L. Brinster, and R.D. Palmiter. 1994b. Lipid composition of carcass tissue from transgenic pigs expressing a bovine growth hormone gene. *J. Animal Sci.* 72:1242–1246.

Spencer, G.S.G. and K.G. Hallett. 1985. Somatostatin antagonists analogue stimulates growth in rats. *Life Sci.* 37:27–30.

Spencer, G.S.G. 1986. Hormonal manipulation of animal production by immunoneutralization. In: *Control and Manipulation of Animal Growth*. P.J. Buttery, D.B. Lindsay, and N.B. Haynes, eds. Pages 279–291. Butterworths, London.

Tannenbaum, G.S. 1985. Physiological role of somatostatin in regulation of pulsatile growth hormone secretion. In: *Advances in Experimental Medicine and Biology, Vol. 188, Somatostatin*. Y.C. Patel and G.S. Tannenbaum, eds. Pages 229–259. Plenum, New York.

Turman, E.J., and F.M. Andrews. 1955. Some effects of purified anterior pituitary growth hormone on swine. *J. Animal Sci.* 14:7–15.

Turton, J.D. 1969. The effect of castration on meat production from cattle, sheep and pigs. In: *Meat Production from Entire Male Animals*. D.N. Rhodes, ed. Pages 1–49. J. and A. Churchill Ltd., London.

USDHHS. 1988. *USDHHS: The Surgeon General's Report on Nutrition and Health*. Public Health Service, USDHHS, PHS, NIH, Publ. No. 88–50210. U.S. Government Printing Office, Washington, DC.

Wang, S.I. and D.R. Beermann. 31. 1988. Reduced calcium dependent proteinase activity in cimaterol-induced muscle hypertrophy in lambs. *J. Anim. Sci.* 66:2545–2550.

Warris, P.D. 1989. The influence of β-adrenergic agonists and exogenous growth hormone on lean meat quality. In: *Proc. Brit. Soc. Anim. Prod.* Winter meeting, p. 52.

Glossary

Absorption: The process of taking digested food components up from the gastrointestinal tract.

Acaracide: Compound intended to kill mites and ticks.

Actin: Myofibrillar protein and main component of the I-band or thin filament of the sarcomere.

Action potential: Electrical impulse carried through nerve and muscle cells by polarization/depolarization of the plasma membrane.

Adipose: Lipid-containing tissue distributed throughout the carcass between muscle fiber bundles (marbling) and muscles (seam fat) internally and subcutaneously.

Adulterated: The inclusion of a deleterious substance that may render the food product injurious to health. It may be a substance that occurs in the raw materials or enters the food during the processing phase. It may be biological, chemical, or physical.

Aged beef: Beef held for relatively long periods of time at controlled temperature and humidity for purposes of improving tenderness and changing flavor.

Aging or conditioning: Tenderization of meat following rigor and storage due to disruption of cross-links in collagen fibers of connective tissues, Z-line disruption in myofibrillar proteins and proteolysis of soluble sarcoplasmic proteins.

Agricultural Marketing Service (AMS): Agency of the United States Department of Agriculture with the responsibility of developing and applying the federal meat and poultry grading systems.

Antemortem: Before death, USDA-FSIS inspectors review livestock and poultry, commonly referred to as antemortem inspection.

ATP: Adenosine triphosphate, a high-energy molecule that is the carrier of free energy used in animal systems.

Attrition: Inactivation of a virus, similar to death in a living organism.

β-Agonist: Compound causing relaxation of smooth muscles by blocking β-adrenergic receptors.

Backfat: Term used in the pork industry to refer to the layer of subcutaneous fat along the back of the carcass.

Ballot: A sensory instrument that defines attributes or asks questions for a sensory test.

Basement membrane: Specialized connective tissue layer surrounding each muscle cell lying above the plasma membrane and beneath the endomysium.

Bioavailability: A measure of a nutrient's utilization by the body in a biologically meaningful way.

Biogenic amine: Any of a group of amines produced by microorganisms from amino acids and having an adverse physiological effect in susceptible humans.

Biotechnology: Implementation of biological sciences for the improvement of life-style.

Blind controls: Controls used in a sensory study where the sensory panelists are not aware which samples are controls.

Bloom: Process of the development of the bright color associated with the formation of oxymyoglobin on lean surface of a muscle cross section when it is exposed to oxygen. The color generally turns from a purple color to a bright reddish color.

Bloom (microbial): Population explosion of dinoflagellates, often manifested by a colored area on the surface of bodies of water.

Bloom time: Time required for the color to bloom or turn a bright red color. The time between the carcass ribbing and evaluation by a grader is important in order for the ribeye to fully bloom (usually more than 20 min).

Boxed beef: A relatively recent (since 1970) innovation in the shipping of beef in boxes, by primal or wholesale cuts rather than as quarters (fore and hind). The beef carcass is cut into the wholesale cuts at the packing plant and trimmed of excess fat and possibly boned; these cuts are vacuum-packaged and boxed by like cuts and shipped under refrigeration to wholesalers or retailers directly.

Branded product: Used by processors and distributors to identify for consumers food articles that are of particular quality.

Break joint: Used in lamb grading to aid in estimating skeletal maturity. There is a thin layer of soft cartilage between the metacarpal and the spool joint, making it easy to break the spool joint from the metacarpal; hence, the rough edge of the metacarpal is referred to as the "break joint." As sheep advance in maturity, the thin layer of soft cartilage undergoes ossification and becomes bone, fusing the spool joint and the end of the metacarpal together, making it impossible to break the spool joint from the foreshank of the carcass.

Broilers: Young chickens raised exclusively for meat, usually under 8 weeks of age, producing dressed weights of 3–5 lbs. It is the most popular poultry meat consumed.

Bulk-pack: Individual pieces of by-product placed in a container without any individual wrapping. Usually, when frozen, the pieces cannot be separated.

Butcher shops: Originally a shop offering fresh meat products, some sausage, and cured meat items made from animals slaughtered in the facility by the proprietor.

By-product and or coproduct: Product derived from animals in the animal-to-food production process that is not part of the skeletal muscle associated with the portion of the animal that is typically considered a carcass.

Calpains: Class of calcium-activated muscle proteases with pH optima at 6.6–6.8, responsible for proteolytic degradation of muscle proteins.

Cannibalism: Sacrificing sales of one product for the sales of the new variety.

Case: A single incident of foodborne disease in a single individual.

Casing: Packaging material for sausage meat; often the lining or membrane of an organ from an animal that produces muscle foods.

Cathepsins: Lysosomal-bound, muscle proteases with optimal activity at acidic pH.

Centrifuge: Device that is used to separate oil/fat and water (two-way centrifuge) from each other and also to separate solids, oil/fat, and water (three-way centrifuge) in rendering systems.

Chain speed: Often the carcass is elevated on a rail and a chain moves along the rail transporting the carcass from one location of the plant to another. Chain speed is often described as the number of carcasses slaughtered and/or processed per hour.

Chemical residue: Chemical substance not naturally present or present above natural levels in foods.

Chine button: The white, cartilaginous tip of the spinous processes.

Chub: Type of package formed from a continuous film tube that is completely filled with product and then tied or clipped and cut into individual containers.

Circuit: One or more official USDA-FSIS establishments under the supervision of a circuit supervisor.

Clone: A population of cells descended from a single cell.

Coextrusion: Extrusion processed used to make objects such as films or sheets in which two or more different polymers are simultaneously formed into a multilayered film or object.

Cold-shortening: A physiological occurrence that results from rapid temperature decline during the onset of rigor mortis; this process results in muscle fibers that are contracted to a greater extent than muscle fibers from muscles that were subjected to higher temperatures during the onset of rigor mortis; severe muscle shortening may occur, resulting in reduced meat tenderness.

Collagen: Main structural protein of vertebrate animals, occurring in all tissues; it is a precursor to gelatin and can be processed into a variety of forms; predominant fibril-forming protein of the connective tissue or stomal portion of muscle.

Collagen cross-links: Lysine-derived structures linking together molecules and fibrils of collagen, increasing its tensile strength.

Collagen solubility: The amount or percentage of collagen that breaks down with heat while the meat sample is suspended in Ringer's solution; it is a result of collagen cross-links being susceptible to heat denaturation and is defined as hydrothermal labile collagen.

Colorimeter: Instrument that measures three kinds of light and provides three values that are either the three CIE tristimulus values or that can be used to calculate the CIE tristimulus values.

Combo-bin: A large cardboard container designed to fit on a standard pallet (40 × 48 in.) and 48–60 in. high and designed to hold 200–2500 lbs.

Commercial sterility: The condition achieved by application of heat, alone or in combination with other ingredients and/or treatments, to render the product free of microorganisms capable of growing in the product at nonrefrigerated conditions (over 10°C/50°F) at which the product is intended to be held during distribution and storage. The product may contain spores of thermophilic bacteria.

Comminution: A process of particle size reduction that includes grinding, flaking, chopping, and milling.

Competitor match: Products are developed to compete with specific competitors currently in the marketplace.

Compliance Division: Division of the USDA that monitors USDA-inspected meat products in food service and retail establishments.

Conformation: Structure and shape of a carcass thought to denote muscularity; considered in lamb and poultry grading.

Connective tissue: Forms cellular structure of most animal tissues including bone, fat, and muscle. The two major proteins are collagen and elastin.

Consumer panel moderator: The individual who acts as the facilitator in conducting consumer sensory evaluation; this individual interacts directly with consumers.

Consumer intercepts: Market research technique used to gain consumer information where consumers are randomly stopped and interviewed.

Consumer sensory panel: A sensory test that assesses the personal response (preference and/or acceptance) by a current or potential customer of a product, a product idea, or specific product characteristics.

Controlled atmosphere: The intentional alteration of the natural gaseous environment and maintenance of that atmosphere during the distribution cycle.

Convenience: Labor saving or comfort giving.

Convenience stores: Small stores located throughout a trading area, easily accessible, offering a limited variety of items commonly used in the home. Many offer self-service gasoline pumps.

Copolymers: Polymers formed from two or more different types of monomers.

Corned beef: One method of preserving beef is that of curing such cuts as boneless brisket or top or bottom round cuts in a brine (salt and water) solution for a few days to a few weeks. These cuts may also be injected with the brine, which may also contain some sugar and spices.

Creatine kinase: An enzyme that catalyzes the transfer of phosphate from creatine phosphate to adenosine diphosphate to yield adenosine triphosphate.

Creatine phosphate: A "reserve" of high-energy phosphate used to regenerate ATP from ADP.

Critical control points (CCP): Part of HACCP, a CCP is a location, practice, procedure, or process at which control can be exercised over one or more factors that, if controlled, could minimize or prevent a hazard. An example of a CCP in a cooked ham product would be the cooking phase in the smokehouse, which should be controlled to assure the destruction of microbiological pathogens.

Critical limits (CL): Part of HACCP, critical limits may be defined as physical (e.g., time or temperature), chemical (e.g., salt or acetic acid), or biological (e.g., sensory evaluation) minimum or maximum levels that a product must meet in order for a hazard to be controlled. The critical limits for the cooking of meat patties would include oven temperatures, cooking times, and patty thickness.

Cross-contamination: Process whereby a food acquires a pathogen from contact with another contaminated food or from contact with an object (e.g., cutting board) used first with a contaminated food.

Cutability: Percentage of edible portion or closely trimmed, boneless retail cuts that can be obtained from a carcass.

Cytoskeleton: Network of intracellular structural proteins responsible for maintaining orientation and alignment of sarcomere structures; may be involved in contraction.

Dark cutter: Dark-cutting is a term used in beef to denote darker than normal lean color. It is a result of a reduced glycogen content in the muscle prior to slaughter and is often associated with stress prior to slaughter. The muscle pH of a dark cutter is high, generally above 6.0, which results in a higher water-holding capacity and more light absorbency than normal thus causing a dark lean color.

Dark, firm, and dry (DFD): Term used in pork to refer to a dark lean color that is a result of the same condition described as dark cutter.

Delaney Clause: Section of the Food, Drug and Cosmetic Act of 1959 requiring that foods be free of residues of pesticides found to cause cancer in experimental animals.

Demersal fish: Species that live near the ocean floor and also are known as ground fish or bottom fish. These fish are limited to the continental shelf and include haddock, cod, flounder, and sardine.

Deoxymyoglobin: The reduced form of the meat pigment myoglobin where a compound is not bound to the ligand and the heme portion of the myoglobin is in the Fe^{2+} or ferrous state; this pigment is purple-red in color.

Depuration: Process whereby shellfish are removed from a contaminated body of water and placed in pools of treated water that is free of pathogens.

Descriptive analysis: A sensory method that involves the detection or discrimination and the description of the qualitative and quantitative sensory attributes of a product by trained panelists.

DFD: *See* Dark, firm, and dry meat.

Difference testing: A sensory method that evaluates the difference of a single or multiple attributes where one or several samples are compared.

Digestion: The process of breaking foods into molecular or chemical components.

Dinoflagellate: Any of an order of chiefly marine planktonic microorganisms related to algae and protozoa.

Dressing percentage: An expression of yield of carcass as a proportion of live animal weight. Calculated as warm carcass weight divided by live weight multiplied by 100.

Ectomorph: A lightly muscled animal that is angular, concave, and narrow in appearance due to a low muscle/bone.

Edible: Products processed and sold with the intent that people will consume it. Also one of the three main classifications for coproducts.

Edible unit: Portion eaten as food, excluding trimmable fat and bone.

Elastin: Rubberlike connective tissue protein noted for its elasticity and extreme insolubility.

Elastography: The process of evaluating the elastic properties of biological tissues using ultrasonic waves.

Electrical stimulation: The application of an electrical current to carcasses for the purpose of hastening the onset of rigor mortis.

Endogenous: Originating within the body.

Endomysium: Connective tissue surrounding each muscle fiber overlying basement membrane.

Endotoxin: A toxin produced in an organism and liberated only when the organism disintegrates (term often refers to the lipopolysaccharide components of bacteria).

Epimysium: Connective tissue sheath surrounding muscle and contiguous with tendon and perimysium.

Establishment number: The number assigned a slaughter facility and/or processing facility by USDA-FSIS to identify the source of the meat or poultry product.

Exogenous: Introduced from or produced outside.

Exotoxin: A toxin produced by a microorganism and excreted into the surrounding environment.

Expert sensory panel: A sensory panel that uses sensory panelists who have been extensively trained, usually 6 months to 1 year on sensory procedures prior to conducting a sensory study.

Exsanguination: The process of removing blood from an animal.

Extracellular matrix: Connective tissue composed of collagen and ground substance surrounding muscle fibers, bundles of fibers, and muscles.

Extrinsic factors: Factors that are environmental in origin.

Failure fees: Money paid to the retail store when a product fails to perform at expected levels.

Fast-food outlets: Small food establishments providing a menu with limited selections and emphasis on fast service.

Fat-reduced beef/pork: Material produced from fatter tissue produced in the breakdown of beef and/or pork carcasses. The material is subjected to a low-temperature process that separates fat and oil from the protein matrix surrounding it, leaving a low-fat, functional lean material that can be used as a raw material for sausage, ground beef/pork, and other meat products.

Fat-free muscle: An expression of muscle composition in which adipose tissue is excluded.

FDA: *See* Food and Drug Administration.

Federal Code of Regulations: Government regulations published by the Office of the Federal Register National Archives and Records Administration of the Government Printing Office. Animal and Animal Product code of federal regulations can be found in Title 9 part 1 to part 399.

Fiberboard: Packaging materials made from matted cellulose fibers in a manner similar to paper but in sufficient thickness to give rigidity to the container.

Fill: Contents of the stomach (ingesta) and the intestines (excreta).

Fish frame: The remaining portion of a fish after the head, viscera, and fillets have been removed.

Flank streaking: Flank streaking is the streaky appearing fat deposits, often shoe-lace shaped, in the primary flank muscle and the secondary flank muscle; also called flank lacing and fat streaking in the flank.

Fleshing: The thickness of the muscle covering over the back, breast, drumstick, and thigh of a chicken carcass. Also considered in chicken part grading.

Focus group: Market research technique involving a round-table question/answer and discussion among a small, select group of consumers using a story board and product example.

Food and Drug Administration (FDA): agency of the U.S. Government Health and Human Services with the responsibility of monitoring the processing and safety of nonmeat and nonpoultry foods. They are also responsible for the review, approval, and use of approved human and animal health products. The agency also approves chemicals used in food directly and indirectly.

Food Safety and Inspection Service (FSIS): Agency of the United States Department of Agriculture that has the predominant responsibility of the safety and wholesomeness of meat and poultry.

Foodservice: Refers to establishments that produce and serve food to a large number of people.

Freezer burn: Dehydration of loosely packaged fresh meat items while in frozen storage. Moisture is lost from the product, causing localized tissue dehydration as ice crystals form on the inner surface of the packaging material due to improper frozen storage.

Fresh by-product: By-product that is sold in nearly the same form that it occurred naturally in the animal. It is a recognizable portion of the animal and has only been minimally processed.

FSIS: *See* Food Safety Inspection Service.

Further processing: Smoking, cooking, canning, curing, refining, or rendering in an official establishment of product previously prepared in an official establishments.

Gas chromatograph: An instrument that carries a chemical mixture in a gas and then separates out compounds contained in the mixture as they flow around or over a stationary liquid or solid phase.

Gel bone: Processed beef bone that has been heated to remove fat and then dried and separated from other nonbone solids in a rendering system. It is the raw material for one type of gelatin.

Gissen: The wetting down of meat within 72 h after slaughter before salting in the Jewish ritual process of koshering.

Glycogen: Polysaccharide of glucose.

Glycolysis: As applied to anaerobic metabolism, glycolysis is the conversion of glucose to lactate and hydrogen ion, with concurrent regeneration of high-energy bonds (as ATP).

Grade: System (government or industry) that is used to segment a heterogeneous population of livestock and poultry into smaller, more homogenous groups according to economic, merchandising, and/or production characteristics.

Grain-fed: Feeding animals diets in which the majority of the feedstuff is composed of grains such as corn, wheat, or sorghum; these diets are also referred to as a high-energy diets, as grains are high in energy compared to grass-based diets.

Grass-fed: Feeding of animals grass where the grass fed could be a multitude of grass species that should be defined.

Green weight: Total weight of the meat raw materials (lean meat and fat meat sources); also referred to as meat weight or block weight. Green weight is often used to determine the amount of a specific nonmeat ingredient (e.g., 200 ppm sodium nitrite) to be incorporated into the meat batch.

Gross margin: Determined by subtracting item cost from the selling price, dividing this difference by the selling price and multiplying the quotient by 100.

Growth promotant: FDA-approved pharmaceutical products that enhance animal growth and feed efficiency above normal levels. The use of these products are monitored and policed by FDA and USDA-FSIS.

Hadacha: Soaking of red meat or poultry after slaughter prior to salting.

Halal: Moslem dietary laws and ritual practices.

Hardbone: Common term used in the beef industry to refer to those carcasses that would qualify for the USDA Commercial, Utility, Cutter, and Canner grades.

Hazard analysis and critical control point (HACCP): The HACCP concept is a systematic approach to hazard identification, assessment, and control.

Health: State of being whole or sound in body, mind, or soul.

Heat-labile: The breakdown of chemical cross-links due to the influence of heat.

Hedonic scales: Scales that ask acceptance, such as like or dislike, for a product.

Hemoglobin: A protein in blood that is the major pigment of blood.

Herbicide: Compound intended to kill plants.

High-temperature conditioning: The holding of carcasses at 59°F (15°C) or greater to improve meat tenderness.

Homopolymers: Polymers formed from a single type of monomer.

House grades: Grading system developed and applied within a processing plant and not associated with a governmental applied grading system.

HPLC: High-performance liquid chromatography.

HRI: Hotel, restaurant, and institutional trade.

Humane Slaughter Act: Handling and immobilization of livestock and poultry is done in a humane manner as described in 1958 Humane Slaughter Act (amended in 1978), overseen by the USDA-FSIS inspector.

Hurdle concept: Superimposing several growth limiting parameters to bacteria to achieve food preservation intermediate moisture foods: products that contain 15–50% moisture and have a_w between 0.60 and 0.85.

Hydrogenation: Process that saturates (adds hydrogen atoms to fatty acid chains) fats, increasing melting points and improving frying characteristics of lard and other unsaturated fats.

Hyperplasia: Increase in the number of cells.

Hypertrophy: Increase in the size of the cell.

Hypothesis: A tentative assumption made in order to test its logical or empirical consequences.

Improved product: A product that is reformulated or modified to meet the needs of the business.

Index of Nutritional Quality (INQ): The ratio of percentage of recommended daily allowance to percentage of calories provided by a specific amount of a given food for a given nutrient.

Inedible: Food defined as inedible by U.S. law; one of the three major classifications of by-products; adulterated, uninspected, or meat and poultry not processed for human consumption.

Infection: The state produced by the establishment of an infective agent in a suitable host; also, the establishment of a pathogen in its host after invasion.

Infectious dose: Number of microorganisms necessary to cause illness in a particular host.

Ingredient Statement: Listing on the food product label of the ingredients. The listing is in descending order based on the quantities used.

Insecticide: Compound intended to kill insects.

Inspector-in-Charge: Designated FSIS inspector who is in charge of one or more official establishments within a circuit and is responsible to the circuit supervisor.

Institutional Meat Processing Specifications (IMPS): Developed by the USDA to provide the HRI (foodservice)/meat businesses with a uniform set of cutting guidelines. Each product described has an IMPS designated number along with the product name. Most boxed beef is marketed using the standards found under the IMPS guidelines.

Intermuscular Fat: A deposit of fat located between muscles; also called seam fat.

Intoxication: The state produced by ingestion of a poisonous (toxic) substance.

Intramuscular fat: Amount of fat within the lean of a meat cut, referred to as marbling.

Intrinsic factors: Inherent chemical or physiological factors.

Ionomer: A polymer that is similar to polyethylene but also contains ionic groups (i.e., acids) which are complexed with metal ions. Ionomers are often used as heat-seal layers.

K-Value: A measure of freshness that represents the concentration of hypoxanthine and inosine relative to total nucleotide present.

Kosher: Jewish dietary laws and ritual practices.

Koshering: Preparation of food and utensils to be fit for consumption by Jews described by a set of Jewish ritual laws. For the preparation of meat and poultry, koshering means schita, hadacha, melicha, niputs, and shetifa immediately after or no later than 72 h after slaughter.

Laminate: Bonding two or more materials together to form a multilayered single material.

Lead time: Time necessary to receive or produce a product once an order is placed for the product.

Lean cuts: The ham, loin, Boston butt, and picnic shoulder of a pork carcass.

Lean maturity: Physiological maturity as indicated by the color of the lean of a carcass; as maturity increases, the lean usually becomes darker.

Leanness: A term describing live animals or carcasses depicting the degree of fat deposited. It is not affected by muscling per se.

Lexicon: A group of words or descriptors used to describe the sensory attributes of a product.

Ligand: A portion of a molecule coordinated to a central atom in a complex.

Line extensions: New varieties of existing products.

Lipid oxidation: A chemical reaction involving unsaturated lipid with oxygen to yield hydroperoxides; degradation of the hydroperoxides yields a variety of products including alkanols, alkenols, hydroxyalkenols, ketones, alkenes, etc.

Lipogenesis: Formation of fatty acids from acetyl coenzyme A in the living body.

Lipolysis: The hydrolysis (decomposition) of fat.

Locker plant: Generally located in rural areas providing custom slaughter, cutting, and packaging for freezer storage and some processing. Proprietor does not take ownership. All services are performed for a fee along with monthly charges for locker rentals.

Loin eye area: Size or area of a cross section of the longissimus muscle at the 10th rib cross section in pork carcasses.

Low-moisture foods: Foods that contain less than 25% moisture and have $a_w < 0.6$.

Lysosomes: Subcellular vesicles that contain digestive enzymes important to turnover of cell constituents.

Maillard reaction: A nonenzymatic chemical reaction involves condensation of an amino acid group and a reducing group, resulting in formation of intermediates that ultimately polymerize to form brown pigments.

Marbling: Amount of visible fat within the lean of a meat cut; also referred to as intramuscular fat.

Mashgiach: Overseer of the koshering process in the slaughterhouse or packing house; also one who sees to it that kosher practices are strictly followed in establishments such as factories and restaurants that have kosher certification from the rabbinate.

Mass spectrometer: An instrument that can measure the mass of a compound.

Mature market: Product market in which the sales are well established and more predictable.

MDM: Mechanically deboned meat; meat removed from its skeletal matrix by mechanical means.

Meat factor: The phenomenon in which the bioavailability of nutrients (specifically iron) in a food is higher when that food is consumed along with meat.

Melicha: Salting of the drained flesh of an animal after soaking in water for 30 min following slaughter. Salting is done over the entire skin surface and any open cuts. A special size salt grain is used which is not large enough to fall off by itself. The salt is left on for 1 h, then shaken off, and the flesh dipped twice in ice-cold water to remove remaining salt.

Mesomorph: A heavily muscled animal that is bulging, convex, and thick in appearance due to a high muscle/bone.

Metabolism: The sum of all processes by which a specific substance is handled by the body.

Metmyoglobin: The oxidized form of the meat pigment myoglobin where water is bound to the ligand and the heme portion of the myoglobin is in the Fe^{3+} or ferric state; this pigment is brown in color.

Microsomes: Membranous structures of subcellular organelles.

Mitochondria: An organelle of the cell that contains enzyme systems responsible for producing energy for the cell; often referred to as the "power house" of the cell.

Modified atmosphere: Initial alteration of the gaseous environment in the immediate vicinity of the product, allowing the packaged product interactions to naturally vary their immediate gaseous environment.

Modified atmosphere packaging (MAP): The alteration of the gas atmosphere inside a package to something other than ambient air for the purpose of extending the shelf-life of the product.

Monomers: Individual small molecules that are linked together to form large molecules or polymers.

Monounsaturated: A fatty acid with only one double bond in the carbon chain.

Morbidity: The relative incidence of disease.

Mortality: The relative incidence of death.

Muscle fiber: Elongated, syncytial fibrous cell containing multiple nuclei and the specialized contractile organelles, myofibrils.

Muscle meat: Striated (skeletal) muscle from the carcasses of meat animals, poultry, and fish. Some variety or glandular meats may be nonstriated (involuntary) muscle tissue. All muscle has a high protein content (15–25%).

Muscle score: Score used in the pork grading system to denote the muscularity of a pork carcass.

Muscling: A term describing live animals or carcasses depicting varying degrees of shape due to magnitude of muscle/bone. It is not affected by fatness per se.

Myofiber: A skeletal-muscle cell

Myofibril: Long rodlike (70–80% of volume) organelle of the muscle cell, composed of sarcomeres held in register laterally and longitudinally; the main component of the muscle fiber that constitutes the contractile apparatus.

Myofibrillar component: The portion of a muscle cell that contains the muscle proteins responsible for muscle contraction.

Myofilaments: Main protein filaments of sarcomere, consisting mostly of actin (thin filament) or myosin (thick filament).

Myoglobin: Sarcoplasmic, heme-containing protein in muscle that binds and delivers oxygen in the myofiber; this major pigment is responsible for red meat color.

Myokinase: Enzyme that catalyzes the conversion of 2 ADP molecules to 1 AMP molecule plus 1 ATP molecule.

Myosin: Major myofibrillar protein and predominant salt-soluble muscle protein that comprises 50–60% of the myofibrillar contractile proteins; it is the main component of the A-band or thick filament of the sarcomere and has a molecular weight of 475,000 D.

Myotendinous junction: Junction of actin filaments of myofibrils with connective tissue at the tapering ends of muscle fibers.

National Association of Meat Purveyors: (NAMP) The organization that provides a Meat Buyer's Guide with colored pictures and cutting instructions for fresh meat items described by IMPS.

New product: Product that is unique and puts the company into a new business area, usually as a result of new technology or innovation.

Nutrient density: Expressed as the units of a specific nutrient contained in meat per calorie of energy. The nutrient density of riboflavin would be low in a muscle containing abundant marbling.

Nutritional Label: Label that specifies the quantity of calories, fat, saturated fatty acids, cholesterol, and other selected minerals and vitamins. The label is mandatory for all further processed food items, meat, and poultry products that make a health claim and is voluntary for raw meat and poultry products.

Official establishment: Meat or poultry processing or packing facility under USDA-FSIS or State Inspection jurisdiction.

Olfactory nerve: Nerves located in the nasal mucous membrane that pass to the anterior part of the cerebrum in the brain; these are the nerves that are involved in the sense of smell.

Omega-3-fatty acid: Fatty acid with its first carbon–carbon double bond starting at the third carbon atom from the CH_3 end.

Onset time: Time interval between ingestion of an infectious organism or toxin and manifestation of illness (often referred to as "incubation period" when an infectious microorganism is involved).

Open-ended questions: Questions used on a sensory ballot where the respondent has to write a response; these questions do not have predetermined scales or answers associated with them.

Ossein: Demineralized gel bone. This material is soft when hydrated, off-white to pale yellow in color, and is the intermediate stage in the production of gelatin from beef bones.

Ossification: Biological process of skeletal cartilage converting to bone, often occurring between joints and on the end of the spinous process.

Outbreak: An occurrence of two or more cases of foodborne disease (other than that caused by botulism or chemical poisoning, wherein a single case constitutes an outbreak), associated in time and place so as to implicate a common food.

Oxymyoglobin: The reduced form of the meat pigment myoglobin in which oxygen is bound to the ligand and the heme portion of the myoglobin is in the Fe^{2+} or ferrous state; this pigment is bright red in color.

Packer: The company or plant that slaughters animals for meat is generally referred to as a meat packer. This name is derived from the fact that this process originally involved "packing" meat in barrels for shipment. Today, the packer often performs other services such as by-product processing and utilization, further processing of carcasses, and dealing in other refrigerated items, such as dairy and poultry products.

Pale, Soft, Exudative (PSE): or pale, soft, watery (PSW) condition of muscle of low (<5.3 pH), especially noticeable in pork; often stress related.

Pasteurization: (Heating) resulting in destruction of all pathogenic bacteria; spores and some thermoresistant spoilage organisms may survive.

Pathogenicity: Capability of a microorganism to cause disease.

Pelagic fish: Those species of fish that normally occur in the upper part of the water column.

Per capita consumption: Refers to the annual disappearance of meat, poultry, and seafood on a per person basis. In the United States, for example, it represents the total tonnage of the products divided by approximately 250 million (U.S. population). However, these figures should be expressed specifically, such as on a carcass basis, or on a trimmed, boneless basis, or on a cooked, ingested-meat basis.

Perimysium: Intramuscular connective tissue surrounding primary and secondary bundles of muscle fibers.

Perirenal fat: Adipose tissue surrounding the kidneys that is commonly called kidney fat.

Permeation: The property of materials, particularly polymers, that permits the diffusion of gases and vapors through the continuous material.

Pharmaceutical: One of the three major types or classifications for coproducts; refers to coproducts that are intended for medical uses.

Phospholipid: A class of lipids in which glycerol is linked to fatty acids at the C-1 and C-2 positions, the C-3 position being occupied by phosphate and usually a nitrogen-containing chemical group, a lipid that contains phosphoric acid, and fatty acids esterified to glycerol; found in all living animal cells in the bilayer of the plasma membrane.

Plant test: The manufacture of a new product under actual production conditions to determine changes necessary in product and process when making the transition to a larger scale and volume.

Plasma membrane: Lipid bilayer membrane encompassing each muscle cell; also called sarcolemma.

Plastic: Any one of a large group of materials made up of large high-molecular-weight organic molecules that are liquid at some point in manufacture but solids in the finished state.

Pleiotropic: Producing more than expected from inherited disposition.

Polymer: A material made from very large molecules that are formed by linking small identical molecules together in large chains that can be branched or linear.

Polyolefin: A general class of polymers that are made from monomers that contain an olefinic or carbon–carbon double bond group.

Polyunsaturated: Fatty acids with two or more double bonds in the carbon chains.

Porging: Removal of the sciatic nerve with all its branchings and the fat around it from the hindquarters of cattle following slaughter. Porging also includes removal of the blood vessels from the hindquarter, or at least cutting them and the removal of the suet, especially around the kidneys.

Postmortem: After death, the USDA-FSIS inspectors reviews livestock and poultry carcasses, carcass parts and viscera after slaughter, commonly referred to as postmortem inspection.

Precooked: Meat products cooked and packaged for sale or use prior to selling to an HRI business or offering for retail sale. They may be reheated prior to serving.

Preferred cuts: The leg, loin, rack, and shoulder of a lamb carcass.

Pre-rigor muscle: Soft, pliable, and extensible compared to muscle in which rigor is complete, and permanent crossbridges between actin and myosin have formed. Pre-rigor muscle has a high pH and sufficient levels of metabolizable substrates necessary to remain physiologically active.

Price: Quantity of one thing, usually money, exchanged or demanded in barter or sale for another.

Primal cuts: The round, loin, rib, and chuck of the beef carcass. The ham, loin, Boston butt, picnic shoulder, and belly of the pork carcass.

Primary ingredient supplier: Manufacturer, distributor, wholesaler, or broker who has proven to be the best source of supply when considering quality, performance, and cost of raw ingredients.

Primary package: Outermost container in which a coproduct is packed; this is the container the product in which it is shipped and stored.

Processed meats: Includes items such as bacon, hams, sausages, and luncheon meats.

Processing: Anything done to a coproduct other than removing it from the animal, washing (water only), trimming to specification, and packaging.

Protease: A protein that in involved in breaking down or degrading other proteins.

Protein denaturation: The process of modifying the molecular structure of a protein; alteration of the protein from its original state.

Protein functionality: Numerous chemical and physical properties of the protein. Properties include the ability to form protein to protein binding and/or to stabilize water and fat relative to texture formation and appearance, organoleptic texture (bind, mouthfeel, hardness, cohesiveness), hydration, gelation, and moisture retention (cooking yields, purge).

Protein turnover: The combination of protein synthesis and protein degradation; often used to refer to the combined process where new proteins are produced and other proteins are degraded and replaced.

Proteolysis: The breakdown of protein.

Protocol: A document that defines the correct procedure used to conduct a study.

Prototype: The preliminary design or formulation of a product that best represents the idea in physical form. Often, it is manufactured by hand from nonindustrial recipes or formulations.

Proximate analysis: Methodology used to assess the proportionate content of protein, water, lipid, mineral, and carbohydrate of meat.

PSE: Acronym that stands for pale, soft, and exudative (watery) lean meat.

PSS: Porcine Stress Syndrome; pigs afflicted with this genetic condition are especially susceptible to stress and produce carcasses with undesirable meat quality that often results in a PSE appearance of the lean carcass.

Psychrophilic: Microorganisms that are capable of growing at refrigeration temperatures.

Purge accumulation: The fluid that migrates out of a muscle food that may collect in the package.

Purveyor: Wholesalers or companies that purchase, warehouse, and resell or distribute meat to retailers and restaurants are often termed purveyors. A specialized segment is referred to as HRI or hotel, restaurant, and institution service.

Quality grade: Grading system that utilizes carcass factors (e.g., marbling, maturity, and color) to estimate the palatability of the meat.

Radappertization: Equivalent to radiation sterilization.

Radiation: Emission and propagation of energy through space or through a material medium.

Radicidation: Equivalent to radiation pasteurization.

Radurization: Process equivalent to pasteurization with respect to spoilage bacteria.

Receptor gene: Gene that has a specific affinity for a particular stimulus.

Recombinant: Exhibiting genetic recombination.

Red meat: Generally includes meat from large farm animals such as cattle (beef), sheep (lamb and mutton), pigs or hogs (pork), and calves (veal). Poultry and seafood are much paler in color and are referred to as "white" meat. However, pork and veal are more "white" than "red" and often are classified as "white" meats.

Reducing equivalents: The degree to which certain ingredients or inherent components in the meat can provide a source of electrons (donors).

Refining: Process that steam strips volatiles from oil, and also filters out impurities and fatty acids from raw (unrefined) oil.

Reflectance: The amount of light that is reflected from a sample.

Relaying: The process whereby shellfish are removed from a contaminated body of water and placed in unpolluted water for decontamination.

Rendering: Process that separates the by-products of muscle food production into fat/oil, water, gel bone, and meat and bone meal.

Restructured: Meat may be ground, flaked, diced, etc., and formed into steaks, chops, and roastlike items.

Retorted: Thermally processing a packaged food item at pressures and temperatures above 121°C (212°F) in order to sterilize the food.

Rib eye area: Size or area of a cross section of the longissimus muscle at the 12th rib cross section in beef and lamb carcasses.

Rigor mortis: General stiffening of muscle that accompanies the loss of ATP in myofibers postmortem.

Safe Food Handling Label: Label placed on every raw (uncooked) meat and poultry product describing the storage, cooking, and re-storage required to ensure safe foods through the preparation stage.

Sarcomere: Basic repeating contractile structural unit of the myofibril; in-register laterally and longitudinally and responsible for striated appearance of skeletal muscle.

Sarcomere length: A measurement of the distance between the two Z-lines (outer boundaries) of one sarcomere.

Sarcoplasm: Cytoplasm of muscle cell; major water depot of muscle cell containing soluble proteins.

Sarcoplasmic reticulum: A specialized form of endoplasmic reticulum in myofibers. This membrane system surrounding myofibrils is responsible for storing, releasing, and resequestering calcium in response to an action potential.

Scribe line: Shallow cut that remains in the fresh pork belly as a result of cutting through the costal (rib) bones in order to remove the spareribs and loin.

Seafood: The term used to identify fish and crustaceans sold for consumption. Generally the two types are referred to as finfish and shellfish. A few others such as octopus and squid do not fall into either description but are definitely seafood.

Seam fat: Deposits of fat located between muscles; also called intermuscular fat.

Secondary ingredient supplier: Secondary source to obtain an ingredient, packaging material, or other supply. The ingredient may not be an exact duplicate of the one currently used.

Secondary package: Sometimes this package is the product package, in other cases it is another barrier to the environment, providing additional protection to the product. An example of this type of package is a shelf pack that is placed inside a master carton (primary package) for shipping.

Self-service deli: Processed meat items are prepackaged into small convenient size packages and priced for consumer purchase. These are usually offered for sale in open display cases or from pegboard displays.

Sensory acuity: The keenness of perception for sensory attributes.

Sensory characteristics: Something that distinguishes or identifies a product; due to stimulating nerves from the sense organs to the nerve centers.

Sensory evaluation (analysis): To determine or fix the value of sensory characteristics by careful appraisal and study.

Sensory panel leader: The individual that acts as a leader or facilitator during sensory evaluation.

Sensory program: Rating of a food for visual, flavor, aroma, texture, and mouthfeel characteristics in comparison to desired qualities.

Sensory session: A continuous segment of time in which sensory properties are determined by sensory panelists.

Shechita: Ritual slaughter of an animal according to the Jewish practice. Shechita means the cutting in one or two swift passes of the soft parts of the neck from front to back including the trachea and the oesophagus as well as the carotid arteries, jugular veins, and the sympathetic and vagus nerve.

Shelf-life: The amount of time the product can be expected to remain consumable as a food based on microbiological considerations, nutritional parameters, and organoleptic properties. The time from processing meat items to the time the products are not considered acceptable for sale.

Shelf presence: Retail sales space allotted for a particular brand or company.

Shipping test: Transportation of the product in the final packaged form, on pallets or stacked, over a normal hauling distance followed by package and product evaluation to better assure quality products at destination points.

Shochet: Person trained and certified by rabbis to carry out kosher shechita of large animals or poultry or both.

Skeletal maturity: Physiological maturity of a carcass as indicated by changes in the skeleton, such as cartilage turning to bone and ribs becoming wider and flatter with advanced maturity.

Slotting fees: Money paid to the retail store to place a new product on the shelf to cover the increased handling involved.

Somatostatin: A hormone that inhibits the release of somatotropin.

Somatotropin: Growth hormone; a polypeptide hormone that is secreted by the anterior lobe of the pituitary gland and regulates growth.

Sous-Vide: Under vacuum; a process by which meal components are packaged in a barrier bag or rigid container, from which the air is evacuated, with or without compensation by neutral or bacteriostatic gasses. It is then cooked in a high-moisture environment to a temperature sufficient to cook and pasteurize but not sufficient to sterilize. The package is rapidly chilled and refrigerated until eaten.

Specialty meat shops: Many are located in shopping malls offering an assortment of cured and processed meat items with highly recognized brand names.

Specific gravity: Expression of tissue density in relation to density of water. Because density of fat is lower than density of either muscle or bone, the specific gravity of a lean carcass is higher than a fat carcass when muscle/bone is constant.

Specifications: Specific parameters that define a process or product such as shape, color, weight, and ingredients.

Spectrophotometer: An instrument that can evaluate light reflectance or transmission at varying wavelengths across the color spectrum.

Sphingomyelin: Member of crystalline phosphatides found primarily in the nerve tissue. Products of sphingomyelin hydrolysis are fatty acid, sphingosine or its saturated dihydro derivative, choline, and phosphoric acid.

Spoilage: Result of the process that causes a food product to become unfit for human consumption.

Spool joint: On the end of the metacarpal, it is a joint that resembles the appearance of a spool of thread ("spool joint"). As sheep advance in maturity, the thin layer of soft cartilage between the break joint and spool joint undergoes ossification and becomes bone. The ossification process fuses the spool joint and the end of the metacarpal together, making it impossible to break the spool joint from the foreshank of the carcass.

Stick water: Water phase that is discharged from a rendering centrifuge. This material may be reprocessed to remove oil and or proteins or sent to the plant discharge.

Super Stores: Large stores offering a large range of hard line and soft line items combined with items found in normal supermarkets and containing a complete pharmacy.

Supermarkets: Food stores offering an assortment of food and related products using the principle of self-service to reduce costs.

Surimi: Washed fish protein that can be used in a variety of products.

Table-ready: Meats that are cut, trimmed, and often boneless, prepared for direct use by the consumer without additional preparation. These cuts may need to be cooked or they may be precooked and ready to serve. Other similar terms such as "oven-ready" or "knife-ready" fall into equivalent categories but need to be cooked (i.e., oven-ready rib) or cut into steaks or roasts (i.e., boneless short loin or top round).

Taste adaptation: The temporary loss of power to respond, induced in a sensory receptor by continued stimulation.

Taste panel: Group of people that can discriminate and detect sensory differences in product qualities.

Teleological: Developments in nature that come about by design in order to meet a specific purpose.

Tenth rib cross section: Area of the carcass that is exposed when a pork carcass is ribbed by dividing the carcass between the 10th and 11th ribs.

Tenth rib fat depth: Linear measurement of the subcutaneous fat layer adjacent to the loineye at the 10th rib cross section, taken on pork carcasses.

Tertiary package: The final, or product contact package in retail packages. For example, the wrapper on a candy bar that is in a display pack (secondary package) that was shipped in a master carton (primary package).

Test market: Limited introduction of product into selected markets in an effort to determine consumer response and project success in commercialization.

Thaw shortening: A phenomenon observed in muscle previously frozen prior to the onset of rigor mortis; subsequent thawing results in massive shortening and loss of tenderness.

Thermoform: A process of forming objects from sheets of plastic by softening a sheet and forcing it into or over a mold by vacuum of gas pressure.

Trained sensory panel: A sensory panel that uses panelists who have had extensive instruction on sensory procedures prior to conducting a sensory study; this level of instruction implies a high level of ability to discern differences in sensory attributes; sensory panels that have been subjected to orientation sessions that may be 15–30 min in length do not qualify as trained panels.

Tref: Animals not fit for kosher slaughter or that are shown after slaughter to have diseased or physically or physiologically abnormal internal organs. Tref also refers to fish without scales and all other forms of seafood.

Triacylglycerol: Class of lipids in which three fatty acids are attached to a glycerol molecule.

Trichina: Microscopic parasite found in rats, dogs, cats, swine, bears, horses, and humans. It is not a widespread problem among domestic hogs and is controllable through proper curing, freezing, or heating. USDA-FSIS has set regulation (in 9 CFR 318.10) for curing, heating, and/or freezing that must be performed on all pork products that are not necessarily cooked at home or in a foodservice establishment before consumption.

Trichinella spiralis: Microscopic parasite that can cause the illness trichinosis. This parasite is a nematode worm sometimes found in pork muscle in the larva stage. Ingestion of insufficiently cooked infected pork can cause nausea, vomiting, diarrhea, and muscle soreness.

Triglyceride: Ester of glycerol that contains three ester groups and either one, two, or three fatty acids attached to the ester group.

Tunnel boned hams: Hams in which the bones (femur, tibia, fibula) have been removed in such a way as to minimize separation of the muscles at the seams.

Twelfth rib cross section: Area of the carcass that is exposed when a beef or lamb carcass is ribbed by dividing the carcass between the 12th and 13th ribs.

Twelfth rib fat thickness: Linear measurement of the subcutaneous fat layer adjacent to the ribeye at the 12th rib cross section; taken on beef and lamb carcasses.

USDA: United States Department of Agriculture.

USDA Inspected: USDA-inspected products that have been prepared in an establishment operating under the meat inspection guidelines of the United States Department of Agriculture designed to provide the consuming public with wholesome meat and meat products. All meat products, covered in the guidelines, crossing state lines must carry the USDA inspection legend which identifies by an inspection number the plant of origin. Inspection is a federally funded program supported by U.S. tax dollars.

USDA Quality/Yield grades: Voluntary service which groups carcasses into expected palatability levels (quality grades) and/or expected yield of retail cuts (yield grades).

U.S. Inspected and Condemned legend: Meat and poultry product that has been inspected by USDA-FSIS and condemned under the regulation described in the Federal Code of Regulations and then marked as such.

U.S. Inspected and Passed legend: Meat and poultry product has been inspected by USDA-FSIS and passed under the regulation described in the Federal Code of Regulations and then marked as such.

U.S. Retained: Carcass, carcass part, viscera, or other product that has been held for further examination by a USDA-FSIS inspector to determine if it should be passed or condemned.

U.S. Suspect: Live animal that suspected of having a disease or other condition that may require its condemnation, in whole or in part, and is subject to further examination by a USDA-FSIS inspector to determine its disposal.

Uniform Product Code (UPC): Computer bar code used to identify products. This would be a universal code for that product.

Uniform Retail Meat Identity Standards (URMIS): Many states require that retail meat products must be prepared and labeled for retail sales following these standards. Each package label must list (a) species, (b) wholesale cut, and (c) recommended retail name.

Variety: Something varying or different from others of the similar kind, subgroup within a group.

Venture group: Group of individuals from varied backgrounds that are removed from their normal assignments and are formed into a product development group for new projects.

Virulence: Degree of pathogenicity associated with a microorganism.

Volatiles: Compounds that are vaporized from a product.

Volative residue: The FDA sets the tolerance residue levels permitted to be found in the tissue after slaughter. These tolerance levels are set by the FDA to provide a 100-fold margin of safety to the public. A residue level found above the approved tolerance level is referred to as a volative residue.

Warehouse club stores: Large stores offering limited lines of food items at reduced costs. A club membership is required to obtain standard discounts.

Warehouse supermarket stores: Large stores with items generally found in supermarkets offered at reduced prices with limited customer services.

Warmed-Over Flavor (WOF): Undesirable rancid or stale flavor that can occur in precooked, uncured meats after short periods of storage time. WOF is caused by the oxidation of unsaturated fatty acids and phosopholipids and iron-catalyzed lipid oxidation.

Web: The roll of material (paper, plastic, foil, or a combination) that is processed into a finished container.

Yield grade: Numerical representation (e.g., 1, most desirable, to 5, least desirable) of the percentage of boneless, closely trimmed retail cuts from the major wholesale cuts of a carcass.

Z-disk: Component of sarcomere consisting mostly of actin and defining length of the sarcomere.

Zero-tolerance: Part of the USDA-FSIS Pathogen Reduction Program; carcasses and carcass parts are closely scrutinized for any form of contamination, and when found, the contamination must be totally eliminated and discarded before that carcass or carcass part is able to be further processed.

Zoonosis: A disease transmissible from animals to humans under natural conditions.

Index